Lecture Notes in Mechanical Engineering

Lecture Notes in Mechanical Engineering (LNME) publishes the latest developments in Mechanical Engineering—quickly, informally and with high quality. Original research reported in proceedings and post-proceedings represents the core of LNME. Volumes published in LNME embrace all aspects, subfields and new challenges of mechanical engineering. Topics in the series include:

- Engineering Design
- Machinery and Machine Elements
- Mechanical Structures and Stress Analysis
- Automotive Engineering
- Engine Technology
- Aerospace Technology and Astronautics
- Nanotechnology and Microengineering
- Control, Robotics, Mechatronics
- MEMS
- Theoretical and Applied Mechanics
- Dynamical Systems, Control
- Fluid Mechanics
- Engineering Thermodynamics, Heat and Mass Transfer
- Manufacturing
- Precision Engineering, Instrumentation, Measurement
- Materials Engineering
- Tribology and Surface Technology

To submit a proposal or request further information, please contact the Springer Editor of your location:

China: Ms. Ella Zhang at ella.zhang@springer.com
India: Priya Vyas at priya.vyas@springer.com
Rest of Asia, Australia, New Zealand: Swati Meherishi at swati.meherishi@springer.com
All other countries: Dr. Leontina Di Cecco at Leontina.dicecco@springer.com

To submit a proposal for a monograph, please check our Springer Tracts in Mechanical Engineering at http://www.springer.com/series/11693 or contact Leontina.dicecco@springer.com

Indexed by SCOPUS. All books published in the series are submitted for consideration in Web of Science.

More information about this series at http://www.springer.com/series/11236

Rajeev Agrawal · Jinesh Kumar Jain ·
Vinod Singh Yadav · Vijaya Kumar Manupati ·
Leonilde Varela
Editors

Recent Advances in Smart Manufacturing and Materials

Select Proceedings of ICEM 2020

Springer

Editors
Rajeev Agrawal
Department of Mechanical Engineering
Malaviya National Institute of Technology
Jaipur, Rajasthan, India

Jinesh Kumar Jain
Department of Mechanical Engineering
Malaviya National Institute of Technology
Jaipur, Rajasthan, India

Vinod Singh Yadav
Department of Mechanical Engineering
National Institute of Technology
Srinagar, Uttarakhand, India

Vijaya Kumar Manupati
Department of Mechanical Engineering
National Institute of Technology Warangal
Hanamkonda, Telangana, India

Leonilde Varela
Department of Production and Systems
University of Minho
Braga, Portugal

ISSN 2195-4356 ISSN 2195-4364 (electronic)
Lecture Notes in Mechanical Engineering
ISBN 978-981-16-3032-3 ISBN 978-981-16-3033-0 (eBook)
https://doi.org/10.1007/978-981-16-3033-0

This Springer imprint is published by the registered company Springer Nature Singapore Pte Ltd.
The registered company address is: 152 Beach Road, #21-01/04 Gateway East, Singapore 189721, Singapore

Contents

About the Editors

Dr. Rajeev Agrawal is currently working as an Associate professor at the Department of Mechanical Engineering, Malaviya National Institute of Technology Jaipur, (Rajasthan). He obtained his B.E. (Mechanical) from Govt. Engg. College, Jabalpur (M.P.) and M.E. (Prod. Engg.), Ph.D. from the MNNIT, Allahabad (U.P.) and Birla Institute of Technology, Mesra (Ranchi) respectively. Dr. Rajeev Agrawal is having more than 20 years of Professional Experience. Dr. Rajeev Agrawal currently managing Research projects includes engaged in Soft Computing (GA, ANN, Fuzzy) applications to Manufacturing, Modelling, and Simulation of manufacturing systems, Sustainable Manufacturing, Lean Six Sigma, Supply chain Design, and Reconfigurable manufacturing system (RMS). He serves on the editorial board of three international journals and has guest edited special issues in around 8 journals and conference proceedings. Several SCI Journal and International Conference articles have been credited into his account. He has research funding from major funding agencies. He has also published book chapters on Industry 4.0, Smart Manufacturing, Integration of Process Planning and Scheduling, Smart Actuators, Sustainable Manufacturing, and Reconfigurable Assembly Systems, etc. with renowned international publishers. He is a recognized reviewer of reputed international SCI Journals and Conferences. He is an active member of several dynamic societies and research organisations worldwide.

Dr. Jinesh Kumar Jain is currently working as Associate Professor at Malaviya National Institute of Technology Jaipur (MNIT). His research interests include joining and welding, additive manufacturing and bio-fabrication, design and manufacturing technology, operation management, and sustainable manufacturing. He completed his Ph.D. and M.Tech. from MNIT and B.E. from MBM Engineering College Jodhpur. He has more than twenty years of teaching, research and industrial experience; published many technical papers in journals and conference proceedings; authored number of books, chapters in reputed publications and delivered lectures in reputed institutions in India and abroad. Prior to joining MNIT, he has worked in several renowned organizations. Dr. Jain is a life member of many reputed professional bodies and board member in the number of Indian universities. Dr Jain successfully guided many M.Tech. dissertations and presently guiding many doctoral and masters scholars. He successfully organized many short-term courses and faculty development programmes to impart, encourage the young fellows to proceed by sharing knowledge and experiences from eminent researchers.

Dr. Vinod Singh Yadav is currently an Assistant Professor at the Department of Mechanical Engineering, National Institute of Technology Uttarakhand, Uttarakhand. He obtained his Bachelor of Engineering (Mechanical Engineering) from University of Rajasthan, and M.Tech. (Energy Engineering) and Ph.D. from Malaviya National Institute of Technology, Jaipur. His major interest of research areas are alternate fuels in IC engines, renewable energy technologies, thermal storage techniques and their applications. He has published more than 35 research papers in reputed journals and international conferences. He is the author of five text and reference books. He is also a reviewer of more than five SCI Elsevier journals. He is Technical Editor of *Journal of Automotive Mechanical & Aerospace Engineering Research* and Consulting Editor of Hon. Editorial Board Member recognized by Innovative Scientific Research Professional Malaysia.

Dr. Vijaya Kumar Manupati is currently working as an Assistant Professor in the Department of Mechanical Engineering, NIT Warangal. He received his Ph.D. in the Department of Industrial and Systems Engineering from the Indian Institute of Technology Kharagpur. His current research interests include intelligent manufacturing systems, cyber-physical systems, sustainable supply chain, and health care systems. He has published more than 80 publications which include prestigious journals like International Journal of Production Research, Computers, and Industrial Engineering, International Journal of Advanced Manufacturing Technology, Journal of measurements, International Journal of Computer Integrated Manufacturing, etc. He is acting as an international reviewer for more than 30 peer-reviewed journals. Currently, he is acting as an Editorial Review Board member of *International Journal of Sustainable Entrepreneurship and Corporate Cocial Responsibility*, *International Journal of Web Portals* IGI Global publications. He received an Early Carrier Research Grant from Department of Science and Technology (DST) for his research work on Telefacturing Systems. He is a member of the Institute of Industrial and Systems Engineering (IISE), Institute of Engineers (IEI) India, Life member of International Association of Engineers, US and also acting as a technical committee member of various International conferences.

Dr. Leonilde Varela received her Ph.D. degree in Industrial Engineering and Management from the University of Minho, Portugal in 2007. She is Assistant Professor at Department of Production and Systems of University of Minho. Her main research interests are in manufacturing management, production planning and control, optimization, artificial intelligence, meta-heuristics, scheduling, web based systems, services and technologies, mainly for supporting engineering and production management, collaborative networks, decision making models, methods and systems, and virtual and distributed enterprises. She has published more than 150 refereed scientific papers in international conferences and in international scientific books and journals, indexed in the Web of Science and/or in the Scopus databases. She coordinates R&D projects in the area of production and systems engineering, namely concerning the development of web-based platforms and decision support models, methods and systems. She is a frequent paper reviewer for several journals, such as, *Journal of Computer Integrated Manufacturing, Engineering Applications of Artificial Intelligence, IEEE Transactions on Neural Networks and Learning Systems, Journal of Decision Systems, Sustainability, International Journal of Management and Fuzzy Systems, International Journal of Sustainability Management and Information Technologies, International Journal of Intelligent Enterprise, International Journal of Decision Support Systems Technology, Management and Production Engineering Review, Mathematical Problems in Engineering, Journal of Robotics*. Moreover, she has been increasing international scientific network and collaborates with more than twenty institutions worldwide, and is a member of several international networks, such as: Machine Intelligence Research Labs, Scientific Network for Innovation and Research Excellence (MirLabs), Euro Working Group of Decision Support Systems (EWG-DSS), Institute of Electrical and Electronics Engineers (IEEE); System, Man, and Cybernetics Society (IEEE SMC), Industrial Engineering Network (IE Network), and Institute of Industrial and Systems Engineers (IISE).

Industry 4.0 and Sustainable Manufacturing: A Bibliometric Based Review

Anbesh Jamwal, Rajeev Agrawal, Monica Sharma, Vijaya Kumar Manupati, and Akshay Patidar

1 Introduction

These days both the technologies Industry 4.0 and sustainable manufacturing (SM) are reshaping the new industrial waves in the world [1, 2]. At present, both Industry 4.0 and sustainability are considered as major emerging trends in the production systems of industries [3, 4]. Now researchers are integrating Industry 4.0 with the sustainability concepts to revolutionize the new industrial wave [5]. Higher level of digitalization in the global industries has transformed the industrial production system in the last few years [6]. This transformation has changed the production from traditional to intelligent and decentralized production system which can be termed as the fourth industrial revolution or Industry 4.0. Industry 4.0 is the digital transformation of the production or manufacturing firms and value creation processes [3]. Both sustainability and Industry 4.0 are considered as the major research areas in the current production system. Industry 4.0 will help to unlock the sustainability in industries by its advanced technology and moving towards social sustainability [7]. Guidelines for Industry 4.0 are driven by different government programs such as in Germany it is driven by Industry 4.0, in the United States it is driven by "Advanced manufacturing partnership", in the United Kingdom it is driven by "Future of manufacturing", in France it is driven by "La Nouvelle France Industrielle", in Sweden it is driven by "Smart Industry", in Japan "Super smart society" and in Europe

A. Jamwal (✉) · R. Agrawal · M. Sharma
Department of Mechanical Engineering, Malaviya National Institute of Technology, J.L.N. Marg, Jaipur, Rajasthan 302017, India

M. Sharma · A. Patidar
Department of Management Studies, Malaviya National Institute of Technology, J.L.N. Marg, Jaipur, Rajasthan 302017, India

V. K. Manupati
Department of Mechanical Engineering, NIT Warangal, Warangal, Telangana, India

R. Agrawal et al. (eds.), *Recent Advances in Smart Manufacturing and Materials*, Lecture Notes in Mechanical Engineering,
https://doi.org/10.1007/978-981-16-3033-0_1

"Factories for future" [8]. At present manufacturing industries are focusing towards higher investment returns and lesser environment impacts due to their manufacturing processes which requires an attractive workplace focused on collaboration and learning [9]. In the past few years, authors have discussed the opportunities of SM in Industry 4.0 but the no. of articles considered for study mapping are in limited number. This study aims at the consideration of Scopus database for the study. All the articles available from January 2000-May 2020 were considered for study mapping with Industry 4.0 and SM as the search terms. This article is an effort to know the opportunities of SM in Industry 4.0. In the present study following research questions are raised which are:

RQ1: Which authors, countries, and Institutes are working in the area of SM in Industry 4.0?
RQ2: Which countries or authors are working together in the area of SM in Industry 4.0?
RQ3: What is the current research status and future opportunity of work?

The paper divided into four sections. Section 1 of paper discuss about the Introduction and various research questions of the study. Section 2 discuss about the SM, Industry 4.0, and relation between them. Section 3 of the paper discuss about the bibliometric analysis of SM in Industry 4.0. Section discuss about the conclusion and implications of the study.

1.1 Sustainable Manufacturing and Industry 4.0

SM focus on the minimizing or elimination of negative impact of manufacturing processes by adoption of eco-efficient practices which includes waste minimization and new technologies [10]. Over the years researchers have proposed many definitions in the area of SM which is as: [11] defines SM as the manufacturing which minimizes the negative impact of manufacturing on environment and increase the resource efficiency.

The need for fast adaption of new technologies and never-ending customer demands are also responsible for the growth of industries. Now the industries are more competitive and these customer requirements can be completed by adopting new technologies over the existing production systems. Industry 4.0 is a promising choice which is the integration of manufacturing and business processes as well as integration of customers and suppliers. Technical aspects of Industry 4.0 can be addressed by the application of IoT and Cyber physical systems. Industry 4.0 is the connection of CPS blocks [12]. The broader aspect of Industrial Internet can be seen in Industry 4.0 which covers different industry sectors such as mining, healthcare, power generation, and manufacturing sector [13]. The different pillars for the Industry 4.0 are: Autonomous robots, Cyber security, Horizontal and vertical systems integrations, simulations, Industrial IoT, additive manufacturing, cloud computing, Big data analytics, and augmented reality.

Table 1 Studies on SM in Industry 4.0

Title or paper and author	Objectives of study	Outcomes of study
[7]	State of art review in Industry 4.0 Opportunities for SM in Industry 4.0 are identified	Both macro perspective and macro perspective of Industry 4.0 is discussed Retrofitting of machine tool as a opportunity for SM in Industry 4.0 is discussed
[3]	Synergy between the SM and Industry 4.0 is discussed	Integrated framework for Industry 4.0 and SM is proposed Industry 4.0 technologies will unlock the GM in the industries
[8]	Systematic literature review with 35 papers from 2008 to 2018	Top research areas and authors were found out working in the area of SM in Industry 4.0. Drivers for SM in Industry 4.0 are discussed

At present digitalization and sustainability concepts in industries are increasing global competitiveness [14, 15]. Both the Industry 4.0 and SM approaches present practices convergence e.g. remanufacturing, life cycle, recycling, reverse logistics for the circular economy, resource efficiency, lean and green philosophies, elimination of toxic parts in the product and production processes, and sustainable designs which minimize the health risks for workers [16]. Sustainability in Industry 4.0 results in improving efficiency, waste reduction; reduce energy consumption, optimization of processes, minimizing overproduction, job opportunities, predictive maintenance, and employment generation for disabled workers. The various studies which integrate the SM with Industry 4.0 have been discussed in Table 1:

2 Methodology

Bibliometric analysis helps to evaluate the published books, articles, conference papers, or chapters statistically [17]. It is an effective way to measure the influence of a scientific document in the scientific community [18]. Over the years researchers have used various databases such as Google scholar, Web of science, and Scopus for bibliometric analysis [19]. In which Scopus and Web of Science are more popular in terms of quality. In the present study, authors have used the Scopus database. In the study search terms "Industry 4.0" OR "4th Industrial revolution" OR "fourth industrial revolution" AND "Sustainable Manufacturing" areused for "Articles, Keywords, and Titles". In the search terms, 612 documents were found.

Table 2 Document wise analysis

S. No.	Description	Results
1	No. of documents	317
2	Total sources	64
3	Author keywords	1024
4	Time period	2000–2020

3 Results

3.1 Document Wise Analysis

Table 2 shows the document wise analysis of the SM in Industry 4.0. There are total of 317 documents were found after the analysis. The conference papers are more in the Scopus database with 137 publications.

3.2 Publication Trend

Figure 1 shows the publication trend of SM in Industry 4.0. The number of articles in this area was less which increases in 2016 and 2017. In 2018 articles were two

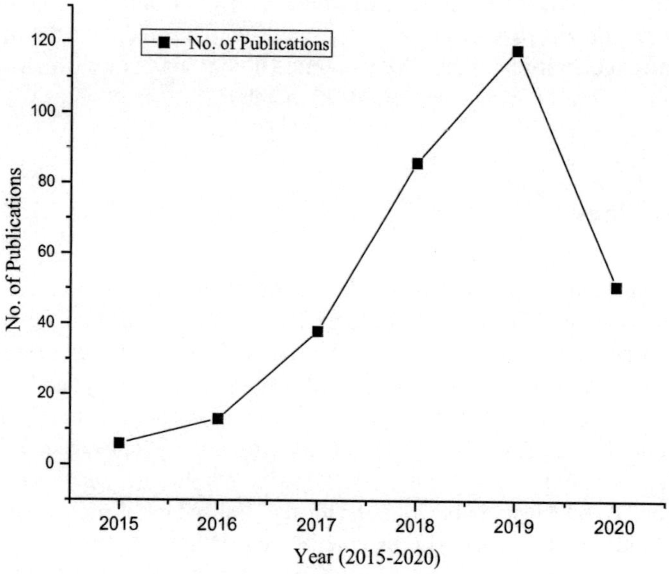

Fig. 1 Publication trend (2015-2020)

Table 3 Top cited Articles with their source

S. No.	Article	Citations
1	Stock and Seliger (2016) [7]	397
2	Kagermann (2015) [20]	192
3	Kamble et al. (2018) [21]	93
4	Jabbour et al. (2018) [3]	90
5	Machado et al. (2020) [8]	86

times as compared to 2015-2017. In 2020, till May, 2020 the number of articles are more than articles till 2017.

3.3 Most Cited Articles

Table 3 shows the most cited articles in the area of SM and Industry 4.0. It is found that Stock and Seliger (2016) [7] is most cited article with 397 citations.

3.4 Most Productive Authors and Journals

Table 4 shows the productive authors with their total numbers of publications, in particular, area. It is found that Romero, D. is the most productive author with 6 documents. In the top 5 most productive areas most of the authors have different research areas but few authors such as Silva, E. J. and Cho, S. has very specific research area and contributing the productive research work in the area of SM in Industry 4.0.

Top 5 Sources is shown in the Table 4 with the total publications. In analysis, it is found that Procedia manufacturing having most of the documents with 22 articles. In some bibliometric studies, authors have excluded the conference papers to maintain the quality of study. But it can be seen that in this analysis conference proceedings shouldn't be excluded while mapping the studies. As in this study, both top sources and top-cited articles are from the conference proceedings. Reputed publishers

Table 4 Top productive authors

S.No.	Author Name	TP	Journal/ Source Name	TP
1	Romero, D	6	Procedia Manufacturing	22
2	Wuest, T	5	Procedia CIRP	16
3	Kiritsis, D	4	IFIP Advances In Information And Communication Technology	13
4	May, G	4	Sustainability Switzerland	12
5	Stahre, J	4	IFAC Papersonline	10

Table 5 Top productive countries

Country name	TP	Category	Institute name	TP	Country
Germany	43	Developed	Tecnologico de Monterrey	10	Mexico
Italy	40	Developed	University of Zilina	8	Slovak Republic
United Kingdom	38	Developed	The Institute of Technology and Business in Ceske Budejovice	6	Czech Republic
United States	34	Developed	Chalmers University of Technology	6	Sweden
India	23	Developing	Università degli Studi di Modena e Reggio Emilia	6	Italy

having Sustainability, International Journal of Production research, Journal of cleaner production and Journal of manufacturing systems still having less documents on SM in Industry 4.0 area which is expected to be increased in upcoming years.

3.5 Top Countries and Institutes Working on Industry 4.0 and SM

Top 5 countries working on SM in Industry 4.0 is shown in the Table 5. It is found that Germany having most of 43 documents which is followed by Italy with 40 documents, United Kingdom with 38 documents, United States with 34 documents and India with 23 documents. It is also found that in top 5 countries only one developing country India is working on the SM in Industry 4.0 all other countries are developed nations.

Top 15 institutes working on SM in Industry 4.0 is shown in the Table 5 with total publication by institute and country. It is found that Tecnologico de Monterrey having most of publications in the research area which is in Mexico country. University of Zilina having 8 publication in the developed nations. It is observed that mostly the institutes working on SM and industry 4.0 are in developed nations which provides them a better platform for industry academia collaboration. United States institutes are more focused on this research area as compared to other countries.

3.6 Top Keywords

Top keywords used in the documents of sustainable manufacturing in Industry 4.0 are shown in the Table 6 with their occurrence. It is found that Industry 4.0 is mostly used keyword in the documents with 203 occurrences followed by "sustainable development" with 99 occurrences, "Manufacture" with 65 occurrences, "Sustainable manufacturing" with 55 occurrences, and "Sustainability" with 51 occurrences. However,

Table 6 Top keywords with total occurrence

S. No.	Keyword name	Total cccurrence
1	Industry 4.0	203
2	Sustainable development	99
3	Manufacture	65
4	Sustainable manufacturing	55
5	Sustainability	51
6	Internet of things	41
7	Embedded systems	39
8	Industrial research	38
9	Decision making	28
10	Smart manufacturing	27
11	Manufacturing	25
12	Cyber physical system	24
13	Industrial revolutions	23
14	Big data	22
15	Life cycle	22

other popular keywords in Industry 4.0 and sustainable manufacturing such as "Cyber physical systems", "Industrial Revolution" and "Big data" still having less occurrence in the documents.

3.7 Co Authorship Between the Authors

Figure 2 shows the co-authorship network between the authors working in the SM in Industry 4.0. For the analysis different parameters are set such as maximum authors per document is considered as 25, minimum no. of documents by author is set to 2, and minimum no. of citations is set to 5. These all parameters are set in the VOSviewer software. The results indicate that three clusters of co-authorship between the authors are found. In cluster-I, there are 6 authors which is represented by red color. In Cluster-II, three authors are working together which is represented by green color. In Cluster-III, two authors are working in collaboration which is represented by blue color.

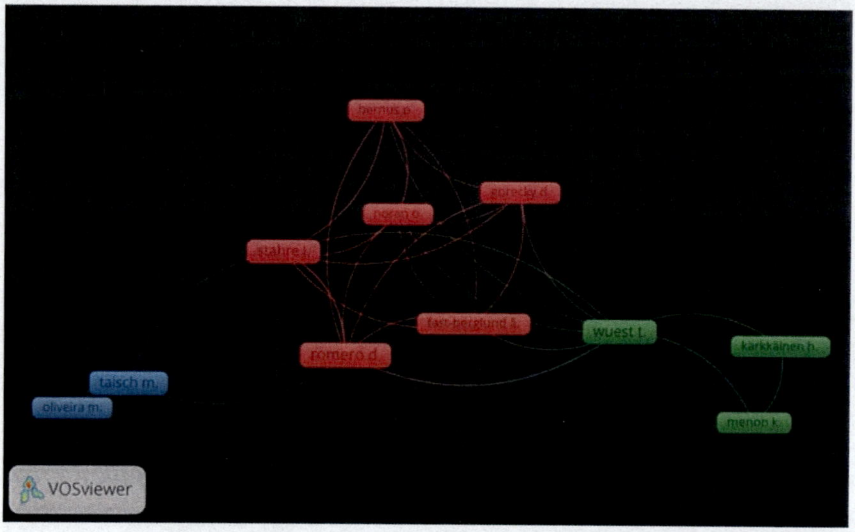

Fig. 2 Co-authorship between authors

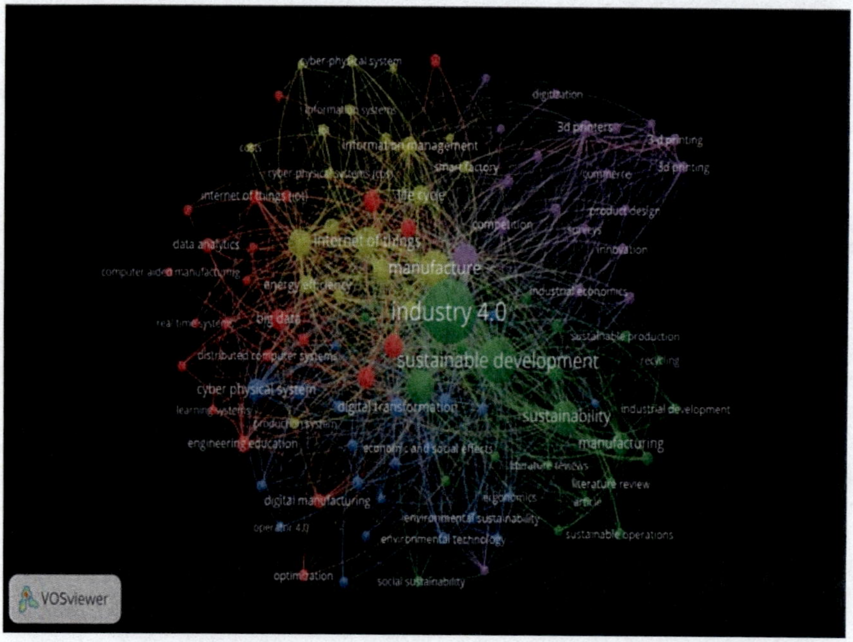

Fig. 3 Keyword network analysis

3.8 *Keyword Occurrence Network*

Figure 3 shows the keyword occurrence network in the SM in Industry 4.0 research area. For the analysis different parameters are set such as minimum keyword occurrence is set to 5. Total of 5 clusters with 109 items is found. In Cluster-I (23), Cluster-II (23), Cluster-III (22), Cluster-IV (21), and Cluster-V (20) items are found. In which Industry 4.0, sustainable development, Internet of things, sustainability and manufacturing are the keywords with maximum node strength.

As discussed above, this study mapping will help the researchers to work on the sustainability issues in Industry 4.0. Now the researchers are more focused towards the sustainability in Industry 4.0. Based on the study mapping following research questions for future research directions have been proposed:

RQ1: How can a CPS or IoT-based framework will help to enable the sustainability in SMEs?
RQ2: What are different common barriers in the adoption of sustainable practices in Industry 4.0?
RQ3: What are the different common enablers in the adoption of sustainable practices in Industry 4.0?
RQ4: How the Industry 4.0 will impact the society by job creation and employment?
RQ5: What will be effect of man-machine collaboration on the economic aspects of sustainability?

These are some future research questions which need to address in the future studies.

4 Conclusion and Future Implications

This study helps to contribute various research scopes in SM in Industry 4.0. SM in Industry 4.0 is an emerging research area for most of the authors. It is found that research articles in this area are increasing rapidly. In 2020, no. of articles is more than the no. of articles published till 2017. This shows that researchers are now more focused towards the sustainability in Industry 4.0. Total of 317 documents were found out in the analysis in which conference proceedings papers are more. It is suggested that in future studies conference papers shouldn't be excluded from the bibliometric analysis. Procedia CIRP and Procedia manufacturing having most of publications while other reputed journals have fewer publications which is expected to increase in upcoming years. In top 5 most productive authors, the authors are from various research areas having a strong research background. While some of the authors have started their research career in this area and doing productive research work. Most of research work is in the developed nations. India is the only developing nation in the top 5 countries working on the SM in Industry 4.0. Developing nations are working collaboratively with the developed nations to promote and explore the research are of sustainability in Industry 4.0. The research questions raised in the present study will help the researchers to decide their future research directions.

References

1. Enyoghasi, C., & Badurdeen, F. (2021). Industry 4.0 for sustainable manufacturing: Opportunities at the product, process, and system levels. *Resources, Conservation and Recycling, 166*, 105362.
2. Ivascu, L. (2020). Measuring the implications of sustainable manufacturing in the context of industry 4.0. *Processes, 8*(5), 585.
3. de Sousa Jabbour, A. B. L., Jabbour, C. J. C., Foropon, C., & Filho, M. G. (2018). When titans meet—Can industry 4.0 revolutionise the environmentally-sustainable manufacturing wave? The role of critical success factors. *Technological Forecasting and Social Change, 132*, 18–25. https://doi.org/10.1016/j.techfore.2018.01.017
4. Jamwal, A., Agrawal, R., Sharma, M., Kumar, A., Kumar, V., & Garza-Reyes, J. A. A. (2021). Machine learning applications for sustainable manufacturing: A bibliometric-based review for future research. *Journal of Enterprise Information Management.*
5. Jamwal, A., Agrawal, R., Sharma, M., & Kumar, V. (2021). Review on multi-criteria decision analysis in sustainable manufacturing decision making. *International Journal of Sustainable Engineering*, 1–24.
6. Jamwal, A., Agrawal, R., Manupati, V. K., Sharma, M., Varela, L., & Machado, J. (2020, December). Development of cyber physical system based manufacturing system design for process optimization. In *IOP Conference Series: Materials Science and Engineering* (Vol. 997, No. 1, p. 012048). IOP Publishing.
7. Stock., T, & Seliger, G. (2016). Opportunities of Sustainable Manufacturing in Industry 4.0. In G. Seliger, H. & M. J. Kohl, (Eds.), *Procedia CIRP* (pp 536–541). Elsevier B.V.
8. Machado, C. G., Winroth, M. P., & da Silva, E. H. D. (2020). Sustainable manufacturing in Industry 4.0: an emerging research agenda. *International Journal of Production Research, 58*, 1462–1484. https://doi.org/10.1080/00207543.2019.1652777
9. Ghobakhloo, M. (2020). Industry 4.0, digitization, and opportunities for sustainability. *Journal of Cleaner Production*, 252. https://doi.org/10.1016/j.jclepro.2019.119869.
10. Bhatt, Y., Ghuman, K., & Dhir, A. (2020). Sustainable manufacturing. Bibliometrics and content analysis. *Journal of Cleaner Production*, 260. https://doi.org/10.1016/j.jclepro.2020.120988.
11. Melnyk, S. A., & Smith, R. T. (1996). Green manufacturing. *Computer Automated Systems of the Society of Manufacturing Engineers*
12. Ejsmont, K., Gladysz, B., Kluczek, A. (2020). Impact of industry 4.0 on sustainability-bibliometric literature review. *Sustainability*, 12. https://doi.org/10.3390/su12145650.
13. Yadav, G., Kumar, A., Luthra, S., et al (2020). A framework to achieve sustainability in manufacturing organisations of developing economies using industry 4.0 technologies' enablers. *Computers & Industrial Engineering*, 122. https://doi.org/10.1016/j.compind.2020.103280.
14. Jamwal, A., Agrawal, R., Sharma, M., Kumar, V., & Kumar, S. (2021). Developing A sustainability framework for Industry 4.0. *Procedia CIRP, 98*, 430–435.
15. Leng, J., Ruan, G., Jiang, P., Xu, K., Liu, Q., Zhou, X., & Liu, C. (2020). Blockchain-empowered sustainable manufacturing and product lifecycle management in industry 4.0: A survey. Renewable and Sustainable Energy Reviews, 132, 110112.
16. Jamwal, A., Agrawal, R., Sharma, M., & Giallanza, A. (2021). Industry 4.0 Technologies for Manufacturing Sustainability: A systematic review and future research directions. *Applied Sciences, 11*(12), 5725.

17. Jamwal, A., Agrawal, R., & Sharma, M. (2021). Life cycle engineering: past, present, and future. In Sustainable Manufacturing (pp. 313-338). Elsevier.
18. Muhuri, P. K., Shukla, A. K., & Abraham, A. (2019). Industry 4.0: A bibliometric analysis and detailed overview. Engineering applications of artificial intelligence, 78, 218-235.
19. Jamwal, A., Agrawal, R., Sharma, M., Dangayach, G. S., & Gupta, S. (2020). Application of optimization techniques in metal cutting operations: A bibliometric analysis. *Materials Today: Proceedings*.
20. Kagermann, H. (2015). Change through digitization—Value creation in the age of Industry 4.0. In *Management of permanent change* (pp. 23–45). Springer.
21. Kamble, S. S., Gunasekaran, A., & Gawankar, S. A. (2018). Sustainable Industry 4.0 framework: A systematic literature review identifying the current trends and future perspectives. *Process Safety and Environmental Protection, 117*, 408–425. https://doi.org/10.1016/j.psep.2018.05.009

Joining of Similar and Dissimilar Metals Through Microwave Hybrid Heating & Its Characterization: A Review

B. G. Koujalagi, Amith Gadagi, C. V. Adake, and Ramesh Katti

1 Introduction

Microwave joining/processing is a nonconventional joining process in which a electromagnetic radiation of wavelength in the range of 1 m to 1 mm & frequency range of 300 MHz to 300 GHz are used to heat the metals & alloys. The frequency for microwave hybrid heating is reserved by federal communication commission (FCC) for metal & alloy joining and commonly used frequency for industrial purpose is 2.45 GHz. The microwave energy can be utilized in variety of applications such as brazing of low melting point metals, metal coating on bulk metal, joining of similar & dissimilar bulk metal, microwave drilling of metal sheet & microwave casting of bulk metals. The joining of bulk metals is challenging as the metals does not absorb the electromagnetic radiation which when incident on the surface of metals gets reflected. The microwave joining of bulk metals is obtained with new hybrid technique using a susceptor powder called "microwave hybrid heating (MHH)". The susceptor powder either charcoal or silicon carbide is used for heating metal powders which is used as interface layer between base metals. Conventional & nonconventional welding techniques are used to join different similar and dissimilar metals. Microwave hybrid heating is an emerging non-conventional welding technique and has many advantages such as fewer defects, volumetric heating, better tensile strength, greater hardness at the joint, and smaller heat affected zone(HAZ) compared to conventional welding techniques [1]. Laser welding, TIG, and FSW are some conventional welding techniques are used to weld the super alloys which has its own advantages and disadvantages as these processes leads to enormous amount of heat input which leads to change in mechanical and chemical properties and reduction

B. G. Koujalagi (✉) · A. Gadagi · C. V. Adake · R. Katti
Department of Mechanical Engineering, KLE Dr. M S .Sheshgiri College of Engineering and Technology, Belagavi, Karnataka 590008, India

R. Agrawal et al. (eds.), *Recent Advances in Smart Manufacturing and Materials*,
Lecture Notes in Mechanical Engineering,
https://doi.org/10.1007/978-981-16-3033-0_2

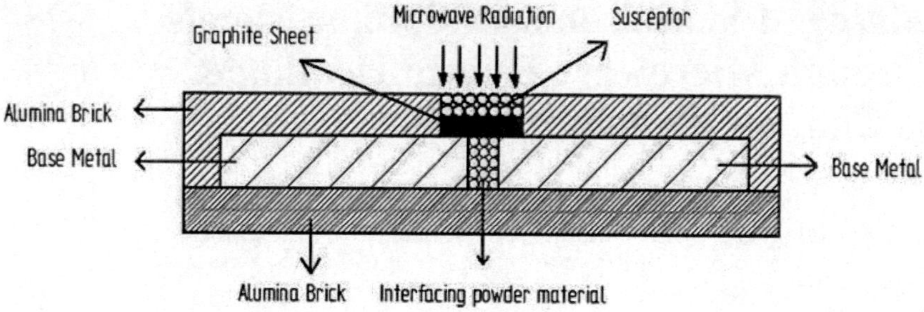

Fig. 1 Schematic diagram of microwave joining of metals

in corrosion resistance. Researchers have made an effort to use microwave hybrid heating to process the super alloys as it involves volumetric heating [2].

Experiment is carried out in a 1 KW multimode microwave applicator and metal plates are exposed to microwave radiation at the frequency of 2.45 GHz. Before joining, the metal plates are cleaned with emery paper of different grades and acetone to remove dirt from the surface of metal plates. Interface powder material of microsize particles are used to fill the gap between two metal plates. When microwave electromagnetic radiation is incident on the surface of metal plates it reflects the microwave electromagnetic radiation. The initial temperature of metal plates is increased by using the susceptor powder which absorbs microwave radiation and has a high dielectric loss. The susceptor powder converts the absorbed radiation into heat and transfers it to a separator sheet. The separator sheet either graphite sheet or silicon carbide sheet of 1 mm thickness is used between interface powder and susceptor powder to avoid the mixing of both. It absorbs the heat released by susceptor powder and transfers it to an interface material, which melts to form the joint. The metal plates with interface layer are placed inside the cavity made in alumina brick. Figure 1 shows the schematic view of microwave hybrid heating.

2 Materials and Method

Joining of similar and dissimilar metals through Microwave hybrid heating (MHH) and the joint characterization is explained in this section.

Joining of Stainless Steel-316 in bulk form has been carried out by M S Srinath et al. and joined the SS-316 plates of dimensions 25 mm × 12 mm × 6 mm prepared in form of square butt. The interface powder material, Ni-based powder, with an average particle size of 40 μm was used in the gap of a butt joint. The epoxy resin was mixed with nickel-based powder to hold the powder particles together during the joining of plates. At high temperature, the epoxy resin gets evaporated without affecting the strength of the joint. An experiment was carried out under atmospheric conditions in 1KW microwave system. The Stainless Steel-316 bulk metal plates were exposed to electromagnetic radiations for upto 390 s at 2.45 GHz & at power level of 900 W. Characterization of the joint was carried out to examine tensile strength, hardness, porosity, XRD & SEM [3]. M S Srinath et al. also developed the joint between the SS-316 & MS and investigated the microstructural & mechanical properties of joint obtained through MHH. Experiments were carried out on Stainless Steel-316 & Mild Steel having the dimensions 25 mm × 12 mm × 6 mm using the 1KW multimode microwave system and metals were exposed to 450 s at a frequency of 2.45 GHz and power level of 900 W with charcoal powder as susceptor and Ni-based powder with average particle size of 40 μm as interface powder material. The joint exhibits the tensile strength of 346.6 Mpa with 13.5% of elongation. and Vickers microhardness of 133 Hv with measured porosity of the dissimilar joint of approximately 0.58% [4]. Ravindra badiger et al. achieved the joining of super alloys like inconel-625 through MHH. Joined the Inconel-625 alloy plates of dimensions 12 mm × 102 mm × 6 mm with the epoxy resin mixed with Ni-based powder as an interfacing powder. The experiment was carried out in multimode microwave oven at 900 W power level and a frequency of 2.45 GHz. Characterization of the obtained joint was carried out through XRD, Vickers microhardness, SEM & UTM. Reported that the formation of carbides of, molybdenum, nickel, and chromium is observed in microstructure and also the presence of titanium molybdenum oxide and nickel oxide were identified. Due to presence of carbides like chromium at the weld zone increased the hardness in the joint region. The tensile strength obtained was 35% of the base metals [2]. Shivinder singh et al. reported the joining of aluminum 6061 plate through MHH using epoxy resin and aluminum powder as a interface layer. Joining of aluminum 6061 alloy was carried out using the microwave multimode applicator with 10 min of exposure time at 2.45 GHz and power level of 900 W. Silicon carbide was used as a susceptor material at both top and bottom of plates to facilitate the microwave joining. The SEM images show good metallurgical bonding of aluminum powder with base metal. From the joint characterization it was reported as joint microhardness at the interface and on the substrate was measured 50.3 ± 10 Hv and 45.2 ± 10 Hv resp. and maximum value was measured at joint about 72.4 ± 10 Hv [5]. Ravindra I Badiger et al. studied the mechanical properties & microstructure of inconel-625 joint processed through MHH. Using the Ni-based powder as interface layer with coal powder as susceptor material. Metallurgical and mechanical characterization reveals that the good metallurgical bonding of interfacing powder particles with base metal and X-Ray Diffraction study shows the formation of carbides like chromium at the interface between base metal and fusion zone and also formation of Niobium carbide at the grain boundaries in the fusion zone. Microhardness test results shows that the maximum hardness is noted at the weld zone is 375 Hv, and porosity at the weld zone is 0.7%. The joint obtained through microwave hybrid heating was subjected to UTS and 3-point bend test where average UTS of the joint found to be 375Mpa with elongation of 9.2% and maximum flexural strength is calculated as 375 Mpa. Fractography study reveals that there was a mixed mode type of fracture [6]. Ravindra I Badiger et al. also joined the inconel-625 through the MHH with two power levels 600 and 900 W using Ni-based powder EWAC as the interfacing powder material. The joint produced with 600 W power yielded a fine grain structure compared to joint produced with 900 W power level and also the joint produced with 600 W power level showed 8–10% higher tensile strength and flexural strength resp [7]. Lucky Bagha et al. studied the comparative analysis of Stainless Steel 304-Stainless Steel 304 plates joined through MHH using different interfacing materials. 99.9% pure nickel-based powder and nickel-based EWAC powder were used to join the SS304 plates and characterization of joint was done through SEM, Rockwell hardness test, and XRD. The characterization study showed that Ni-based EWAC powder exhibited better weldability than 99.9% pure Ni-based powder [8].

Lucky Bagha et al. Joined the SS304-SS304 plates using the nickel-based as interface powder of 50 μm, 40 μm, 30 μm & 20 μm size through MHH and carried out the characterization through SEM, EDS, Microhardness and Micro Tensile test. The nickel-based powder is mixed with blumer 1400XX in the ratio of 75:25. It was observed that the ratio of nickel-blumer of 75:25 produces a good joint. The experiment results show that smaller size of nickel-based powder increases hardness of joint and HAZ [9]. Pawansoni et al. Joined the SS316-SS316 plates through Microwave hybrid heating using the nickel nanopowder as interfacing material. Experiment was carried out with 800 W power level at 2.45 GHz frequency and whole assembly was exposed to microwave radiation for 5 min. The characterization study showed that the joint was free from microcracks due to volumetric heating. The joint produced using nanosize particles as interface powder produces better tensile strength and hardness than joints produced with microsize particles as interface powder [10]. Rahul Samyal et al. conducted the experiments using different grades of stainless steel to predict the exposure time for microwave-based joining. 16 experiments were carried out for Stainless Steel-304, Stainless Steel-202, and Stainless Steel-316 at different dimensions with different thickness of separator sheet, exposure time and also with different dimensions of upper and lower brick with varying percentage content of alumina. For all the experiments, a 99.9% pure nickel-based powder with blumer 14500XX having a ratio of 75:25 was used as interface layer. From the experimental results it was reported that a sample size 10 mm × 10 mm × 5 mm for SS202-SS202 the exposure time to obtain a strong joint is 270 s and for SS304-SS304, SS316-SS316 with same sample size the exposure time to obtain a strong joint is 370 s and 330 s [11]. Amit Bansal et al. used microwave energy to join the mild steel plates. They carried out the experiment using microwave oven at a frequency of 2.45 GHz and power level of 900 W. The plates were joined using Ni-based powder as an interfacing layer, with a 0.3 mm gap between interfacing surfaces. It was reported through SEM images that, there was a complete melting of interfacing powder particles and metallurgical bonding with interfacing powder material and Mild Steel plates. The average microhardness measured was 420 ± 30 Hv, which is 1.82% of base metal. The joined specimen was subjected to a standard tensile test using UTM, which exhibited a tensile strength of 250 Mpa with percentage elongation of 6% [12]. Amit Bansal et al. also carried a metallurgical and mechanical characterization of MS–MS joined with different dimensions specimens through MHH and experimental study revealed that the joint exhibited a tensile strength of 50% of base material and microhardness of the joint is more at the joint than base material due to formation of various carbides like chromium carbide. The SEM images revealed that there was good metallurgical bonding between the interfacing powder and base material [13].

Shashiprakash dwivedi et al. Reported the parameters affecting the tensile strength of 1018 MS plates joined through microwave hybrid heating. The two plates were used to join with different rated output power, exposure time, and temperature. Based on experiment results the tensile strength of a 1018 MS joint decreased as the rate output power increased while it increased with increasing exposure time and temperature from lower to upper limit[14]. S Tamang et al. joined Al 6061-T6 and A231B by MHH using the active braze alloy of silver, copper, and titanium paste as the interfacing layer between the Mg and Al alloy. The joint was investigated using Scanning electron microscope, EDX, and X-Ray Diffraction. The EDX and XRD images reveal the formation of oxides like $MgAl_2O_4$, Tio and $Al_{60}Mg_{38}Ag_2$ intermetallic compound and reported that the formation of these oxides are the reason for the low strength of the joint[15]. M S Srinath et al. joined the bulk copper through MHH using the copper powder particles of 2 μm as the interface material and from the experimental results, it was revealed that better microhardness, porosity, and tensile strength of bulk copper joint were reported processed through MHH compared to TIG welded joints [16]. Satnam Singh et al. joined the cast iron using nickel-based powder as interlayer, charcoal as the susceptor, power level of 900 W, and reported that the SEM images reveal the formation of fine grain structure in the joint region. The joint exhibited a tensile strength of 90% of base material. Table 1 shows the process parameters for Microwave hybrid heating of different materials [17].

3 Characterization

The most important aspect after joining the metal plates through microwave hybrid heating is characterization of joint to investigate the change in properties and microstructure of material after joining. The characterization is done to determine the strength of material after the joining [18]. The characterization carried out by earlier researchers are SEM, EDS, XRD, Microhardness, charpy test, micro tensile test, 3-point bend test, porosity, fractography, and fatigue life.

3.1 Scanning Electron Microscope

SEM is a type of electron microscope that produces images of a sample by scanning the surface with a focused beam of electrons which reveals information about morphology, chemical composition, and crystalline structure [19, 20]. 1 nm of resolution can be achieved through SEM. Figure 2a and b shows the complete melting of interface powder particles microstructure indicates good metallurgical bonding with base metals [5].

Table 1 Process parameters for MHH of different materials

Sr. No	Materials	Interface powder	Susceptor	Power level (W)	Exposure time (S)
1	SS316 [3]	Ni-based powder	Charcoal powder	900	390
2	SS316-MS [4]	Ni-based powder	Charcoal powder	900	450
3	Inconel625 [2]	Ni-based powder	Charcoal powder	900	1260
4	Aluminum6061 [5]	Aluminum powder	SiC powder	900	600
5	Inconel625 [6]	Ni-based powder	Coal powder	900	1260
6	Inconel625 [7]	Ni-based powder	Coal powder	600 and 900	–
7	SS304 [8]	Ni-based powder	Charcoal powder	800	360,420 and 450
8	SS304 [9]	Ni-based powder	Charcoal powder	800	570
9	SS316 [10]	Ni-based powder	Charcoal powder	800	300
10	SS304 [11]	Ni-based powder	Charcoal powder	900	370
11	SS202 [11]	Ni-based powder	Charcoal powder	900	270
12	SS316 [11]	Ni-based powder	Charcoal powder	900	330
13	MS [12]	Ni-based powder	Charcoal powder	900	-
14	MS [13]	Ni-based powder	Charcoal powder	900	600
15	MS1018 [14]	Ni-based powder	Charcoal powder	800	1000
16	AL6061T6-AZ31B [15]	TiCuSil paste	Graphite powder	700	525
17	Copper [16]	Copper powder	–	900	300
18	Cast iron [17]	Ni Based powder	Charcoal powder	900	420

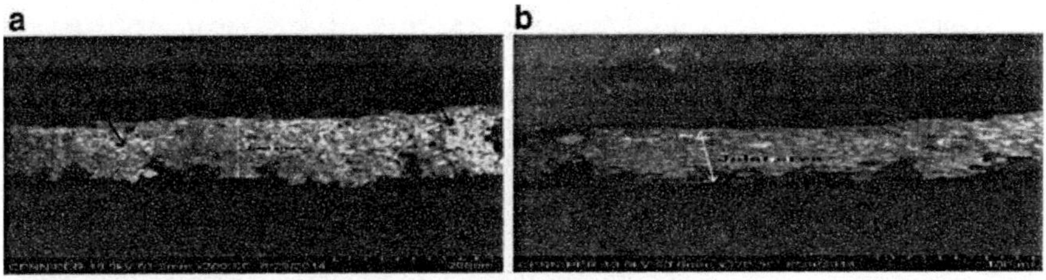

Fig. 2 **a** SEM image of the joint **b** Joint zone at higher magnification[5]

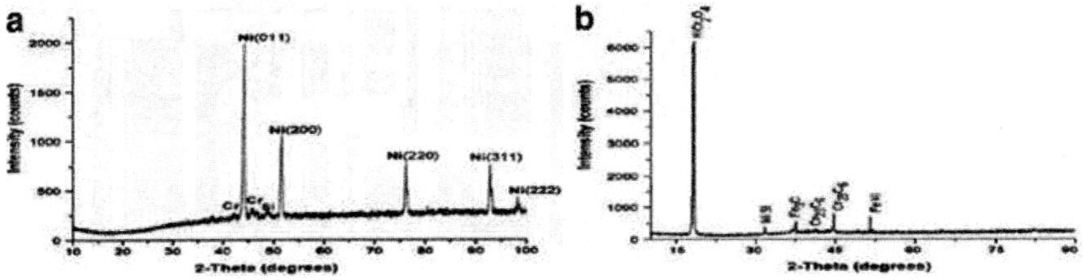

Fig. 3 Typical XRD spectra of **a** Ni-based powder; **b** Dissimilar joint developed through microwave hybrid heating [4]

3.2 XRD

X-Ray Diffraction is an experimental technique used for material characterization and phase identification of a crystalline material [21]. In XRD, generally results are reported as peak position at 2Θ Fig. 3a and b shows the presence of nickel-chromium carbide at angle 2Θ = 18.2° and formation of chromium carbide at the peak corresponding to 2Θ = 44.2° [4].

3.3 Microhardness

Microhardness test is carried out to determine the change in the hardness, when the test samples are very small or thin at microscopic level [22]. Vickers hardness test and Knop hardness test are commonly used test methods to determine hardness. Figure 4 obtained through SEM shows the indentation marks at the different point of joint area and Fig. 5 shows the hardness is decreased away from the joint area [2].

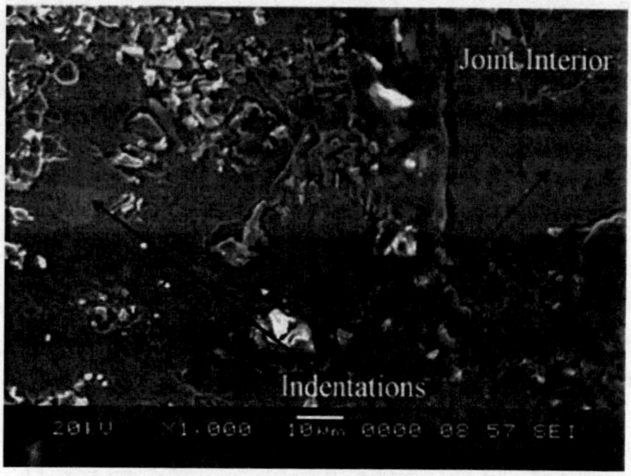

Fig. 4 Typical SEM microstructure showing the indentation marks [2]

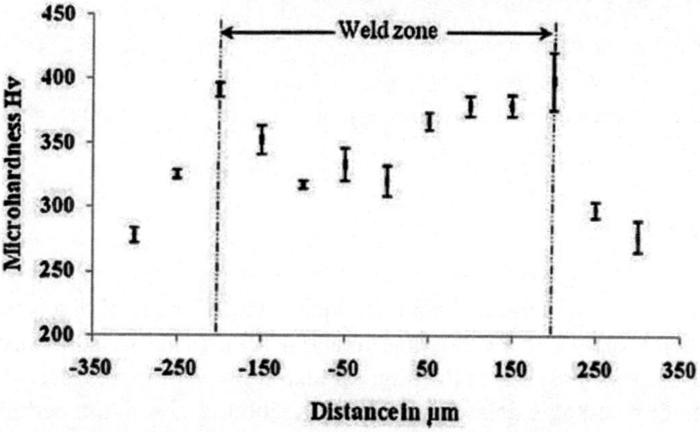

Fig. 5 Variation in microhardness profile from the center of the joint [2]

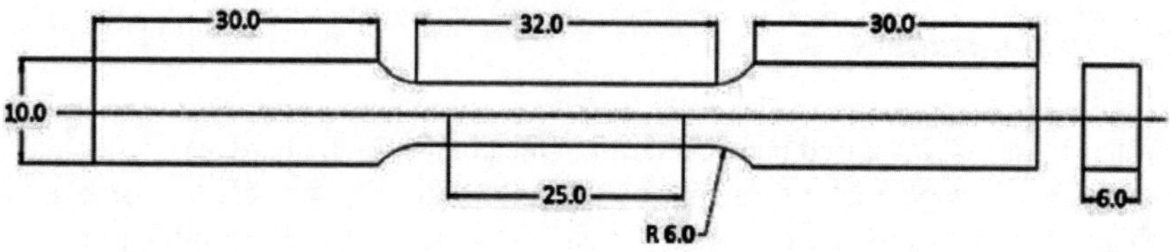

Fig. 6 ASTM E-8 standard tensile test specimen with dimensions in mm [2]

3.4 *Micro Tensile Test*

Micro tensile tests are carried out to determine the tensile strength of the joint through MHH, since the joint formed by MHH are very small. The joined specimen are prepared as per the ASTM standards using the wire cut EDM. Figure 6 shows the specimens prepared as per ASTM E-8 with dimensions.[2]

4 Conclusion

Similar and Dissimilar metals are joined through MHH using different interface powder and susceptor. The following conclusions are drawn based on the review study.

1. From the review study it is clear that hardness of the joint increases than the base metal through microwave hybrid heating.
2. Most of similar and dissimilar metals are not processed through MHH using nanoparticles as interfacing material.
3. The joint produced with nanosize powder interfacing material produces better tensile strength & hardness than the joints produced with microsize powder interfacing material.
4. Aluminum alloys were not being processed through MHH.
5. Fatigue life estimation of the joint formed through MHH is not studied properly.
6. From the literature study, it is clear that only small size specimens are processed through MHH.

References

1. Mishra, R. R., & Sharma, A. K. (2016). A review of research trends in microwave processing of metal-based materials and opportunities in microwave metal casting. *Critical Reviews in Solid State and Materials Sciences, 41*(3), 217–255.
2. Badiger, R. I., Narendranath, S., & Srinath, M. S. (2015). Joining of Inconel-625 alloy through microwave hybrid heating and its characterization. *Journal of Manufacturing Processes, 18*, 117–123.
3. Srinath, M. S., Sharma, A. K., & Kumar, P. (2011). A novel route for joining of austenitic stainless steel (SS-316) using microwave energy. *Proceedings of the Institution of Mechanical Engineers, Part B: Journal of Engineering Manufacture*. https://doi.org/10.1177/2041297510393451.
4. Srinath, M. S., Sharma, A. K., & Kumar, P. (2011). Investigation on microstructural and mechanical properties of microwave processed dissimilar joints. *Journal of Manufacturing Processes, 13*(2), 141–146.
5. Singh, S., Suri, N. M., & Belokar, R. M. (2015). Characterization of joint developed by fusion of aluminum metal powder through microwave hybrid heating. *Materials Today: Proceedings, 2*(4–5), 1340–1346.
6. Badiger,R. I., Narendranath, S., & Srinath, M. S. (2018). Microstructure and mechanical properties of Inconel-625 welded joint developed through microwave hybrid heating, *Proceedings of the Institution of Mechanical Engineers, Part B: Journal of Engineering Manufacture 232*(14), 2462–2477.

7. Badiger, R. I., Srinath, S. N. M. S., & Hebbale, A.M. (2019). Effect of power input on metallurgical and mechanical characteristics of Inconel-625 welded joints processed through microwave hybrid heating. *Transactions of the Indian Institute of Metals.*
8. Bagha, L., Sehgal, S., & Thakur, A. (2016). Comparative analysis of microwave based joining/welding of SS304-SS304 using different interfacing materials. *MATEC Web of Conferences, 57*, 03001.
9. Bagha, L., Sehgal, S., Thakur, A., & Kumar, H. (2017). Effects of powder size of interface material on selective hybrid carbon microwave joining of SS304–SS304. *Journal of Manufacturing Processes, 25*, 290–295.
10. Soni, P., Sehgal, S., Kumar, H., & Singh, A. P. (2018). Joining of SS316-SS316 through microwave hybrid heating by using Nickel nano-powder. 13(8), 6446–6449
11. Samyal, R., Bagha, A. K., Bedi, R. (2019) An experimental study to predict the exposure time for microwave based joining of different grades of stainless steel material. *Materials Today: Proceedings*
12. Bansal, A., Sharma, A. K., Kumar, P., & Das, S. (2012). Joining of mild steel plates using microwave energy. *Advances in Materials Research, 585*, 465.
13. Bansal, A., Sharma, A. K., & Das, S. (2013). Metallurgical and mechanical characterization of mild steel-mild steel joint formed by microwave hybrid heating process. *Sadhana—Academy Proceedings in Engineering Sciences, 38*(4), 679–686.
14. Dwivedi, S. P., & Sharma, S. (2014). Effect of process parameters on tensile strength of 1018 mild steel joints fabricated by microwave welding. *Metallography, Microstructure, and Analysis, 3*, 58.
15. Tamang, S., & Aravindan, S. (2017). An investigation on joining of Al6061-T6 to AZ31B by microwave hybrid heating using active braze alloy as an interlayer. *Journal of Manufacturing Processes, 28*, 94–100.
16. Srinath, M. S., Sharma, A. K., & Kumar, P. (2011) Microwave processing of metallic joints and their characterization. *I-manager's Journal on Mechanical Engineering, 1*(1).
17. Singh, S., Gupta, D., & Jain, V. (2016). An energy efficient processing route for advance ceramic composites using microwaves. *Advanced Ceramic Materials*, 97–143.
18. Garg, P., Jamwal, A., Kumar, D., Sadasivuni, K. K., Hussain, C. M., & Gupta, P. (2019). Advance research progresses in aluminium matrix composites: manufacturing & applications. *Journal of Materials Research and Technology, 8*(5), 4924–4939.
19. Jamwal, A., Seth, P. P., Kumar, D., Agrawal, R., Sadasivuni, K. K., & Gupta, P. (2020). Microstructural, tribological and compression behaviour of copper matrix reinforced with Graphite-SiC hybrid composites. *Materials Chemistry and Physics, 251*, 123090.
20. Thamaphat, K., Limsuwan, P., & Ngotawornchai, B. (2008). Phase characterization of TiO2 powder by XRD and TEM. *Agriculture and Natural Resources, 42*(5), 357–361.
21. Seth, P. P., Singh, N., Singh, M., Prakash, O., & Kumar, D. (2020). Formation of fine Mg2Si phase in Mg–Si alloy via solid-state sintering using high energy ball milling. *Journal of Alloys and Compounds, 821*, 153205.
22. Badra, V. V., Faraoni, J. J., Ramos, R. P., & Palma-Dibb, R. G. (2005). Influence of different beverages on the microhardness and surface roughness of resin composites. *Operative Dentistry, 30*(2), 213–219.

Evaluation of Cylindricity Deviation from Coordinate Measurement Data using a Volume Hunting Method

G. Rajamohan, G. Sai Krishna, and Shariqul Hoda

1 Introduction

Form tolerances, which control the shape deviations of the individual features, may be measured using the form measuring instruments or the coordinate measuring machine (CMM). The cylindricity form tolerance is considered in the present work. Cylindricity tolerance specifies a tolerance zone enclosed by two coaxial cylinders within which the specified surface must lie [1]. Several researchers have worked on the cylindricity evaluation using form data and CMM data. The literature review is limited to research works related to CMM data.

Roy and Xu exploited the properties of convex hulls and Voronoi diagrams to fit reference features with minimum radial separation [2]. Lai and Chen transformed a cylinder into plane using a nonlinear transformation and solved it by using a flatness evaluation scheme [3]. Yau and Menq proposed a geometric evaluation algorithm that inversely transforms the measured coordinates to best-fit a nominal geometry instead of reference feature [4]. Radhakrishnan et al. proposed an optimal characterization of minimax out-of-cylindricity problem and an iterative cyclic coordinate procedure that offers a near-optimal solution [5]. Hodgson et al. proposed a minimum shell method that searches for six points of MZC from given data [6]. Lai et al. proposed a genetic algorithm-based heuristic method for the measurement data comprising both random and deterministic elements [7].

Weber et al. used the Taylor expansion to develop a unified linear approximation technique and solved as linear equations [8]. Endrias and Feng defined minimum zone form evaluation problems based on the parameters of a rigid-body coordinate transformation and minimized them using the downhill simplex search [9]. Cheraghi et al. proposed a cylindricity evaluation method based on circularity evaluation

G. Rajamohan (✉) · G. Sai Krishna · S. Hoda
National Institute of Foundry and Forge Technology, Ranchi 834003, India
e-mail: grajamohan.nifft@gov.in

R. Agrawal et al. (eds.), *Recent Advances in Smart Manufacturing and Materials*, Lecture Notes in Mechanical Engineering,
https://doi.org/10.1007/978-981-16-3033-0_3

[10]. Zhu and Ding applied the kinematic geometry optimization by formulating and solving the LSC, MIC, MCC, and MZC as the problems of nonlinear constrained optimization and nonlinear least squares [11]. Gosavi and Phatakwala proposed a mathematical model for cylindricity evaluation using the theory of finite-differences derivative descent and Nelder-Mead search [12]. Shunmugam and Venkaiah used equidistant lines and diagrams to establish the MIC, MCC, and MZC and proposed a heuristic algorithm to find a unique convex inner hull [13].

Wen et al. proposed a particle swarm optimization (PSO)-based algorithm for simultaneous evaluation of MZC, MCC, and MIC [14]. Lee et al. proposed a support vector machines-based algorithm with a specific kernel function [15]. Zhang et al. proposed a hybrid of PSO and differential evolution (DE) that takes advantage of the fast convergence of PSO and ability of DE to avoid local minima [16]. Liu et al. used an adaptive ant colony optimization algorithm for MZC [17]. Lei et al. proposed a geometry optimization searching algorithm that collocates a hexagon each at starting and end measured sections and computes the radius using the lines between vertices of hexagons as ideal axes [18]. The MZC, MCC, and MIC are found using a simple procedure. Wang and Xu stated the constrained problem of MZC as unconstrained problem using coordinate transformations and solved using computational geometry [19].

Zheng et al. proposed linear programming models for cylindricity evaluation and a modified simplex method to solve them [20]. Pathak and Singh proposed an improved PSO algorithm for minimum zone form evaluation and a greedy selection procedure for selecting the best candidates [21]. Yang et al. proposed an improved harmony search algorithm, initialized by logistic chaotic initialization [22]. Cauchy mutation strategy was used with the best solution to improve the precision. Zheng et al. proposed a kinematic geometry optimization algorithm [23]. They extracted the feature points using a projective transformation and convex set construction methods. Wu et al. proposed an improved PSO algorithm by adding local search based on Latin hypercube sampling [24]. A control method has been designed to set the hypercube size.

The existing evaluation algorithms have resulted only in a moderate to very small improvement in results for the same dataset. The problem of finding the lowest cylindricity deviation, therefore, still remains unanswered and hence worth exploring. With this as motivation, this paper presents a volume hunting method for cylindricity evaluation based on MZC. The proposed method is compared with several existing algorithms using coordinate measurement data taken from the literature. Rest of the paper is organized as follows: Sect. 2 presents the problem formulation, Sect. 3 describes the proposed method, Sect. 4 presents and discusses the results obtained, and finally, Sect. 5 presents the conclusions and future scope.

2 Problem Formulation

The axis of two coaxial cylinders defining the cylindricity tolerance zone can be specified by any arbitrary point P_O (x_0, y_0, z_0) and directional cosines (l, m, n), as follows:

$$\frac{(x - x_0)}{l} = \frac{(y - y_0)}{m} = \frac{(z - z_0)}{n} \tag{1}$$

The normal distance e_i between any measured point P_i (x_i, y_i, z_i) on the surface of the cylinder and its axis is given by:

$$e_i = \sqrt{\left(a^2 + b^2 + c^2\right)/\left(l^2 + m^2 + n^2\right)} \tag{2}$$

In Eq. (2), a, b, and c are given as:

$$\begin{aligned} &a = (y_i - y_0) \cdot n - (z_i - z_0) \cdot m; b = (z_i - z_0) \cdot l - (x_i - x_0) \cdot n; \\ &c = (x_i - x_0) \cdot m - (y_i - y_0) \cdot l \end{aligned} \tag{3}$$

The minimum zone cylindricity deviation (s) is computed as follows:

$$s = \max(e_i) - \min(e_i) \tag{4}$$

3 The Proposed Method

Cylindricity is usually measured as circularity measurements in several sections along the workpiece length. The axis of two coaxial cylinders enclosing the cylindricity tolerance zone may be found by taking a point each from first and last measured sections by search methods, typically centered on least square solution. The other initialization methods have used certain approximations that require user interaction to get the initial axis. The proposed method uses a simple approximation to fix the endpoints (x_{L0}, y_{L0}, z_L) and (x_{R0}, y_{R0}, z_R) of the initial axis of reference cylinder as given below:

$$\begin{aligned} &x_{L0} = (x_{\max} + x_{\min})/2; \; y_{L0} = (y_{\max} + y_{\min})/2; \; z_L = z_{\min} \\ &x_{R0} = (x_{\max} + x_{\min})/2; \; y_{R0} = (y_{\max} + y_{\min})/2; \; z_R = z_{\max} \end{aligned} \tag{5}$$

where $x_{\max}$ and $x_{\min}$ are the maximum and minimum values of x-coordinates, $y_{\max}$ and $y_{\min}$ are that of y-coordinates, and $z_{\max}$ and $z_{\min}$ are that of z-coordinates. The points of first and last sections are used in Eq. (5) to get the respective initial centers. The z_L and z_R values do not change during the search. With non-sectional measurements, the overall coordinate values may be considered instead of first and last sections.

3.1 The Search Volumes

The volume hunting method is an iterative search procedure carried out using a search volume (*prismatic* or *cylindrical*), set on two initial centers obtained using Eq. (5).

Prismatic Search Volume. Initial centers of prismatic search volume are found using Eq. (5). The line joining (x_{L0}, y_{L0}, z_L) and (x_{R0}, y_{R0}, z_R) is taken as the initial axis. The cross-sectional length of side of square "*a*" is then divided into $n \times n$ units. The value of *n* is taken as 3 in this work (Fig. 1), which results in 4×4 intersection points (x_{Lj}, y_{Lk}, z_L) in the first section and (x_{Rp}, y_{Rq}, z_R). These are considered as auxiliary centers.

$$x_{Lj} = x_{L0} - \frac{a}{2} + j\left(\frac{a}{n}\right); \quad y_{Lk} = y_{L0} - \frac{a}{2} + k\left(\frac{a}{n}\right); \quad j, k = 0, \ldots, n \tag{6}$$

The x_{Rp} and y_{Rq} values can be found by substituting for x_{R0} and y_{R0}, and varying *p* and *q* from *0* to *n*. A total of 256 auxiliary axes are formed by taking the combinations of all auxiliary centers in first and last sections. The initial and auxiliary axes are used to compute the respective cylindricity values and used in the proposed method.

Cylindrical Search Volume. The initial axis and the corresponding cylindricity of cylindrical search volume are found in the same manner as detailed for prismatic search volume. The diameter of circle "*d*" is divided into *n* segments. A value of $n = 8$ leads to 8 intersection points (x_{Lj}, y_{Lj}, z_L) in first section and (x_{Rp}, y_{Rp}, z_R) in last section (Fig. 2). The auxiliary centers are calculated as follows:

$$x_{Lj} = x_{L0} + j\left(\frac{d}{2}\right)\cos\left(\frac{360}{n}\right); \quad y_{Lj} = y_{L0} + j\left(\frac{d}{2}\right)\sin\left(\frac{360}{n}\right);$$
$$j = 0, \ldots, (n-1) \tag{7}$$

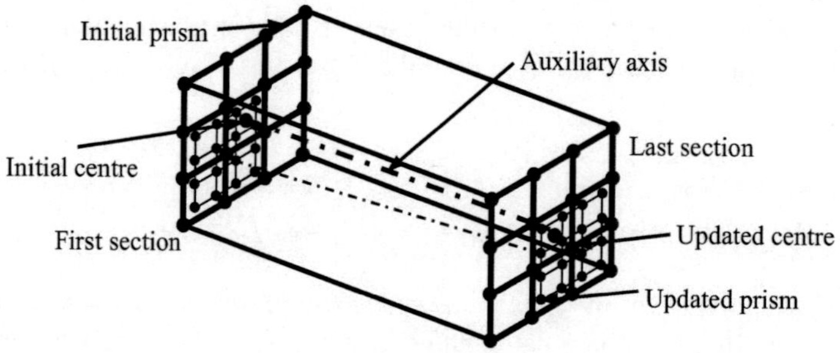

Fig. 1 Prismatic search volume

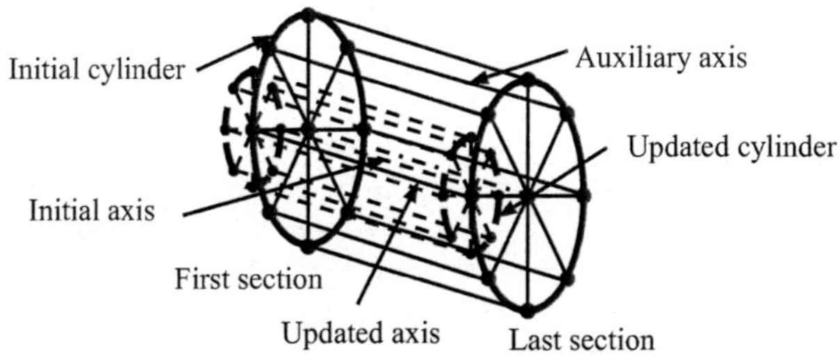

Fig. 2 Cylindrical search volume

The x_{Rk} and y_{Rk} values are found by substituting for x_{R0} and y_{R0}, and varying p from 0 to $(n - 1)$. The 64 auxiliary axes, formed by taking the combinations of all auxiliary centers in first and last sections, are used similar to the prismatic search volume.

3.2 *The Algorithm*

Figure 3 shows the steps involved in the proposed volume hunting method. The initial centers of first and last sections, i.e., (x_{L0}, y_{L0}, z_L) and (x_{R0}, y_{R0}, z_R), are found using Eq. (5). The cylindricity deviation (s_{cur}) is then computed using Eq. (2) and (4), with the line joining those points as the initial axis of reference cylinder. The search volume is divided to get the auxiliary centers, auxiliary axes, and cylindricity values. The lowest cylindricity value corresponding to auxiliary axes (s_{new}) is found and its location (j, k) in first section and (p, q) in last section (prismatic) or j in first section and p in last section (cylindrical) are saved. If the current value of a (or d) is less than ε, iterations will be stopped, and s_{new} will be the minimum deviation. Otherwise, the s_{new} and s_{cur} values are compared. If s_{new} is less than s_{cur}, iteration will be repeated by taking the coordinates of saved location as the new initial centers and retaining the size of search volume. If s_{new} is greater than s_{cur}, iteration will be repeated by retaining the initial center and reducing a (or d) by half.

4 Results and Discussion

The proposed MZC cylindricity evaluation method has been implemented using C++ language under Visual Studio 2015® Environment. Through several trials, the search volume dimensions (a and d) and iteration accuracy were identified as 0.40 units and 0.000001, respectively. Ten coordinate datasets have been taken from literature for testing and validation. Table 1 shows the literature results and those yielded by

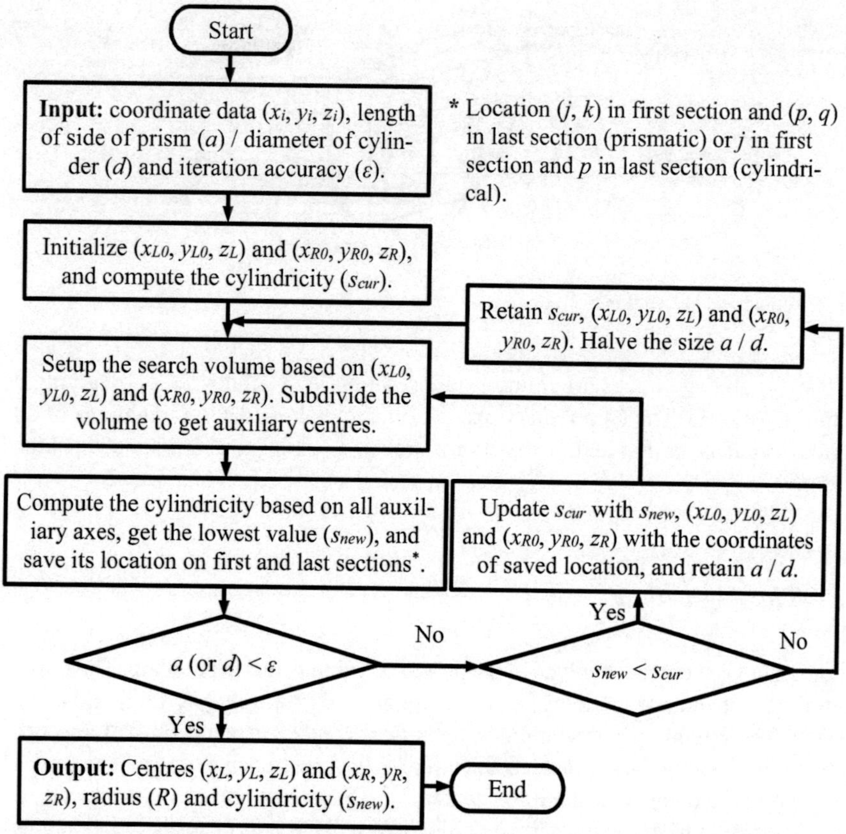

Fig. 3 Flowchart showing the proposed method

proposed algorithm. The acronym PP stands for proposed method with prismatic search volume and PC for proposed method with cylindrical search volume.

Table 1 reveals that the proposed algorithm has yielded smaller cylindricity deviations, except for datasets 4 and 7. These values are 0.003091 (PP), 0.000438 (both), 0.012663 (PP), 0.001801 (both), 0.000375 (PP), 0.106724 (PP), 0.026280 (both), and 0.031830 (PP), respectively. The respective lowest literature values are 0.0042 [8], 0.0005 [8], 0.014177 [11], 0.0019289 [13], 0.0004 [8], 0.1107 [6], 0.0288134 [2], and 0.03188 [22]. The proposed algorithm nearly equals the literature results of 0.183959 (PP) against 0.173520 [21] for dataset 4 and 0.019397 (PP) against 0.01938 [22] for dataset 7. Therefore, it can be stated that the proposed algorithm meets the set goals. The datasets 3 and 4 are not from section-wise measurements. Hence, the proposed algorithm works with non-sectional datasets also.

Though PP has resulted in smaller cylindricity values than the PC, the differences are mostly in the order of about 1×10^{-4} or better. Hence, the PC is equally effective. It may be noted that the prismatic search volume has 256 auxiliary axes, while the

Table 1 Results obtained from different datasets from literature

Dataset	Points	Method	Radius	Cylindricity	Iterations
1	16	Weber et al. [8]	–	0.0042	–
		PP	0.535027	0.003091	42
		PC	0.535034	0.003094	34
2	18	Weber et al. [8]	–	0.0005	–
		PP / PC	0.494673	0.000438	46 / 40
3	20	Zhu and Ding [11]	–	0.014177	–
		PP	34.999657	0.012663	68
		PC	34.999655	0.012669	52
4	20	Endrias and Feng [9]	–	0.183960	–
		Cheraghi et al. [10]	–	0.183950	–
		Gosavi and Phatakwala [12]	–	0.188621	–
		Wen et al. [14]	–	0.174635	–
		Zhang et al. [16]	–	0.183958	–
		Pathak and Singh [21]	–	0.173520	–
		Yang et al. [22]	–	0.183957	–
		PP	59.989505	0.183959	54
		PC	59.989414	0.184183	47
5	24	Shunmugam and Venkaiah [13]	–	0.0019289	–
		PP / PC	25.002600	0.001801	36 / 35
6	24	Weber et al. [8]	–	0.0004	–
		PP	0.199864	0.000375	36
		PC	0.199685	0.000376	33
7	32	Yang et al. [22]	–	0.019380	–
		PP	12.866333	0.019397	37
		PC	12.866332	0.019398	39
8	40	Hodgson et al. [6]	–	0.1107	–
		PP	32.592394	0.106724	55
		PC	32.592484	0.106911	41
9	64	Roy and Xu [2]	–	0.0288134	–
		PP / PC	5.000067	0.026280	51 / 44
10	80	Lei et al. [18]	–	0.0319	–
		Yang et al. [22]	–	0.03188	–
		PP	34.966987	0.031830	48
		PC	34.966987	0.031832	47

cylindrical search volume has just 64 auxiliary axes. It means that the prismatic search volume will need 257 × n (1 initial axis and 256 auxiliary axes) calculations in first iteration and 256 × n (256 auxiliary axes) calculations in subsequent iterations, where n is the number of measured points. The corresponding figures for the cylindrical search volume are 65 × n (1 initial axis and 64 auxiliary axes) and 64 × n (8 auxiliary axes), respectively. From the last column (Table 1), a comparison based on these facts clearly indicates that PC is very efficient and almost equally effective in comparison to PP.

5 Conclusions and Future Scope

A volume hunting method has been proposed for cylindricity evaluation based on the MZC. Two search volumes, viz. prismatic and cylindrical, have also been proposed. Ten example datasets and their corresponding published results have been taken from the literature for validating the proposed method. Both the search volumes have been found to yield nearly same results that are smaller than the literature results in eight of the ten datasets and at par in remaining two datasets. A comparison between prismatic and cylindrical search volumes revealed that the latter is faster as it requires far lesser computations, while being equally effective as the former. The proposed method using cylindrical search volume can therefore be of practical use. Future work may involve an extension of the proposed method to MIC and MCC-based cylindricity evaluation and optimal selection of search volume so as to reduce the number of iterations without compromising the accuracy.

References

1. ASME Y14.5–2009. (2009). Dimensioning and tolerancing—engineering drawing and related documentation practices. American Society of Mechanical Engineers, New York.
2. Roy, U., & Xu, Y. (1995). Form and orientation tolerance analysis for cylindrical surfaces in computer-aided inspection. *Computers in Industry, 26*(2), 127–134.
3. Lai, J. Y., & Chen, I. H. (1996). Minimum zone evaluation of circles and cylinders. *International Journal of Machine Tools and Manufacture, 36*(4), 435–451.
4. Yau, H. T., & Menq, C. H. (1996). A unified least-squares approach to the evaluation of geometric errors using discrete measurement data. *International Journal of Machine Tools and Manufacture, 36*(11), 1269–1290.
5. Radhakrishnan, S., Ventura, J. A., & Ramaswamy, S. E. (1998). The minimax cylinder estimation problem. *Journal of Manufacturing Systems, 17*(2), 97–106.
6. Hodgson, T. J., Kay, M. G., Mittal, R. O., & Tang, S. Y. (1999). Evaluation of cylindricity using combinatorics. *IIE Transactions, 31*, 39–47.
7. Lai, H. Y., Jywe, W. Y., Chen, C. K., & Liu, C. H. (2000). Precision modelling of form errors for cylindricity evaluation using genetic algorithms. *Precision Engineering, 24*(4), 310–319.
8. Weber, T., Motavalli, S., Fallahi, B., & Cheraghi, S. H. (2002). A unified approach to form error evaluation. *Precision Engineering, 26*(3), 269–278.

9. Endrias, D. H., & Feng, H. Y. (2003). Minimum-zone form tolerance evaluation using rigid-body coordinate transformation. *ASME Journal of Computing and Information Science in Engineering, 3*(1), 31–38.
10. Cheraghi, S. H., Jiang, G., & Ahmad, J. S. (2003). Evaluating the geometric characteristics of cylindrical features. *Precision Engineering, 27*(2), 195–204.
11. Zhu, L. M., & Ding, H. (2003). Application of kinematic geometry to computational metrology: distance function based hierarchical algorithms for cylindricity evaluation. *International Journal of Machine Tools and Manufacture, 43*(2), 203–215.
12. Gosavi, A., & Phatakwala, S. (2006). A finite-differences derivative-descent approach for estimating form error in precision-manufactured parts. *Journal of Manufacturing Science and Engineering, 128*(1), 355–359.
13. Shunmugam, M. S., & Venkaiah, N. (2010). Establishing circle and circular-cylinder references using computational geometric techniques. *International Journal of Advanced Manufacturing Technology, 51*, 261–275.
14. Wen, X., Huang, J., Sheng, D., & Wang, F. (2010). Conicity and cylindricity error evaluation using particle swarm optimization. *Precision Engineering, 34*(2), 338–344.
15. Lee, K., Cho, S., & Asfour, S. (2011). Web-based algorithm for cylindricity evaluation using support vector machine learning. *Computers & Industrial Engineering, 60*(2), 228–235.
16. Zhang, X., Jiang, X., & Scott, P. J. (2011). A reliable method of minimum zone evaluation of cylindricity and conicity from coordinate measurement data. *Precision Engineering, 35*(3), 484–489.
17. Liu, J., Wang, G., & Pan, X. (2011). Minimum-zone form tolerance evaluation for cylindrical surfaces using adaptive ant colony optimization. *Journal of Computational Information Systems, 7*(12), 4480–4490.
18. Lei, X., Song, H., Xue, Y., Li, J., Zhan, J., & Duan, M. (2011). Method for cylindricity error evaluation using geometry optimization searching algorithm. *Measurement, 44*(9), 1556–1563.
19. Wang, C., & Xu, B. S. (2014). Evaluation of cylindricity geometrical error based on calculational geometry. *Applied Mechanics and Materials, 722*, 359–362.
20. Zheng, P., Wu, J. Q., & Zhang, L. N. (2017). Research of the on-line evaluating the cylindricity error technology based on the new generation of GPS. *Procedia Engineering, 174*, 402–409.
21. Pathak, V. K., & Singh, A. K. (2017). Effective form error assessment using improved particle swarm optimization. *Journal of Metrology Society of India, 32*(4), 279–292.
22. Yang, Y., Li, M., Wang, C., & Wei, Q. Y. (2018). Cylindricity error evaluation based on an improved harmony search algorithm. *Scientific Programming*, 1–13.
23. Zheng, P., Liu, D., Zhao, F., & Zhang, L. (2019). An efficient method for minimum zone cylindricity error evaluation using kinematic geometry optimization algorithm. *Measurement, 135*, 886–895.
24. Wu, Q., Zhang, C., Zhang, M., Yang, F., & Gao, L. (2019). A modified comprehensive learning particle swarm optimizer and its application in cylindricity error evaluation problem. *Mathematical Biosciences and Engineering, 16*(3), 1190–1209.

Investigations of Laser Machining Parameters Using Taguchi Approach for Response Kerf Ratio

Sagar Hiwale and Rajiv Basavarajappa

1 Introduction

In laser machining process, a high-intensity beam is directed towards the surface of the workpiece material which heats the surface rapidly result in melting substance that blows away using assisted gas. This process is basically a fabrication process which is used to cut the parts made of different material. The cutting of material using laser process is complicated as it involves many factors like environment moisture, thermal conductivity, composition, and internal bond strength. Further, the selection of laser process parameters plays an important role to obtain the desired output characteristics. In this work, an attempt is made to obtain the optimum level of selected laser parameters that enhance desired output characteristics.

2 Literature Survey

The broad literature survey is attempted to acquire basic understanding of the laser cutting process and to identify literature gap for the current research work. Stournaras et al. [1] have evaluated the performance characteristics "*kerf width*", "*edge roughness*" and "*heat affected zone* (HAZ)" with respect to process parameters "laser power", "scanning speed", "pulsing frequency" and "gas pressure" using statistical analysis during machining of material aluminium alloy-5083. Pandey, and Dubey [2] have investigated laser process parameters, i.e., "gas pressure", "pulse width", "pulse frequency" and "cutting speed" to observe effects on *surface roughness* during machining of duralumin sheet using ANOVA analysis. They have used artificial

S. Hiwale (✉) · R. Basavarajappa
Department of Manufacturing Engineering and Industrial Management, College of Engineering Pune, Pune 411005, India

R. Agrawal et al. (eds.), *Recent Advances in Smart Manufacturing and Materials*, Lecture Notes in Mechanical Engineering,
https://doi.org/10.1007/978-981-16-3033-0_4

neural network to obtain the predicted model for *Ra* and applied genetic algorithm to obtain the optimum parameter setting. In the other paper, Pandey and Dubey [3] have investigated laser process parameters to observe effects on "*kerf taper*" and "*surface roughness*" during machining of titanium sheet (grade 5). They have attempted multi-objective optimization using genetic algorithm using the predicted model for considered characteristics. Sharma and Yadava [4] have attempted modeling and optimization for the considered performance characteristics "*kerf taper*" and "*average surface roughness*" using hybrid approach of Taguchi, RSM and Grey relational analysis. They have reported the influence of considered output characteristics with respect to input laser process parameters like, "oxygen pressure, pulse width, pulse frequency, and cutting speed". Eltawahni et al. [5] have investigated the laser parameters like, "laser power, cutting speed, air pressure and focal point position" on the performance characteristics kerf width ratio, surface roughness, operating cost for different workpiece material thickness of medium density fiberboard. In other paper, Eltawahni et al. [6] have investigated the same laser parameters on the performance characteristics *ratio of "kerf width"* and "*operating cost*" for different material thickness of plywood.

Hascalik and Ay [7] has performed the experimental investigation to analyze the effects of CO_2 laser cutting parameters "laser power, cutting speed and assisting gas pressure" on the cut quality parameter like, "*surface roughness, kerf geometry* and *recast layer thickness*" for the selected Inconel718 nickel-based superalloy. Adalarasan et al. [8] have reported the influence of laser parameters, i.e., "laser power", "pulsing frequency", "cutting speed" and "assist gas pressure" on "*kerf width*", "*surface roughness*" "*cut edge slope*" during machining of aluminium based Al6061/SiCp/Al_2O_3 composite. They have used grey-based RSM approach to obtain the optimum parameter setting for considered characteristics. In the other paper, Adalarasan et al. [9] have obtained the optimum setting for laser parameters using RSM approach and studied EDX plot along with P- profile of cutting edges using scanning electron microscope for Al6061/Al_2O_3 composite. Moradi et al. [10] have reported the effects of CO_2 laser cutting process parameters, i.e. "laser focal plane position, cutting speed and the laser power" on the "geometry" and "surface roughness" in specimen of polycarbonate injection molded sheets having thickness of 3.2 mm. Muthuramalingam et al. [11] have performed experimental investigation to obtain the influence of laser process parameters to enhance the *machinability* for machining of selected work piece material Titanium alloy. It was found that laser power has great influence on the response machinability. From the literature, it is observed that researcher has reported the effects of laser process parameters, optimized laser process parameters to enhance *kerf geometry* and studied surface characteristics during machining of *kerf*.

The objective of the present research paper is to investigate into Prima 4000 W CW-CO_2 laser cutting process of Hastelloy C-276 material. The inspiration for the current investigation is within the fact that the CO_2 based laser sources are frequently found in manufacturing shops and therefore, used for cutting engineering materials. As for assembly tight tolerance is prime requirement, quality of cutting is significant in machining of workpiece. In the present work, material Hastelloy C-276 is selected as it has corrosion resistance properties. The effect of selected laser process parameters on quality characteristics, *kerf ratio* for cutting considered material is evaluated using statistical analysis to obtain the contribution of each process parameter.

3 Experimental Details and Methodology

An experimental investigation is conducted on "prima 4000 W CW CO_2 laser process" during machining of material "Hastelloy C276" as depicted in Fig. 1. The material thickness is 3.7 mm. It investigates the influence of the selected laser parameters like, "laser power, cutting speed, gas pressure, working distance, and focal position" on *kerf ratio.* The range of process parameters is shown in Table 1 which is selected on the basis of trial experiments. All the runs are carried out using same size of nozzle at constant flow rate of nitrogen gas.

In the present work, long straight cut having size of 15 mm is made on each specimen to obtain the desired kerf characteristics. The schematic view of kerf characteristics is depicted in Fig. 1b. To evaluate the average of kerf width at top and

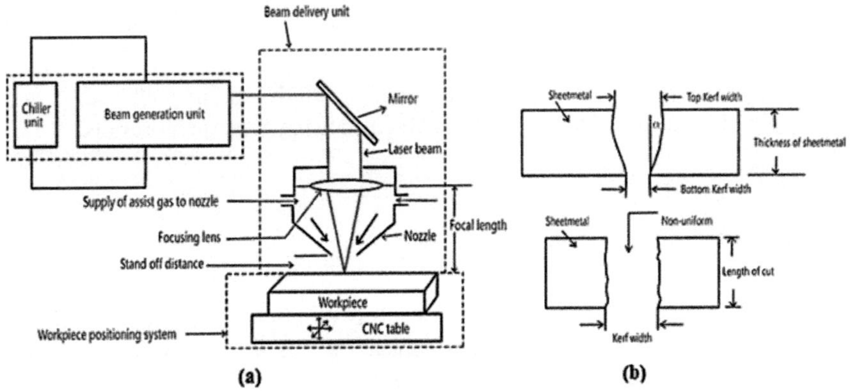

Fig. 1 **a** Line diagram experimental setup **b** Schematic view of kerf characteristics

Table 1 Range and Levels of the laser process parameters

Process parameters	Unit		Level	
		−1	0	+1
Laser power	watt	2400	3100	3800
Cutting speed	mm/min	1125	1762	2400
Gas pressure	bar	11	14	17
Working distance	mm	0.5	1	1.50
Focus position	mm	-4.2	-1.85	0.5

bottom surface, set of locations were selected along the cut on the specimen. An equipment stereo zoom microscope having trinocular body with working distance up to 100 mm and zoom range of 0.65 X- 4.5 X was utilized to measure kerf characteristics. After the measurement of the kerf width at the upper and lower surface, a desired output characteristics "*kerf ratio*" is computed using the Eq. (1).

$$kerf\ ratio = \text{Upper kerf width/Lower kerf width} \tag{1}$$

The design of experiments is based on Taguchi orthogonal array L_{27} [12–14]. In the present work, five process parameters (i.e., "laser power", "cutting speed", "gas pressure", "working distance" and "focal position") are considered and each parameter is varied at three levels. The parameter nozzle diameter is kept constant with diameter 2.5 mm. The range is selected which reflects the wide parameter range available on the set up of laser machine as given in Table 2. The objective of the present experimental work is to obtain minimum *kerf ratio* under the influence of Laser process parameters. Therefore, a quality characteristic of smaller-the-better (SB) for *kerf ratio* is applied in current study. The signal-to-noise (S/N) ratio [12, 13] is computed using the Eq. (2). As per rule, higher value of SN ratio is preferred for lower the better characteristics like *kerf ratio.*

$$S/N_{LB} = -10\log\left(y_{ij}^2\right) \tag{2}$$

where, y_{ij} is the output of *i*th quality characteristic at *j*th experimental trial. The performance parameter *kerf ratio* is obtained using experiment results.

4 Result and Discussion

The effect plot based on the mean value of the S/N ratio for each selected process parameter at given levels for *kerf ratio* is depicted in Fig. 2. The analysis of S/N ratio reveals that the optimal value of *kerf ratio* is obtained at laser power 3800 W (level 3), cutting speed 2400 mm/min (level 3), gas pressure 11 bar (level 1), working distance 1.5 mm (level 3) and focus position −4.20 mm (level 1). Hence, the optimal combination of laser process parameters for *kerf ratio* is A3, B3, C1, D3, and E1 as "laser power", "cutting speed", "gas pressure", "working distance" and "focal position" respectively.

The effects of selected laser process parameters on *kerf ratio* at each level are shown in Fig. 3. It is observed from Fig. 3 that the response "*kerf ratio*" is decreasing with respect to increase value of laser power and cutting speed, while decreases with respect to focus position. As the laser power and cutting speed increases the response *kerf ratio* is decreased. This is due to the fact that as the intensity of laser power increases the *bottom kerf width* would tend to be equal to its *top kerf width* and therefore, high laser power is required to have *kerf ratio* to be minimum or tends to 1. Furthermore, to reduce the machining time and cost, the user may select the

Table 2 Experimental result using L27 orthogonal array of Taguchi approach

Laser power (watt)	Cutting speed (mm/min)	Gas pressure (bar)	Working distance (mm)	Focus Position (mm)	Upper kerf width (mm)	Lower kerf width (mm)	Ratio of kerf width (U/L)	Taguchi S/N value
2400	1125	11	0.5	−4.2	0.391	0.332	1.178	−1.4229
2400	1125	11	0.5	−1.85	0.561	0.491	1.143	−1.1609
2400	1125	11	0.5	0.5	0.339	0.249	1.361	−2.6771
2400	1762	14	1	−4.2	0.675	0.353	1.910	−5.6206
2400	1762	14	1	−1.85	0.832	0.385	2.161	−6.6931
2400	1762	14	1	0.5	0.785	0.321	2.445	−7.7655
2400	2400	17	1.5	−4.2	0.339	0.398	0.852	1.3912
2400	2400	17	1.5	−1.85	0.362	0.401	0.903	0.8862
2400	2400	17	1.5	0.5	0.492	0.378	1.300	−2.2788
3100	1125	14	1.5	−4.2	0.881	0.589	1.490	−3.4637
3100	1125	14	1.5	−1.85	0.278	0.171	1.626	−4.2224
3100	1125	14	1.5	0.5	0.691	0.440	1.570	−3.9179
3100	1762	17	0.5	−4.2	0.510	0.452	1.128	−1.0461
3100	1762	17	0.5	−1.85	0.459	0.352	1.304	−2.3055
3100	1762	17	0.5	0.5	0.502	0.431	1.165	−1.3265
3100	2400	11	1	−4.2	0.468	0.361	1.296	−2.2521
3100	2400	11	1	−1.85	0.501	0.295	1.698	−4.5987
3100	2400	11	1	0.5	0.582	0.394	1.477	−3.3876
3800	1125	17	1	−4.2	0.501	0.321	1.561	−3.8680
3800	1125	17	1	−1.85	0.571	0.410	1.393	−2.8790
3800	1125	17	1	0.5	0.965	0.512	1.885	−5.5062
3800	1762	11	1.5	−4.2	0.845	0.772	1.090	−0.7485
3800	1762	11	1.5	−1.85	0.388	0.453	0.857	1.3403
3800	1762	11	1.5	0.5	0.331	0.375	0.883	1.0807
3800	2400	14	0.5	−4.2	0.543	0.414	1.312	−2.3586
3800	2400	14	0.5	−1.85	0.542	0.391	1.386	−2.8352
3800	2400	14	0.5	0.5	0.521	0.331	1.574	−3.9400

cutting feed rate as high as possible but it increases the level of inaccuracy in the desired response. Similarly, the response *kerf ratio* has sudden rise and fall with respect to process parameter gas pressure and working distance as depicted in Fig. 3. This is due to the fact that along the thickness of material as the certain range of considered process parameters, i.e., working distance and gas pressure limit crossed it has adverse effect to the *bottom kerf width zone* (due to reduced intensity of laser beam) which creates such sudden rise and fall pattern in response *kerf ratio.*

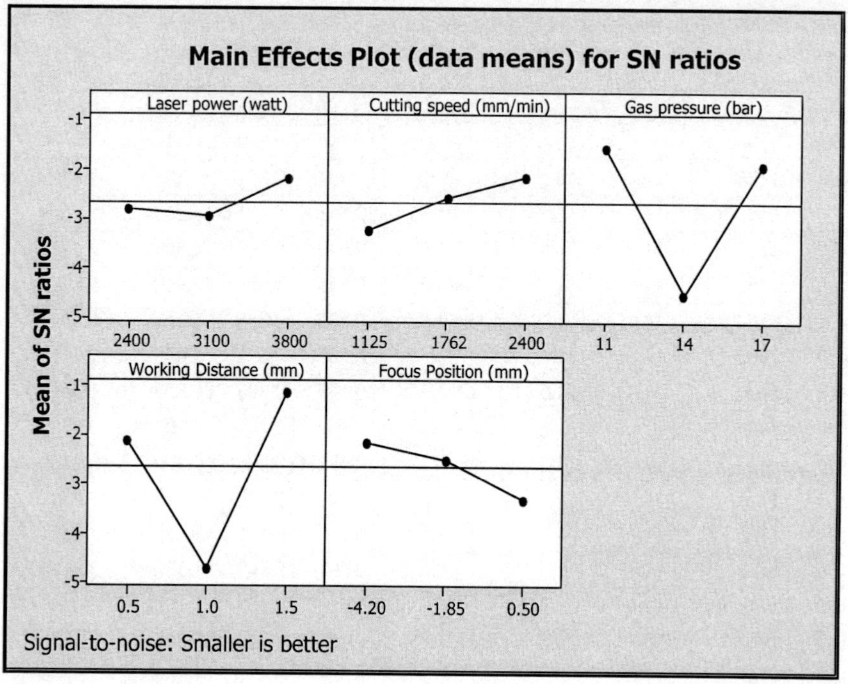

Fig. 2 S/N plot

In the present study, ANOVA analysis [12–14] is performed to obtain significant laser process parameters that affect the response *kerf ratio* as depicted in Table 3. It is observed from Table 3 that P-value of factor C, D, and E (i.e., gas pressure, working distance, focal position) is less than 0.05. Therefore, these laser process parameters have statistically significant effect on *kerf ratio* at 95% confidence level. The obtained % P values reveal that working distance has most significant effect followed by gas pressure on *kerf ratio*, if only % P is observed then it is clearly observed that focus position is significant but its percentage contribution is very less compared to working distance and gas pressure The result observed from the F-value of the significant parameters shows that working distance is significant than other parameters. The rank in the manner that affects the most to the *kerf ratio* for considered laser process parameters is depicted in Table 4.

Interaction plot of laser process parameter with respect to *kerf ratio* is depicted in Fig. 4. Here, the level of each process parameter is indicated on the horizontal axis and others laser process parameters with colored lines in matrix plot. It is observed from Fig. 4 that each row depicts the average mean of the response *kerf ratio* at each level for considered parameters. To plan the interaction plot gas pressure [11, 14, 17] in bar and working distance [0.5, 1, 1.5] in mm, the average mean of the response *kerf ratio* is computed at each level for the considered parameters using the

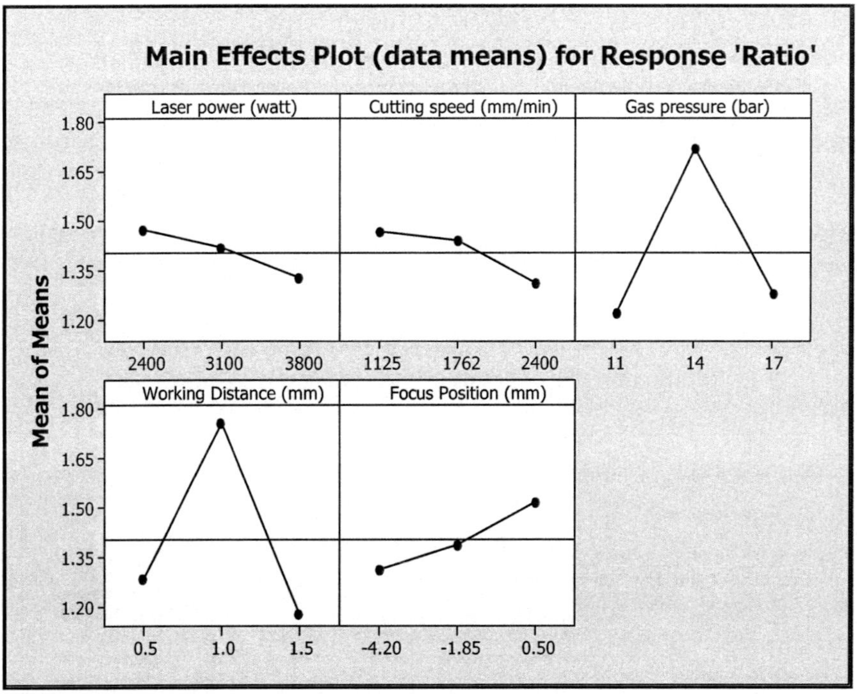

Fig. 3 Main effect plot

Table 3 ANOVA for Ratio of kerf width

Factor	Degree of freedom	Sum of square	Mean square	*F* value	*P* value	% P
A: Laser power (watt)	2	0.09746	0.04873	1.95	0.1740	2.503
B: Speed (mm/min)	2	0.12467	0.06233	2.50	0.1130	3.201
C: Gas pressure (bar)	2	1.34413	0.67206	26.96	0.0001	34.52
D: Working Distance (mm)	2	1.73522	0.86761	34.81	0.0001	44.56
E: Focus Position (mm)	2	0.19400	0.0970	3.89	0.042	4.982
Error	16	0.39882	0.02493			10.24
Total	26	3.89430				

$S = 0.15788$ $R^2 = 89.76\%$ R^2 (adj) $= 83.36\%$

Table 4 Rank of process parameter for response kerf ratio

Level	Laser power (watt)	Cutting speed (mm/min)	Gas pressure (bar)	Working distance (mm)	Focus position (mm)
1	1.473	1.467	1.220	1.283	1.313
2	1.417	1.438	1.719	1.753	1.386
3	1.327	1.311	1.277	1.175	1.518
Delta	0.146	0.157	0.499	0.580	0.205
Rank	**5**	**4**	**2**	**1**	**3**

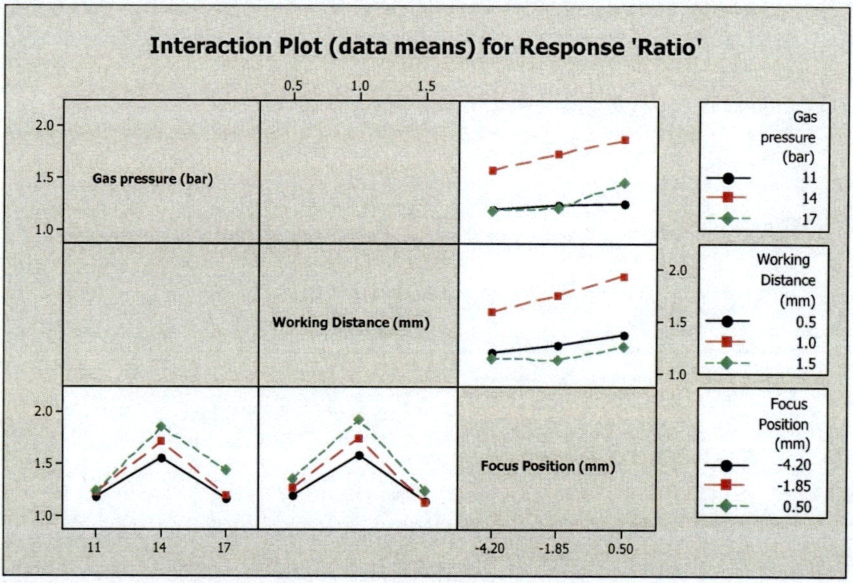

Fig. 4 Interaction plot of process parameter for kerf ratio

experimental values. So, combinations of mean response are obtained for considered process parameters.

4.1 Confirmatory Test Results

In order to confirm the results that are obtained using the Taguchi approach, confirmatory test is performed. From the Taguchi analysis, the optimal combination of process parameter is found at level of laser process parameters for *kerf ratio* is A3, B3, C1, D3, and E1. The corresponding process parameter values are "laser power" as 3800 W (level 3), "cutting speed" as 2400 mm/min (level 3), "gas pressure" as 11 bar (level 1), "working distance" as 1.5 mm (level 3) and "focus position" as –

Table 5 Confirmatory test result

Sr. No	Theoretical value	Confirmatory test result	Error	AverageError
1	1.091	1.080 (Trail–1)	0.011	0.012
2		1.082 (Trial–2)	0.009	
3		1.076 (Trial–3)	0.015	

4.20 mm (level 1) with response kerf ratio as 1.091. On performing the confirmatory test with same reading obtained using Taguchi analysis, the best value of response kerf ratio obtained is 1.082. The corresponding value of average percentage error obtained is 1.1%. As given in Table 5 Therefore, it is found from the confirmatory test that test result shows the percentage error is within 5%.

5 Conclusion

- The considered Taguchi-based approach is successfully applied to analyze the effect of "laser power", "cutting speed", "gas pressure", "working distance" and "focal position" on response kerf ratio.
- The significance of these considered laser process parameters is checked and it is observed that working distance followed by gas pressure has the most influence on the response *kerf ratio.*
- Furthermore, the percentage contribution of process parameter working distance and gas pressure on the kerf ratio is 44.56% and 34.52%, respectively.
- Using Taguchi approach, the optimal combination of process parameter is found at "laser power" as 3800 W, "cutting speed" as 2400 mm/min, "gas pressure" as 11 bar, "working distance" as 1.5 mm and "focus position" as −4.20 mm with response "*kerf ratio*" as 1.091. On performing the confirmatory test, the value of response *kerf ratio* obtained is 1.082.
- The corresponding value of error obtained between Taguchi approach and confirmatory test is as 1.1%. Therefore, the percentage is within 5%.

References

1. Stournaras, A., Stavropoulos, P., Salonitis, K., & Chryssolouris, G. (2009). An investigation of quality in CO2 laser cutting of aluminum. *CIRP Journal of Manufacturing Science and Technology, 2*, 61–69.
2. Pandey, A. K., & Dubey, A. K. (2013). Modeling and optimization of surface roughness in pulsed laser cutting of duralumin sheet. In 3rd *International Conference on Production and Industrial Engineering* (pp. 1033–1039).

3. Pandey, A. K., & Dubey, A. K. (2012). Simultaneous optimization of multiple quality characteristics in laser cutting of titanium alloy sheet. *Optics & Laser Technology, 44*, 1858–1865.
4. Sharma, A., & Yadava, V. (2012). Modelling and optimization of cut quality during pulsed Nd: YAG laser cutting of thin Al-alloy sheet for straight profile. *Optics & Laser Technology, 44*, 159–168.
5. Eltawahni, H. A., Olabi, A. G., & Benyounis, K. Y. (2011). Investigating the CO_2 laser cutting parameters of MDF wood composite material. *Optics & Laser Technology, 43*, 648–659.
6. Eltawahni, H. A., Olabi, A. G., & Benyounis, K. Y. (2013). Evalaution and optimization of laser cutting parameters for plywood materials. *Optics and Lasers in Engineering, 51*, 1029–1043.
7. Hascalik, A., & Ay, M. (2013). CO_2 laser cut quality of Inconel 718 nickel—based superalloy. *Optics Laser Technol., 48*, 554–564.
8. Adalarasan, R., Santhanakumar, M., & Rajmohan, M. (2015). Optimization of laser cutting parameters for Al6061/SiCp/Al_2O_3 composite using grey based response surface methodology. *Measurement, 73*, 596–606.
9. Adalarasan, R., Santhanakumar, M., & Thileepan, S. (2016) Selection of optimal machining parameters in pulsed CO_2 laser cutting of Al6061/Al_2O_3 composite using Taguchi-based response surface methodology (T-RSM). *International Journal of Advanced Manufacturing Technology*. http//doi:https://doi.org/10.1007/s00170-016-8978-5
10. Moradi, M., Mehrabi, O., Azdast, T., & Benyounis, K. Y. (2017). Enhancement of low power CO_2 laser cutting process for injection molded polycarbonate. *Optics & Laser Technology, 96*, 208–218.
11. Muthuramalingam, T., Moiduddin, K., Akash, R., Krishnan, S., Mian, S. H., Ameen, W., & Alkhalefah, H. (2020). Influence of process parameters on dimensional accuracy of machined Titanium (Ti-6Al-4V) alloy in Laser Beam Machining Process. *Optics & Laser Technology, 132*, 106494.
12. Taguchi, G., Chowdhury, S., & Wu, Y. (2004). *Taguchi's quality engineering handbook*. Newyork: Wiley.
13. Montogomery, D. C. (2010). *Design and analysis of experiments*. NJ, USA: Wiley.
14. Paneerselvam, R. (2012). *Design and analysis of experiment*. PHI Publisher, India.

A Review on Importance of Dielectric Fluids for Electro Discharge Machining (EDM)

Shatarupa Biswas, Yogesh Singh, and Manidipto Mukherjee

1 Introduction

English physicist Joseph Priestley first invented the erosive effect of the EDM process in 1770. After that, in 1940 two Russian Scientists B.R Lazarenko and N.I Lazarenko discovered the EDM method [1]. EDM is a non-traditional method [2]. The tool (electrode) and work material are submerged in the dielectric fluid when machining operation is ongoing. Work material and tool both are connected to an appropriate power supply (DC power supply). When the power supply is ON i.e. then dielectric fluids break down into positive and negative ions. At that time metal ions are released from the work material, and electrons are released from the tool. That's why creates a spark (hundreds of thousands of sparks produced in one second) between tool and work material. The ions (positive–negative) are combined together and strike the base metal. During the sparking process, the temperature reaches 8000–12,000 °C to erode the metal from the surface. EDM is mainly used for machining of hard metals (such as Inconel, Cover, Hastelloy etc.) and for those which are difficult to machine compared to the traditional method (such as milling, boring etc.) [3, 4]. In EDM those metals are electrically conductive (such as silver, copper etc.) can be machined. EDM machining performance is depended on process parameters (such as pulse on time, pulse off time, current, etc.) [5], and different types of dielectric fluids (such as kerosene, EDM oil, deionized water, etc.). In general, dielectric fluids are used to enhance the performance parameters (such as Material removal rate (MRR), Surface roughness (SR), Tool wear rate (TWR), Recast layer thickness (RLT), Surface hardness (SH), Microstructures, etc.) of machining. As dielectric fluid kerosene, EDM oil, vegetable oil, etc. can be used [6]. When these liquids are used as a dielectric medium, then the

S. Biswas (✉) · Y. Singh
NIT Silchar, Silchar, Assam 788010, India

M. Mukherjee
CAMM, CSIR-CMERI, Durgapur, West Bengal 713209, India

R. Agrawal et al. (eds.), *Recent Advances in Smart Manufacturing and Materials*,
Lecture Notes in Mechanical Engineering,
https://doi.org/10.1007/978-981-16-3033-0_5

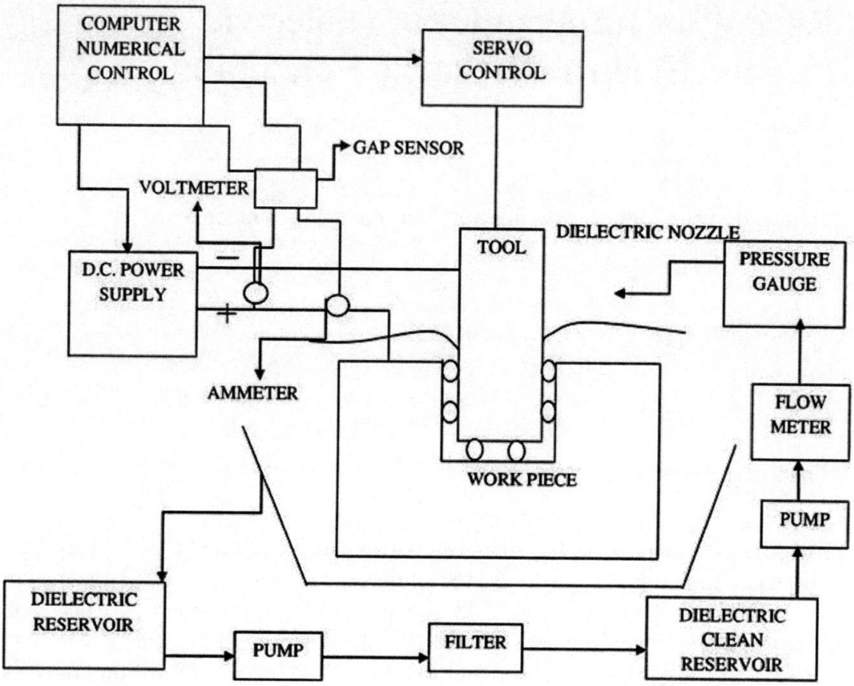

Fig. 1 Block diagram of EDM

process is called wet EDM machining. The main function of dielectric fluids are to create a spark between the tool or electrode, act as a semiconductor, controlled spark gap, cool the work material, and flush away the debris in the time of machining. Figure 1 shows the block diagram of EDM.

In this article it is reviewed that the uses of various dielectric fluids in EDM for improving measured responses.

2 Parameters in EDM

Generally, in EDM two types parameters are used. These are process parameters and response parameters. But response parameters are depended on process parameters. Some important process and response parameters are described below.

Table 1 Process parameters on EDM

Process parameters	Description
Pulse on time (T_{on})	It is expressed as the total duration of discharge time between the tool and work material. It's measured in a microsecond (μs)
Pulse off time (T_{off})	It is expressed as the non-discharge time between the tool and work material. It's also measured in a microsecond (μs)
Voltage (V)	During T_{on}, specific voltage drops and discharge voltage (V_d) is significantly influenced by the dielectric strength
Current (I)	The current, measured in amp, is the measure of the number of electrical charges flowing between the tool and work material
Electrode polarity	The polarity is either positive (+) or negative (−), based on the necessity of work

2.1 Process Parameters

Responses are controlled by using process parameters. The most significant process parameters are described briefly in Table 1.

2.2 Response Parameters

Some responses are described in Table 2.

3 Dielectric Fluids

In EDM dielectric fluid is important for machining. The dielectric fluid acts as a semiconductor and insulator between the tool and the work material to facilitate a stable and regulated spark gap when ionization occurs. Throughout machining operations, it also eliminates the dust content from the machining surface. The four key functions of dielectric fluids are insulation, ionization, cooling, dust particle removal [14]. The dielectric fluid has certain properties (such as high flash point, low viscosity, light coloured, rust-resistant etc.) and it increases the tool life, improves efficiency and enhances surface texture consistency. Some examples of dielectric fluids are kerosene, hydrocarbon oil, jatropha oil, EDM oil, vegetable oil etc. which are briefly described below.

3.1 Kerosene

Kerosene is easily available, and compared to other dielectric oils it is very reasonable as well [15]. It's also called paraffin oil or coal oil. The important properties of kerosene are, it has no defined cetane-number, less lubricity, and large delays in the

Table 2 Response parameters on EDM

Response parameter	Description
Material removal rate (MRR)	The volumetric removal of material per unit time can be described as MRR. It is closely related to the dielectric fluid used for machining purposes [7, 8]. The estimation of MRR takes the following equation [9] $\text{MRR} = \frac{\text{volume of the material remove}}{\text{machining time}} = \frac{w_{sb} - w_{sa}}{\rho \times t_m}$ where, w_{sb} denotes the work material's weight before machining and w_{sa} is the work material's weight after machining. ρ is the density of work material and t_m is the machining time
Tool wear rate (TWR)	The corrosion of cutting tools due to any machining operation in the EDM process is called TWR. TWR is calculated by using this equation [9] $\text{TWR} = \frac{\text{volumetric removal of the tool material}}{\text{total machining time}} = \frac{w_{tb} - w_{ta}}{\rho \times t_m}$ where w_{tb} denotes the weight of the tool before machining and w_{ta} is the weight of tool after machining. ρ is the density of tool and t_m is the machining time
Wire wear ratio (WWR)	It is defined as the wire weight loss (WWL) to the wire initial weight (IWW). The WWR is calculated using the following equation [10] $\text{WWR} = \frac{\text{WWL}}{\text{IWW}}$
Surface quality (SQ)	It is usually dependent on the texture and on the undulation of the work piece's surface. For that reason, it is also often called specimen topographic property. The SQ is one of the important parameters of performance that is based on the dielectric fluid used for EDM [11, 12]
Recast layer thickness (RLT)	During the machining time between tool and work piece, the spark is made. Resulting in a sudden rise in localized heat which melts the machining surfaces to an extent followed by rapid quenching due to the presence of dielectric fluid. This phenomenon produces a surface layer that resembles the cast state, and is known as RLT [13]

ignition. Its density is 0.80 g/millilitre, flashpoint lies between 37 and 65 °C (100–150 °F), and its auto-ignition is 220 °C (428 °F).

3.2 *Hydrocarbon Oil*

Hydrocarbon oil is found in petroleum. It is a mixture of hydrogen and carbon and its basic properties are high flashing point, and attributable to a petroleum product. But during the machining time of EDM, it can produce different forms of poisonous gases, which is therefore very dangerous for the environment and operators.

3.3 *Jatropha Oil*

Jatropha seeds are very rich in oil (40%). It has 4.2% ash content including minerals, such as Ca (Calcium), Mg (magnesium), K (Potassium), and Na (Sodium). The oil has high levels of unsaturated fatty acids particularly oleic (44.7%) and linoleic (32.8%) [16].

3.4 *EDM Oil*

It is highly refined oil, with outstanding resistance to oxidation. The oil has no colour and odours, and has a high flash point which helps to minimize the risk of fire during the machining process. It is a low viscose fluid, providing good flow through the spark gap between tool and specimen. In general, EDM oil is available in the market in three grades (grade 2, grade 3, grade 4).

3.5 *Vegetable Oil*

Vegetable oils are extracted from a variety of seeds (such as rapeseed, soybean, corn, sunflower, peanut etc.). These kinds of oils are also becoming very popular in EDM because of some of the advantages (not harmful) and availability [17].

4 Literature Review on EDM

Various researches have been done on EDM by using several dielectric fluids, which is shown below in Table 3.

5 Conclusions

The above overall review suggest that the dielectric fluid is an integral part of the EDM process. And it plays a pivotal role in order to find the values of various response parameters (such as MRR, TWR, SR, EWR, RLT etc.). It is also clear that for finding several response parameters, all types of dielectric fluids alone may not perform best. The following observations are taken from previous research work:

- Kerosene is used extensively as a dielectric fluid because of its availability and cheap price compare to other dielectric fluids.

Table 3 Literature review on EDM

Author and Year	Metal	Tool	Dielectric fluid	Finding
Chen et. al. [18]	Ti–6A1–4V	Copper	Kerosene	It is found that after machining the crack density is low and the depth of the white layer is thick. But carbide is formed on the work piece surfaces
Bai and Koo [19]	Haynes 230	Al–Mo composite	Kerosene	In this study, it is shown that the max hardness value is 1086 HK and minimum SR (2.62 μm). For measuring both positive polarity gives better result
Wu et. al. [20]	SKD 61steel	Copper	Kerosene	It is discovered that after cutting a slot, over cut value is 0.016 mm and RLT is 5 μm
Bhattacharya et. al. [21]	H11, HCHCr, AISI 1045	Graphite, W-Cu, Brass	Kerosene	It is observed that I and T_{on} are the main input parameters for finding SR. When kerosene is used as dielectric contributes better cooling effect
Mehta [7]	Inconel 718	Copper	Kerosene	In this study, it is shown that the percentage of the difference between actual value and predicted value of MRR is lies between 0.38 and 2.38

(continued)

Table 3 (continued)

Author and Year	Metal	Tool	Dielectric fluid	Finding
Valaki et. al. [22]	P20 + plastic mould steel	Copper	Kerosene	It is found that the values of MRR (29.21mm^3/min) is max when I value is 18 A, T_{on} value is 200 μs, T_{off} value is 20 μs, V value is 50v. But SR (4.89 μm) is min when I value is 3 A, T_{on} value is 200 μs, T_{off} value is 20 μs, V value is 50 v. MH is (339 HV) max when I value is 9 A, T_{on} value is 11 μs, T_{off} value is 20 μs, V value is 50v
Singh and Sharma [23]	Cobalt-bonded tungsten carbide	Copper	Kerosene	It is observed that maximum MRR (8.72mm^3/min when T_{on} (100 μs), I (9 A), flushing pressure (0.6 kg-cm^{-2}). TWR in minimum (0.678 mm^3/min) when T_{on} (100 μs), I (3 A), flushing pressure (0.6 kg-cm^{-2})
Niamat et. al. [9]	Al 6061 T6 alloy	Copper	Kerosene	It is discovered that the value of MRR is maximum (302.47mm^3/min) when increased I (12 A) and T_{on} (45 μs), EWR is minimum (0.56mm^3/min) when I (6A), T_{on} is 45 μs respectively. When T_{on} and I value increased then shows more crack and micro holes

(continued)

Table 3 (continued)

Author and Year	Metal	Tool	Dielectric fluid	Finding
Li et. al. [24]	Inconel 738	Hollow brass	Kerosene	In this study, measured RLT is 4.8 μm, but the machining speed was low due to the thermal corrosion
Rahang and Patowari [25]	Aluminium	P/M green compacts	Hydrocarbon oil	In this study, it is shown that low I and shorter T_{on} caused small damage to the masking edge with low thermal energy within a short duration
Valaki et. al. [22]	P20 + plastic mould steel	Copper	Jatropha oil	In this literature, it is found that when I (18A), T_{on} (200 μs), T_{off} (20 μs), V (50 v) then shows max MRR (53.33mm^3/min). When SR is minimum (3.66 μm) and MH (354HV) is maximum then I (9 A), T_{on} (11 μs), T_{off} (20 μs),V (50v) respectively
Das et. al. [26]	Ti–6Al–4 V	Copper	Jatropha oil	In this article, it is shown that MRR is improved by 15% compared to other dielectric fluids(canola oil, neem oil)

(continued)

Table 3 (continued)

Author and Year	Metal	Tool	Dielectric fluid	Finding
Pradhan and Biswas [27]	AISI D2	Brass	EDM oil	It is shown when I and T_{on} are increased then WLT, (white layer thickness) also increases. When T_{on} decreases, higher numbers of holes and micro cracks is detected in the work piece
Dewangan et.al. [28]	AISI P20	Copper	EDM oil	In this study, both I and T_{on} are straight proportional to WLT. Tool work time and Tool lift time have no significant effect on WLT
Dhakar et. al. [29]	HSS T2 grade	Copper	EDM oil	In this study it is found that MRR is high (3.58 mm^3/min) compare to water. Wear ratio is 0.94%
Mohanty and Mohanty [30]	D2 steel	Direct metal laser sintered (DMLS) part	EDM oil	In this study, when TWR (4.61mm^3/min) is minimum then I (10A), T_{on} (100 µs), T_{off} (20 µs), flushing pressure (0.9 kg-cm^{-2}). But when MRR is maximum (12.83 mm^3/min) then I (12 A), T_{on} (100 µs), T_{off} (30 µs), flushing pressure (0.6 kg-cm^{-2}). When SR is minimum (3.14 µm) then I(8A), T_{on} (200 µs), T_{off} (30 µs), flushing pressure (0.9 kg-cm^{-2}) respectively

(continued)

Table 3 (continued)

Author and Year	Metal	Tool	Dielectric fluid	Finding
Mishra et. al. [31]	Inconel 625	Copper	EDM oil	It is observed that when T_{on} increase then increased SR, SCD, and WLT respectively and optimum values of parameters are I (15 A), T_{on} (300 μs), V (20v) respectively. Then responses values are MRR (70.32mm^3/min), SR (5.3 μm), SCD (μm/μm^2) and WLT (10.25 μm)
Valaki and Rathod [32]	M238 HH	Copper	Vegetable oil	In this study, it is shown that TWR is minimum when T_{off} (20 μs), and I (9 A)

- When kerosene is used as a dielectric fluid then, for Al 6061 alloy maximum MRR (302.47mm^3/min) shows when I (12 A), T_{on} (45 μs) and EWR is minimum (0.56 mm^3/min) when I (6 A), T_{on} (45 μs) respectively. For Cobalt-bonded tungsten carbide minimum TWR (0.678 mm^3/min) shows when T_{on} (100 μs), I (3 A), flushing pressure (0.6 kg-cm^{-2}). For P20 + plastic mould steel MH is (339HV) max when I (9A), T_{on} (11 μs), T_{off} (20 μs), V (50 v) respectively. When jatropha oil is used as a dielectric fluid then, for P20 + plastic mould steel shows maximum MRR (53.33mm^3/min) when I (18 A), T_{on} (200 μs), T_{off} (20 μs), V (50 v) respectively and SR is minimum (3.66 μm) and MH (354HV) is maximum then I (9 A), T_{on} (11 μs), T_{off} (20 μs), V (50 v) respectively.
- When EDM oil is used as a dielectric fluid then, for Inconel 625 shows maximum MRR (70.32mm^3/min), when I (15 A), T_{on} (300 μs), V (20 v) respectively. For D2 steel shows minimum SR (3.14 μm) when I (8 A), T_{on} (200 μs), T_{off} (30 μs), flushing pressure (0.9 kg-cm^{-2}) and TWR (4.61mm^3/min) is minimum when I (10 A), T_{on} (100 μs), T_{off} (20 μs), flushing pressure (0.9 kg-cm^{-2}) respectively.
- Vegetable oil and Hydrocarbon oil both are fewer used in previous work. But for Aluminum hydrocarbon oil minimized the damage of masking edge.

References

1. Ho, K. H., & Newman, S. T. (2003). State of the art electrical discharge machining (EDM). *International Journal of Machine Tools and Manufacturing, 43*, 1287–1300
2. Singh, G., & Dhiman, D. P.: Review: parametric optimization of Edm machine using Taghuchi & Anova Technique. *International Research Journal of Engineering and Technology*, 783–788.
3. Jamwal, A., Aggarwal, A., Gautam, N., & Devarapalli, A. (2018). Electro-discharge machining: recent developments and trends. *International Research Journal of Engineering and Technology, 5*, 433–448.
4. Nayim, S. T. I., Hasan, M. Z., Jamwal, A., Thakur, S., & Gupta, S. (2019, September). Recent trends & developments in optimization and modelling of electro-discharge machining using modern techniques: a review. In *AIP Conference Proceedings* (Vol. 2148, No. 1, p. 030051). AIP Publishing LLC.
5. Kansal, H. K., Singh, S., & Kumar, P. (2005). Application of Taguchi method for optimisation of powder mixed electrical discharge machining. *International Journal Manufacturing and Management, 7*, 329–341.
6. Yu, Z., & Jun, T. K. (2004). Dry electrical discharge machining of cemented carbide. *Masanori Journal of Material Processing Technology, 149*, 353–357.
7. Mehta,H. N. (2015) Modeling of electrical discharge machining process. *International Journal of Engineering Research and Technology, 4*(6), 153–157
8. Upadhyay, C., Datta, S., Masanta, M., & Mahapatra, S. S. (2018) An experimental investigation on electro discharge machining of Inconel 601. *International Journal of Industrial and Systems Engineering. 29*(2), 223–251.
9. Niamat, M., Sarfraz, S., Aziz, H., Jahanzaib, M., Shehab, E., & Ahmad, W. (2017) Effect of different dielectrics on material removal rate. Electrode Wear Rate and Microstructures in EDM. Procedia CIRP 60, 2–7.
10. Pitayachaval, P., Jittamai, T., & Baothong. (2017) A review of machining parameters that effect to wire electrode wear. In *4th International Conference on Industrial Engineering and Applications* (pp. 1–4). IEEE, Japan

11. Rahul, S., Datta, M., Masanta, B. B., Biswal, S., & Mahapatra., S. (2018). Analysis on surface characteristics of electro discharge machined Inconel 718. *International Journal of Materials and Product Technology, 56*(1/2), 135–168.
12. Mohapatra, K. D., & Sahoo, S. K. (2017) *Microstructural analysis of titanium alloy gear using WEDM process*. World Scientific, 1–20.
13. Patowari, P. K., & Mishra, P. K. (2010). Surface modification of C40 steel using WC-Cu P / M green compact electrodes in EDM. *International Journal of Manufacturing Technology and Management, 21*, 83–98.
14. Chakraborty, S., Dey, V., & Ghosh, S. K. (2014). A review on the use of dielectric fluids and their effects in electrical discharge machining characteristics. *Precision Engineering*, 1–6.
15. Singh, K., Agarwal, A. K., & Yadav, R. (2017). Effect of dielectric fluids used on edm performance : a review. *International Journal of Emerging Technologies in Engineering Research, 5*(10), 10–16
16. Nzikou,J. M., Matos, L., Mbemba, F., Ndangui, C. B., Pambou-Tobi, N. P. G., Kimbonguila, A., Silou, T., Linder, M., & Desobry, S. (2009). Characteristics and composition of Jatropha curcas Oils, Variety Congo-Brazzaville. *Research Journal of Applied Science, Engineering and Technology, 1*(3), 154–159
17. Singaravel, B., Shekar, K. C., Rao, K. M. & Reddy, G. G. (2018). Study of vegetable oil and their properties for as an alternative source to mineral oil-based dielectric fluid in electric discharge machining. *International Journal of Modern Engineering and Research Technology, 5*, 237–244
18. Chen, S. L., Yan, B. H., & Huang, F. Y. (1999). Influence of kerosene and distilled water as dielectrics on the electric discharge machining characteristics of Ti–6A1–4V. *Journal of Materials Processing Technology, 87*, 107–111.
19. Bai, C., & Koo, C. (2006). Effects of kerosene or distilled water as dielectric on electrical discharge alloying of superalloy Haynes 230 with Al–Mo composite electrode. *Surface and Coating Technology, 200*, 4127–4135.
20. Wu, K. L., Hwa, B., Lee, J., & Gian, C. (2009). Study on the characteristics of electrical discharge machining using dielectric with surfactant. *Journal of Materials Processing Technology, 209*, 3783–3789.
21. Bhattacharya, A., Batish, A., & Kumar, N. (2013). Surface characterization and material migration during surface modification of die steels with silicon, graphite and tungsten powder in EDM process. *Journal of Mechanical Science and Technology, 27*(1), 133–140.
22. Valaki, J. B., Rathod, P. P., & Sankhavara, C. D. (2016). Investigations on technical feasibility of Jatropha curcas oil based bio dielectric fluid for sustainable electric discharge machining (EDM). *Journal of Manufacturing Process, 22*, 151–160.
23. Singh,J., & Sharma, R. K. (2016). Assessing the effects of different dielectrics on environmentally conscious powder-mixed EDM of difficult-to-machine material (WC-Co). *Front Mechanical Engineering*, 1–14.
24. Li, C., Xu, X., Li, Y., Tong, H., Ding, S., Kong, Q., Zhao, L., & Ding, J. (2018). Effects of dielectric fluids on surface integrity for recast layer in high speed EDM drilling of nickel alloy. *Journal of Alloys and Compounds*, 1–16.
25. Rahang, M., & Patowari, P. K. (2015). Application of masking technique in EDM for generation of rectangular shaped pattern. *International Journal of Precision Technology, 5*(2), 140–156.
26. Das, S., Paul, S., & Doloi, B. (2019). An experimental and computational study on the feasibility of bio-dielectrics for sustainable electrical discharge machining. *Journal of Manufacturing Processes, 41*, 284–296.
27. Pradhan, M. K., & Biswas, C. K. (2011). Multi-response optimisation of EDM of AISI D2 tool steel using response surface methodology. *International Journal of Machining and Machinability of Materials, 9*(1/2), 66–85.
28. Dewangan, S., Gangopadhyay, S., Biswas, C. K. (2013). Study of surface integrity and dimensional accuracy in EDM using fuzzy TOPSIS and sensitivity analysis. *Measurement, 63*, 364–376.

29. Dhakar, K., Dvivedi, A., & Dhiman, A. (2016). Experimental investigation on effects of dielectric mediums in near-dry electric discharge machining. *Journal of Mechanical Science and Technology*, *30*(5), 2179–2185.
30. Mohanty, S. D., Mohanty, R. C., & Mahapatra, S. S. (2017). Study on performance of EDM electrodes produced through rapid tooling route. *Journal of Advanced Manufacturing System*, *16*(4), 357–374.
31. Mishra, D. K., Datta, S. & Masanta, M. (2018). Through hole making by electro-discharge machining on Inconel 625 super alloy using hollow copper tool electrode. *Journal of Process Mechanical Engineering*, 1–23
32. Valaki, J. B., Rathod, P. P. (2016) Assessment of operational feasibility of waste vegetable oil based bio-dielectric fluid for sustainable electric discharge machining (EDM). *International Journal Advanced Manufacturing Technology, 87*, 1509–1518.

A Comprehensive Review on Single- and Multi-Objective Optimization of Liquid Composite Moulding Process

Anita Zade and Raghu Raja Pandiyan Kuppusamy

1 Introduction

The designing of composite process parameters has become crucial factor in different industries with increase in applications of composite materials in vital sectors. There have been various composite part manufacturing techniques from traditional methods like hand-layup which is cost and labour intensive to the use of automated techniques such as autoclave, injection moulding, extrusion, liquid composite moulding (LCM) process, etc. [1–3]. Although each technique has different manufacturing procedure, the major objectives and design parameters will be same to manufacture the cost-effective and void-free composite part. In this article, we will mainly focusing on the LCM process parameters, as this method is effective in manufacturing properly finished composite parts with complex geometries.

LCM is a closed moulding process; hence after closing the mould, it is difficult to know whether fibre preform has impregnated with resin or there are unsaturated regions where the air has entrapped. Therefore, it is important to identify and optimize the parameters during the process.

This paper addresses following research questions for making the cost-effective and void-free composite part using LCM process.

Research Questions:

a. Which factors need to consider to manufacture a void-free and cost-effective composite part?
b. Why there is need to develop meta-heuristic techniques over traditional numerical techniques?

A. Zade · R. R. P. Kuppusamy (✉)
Department of Chemical Engineering, National Institute of Technology Warangal, Warangal, Telangana 506004, India
e-mail: raghuraj@nitw.ac.in

R. Agrawal et al. (eds.), *Recent Advances in Smart Manufacturing and Materials*, Lecture Notes in Mechanical Engineering,
https://doi.org/10.1007/978-981-16-3033-0_6

c. What advancements needed in meta-heuristic techniques for optimizing LCM process parameters for constrained single- and multi-objective optimization problem?

Following objectives formed to address these questions.

Objectives:

a. To analyze the effect of mould, raw material, and process parameters on mould fill time and cure time
b. To develop proper optimization technique for addressing the full-fledged complex and nonlinear optimization problem
c. Development of effective hybrid MOO algorithm to optimize the objectives related to mould fill time with minimum void content and cure time with given temperature range

In the LCM process, mould filling and curing are the critical steps. The factors which affect during the mould filling stage are air entrapment, transverse flow, race-tracking effect and dual-scale flow. Air entrapment arises due to non-uniform impregnation of fibre preform with resin. Usually, composite structures are very thin; hence, flow along the thickness direction has been considered negligible. However, when the in-plane permeability component of fibre preform along the thickness direction will change significantly, then the transverse flow has to take into account [4]. When air channels will be present between the mould wall and fibre preform, the race-tracking occurs. This effect arises mainly along mould wall edges, joints, inserts and around ribs for complex structures. When there is a difference between the fluid flow within the fibre tows and in between the tows, the dual-scale flow arises [5, 6]. Due to these defects, voids and dry spots form in the composite part [7]. Therefore, proper selection of bulk and tow permeability, gate and vent location, porosity and injection pressure/flow rate needs to take into account.

Curing is an exothermic process where heat transfer occurs between mould and saturated preform. With an increase in temperature of saturated preform, liquid resin will convert to gel and then gel to solid; this phase transition is called as gelation and solidification, respectively. The factors which affect the curing process are temperature overshoot, glass transition temperature during phase change and resin viscosity [8, 9]. This may develop the temperature and cure gradients along thick sectioned parts which results to matrix micro-cracks, residual stresses and geometrical distortions.

Numerous numerical techniques are used by researchers to optimize the mould filling and curing parameters of LCM process. We found few review articles, book chapters and thesis chapters related to numerical optimization of composite process parameters. They are particularly specified to some application based on use of traditional numerical techniques. To the best of our knowledge, we did not found review article in last ten years addressing use of meta-heuristic or hybrid optimization techniques for constrained single- and multi-objective optimization problem in the LCM process.

This article describes the comprehensive review on identification of problem statement, formulation of optimization problem and use of suitable optimization technique based on the type of optimization problem in the composite processing. Depending on the accomplishment of target objectives, design parameters need to set. Then, we will present the different types of optimization techniques used for optimizing the different parameters of LCM processes. After that, to optimize the different conflicting objectives simultaneously the studies on multi-objective optimization (MOO) in composite processing have been reviewed. Here, the combination of different mould parameters, cure parameters and both mould and cure parameters is optimized simultaneously.

2 Numerical Optimization for Composite Processing

2.1 General Formulation of Optimization Problem

Generally, the optimization problem formulated as

$$\max_X / \min f_k(X) \quad k = 1, 2, \ldots, N \tag{1}$$

Subject to,

$$h_l(X) = 0 \quad l = 1, 2, \ldots, L$$

$$g_m(X) \le 0 \quad m = 1, 2, \ldots M$$

$$lb \le X \le ub \quad X = (x_1, x_2, \ldots .x_d)^T$$

here, $f_k(X)$ is the scalar objective function which has to maximize or minimize, $h_l(X)$ represents the J number of equality constraints also called active constraints, $g_m(X)$ represents the M number of inequality constraints and X is the vector of d-dimensional design variables.

For the sake of understanding and correlation, we have discussed the mathematical formulation of one example based on composite process parameter optimization. Jahromi et al. [10] used a dynamic artificial neural network (ANN) for achieving the uniform temperature and degree of cure for thick fibre reinforced composite parts, which is directly dependent on the temperature profile of mould wall. The objective function was formulated to minimize the temperature gradients between two selected points, i.e. central and corner, to indicate the overall cure process subject to constraints on the degree of cure. Mathematically,

$$\min \sum_{i=1}^{n} \left(T_{i(\text{center})} - T_{i(\text{corner})}\right)^2 \tag{2}$$

$$\text{Subject to,} \quad \alpha_{\text{corner}} \geq \alpha_{\text{crit}}$$

where $\alpha_{\text{crit}} = 0.8$

$$298K \leq T_i \leq 430K \quad i = 1, 2, \ldots 5.$$

2.2 *Numerical Studies on Single-Objective Optimization*

The number of research articles is found on the single-objective optimization of gate and vent location using numerous deterministic and stochastic techniques. Particle swarm optimization (PSO) and genetic algorithm (GA) found to be the most used stochastic algorithms with the aim of reducing computational time and increasing accuracy [11]. The mostly formulated objective was minimizing mould fill time with reduced void content [12–21]. To optimize the curing stage, the majorly formulated objective was cure process time with respect to variables on temperature and cure gradients, a constraint on degree of cure. Table 1 reviews the categorization of single-objective optimization studies of composite processing in terms of objectives, parameters and numerical techniques.

2.3 *Numerical Studies on Multi-Objective Optimization*

Multi-objective optimization (MOO) problem contains more than one objective which is conflicting to each other. Hence, there will be more optimal solutions which are represented in terms of Pareto fronts. The best optimal solutions are called non-dominated solutions. Different MOO techniques have been developed to find out the Pareto front between the quality and productivity of the composite part. Evolutionary algorithms like MOOGA and NSGA-II have been mainly used for optimizing the process and design parameters for composite parts. Different objectives are addressed using numerous MOO techniques such as mould fill and cure process time, gate and vent location [32], temperature overshoot, degree of cure, etc. Few algorithms have been developed for thick and ultra-thick components for simple geometries to optimize the trade-off between temperature overshoot and cure process time [33]. Table 2 reviews the categorization of multi-objective optimization studies of composite processing in terms of objectives, parameters and numerical techniques.

Table 1 Objective function categorization for single-objective optimization of composite processing

Objective	Parameters	Numerical techniques	Reference	Year
Least square error between experimental and predicted data from simulation	In-plane 3D permeability components	FEM, golden-section search method	[22]	2011
Flow front progression time	Characterization of 3D permeability	FEM, golden-section search method	[23]	2017
Buckling load, fundamental frequency, structural weight	Geometry, temperature	Meta-heuristic technique	[24]	2018
Stress distribution, critical buckling load, fundamental frequency	Material distribution pattern,	GA, PSO, ANN, ANFIS	[25]	2019
Gate location	Permeability, fill time, mould fill fraction	GA, Gradient-based algorithm	[26]	2007
Injection gates and vents	Permeability	Graph-based two-phase heuristic algorithm (GTPH)	[27]	2008
Fill time	Gate and vent location, distribution media, race-tracking effect	Depth-first search and tree search algorithm	[28]	2015
Total cure cycle time	Central temperature, duration of two steps	ANSYS, Simulated annealing (SA), NM-simplex method	[29]	2014
Process cycle time	Mould temperature profile	GA	[30]	2005
Mould fill time	Gate locations	GA, exhaustive search, centroidal Voronoi diagram (CVD) method	[30]	2016
Mean square error between corner and centre part of temperature	Temperature	ANN, SQP	[10]	2012

3 Summary and Conclusion

This review article has targeted the numerical optimization of mould filling and curing steps in the LCM process. Bulk and tow permeability, gate and vent location, injection pressure and component geometry were found to be the crucial parameters for addressing the multi-phase and multi-scale problems for mould filling stage, whereas mould temperature, initial temperature of resin, degree of cure, heat flux, temperature gradient and temperature overshoot are found to be crucial parameters for optimizing thermal profile in curing stage. To manufacture a void-free composite part

Table 2 Objective function categorization for multi-objective optimization of composite processing

Objectives	Parameters	Numerical techniques	Reference	Year
Crowding distance, filling time	Gate and vent location	NSGA-II	[32]	2009
Warpage, shrinkage rate, short shot possibility	Process parameters: part cooling time, melt temperature, pressure holding time and mould fill time; Geometric parameters: modified edge and round gate	FAHP, TOPSIS	[34]	2018
Cure degree difference, set up cost, fill time, variance of cure degree, cure time	First temperature rise, dwell time, dwell temperature, second temperature rise, hold temperature	NSGA-II	[31]	2019
Setup cost, fill time	Resin temperature, temperature of mould	Hybrid FE/FD method	[35]	2019
Weld line, fill time, wasted resin, dry spot	Gate and vent location	FEA, MOOGA	[36]	2016
Cure time, temperature overshoot	First and second dwell temperature, duration of 1st dwell, heating rate	Surrogate model, Monte Carlo simulator, MOOGA	[33]	2018
Standard deviation and average of degree of cure	Mould temperature, heat flux, cure part temperature	Ant swarm strategy	[37]	2015
Tensile load, flexural strength	Duration and temperature of the 1st curing step, heating rate	MOOGA toolbox	[38]	2017
Residual stresses, cure process time, degree of cure	Temperature profile	MOOGA	[39]	2006
Cure process time, temperature gradient, degree of cure at the end of mould filling, filling time	Cure profile, thermal profile of mould filling, initial resin temperature, gate and vent location	MOOGA	[40]	2014
Cure process time, temperature overshoot	Thermal profile, thickness of composite part, mould geometry	MOOGA	[41]	2016
Warpage, volumetric shrinkage and residual stress	Fibre content, fibre aspect ratio, melt temperature, cooling time, injection pressure	NSGA-II	[42]	2018

(continued)

Table 2 (continued)

Objectives	Parameters	Numerical techniques	Reference	Year
Peak tensile residual stress, roller speed	Degree of bonding, peak residual stress, and thermal degradation	NM-simplex method	[43]	2007
Maximum difference in degree of cure, total cure time and maximum difference in temperature	Temperature, degree of cure	MOOGA	[45]	2018
Weld line, void content, fill time, wasted resin	Void fraction, distance between gate	MOGA	[6]	2019

with the optimized parameters, different optimization techniques have been developed. The number of single-objective and multi-objective optimization techniques has been developed to optimize the objectives like the degree of cure, temperature overshoot, process time, gate and vent location and mould fill time. The Nelder mead simplex method, simulated annealing, GA, NSGA-II and MOOGA are the most used traditional techniques for optimizing the process as well as mould parameters.

Although there are number of articles found on the development of numerical models for addressing multi-scale and multi-phase problems for the complex geometries, the development of an effective optimization technique for addressing such problems is a crucial task. Nowadays, the development of a hybrid multi-objective optimization technique for optimizing the LCM process parameters is an active area of research. Till now, very few articles are found on MOO for both mould filling and curing parameters using appropriate optimization techniques. Hence for optimizing the parameters of complex geometry, thick and ultra-thick component development of an effective optimization technique is needed. This may advance the use of LCM technology in the industry for manufacturing large and complex structures instead of traditional techniques.

References

1. Chang, C. (2015). Simulation of non-isothermal filling process in vacuum assisted compression resin transfer moulding process. *Plastics, Rubber and Composites, 44*, 162–170.
2. Yang, B., Jin, T., Li, J., & Bi, F. (2015). Three-Dimensional Numerical Simulation of Mold Filling Process in Compression Resin Transfer Molding. *Applied Composite Materials, 22*, 209–230.
3. Yang, B., Jin, T., Li, J., & Bi, F. (2014). Simulating the resin flow and stress distributions on mold tools during compression resin transfer molding. *Journal of Reinforced Plastics and Composites, 33*, 1316–1331.
4. Kim, J., Hwang, Y., Choi, K., Kim, H., & Kim, H. (2019). Prediction of the vacuum assisted resin transfer molding (VARTM) process considering the directional permeability of sheared woven fabric. *Composite Structures, 211*, 236–243.

5. Hamidi, Y. K., & Altan, M. C. (2017). Process Induced defects in liquid molding processes of composites. *International Polymer Processing, 32*, 527–544.
6. Oya, Y., Matsumiya, T., Ito, A., Matsuzaki, R., & Okabe, T. (2019). Gate optimization for resin transfer molding in dual-scale porous media: numerical simulation and experiment measurement. *The Journal of Composite Materials 0*, 1–15.
7. El, M. A., Tarfaoui, M., & Lafdi, K. (2019). Additive manufacturing of polymer composites: Processing and modeling approaches. *Compos Part B*. https://doi.org/10.1016/j.compositesb.2019.04.029
8. Kuppusamy, R. R. P., & Neogi, S. (2017). Simulation of air entrapment and resin curing during manufacturing of composite cab front by resin transfer moulding process. *Archives of Metallurgy and Materials, 62*, 1839–1844.
9. Kuppusamy, R. R. P. (2018). Development of liquid composite moulded thermoset composite automotive parts using process simulations: liquid composite moulding process simulations. In *Design and Optimization of Mechanical Engineering Products* (pp. 24–36)
10. Jahromi, P. E., Shojaei, A., & Pishvaie, S. M. R. (2012). Prediction and optimization of cure cycle of thick fiber-reinforced composite parts using dynamic artificial neural networks. *Journal of Reinforced Plastics and Composites, 31*, 1201–1215.
11. Elsheikhi, S. A., & Benyounis, K. Y. (2016). Review of recent developments in injection molding process for polymeric materials *The Reference Module in Materials Science and Materials Engineering*. https://doi.org/10.1016/B978-0-12-803581-8.04022-4
12. Kessels, J. F. A. (2007). Optimising the flow pipe arrangement for resin infusion under flexible tooling. *Composites Part A: Applied Science and Manufacturing, 38*(38), 2076–2085.
13. Montés, N., & Sánchez, F. (2010). A new computational tool for liquid composite moulding process design based on configuration spaces. *Composites Part A: Applied Science and Manufacturing, 41*, 58–77.
14. Maier, R., Rohaly, T., Advani, S. G., & Fickie, K. (1996). A fast numerical method for isothermal resin transfer mold filling. *International Journal for Numerical Methods in Engineering, 39*, 1405–1417.
15. Lin, M., Murphy, M., & Hahn, H. (2000). Resin transfer molding process optimization. *Composites Part A: Applied Science and Manufacturing*. https://doi.org/10.1016/S1359-835X(99)00054-8
16. Li, J., Zhang, C., Liang, R., & Wang, B. (2008). Robust design of composites manufacturing processes with process simulation and optimisation methods. *International Journal of Production Research, 46*, 2087–2104.
17. Samir, J., Echaabi, J., & Hattabi, M. (2011). Numerical algorithm and adaptive meshing for simulation the effect of variation thickness in resin transfer molding process. *Composites Part B: Applied Science and Manufacturing, 42*, 1015–1028.
18. Mathur, R., Fink, B. K., & Advani, S. G. (1999). Use of genetic algorithms to optimize gate and vent locations for the resin transfer molding process. *Polymer Composition, 20.*
19. Jovanovic, V., Manoochehri, S., & Chassapis, C. (2001). For resin transfer molding. https://doi.org/10.1108/02644400110409186.
20. Nielsen, D., & Pitchumani, R. (2001). Intelligent model-based control of preform permeation in liquid composite molding processes, with online optimization. *Composition Part A, 32*, 1789–1803.
21. Chen, X., Xie, H., Chen, H., & Zhang, F. (2010). Optimization for CFRP pultrusion process based on genetic algorithm-neural network. *International Journal of Material Forming, 3*, 1391–1399.
22. Okonkwo, K., Simacek, P., Advani, S. G., & Parnas, R. S. (2011). Characterization of 3D fiber preform permeability tensor in radial flow using an inverse algorithm based on sensors and simulation. *Composition Part A, 42*, 1283–1292.
23. Yun, M., Sas, H., Simacek, P., & Advani, S. G. (2017). Characterization of 3D fabric permeability with skew terms. *Composition Part A, 97*, 51–59.
24. Nikbakt, S., Kamarian, S., & Shakeri, M. (2018). A Review on Optimization of Composite Structures Part I: Laminated Composites. *Composite Structures*. https://doi.org/10.1016/j.compstruct.2018.03.063

25. Nikbakht, S., Kamarian, S., & Shakeri, M. (2019). A Review on optimization of composite structures Part II: Functionally graded materials. *Composite Structures*. https://doi.org/10.1016/j.compstruct.2019.01.105
26. Henz, B. J., Mohan, R. V., & Shires, D. R. (2007). A hybrid global—local approach for optimization of injection gate locations in liquid composite molding process simulations. *Composition Part A, 38*, 1932–1946.
27. Li, J., Zhang, C., Liang, R., & Wang, B. (2008). Robust design of composites manufacturing processes with process simulation and optimisation methods. *International Journal of Production Research*. https://doi.org/10.1080/00207540600806455
28. Advani, S. G., Sas, H. S., & Šimác, P. (2015). A methodology to reduce variability during vacuum infusion with optimized design of distribution media. *Composition Part A, 78*, 223–233.
29. Carlone, P., Palazzo, G. S. (2014). A simulation based metaheuristic optimization of the thermal cure cycle of carbon-epoxy composite laminates. In *AIP Conference on Proceedings*. https://doi.org/10.1063/13589483.
30. Ruiz, E., & Trochu, F. (2005). Comprehensive thermal optimization of liquid composite molding to reduce cycle time and. *Polymer Composites*. https://doi.org/10.1002/pc.20077
31. Wang, J., Simacek, P., & Advani, S. G. (2016). Composites: Part A Use of Centroidal Voronoi Diagram to find optimal gate locations to minimize mold filling time in resin transfer molding. *Composition Part A, 87*, 243–255.
32. Matsuzaki, R., Yokoyama, R., Kobara, T., & Tachikawa, T. (2019). Multi-objective curing optimization of carbon fiber composite materials using data assimilation and localized heating. *Composition Part A*. https://doi.org/10.1016/j.compositesa.2019.01.021
33. Achim, V., Ratle, F., & Trochu, F. (2009). Evolutionary operators for optimal gate location in liquid composite moulding. *Applied Soft Computing, 9*, 817–823.
34. Tifkitsis, K. I., Mesogitis, T. S., Struzziero, G., & Skordos, A. A. (2018). Stochastic multi-objective optimisation of the cure process of thick laminates. *Compos Part A, 112*, 383–394.
35. Moayyedian, M., Abhary, K., & Marian, R. (2017). Optimization of injection molding process based on fuzzy quality evaluation and Taguchi experimental design. *CIRP Journal of Manufacturing Science and Technology*. https://doi.org/10.1016/j.cirpj.2017.12.001
36. Gupta, A., Kelly, P. A., Bickerton, S., & Walbran, W. A. (2019). Simulating the effect of temperature evaluation on clamping force requirements during rigid-tool Liquid Composite Moulding process. *Compos Part A, 43*, 1–9.
37. Oya, Y., Yamamoto, G., Sato, J., Matsumiya, T., Okabe, T., Matsuzaki, R., Yashiro, S., & Obayashi, S. (2016). Multi-objective optimization for resin transfer moulding process. *Compos Part A*. https://doi.org/10.1016/j.compositesa.2016.09.023
38. Tarasov, I. V., Shevtsov, S. N., Evlanov, V., Orozaliev, E. E., Tarasov, V., & Evlanov, V. (2015). Model-based optimal control of polymeric composite cure in autoclave system. In *IFAC-PapersOnLine* (pp. 204–210). Elsevier Ltd.
39. Seretis, G., Kouzilos, G., Manolakos, D., & Provatidis, C. (2018). Multi-objective curing cycle optimization for glass fabric/epoxy composites using poisson regression and genetic algorithm. *Materials Research, 21*.
40. Ruiz, E., & Trochu, F. (2006). Multi-criteria thermal optimization in liquid composite molding to reduce processing stresses and cycle time. *Composition Part A, 37*, 913–924.
41. Struzziero, G. (2015.) Optimisation of the VARTM process. https://doi.org/10.13140/RG.2.1.4994.9921.
42. Struzziero, G., & Skordos, A. A. (2016). Multi-objective optimisation of the cure of thick components. *Composition Part A*. https://doi.org/10.1016/j.compositesa.2016.11.014
43. Li, K., Yan, S., Zhong, Y., Pan, W., & Zhao, G. (2018). Multi-Objective Optimization of the Fiber-reinforced Composite Injection Molding Process using Taguchi method, RSM, and NSGA-II. *Simulation Modelling Practice and Theory*. https://doi.org/10.1016/j.simpat.2018.09.003

44. Sonmez, F. O., & Akbulut, M. (2007). Process optimization of tape placement for thermoplastic composites. *Composition Part A*. https://doi.org/10.1016/j.compositesa.2007.05.003
45. Dolkun, D., Zhu, W., & Xu, Q. (2018). Optimization of cure profile for thick composite parts based on finite element analysis and genetic algorithm. *Journal of Composite Materials*. https://doi.org/10.1177/0021998318771458

Effect of Heat Treatment on Microstructure and Room Temperature Mechanical Properties in 55Si7 Spring Steel

Rahul R. Kulkarni, Sumit Mate, Stephen Joseph, and Nigel D'sa

1 Introduction

55Si7 is the commercial-based spring steel [1]. This steel is used in the automotive industries. Further, this kind of steel may be used as leaf spring, helical spring, Belleville washers. Due to operational safety, springs have to meet increasing performance requirements, which concern mechanical properties as well as fatigue strength. There are many ways by which one can alter the mechanical properties of the spring steels. To change the mechanical properties of the steel in the solid state, heat treatments are considered to be one of the conventional methods. Further, an alloying element plays an important role to enhance the mechanical property during heat treatment. Among the alloying elements, silicon (Si) is a crucial element in spring steel. The study revealed that the addition of Si decreases the diffusion of carbon atoms and increases the stability of agglomerations of the carbon [2]. Further, Si induces the directional covalent bonds in the solid solutions which increase the resistance to the motion of dislocations and susceptible to temper embrittlement [2]. Manganese is another important alloying element which reduces the critical cooling rate during the heat treatment. Microstructural changes found to have occurred after annealing, spheroidizing, hardening and tempering in spring steel, which further alter the flow formability and mechanical properties [3–5]. The strain hardening increases sharply with the increase in the interlamellar pearlite spacing in 60SiMnA steel [6]. Among various room temperature tensile properties, yield strength decides the damping capability in spring steels. Yield strength and tensile strength are dependent on time–temperature characteristics of tempering heat treatment. Further, these

R. R. Kulkarni (✉)
National Institute of Foundry and Forge Technology, Ranchi 834003, India

S. Mate · S. Joseph · N. D'sa
Dom Bosco Institute of Technology, Mumbai 400070, India

R. Agrawal et al. (eds.), *Recent Advances in Smart Manufacturing and Materials*, Lecture Notes in Mechanical Engineering,
https://doi.org/10.1007/978-981-16-3033-0_7

properties are due to the diffusion of carbon atoms during tempering heat treatment [7].

Oil quenching is generally being used in case of Si-Mn-type spring steels during hardening heat treatment. But oil quenching is not environmentally friendly. There may be a need for recycling the process. Also, alloying elements such as phosphorous/tin/Sulfur lead to temper embrittlement during oil quenching due to slow cooing rate. Because of this, the cost of production is reasonably high. So, there is a need of exploring another possibility such as the use the water as a quenching medium without compromising the mechanical properties. Deformability is another aspect that needs to be considered in spring steels since these steels may require to form in different shapes.

By considering the above aspects, it was found that not much work has been reported in these regards for 55Si7 spring steel. This kind of work could help the process designer to decide the process parameters during the processing of these steels. Therefore, the objective of the present work is to study the microstructure evolution during various heat treatments such as annealing, normalizing, hardening and tempering in 55Si7 spring steel. The present study uses water quenching instead of oil quenching for hardening purpose. Further, tempering heat treatment has been studied as a function of time and temperature, and room temperature mechanical properties are evaluated and analyzed. Further, an attempt is made to bring out the correlation between the room temperature mechanical properties and evolved microstructure with the constants of constitutive equations by studying the metallurgical characteristics. This kind of study could help in optimizing the process of heat treatment.

2 Experimental Work

2.1 Material

The material used for this study was 55Si7 spring steel in a wrought form whose chemical composition was measured by weight analysis. This is given in Table 1.

Table 1 Chemical composition of 55Si7 spring steel

Element	C	Si	Cr	P	Mn	S	Fe
Wt%	0.60	1.78	0.031	0.07	0.82	0.093	Balance

Table 2 Specimen dimensions for tensile test

Width (mm)	Thickness (mm)	Cross-sectional area (mm^2)	Gauge length (mm)
12.5	3.05	38.125	50

2.2 Heat Treatment and Microstructure Study

The specimens of the 55Si7 spring steel were subjected to the annealing, normalizing, hardening and tempering heat treatment. At the start of the heat treatment, all the specimens were heated in the muffled furnace at 900 °C for 0.5 h. to get homogenized austenitic structure. One of the specimens was cooled inside the furnace by shutting it for annealing heat treatment. This means the specimen is cooled with the furnace. The other specimen was cooled in the air for normalizing heat treatment. Besides, other specimens were subjected to the hardening treatment by water quenching. Further, tempering was done in three batches. Each batch contains three specimens. Each batch of the specimens was heated in the Muffled furnace at 400, 500, 600 °C for 1 h, 2 h and 3 h and cooled in air. Metallography study was done on as-received specimen as well as on all heat-treated specimens.

For microstructure study, specimens were polished with a series of SiC papers of 300, 600 and 800 microns. The wet polishing was done on these specimens with the help of a mixture of alumina (Al_2O_3) powder and water (H_2O). Further, the specimens were etched with Nital (3% HNO_3, 97% methanol). Microstructure study was done under an optical microscope at different magnifications after drying the specimens. Grain size study was done by using an ASTM comparison method which further converted to the micron.

2.3 Room Temperature Tensile and Hardness Test

Tensile test for annealed and normalized specimens was carried out on MTS machine. Flat specimens were prepared for the tensile test with dimensions which is mentioned in Table 2. Rockwell hardness in C scale (HRC) was carried out on hardened and tempered specimens and reported the HRC number for further analysis.

3 Results and Discussions

3.1 Microstructure Study

The optical microstructure of as-received 55Si7 spring steel is shown in Fig. 1. This shows the presence of pearlite with lamellar packing of ferrite and cementite particles

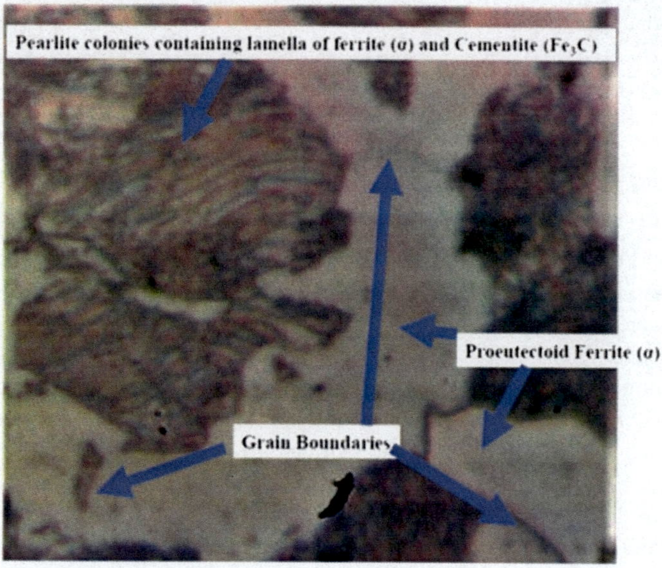

Fig. 1 Optical microstructure of as-received 55Si7 spring steel

also of eutectoid ferrite. Pearlite is observed to be more in weight percentage as compared to proeutectoid ferrite. This observation matches with the Fe–Fe_3C phase diagram and lever rule for 0.6 wt% carbon [8]. In addition, non-uniformity in grain sizes was observed in the present microstructure.

Figure 2a, b, c and d shows the typical optical microstructures of the specimens after annealing, normalizing, hardening and tempering heat treatment. Micrograph of the specimen after annealing heat treatment shows a coarser grain structure of ferrite and larger colonies of pearlite as compared to the micrograph of the specimen subjected to normalizing heat treatment.

Qualitatively, interlamellar spacing is observed to be more in the former case. This is due to the fact that annealing heat treatment is a slow cooling process as compared to normalizing heat treatment. This results in less nucleation rate and more grain growth. During normalizing heat treatment, nucleation rate is higher, and subsequently, grain growth is less due to faster cooling rate as compared to annealing heat treatment. More uniformity in grain sizes was observed after annealing heat treatment. This is because sufficient time has given to solute atoms to redistribute. But this is not the case for normalizing heat treatment in which no time has provided due to faster cooling rate. Grain size numbers were found by ASTM comparison method. Grain size number was found to be 6 in the specimen of annealing heat treatment, and the same was found to be 7 in the specimen of normalizing treatment. This grain size number was taken from three different positions of the specimens at 400X magnification. Hardened and tempered specimens show small needle-like tempered martensite structure.

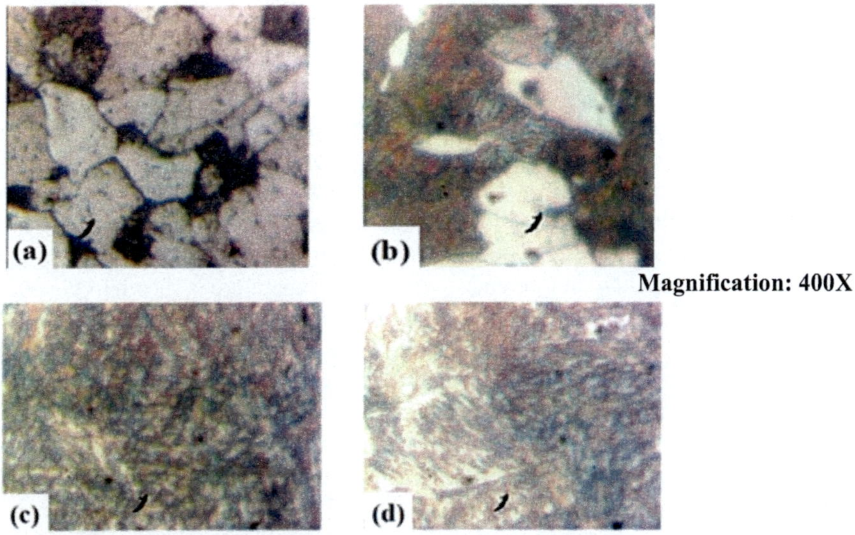

Fig. 2 Optical micrographs of specimens subjected to **a** annealing heat treatment, **b** normalizing heat treatment, hardening and tempering at **c** 400 °C and **d** 600 °C

The ASTM grain size number of proeutectoid ferrite and colonies of ferrite for a given photomicrograph can be found using Eq. (1)

$$N = 2^{n-1} \tag{1}$$

where N = the number of grains observed in an area of 1 in^2 on a photomicrograph taken at a magnification of 100X, and n = the ASTM grain size number [9]. This number is further translated into an average grain diameter via

$$d = \frac{\sqrt{\frac{25.4^2}{2^{n-1}}}}{100^2} \times 1000 \tag{2}$$

where d = the average grain diameter in µm.

Equation 3 is used to adjust the values to the required magnification.

$$N = \left(\frac{M}{100}\right)^2 g_i \tag{3}$$

The average grain size was obtained to be 20.3 µm and 48.1 µm for the specimen subjected to the normalized and annealing heat treatment, respectively. It shows finer grain structure for a normalized specimen as compared to that of the annealed specimen.

3.2 Room Temperature Mechanical Testing

Hardness Test

Rockwell hardness after hardening heat treatment was found to be 57 HRC. This hardness is attributed to the transformation of austenite (FCC: face center cubic crystal structure) to martensite (BCT: body center tetragonal crystal structure) [10]. In addition, this could be due to the internal stresses generated during quenching in water from high temperature, i.e., 900 °C to room temperature. Variation of hardness after tempering treatment with reference to time and temperature is shown in Fig. 3a and b. The marginal decrease in hardness is observed after tempering for the duration of 1 h, 2 h and 3 h for each 400 °C, 500 °C and 600 °C, respectively. When the tempering duration is extended over 2 h. at 500 °C, there is a drastic reduction in the hardness value (Fig. 3a). The marginal decrease could be due to relaxation of internal stresses. The decrease in hardness value as a function of time could be attributed to diffusion of carbon from BCT martensite to the matrix, i.e., relaxation of BCT martensitic cell [10] as well as relaxation of internal stresses. With the increase in tempering temperature, the hardness value at constant time decreases (Fig. 3b). This is due to the fact that the temperature enhances the diffusion process, which leads to the more relaxation of BCT martensitic cell. As tempering temperature reaches to near 723 °C, i.e., eutectoid temperature, the hardness reaches its lowest value.

Further, the rate of decrease in hardness is observed to be faster after 500 °C (Fig. 3b). This means after 500 °C; transformation becomes fast. This is also attributed to the fact that lower tempering temperature is used for relaxation of stresses and higher temperature ranges enhance the diffusion of carbon from the martensitic structure. It is clear from Fig. 3a and b that the optimum strength and ductility can be achieved at 500 °C for 2 h for the present set of investigations.

Tensile Test

Figure 4a and b shows the engineering stress-engineering strain curve for the specimen subjected to (a) annealing heat treatment and (b) normalizing heat treatment

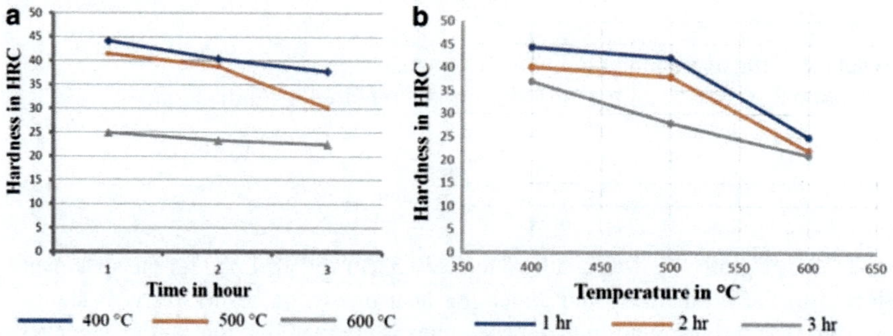

Fig. 3 Variation of hardness as a function of **a** time and **b** temperature

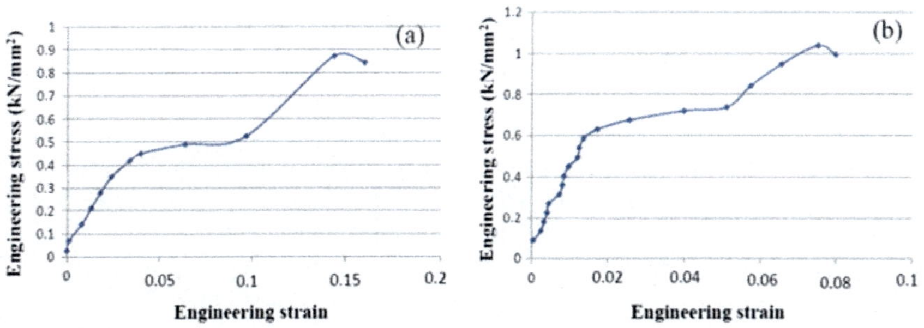

Fig. 4 Engineering stress-engineering strain curves for the specimen subjected to **a** annealing heat treatment **b** normalizing heat treatment under tensile loading

Table 3 Tensile test parameters with reference to the type of heat treatment

Tensile test parameters	Ultimate tensile load (kN)	Ultimate tensile strength (MPa)	0.2% yield strength (MPa)	Percentage elongation to failure (%)	Overall work hardening rate (MPa)
Annealing	33.32	873.96	465.01	16	2556
Normalizing	39.44	1037.895	683.494	8.8	4027

The yield strength of as-received material is 515 MPa and ultimate tensile strength of 919 MPa. Room temperature tensile test was carried out on specimens subjected to annealing and normalizing heat treatment. The parameters like ultimate tensile strength (UTS), yield strength (YS) and elongation to failure of the specimen derived from engineering stress-engineering strain curve with reference to the type of treatment are shown in Table 3.

The engineering stress–strain curve of the tensile specimens subjected to annealing and normalizing heat treatment is shown in Fig. 4a and b, respectively. The yield strength and tensile strength of specimen of normalizing heat treatment are higher than that of the specimen subjected to annealing heat treatment. At the same time, ductility is found to be higher in the specimen subjected to annealing heat treatment (Table 3). The total energy absorbed (shown by area under stress–strain curve) by the annealed material during deformation till breakage is found to be more than that of normalized material, though the strength of the material is more in later cases as shown by stress–strain curve (Fig. 4a and b). The deformability is generally being seen from the uniform strain and the plastic part of the stress–strain curve. An attempt is made to understand the deformability for the given study from engineering stress and engineering strain curves (Fig. 4a and b). The plastic part of the stress strain curve is large for the specimen subjected to the annealing heat treatment (Fig. 4 a). This indicates the deformability is better in the specimen, which was undergone annealing heat treatment. The overall work hardening rate is calculated as $(\sigma_u - \sigma_y)/\varepsilon_f$ where σ_u is the ultimate tensile strength in MPa, σ_y is the yield strength in MPa and ε_f is the total strain to failure. From Table 3, it has been seen that the overall work hardening rate is higher for a normalized specimen than that of the annealed specimen. This is an important parameter to be considered as forming characteristics of this steel. These properties can be related to evolved microstructure. The analysis of optical micrograph shows the finer grain structure for the specimen subjected normalized heat treatment. This results in the higher yield strength. These findings are related to dislocation activities in other steels [11–13]. Finer the grain structure and finer interlamellar spacing lead to the resistance to the movement of dislocation during deformation, which results in higher strength. In addition, non-uniformity in the grain sizes results in less ductility. This is due to accumulation of dislocations at few grains than others. This means that more strain localization is possible in normalized specimen than that of the annealed specimen. This makes the crack to form very quickly results in failure of the specimen. Further, the large toughness value is due to the large ductility. This is because of the large grain size and its uniformity. Large grain size facilitates the movement of dislocation, which means there is less resistance to the movement of dislocation [11–13]. These grain structures are also the cause of work hardening rates of different specimens. In addition, interlamellar spacing affects the strength, ductility and overall work hardening. Qualitatively, lesser interlamellar spacing for normalized specimen leads to higher strength, higher work hardening effect and less ductility. Further, these parameters were related to the total energy absorbed by the specimen during tensile deformation till breakage. Higher strength with less ductility and higher overall work hardening of the specimen of normalizing heat treatment give rise to less absorption of total energy before failure.

The interrelationship between the grain size and tensile yield stress is given by Hall–Petch equation [13–15]. Equation (4) represents the Hall–Petch equation.

$$\sigma_y = \sigma_0 + K d^{(-1/2)} \tag{4}$$

whereas σ_y is the yield strength of the material in MPa, σ_0 is the frictional stress MPa and K is the locking parameter in MPa-m$^{1/2}$. The grain size in micron is represented by d.

Using the tensile stress–strain graph and microstructural analysis, the value of σ_0 and K is calculated to be (from Eq. 4) 59.88 MPa and 2.81 MPa-m$^{1/2}$. These constants can be helpful in predicting the yield strength of this material for a variety of grain sizes.

4 Conclusions

The optical microstructure of 55Si7 spring steel consists of proeutectoid ferrite and lamellar structure of ferrite and cementite. Further, the microstructures after normalizing heat treatment show the finer grain sizes of proeutectoid ferrite and pearlite colonies. More uniform and courser grains are observed to be formed after annealing heat treatment. The hardness of the material is found to be dependent on the tempering temperature as well as on the tempering time. With an increase in the duration of the tempering treatment in all temperatures, hardness value decreases. Further, this rate of decrease is found to be more beyond 500 °C. Higher yield strength is found to be in the specimen subjected to normalizing heat treatment than that of annealing heat treatment. The Hall–Petch constants based on grain sizes are found to be 59.88 MPa and 2.81 MPa-$m^{1/2}$ for the 55Si7 spring steel.

These constants have limitations since yield strength of the material not only depends on grain size, but interlamellar spacing plays a role in this regard. For this, extended study is required.

References

1. Aleksandra, K., et al. (2019). Analysis of grain growth and morphology of bainite in medium-carbon spring steel. *Materials Science and Engineering A, 768*, 138446–138456.
2. MacKenzie, S. (2008). Overview of the mechanisms of failure in heat treated steel components. Failure Analysis of Heat-Treated Steel Component, ASM International.
3. Bikramjit, P., et al. (2012). Effect of preform heat treatment on the flow formability and mechanical properties of AISI4340 steel. *Materials and Design, 37*, 174–181.
4. Barani, A., et al. (2006). Refinement of grain boundary carbides in a Si-Cr spring steel by thermomechanical treatment. *Materials Science and Engineering A, 426*, 194–201.
5. Bo, X., et al. (2014). The research and development of high strength spring steel with long service life. *Key Engineering Materials, 575–576*, 365–369.
6. Chao-Lei, Z., et al. (2014). Influence of pearlite interlamellar spacing on strain hardening behavior in spring steel 60Si2MnA. *Procedia Eng., 81*, 1283–1287.
7. Garg, P., Jamwal, A., Kumar, D., Sadasivuni, K. K., Hussain, C. M., & Gupta, P. (2019). Advance research progresses in aluminium matrix composites: Manufacturing & applications. *Journal of Materials Research and Technology, 8*(5), 4924–4939.
8. Avner, S. H. (1974). *Introduction to physical metallurgy* (2nd edn.). McGraw Hill Book Company
9. ASTM E112–10 (2010) *Standard test methods for determining average grain size.* ASTM International
10. Totten, G. E. (2006). *Steel heat treatment: metallurgy and technologies.* CRC Press, Taylor and Francis.
11. Gang, N. (2019). Strong and ductile steel via high dislocation density and heterogeneous nano/ultrafine grains. *Materials Science and Engineering A, 759*, 1–10.
12. Pierce, D. T., Jiménez, J. A., Bentley, J., Raabe, D., & Wittig, J. E. (2015). The influence of stacking fault energy on the microstructural and strain hardening evolution of Fe–Mn–Al–Si steels during tensile deformation. *Acta Materialia, 100*, 178–190.
13. Hansen, N. (2004). Hall-Petch relation and boundary strengthening. *Scripta Materialia, 51*, 801–806.
14. Hall, E. O. (1951). Deformation and aging of mild steel. *Proceedings of of Physical of Society London, 9*, 747–753.
15. Petch, N. J. (1953). The cleavage strength of polycrystals. *Journal of the Iron and Steel Institute, 173*, 25–28.

A Review on Latest Trends on Different Research Domains of Composite Materials

Aditya Pratap Shahi, Vikas Dwivedi, and Garima Verma

1 Introduction

The term composite materials mainly refer to those materials having composition of at least more than two materials which means that there will hybrid properties of two materials simultaneously. The composite materials can be made up of various materials like carbon, polymers, aramids and natural fibers which are somehow embedded in any kind of polymer matrix. Various types of materials are blended into a single material containing different properties of substances known as the composite materials. But, within this composite composition, one can easily distinguish about the different materials because these materials really do not dissolve or blend into one another. Utilization of composite materials was for the most part started distinctly at airplane business in 1970s; however, these days after just three decades, it is created in many enterprises. Then, the car business, considered as a mother one in every nation, has profited by capacities and qualities of these propelled materials. Alongside progress in innovation, metallic car parts are supplanted by composite ones.

The composite material is arranged by blending at least two distinct components so as to make the subsequent material having prevalent properties from its parental materials. There are two pieces of composite material, grid and filler/fiber (fortifying stage). We can strengthen in different stage like filaments, sheets or particles. It is encompassed in different materials called the framework stage. Metal, clay, non-metal and polymer material can be utilized as fortifying component and lattice material being developed of composites. The fiber/filler utilized in composite is stiffer

A. P. Shahi (✉) · V. Dwivedi
Department of Mechanical Engineerning, B. N. College of Engineering and Technology, Lucknow, India

Garima Verma
Department of Computer Science, Pranveer Singh Institute of Technology, Kanpur, India

R. Agrawal et al. (eds.), *Recent Advances in Smart Manufacturing and Materials*, Lecture Notes in Mechanical Engineering,
https://doi.org/10.1007/978-981-16-3033-0_8

and more grounded than the framework material (called persistent stage), which fills in as burden conveying individuals. Persistent stage (lattice) of composite goes about as the heap move medium between strands/fillers.

In right now, improvement of material tending toward green composite because of difficulties of worldwide ecological concerns like as follows: (1) rising ocean levels, (2) rising normal worldwide temperatures, (3) decreasing polar ice top, (4) rapidly draining oil assets and so on. These issue increased weights on specialists, scientists and industrialists toward assembling and new item configuration utilizing green material halfway or completely. The biodegradable waste transfer issue and benchmarks for more clean just as more secure condition give a copious part of logical research toward eco-composite materials, which effectively have debased or bio-absorbed. The plenteous nearness of characteristic fiber and some other accessible agro- squander has been moreover liable for most recent improvement in inquire about toward green composite material. In Fig. 1, various types of analysis related to durability of some composite materials has been discussed.

Polymer composites are these days broadly utilized in aviation vehicles, cars, marine, sports types of gear, windmills, building, customer businesses for their amazing properties like lightweight, high solidness, high quality, great erosion obstruction, superb plan adaptability, less inclined to natural debasement, incredible warm protection, acoustic damping and great stylish highlights. Polymer composite materials are typically utilized as monocoque structures. The monocoque structure is overwhelming, costly and the laminas of the center segment, near the impartial plane of the structure, are not used to their full quality under flexural stacking. To defeat these impediments of the monocoque structures, sandwich structures are being created. The sandwich structures are generally lighter in weight and more affordable. Objective is to complete the writing survey identified with portrayal

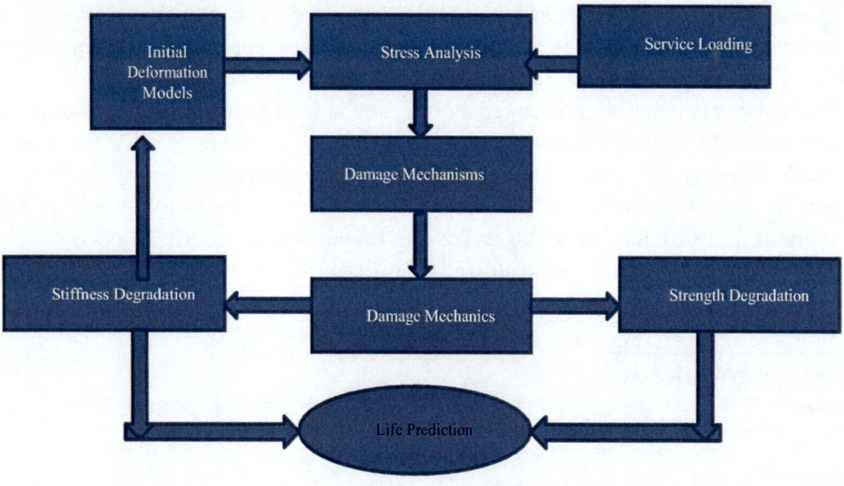

Fig. 1 Durability analysis of composite materials

of polymer composite materials, various materials utilized in polymer composites, disappointment component, fabricating procedure of polymer composite and so forth which will be useful to grow slim polymer sandwich board appropriate for setting up the external spread or external assortment of different machines also, machines. Therefore, composite materials consist of wide range of applications required for manufacturing various products like aluminum-based ceramic matrix.

2 Literature Survey

Basically, there are a lot of materials which possess different types of qualities like hardness, brittleness, roughness and toughness. Various researchers have worked in various domains of metal characteristics and their qualities. One by one, a sequential audit of different types of surveys on composite materials has been discussed in this paper. There are some types of polymers which are very hard in nature and such composite materials are further introduced in different domains of polymers, fillers, applications and so on.

Jayaramudu et al. in [1] have studied the various characteristics of polyalthiacerasoide which gave textures blending in with different types of epoxy composite. These woven textures are helpful in making different composite structures for better use and machinery designs. The hand layup system was used to manufacture of half-breed composite at room temperature.

There are different authors who are searching on different compositions of materials in order to give the methodology for new types of composite materials.

The surface change of woven fiber was finished by the procedure of antacid treatment. The microstructure and morphology considered were finished utilizing Fourier changes infrared spectroscopic (FTIR) and examining electron minute strategies separately. The FTIR investigations speak to minimal estimation of hemi-cellulose and lignin substance of soluble base treated woven texture. The mixture composite is proposed for different applications in building and development enterprises as boards for dividing, flooring, stockpiling tanks, table taps and so forth.

Barnasree et al. [2] contemplated wood dust molecule fortified in epoxy-based composite for examination of mechanical conduct. The sandy wood dust molecule utilized as fortification and LY 556 epoxy for gum. The six distinctive rate of filler molecule was the major concern of this study. Tractable and flexural test were done utilizing UTM and test size dependent on ASTM standard. The diverse structure parameters like filler substance and speed for stacking with tractable and flexural quality utilizing GRA were optimized. Optimization by GRA has the upside of choosing best and most noticeably terrible choices. GRG shows that trial number 13 is the most appropriate, and trial number 3 is the least significant. Epoxy composite with 10 filler contents (wt%) at comparing rate of 1 mm/min shows best execution, and then, again with 0 filler content (wt%) at the speed of 3 mm/min shows the worst execution.

Sarkia et al. [3] inspected the morphology and mechanical conduct of coconut shell particles strengthened filler with epoxy-based composites be assessed. The coconut shell-filled composites manufactured with epoxy polymer network having up to 30 wt.% of the coconut shell fillers. The effect quality was somewhat diminished as contrasted and unadulterated epoxy tar. Ku et al. [4] checked on the malleable conduct of regular fiber strengthened polymer composites. From recent decades, regular filaments drawn consideration by analysts, specialists and researchers as an elective source of fortifying materials in assembling of the composites. Inexhaustible accessibility ease in preparing and ease is the primary issue of fascination and enthusiasm to use as fortified material being developed of composites. It was uncovered that the standard of blend (ROM) anticipated and tentatively got rigidity of various normal filaments fortified HDPE composites were very near one another, and the Halpin–Tsai condition was utilized to anticipate the Young's modulus of composite materials which made of various kinds of characteristic filaments.

Mahindra et al. [5] worked with many other authors on the domain of epoxy-based composite material for knowing and understanding the various degrading properties of the materials.

For experimenting with such work, araldite LY 556 and HY 951 were mostly used as resin and hardener. Further in this work, they discovered that the lifetime of the composite materials tends to vary with various temperature values corresponding to the time function. It was further found that the hot laminate was having good strength as compared to the cold one.

Rao et al. [6] worked on the different wearable properties of coconut dust particle filled with epoxy resin-based composite materials which were both in nature that is treated as well as untreated. The values for the concentration of coir were taken as 10%, 20% and 30% used for the fabrication of composites.

In [7], different characteristics of copper matrix have been discussed involving reinforced SiC–graphite hybrid composites. The authors have described various effects of CNT and TiC hybrid reinforcement on the micro-mechanic-tribo behavior of aluminum matrix composites [8, 9]. In [10], different applications of multi-criteria decision-making techniques in the optimization of mechano-tribological properties of copper-SiC-graphite hybrid metal matrix composites have been discussed. In [11], authors have given characterization of mechanical and tribological properties of graphite and alumina reinforced zinc alloy (ZA-27) hybrid metal matrix composites.

3 Contribution

The authors have discussed various parameters of different types of composites and their types [12, 13]. Along with this, a comparative study of different parameters involved in the composite materials has also been discussed. After that, different research challenges and future scope have also been discussed which makes the study more robust.

4 Types of Fibers

4.1 Materials

The term composite could mean nearly anything whenever taken at face esteem, since all materials are made out of divergent subunits whenever analyzed at close enough detail [14, 15]. In any case, in present-day materials designing, the term for the most part alludes to a "lattice" material that is strengthened with strands. For example, the term "FRP" (Fiber-Reinforced Plastic) normally demonstrates a thermosetting polyester grid containing glass strands, and this specific composite has a lot of the present business advertise [7, 16]. Numerous composites utilized today are at the main edge of materials innovation, with execution and costs proper to ultra-requesting applications, for example, rocket. Yet heterogeneous materials joining the best parts of divergent constituents have been utilized commonly for many years. Old society, impersonating nature, utilized this methodology too: the Book of Exodus talks about utilizing straw to fortify mud in block making, without which the blocks would have no quality.

As found in Table 1, [17] the filaments utilized in current composites have qualities and stiffness's far over those of customary mass materials. The high qualities of the glass strands are because of preparing that stays away from the interior or surface imperfections which regularly debilitate glass, and the quality and solidness of the polymeric aramid fiber is an outcome of the about great arrangement of the

Table 1 Comparative study of various parameters

Parameters	Glass ionomers	Resin ionomers	Composite materials	Amalagam
Durability	Holds good record in non-load bearing but bad in load-bearing mechanisms	Switches from moderate to good in non-load bearing	Good in non-load bearing	Excellent in non-load
Leakage	Low process	Low having proper bonding issues	Low in proper bonding	Moderate in bonding
Recurrent delay	Same as other materials have	Same as other materials have	Have tooth material like bond	Similar like other materials
Wear resistance	Mostly low on occlusal surfaces	Usually low	Usually moderate	High
Operator skills	Materials are unpredictable and unforgiving	Materials are unpredictable and unforgiving	Materials are mostly predictable	Materials are unpredictable and unforgiving
Clinical conditions	Very well controlled operational fields	Very well controlled operational fields	Strong range of tolerance	Very well controlled operational fields

sub-atomic chains with the fiber hub. Obviously, these materials are not commonly usable as strands alone, and ordinarily they are impregnated by a lattice material that demonstrations to move burdens to the filaments, and furthermore to ensure the strands from scraped area and ecological assault. The grid weakens the properties somewhat, yet even so high explicit (weight-balanced) properties are accessible from these materials.

Metal and glass are accessible as grid materials, yet these materials are more expensive and are mainly used for research activities.

4.2 Fibers

The essential capacity of the support in composites strengthened with nonstop strands is to give quality and solidness and to help the auxiliary burden. The motivation behind the grid is to give shape and structure, to shield the filaments from basic harm and unfavorable concoction assault, to circulate stress and to give sturdiness. Strands, otherwise called fibers have limited lengths of in any event multiple times their distance across and are set up by drawing from a liquid shower, by turning, or by concoction fume affidavit on a substrate, for example, tungsten or carbon. They are assembled into packages or strands of 500–12 000 fibers. The packs or strands might be hacked into short filaments or contorted into yarns appropriate for the weaving of textures or three-dimensional performs, utilizing an assortment of weaving designs. The strands might be further joined to shape tows, with upwards of 40,000–300,000 fibers each. Woven textures and tows can be prepared into slashed texture squares or cleaved fiber tows. In the plain weave design, yarns are intertwined in exchanging style over and under each and every other yarn to give greatest texture steadiness and solidness, and least yarn slippage. At the hour of their arrangement, the filaments or yarns are measured, to secure the surface and help the procedure of further dealing with, for example, weaving. Before the strands are at last utilized in the creation of a composite, the size is normally expelled by heat cleaning or washing. Strands are additionally now getting accessible as blends. An propelled maker of composites, Textron Specialty Materials (TSM), has begun to advertise a consistent fiber epoxy gum prepreg tape that contains a blend of enormous boron strands and littler width carbon strands, with a fiber thickness of 70–80%. While carbon is acceptable in strain, it needs great compression properties. This insufficiency is defeated through the utilization of boron filaments that display their best properties in pressure. The joined material has great flexural properties, which are of significance to the assembling of submersible structures, outdoor supplies and restorative gear [18].

5 Types of Composites

5.1 Continuous Phase (Matrix) Material

5.1.1 Metal Matrix Composites (MMC)

There are metal lattice of different materials that are being formed by the fortification process of metal penetrations. After fortification process, the strength of the material is being increased up to several levels. Remelting is also a process of converting or melting down the materials in some form or the other. Metal matrix composites consist of several types of matrix materials resulting in different types of hybrid materials.

In spite of the fact that the grid is liquid, such MMCs are precisely steady up to around 850 °C (Burkhard 1995) and can be dealt with. The respectability of such MMC rings can be affirmed by nondestructive testing and embedded again into the throwing of a cylinder, which has just to accomplish holding between the composite and the grid, as the essential throwing process had just penetrated the preform. Reusing of MMC areas of characterized shape can be imagined for alumina fiber fortified aluminum (or magnesium amalgams gave that recovery of the composite from the liquefy is sufficiently quick to limit interface responses), SiC strengthened aluminum–silicon compounds, carbon fiber strengthened magnesium, and, with certain reservations, for monofilament fortified titanium combinations. In Fig. 2, different composite materials have been categorized into different parts as shown below.

A few properties like higher quality, break sturdiness and firmness are offered by metal networks-based composite material. Metal framework can hold up under raised temperature in destructive environment in contrast with polymer composites [19]. Titanium, aluminum what's more, magnesium are the well-known framework metals as of now in pattern, these materials, especially utilized in airplane applications [20]. On account of these attributes metal framework, composites are getting looked at for wide scope of uses like burning chamber spout (in rocket, space transport), lodgings, tubing, links, heat exchangers, auxiliary individuals and so on.

Molecule fortified metals with particulate volume portions more noteworthy than 40% are additionally much of the time arranged by liquefying invasion of powder preforms. It might be imagined to reshape straightforward geometries of these composites somewhat by shaping procedures; if remelted, the volume portion of support may be expanded if a portion of the network is crushed out during auxiliary weight throwing (Burkhard 1995). On the other hand, these composites could be utilized as a reason for the creation of lower volume portion composites by remolding and weakening by a framework soften (Klier et al. 1991).

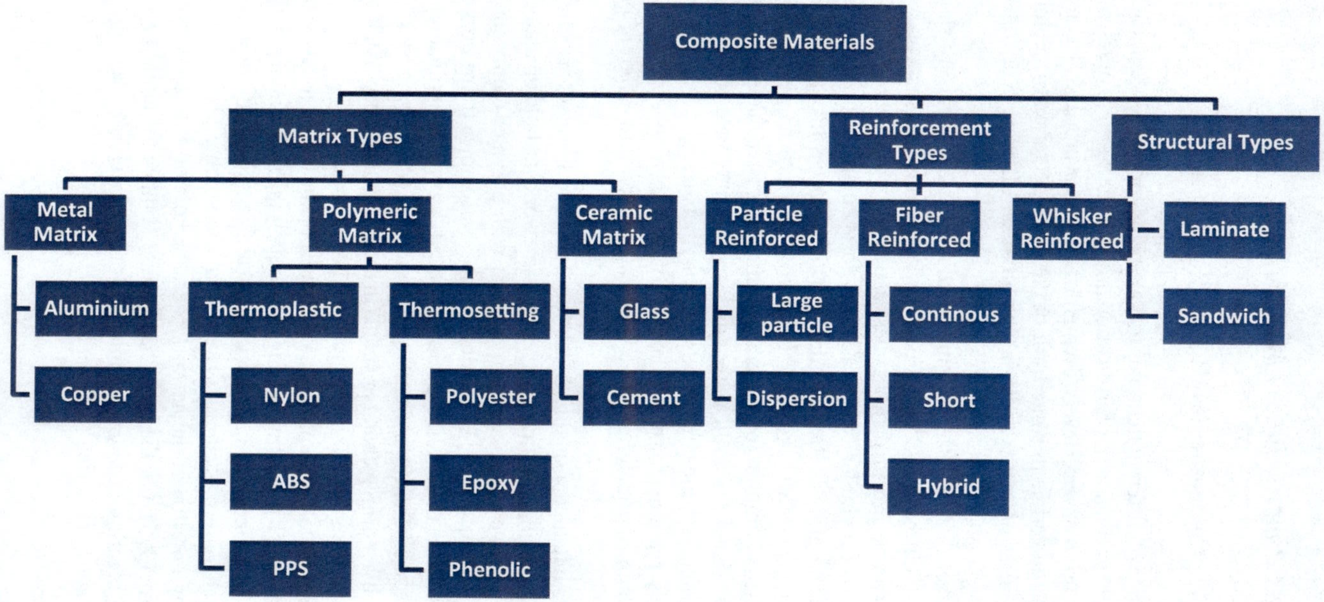

Fig. 2 Various types of composite materials

5.1.2 Ceramic Matrix Composites (CMC)

These composites are basically one of the sub-domains of composite materials and also of ceramics. There is a ceramic matrix which contains ceramic fibers in it. A ceramic material can be composed of both the matrix as well as different fibers. Basically, the concept of ceramic matrix composites (CMC) came in picture in order to overcome the problems of different materials like silicon nitride and silicon carbide. Major disadvantage of such materials is that these materials get fractured easily under any kind of thermo-mechanical loads due to cracks or scratches. To overcome such problems, these materials are being embedded into matrices called as matrix composites. The parameters used in ceramic matrix composites (CMC) generally help in enhancing the toughness of the materials and also help in increasing the strength of the material. It also enhances the value of Young's Modulus of the ceramic material. The major functions of using ceramic matrix composites (CMC) are given below:-

- To increase the stress value of the cracks obtained in the material which leads to increase in the energy of the crack propagation.
- After formation of little cracks, the CMC helps in recovering these cracks with a high ultimate tensile strength (UTS) value.

Majorly used materials in ceramic composites are carbon metals, alumina materials, silicon carbide, etc. Artistic network composites (CMCs) are a subgroup of composite materials just as a subgroup of earthenware production. They comprise of earthenware filaments implanted in an artistic grid. The lattice and strands can comprise of any fired material, whereby carbon and carbon filaments can likewise be viewed as a clay material.

This sort of composite materials comprises of earthenware strands/fillers strengthened in a clay network. Models and silicon carbide strands fixed in a grid produced using a borosilicate glass. One of the principle goals, in creating earthenware grid composites, is to improve the sturdiness of material. Normally, it is normal and certainly.

Found that there is a concurrent improvement in quality and solidness of earthenware network composites [21].

5.1.3 Polymer Matrix Composites (PMC)

The polymer-based matrix composites are basically composed of various types of short fibers which are continuous in nature, and these are bound together by some means of organic matrix of polymers. PMCs are generally designed in order to transfer and manage the loads between the various fibers of matrix. There are various advantages of using polymer matrix composites as they are very lightweight in nature, have high toughness and stiffness and also have high strength to deal with typical metals. Also, they have good corrosion resistance property and better abrasion resistance quality. In this, the materials generally form bonds with one another and then transfer the loads with each other. Polymer matrix composites are generally divided into two main parts as thermosets and thermoplastics. Thermosets are predominant type in today's usage, and they are further classified into resin systems like polymides, polyurethanes and epoxies. There are various types of disadvantages also of using such structure of composite materials as given below:

- Environmental degradation is one of the harnessing factors of degrading qualities of composite materials.
- There is a phenomenon known as moisture absorption by which swelling is caused in the surface of the polymers leading to a decrease in the Tg value.
- There can be a thermal mismatch between various bonds of the polymers like in polymer and fiber which can cause cracking or debonding in the surface of the polymers or materials.
- Due to moisture absorption, sometimes, there is a great increase in the temperature affecting the bonds in the composite materials. This may degrade the quality of the composite materials.

In first kind, fibers are the fortification and the principle wellspring of solidarity while framework sticks every one of the strands together fit as a fiddle and moves worries between the fortifying strands. In the subsequent sort, particles are utilized for strengthening and to expand the modules and to diminish the pliability of the grid. Particles utilized for strengthening incorporate earthenware production and glasses, for example, little mineral particles, metal particles, for example, aluminum and nebulous materials, including polymers and carbon dark. These kinds of composites have application in airplane fabricating, space industry, really taking shape of outdoor supplies and so forth [22].

5.2 *Reinforcing (Fiber/Filler) Material*

5.2.1 Fibrous Composite

Stringy composites are materials comprising of lightweight, high modulus filaments embedded in an encompassing material called the lattice. These composites have properties that change with the course of intrigue [5]. Filaments are partitioned into two gatherings, engineered strands and characteristic strands. Engineered strands are man-made filaments which are an aftereffect of investigation by researchers to improve common happening plant and creature filaments. First engineered fiber was nylon. Strands produced using plant, creature and mineral sources are known as natural filaments. It very well may be additionally characterized by their beginning, for example, organic product strands, best filaments and leaf strands. Natural product strands are removed from the products of the plant; they are light and bushy and permit the breeze to convey the seeds [6]. Best strands are found in the stems of the plant. Strands separated from the leaves are unpleasant and strong which are called as leaf filaments. As the normally inferred filaments are ease, inexhaustible and biodegradable in nature, so from most recent couple of decades, zone of building furthermore, explore has been moving toward the utilization of these filaments as fortifying material in the polymer lattice. Most recent couple of decades different examines led on mechanical properties of normal fiber-based polymer composites with extraordinary unmistakable quality on wood/polymer composites. A point-by-point writing overview is done to discover the extent of composite with the present work [23].

5.2.2 Particulate Composites

A particulate composite is described as being made out of particles suspended in a network. Such composites are utilized for some applications in which quality is definitely not a critical part of the plan. Composites allude to a material comprising of at least two individual constituents. The fortifying constituent is installed in a lattice to frame the composite. One type of composites is particulate fortified composites with concrete being a genuine model. In such kind of composites, the support is of molecule nature fit as a fiddle, for example, circular, cubic, tetragonal, a platelet or then again of other customary or unpredictable shape. They upgrade the solidness of the composite to a constrained degree, are used to improve the properties of framework materials like warm and electrical conductivities, improve execution at raised temperatures, lessen grinding, increment wear and scraped area obstruction, improve machinability, increment surface hardness and decrease shrinkage [24]. In [23], authors have given so many parameters study about the composite materials relating with different domains. Along with this, the composite materials have great stiffness and strength so these materials can be used for designing various types of Iot-enabled [12] devices for developing the robotic arms, hardware equipment, etc. Various kinds of research progressions have been discussed in [25] about advance aluminum matrix composites including various manufacturing aspects and applications. In [26], the authors have discussed various sustainable copper matrix composites. In [15], processing and characterization of pine epoxy-based composites have been discussed. Each particular methodology has its own significance and relevance by which the composite materials are being manufactured. In Fig. 3, various types of parameters like production cost, efficiency of process, and quality of the product which plays a major role in the manufacturing industry of composites has been discussed.

6 Comparative Study

The composite materials have a lot of things in common with other materials but they show greater results when applied and used over different parameters [24]. These parameters include tensile strength, roughness, toughness, stiffness, throughput etc. other types of parameters resulting in developing and designing such hard composite

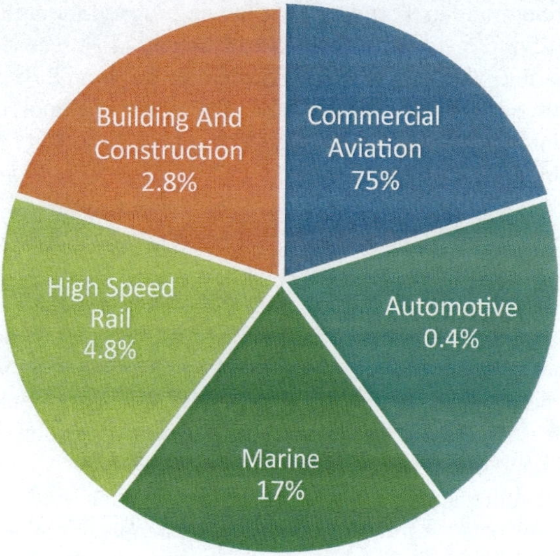

Fig. 3 Division of materials into different sectors

materials. Below given is a comparative study table which is based on different parameters and is compared with other materials as given below:

7 Applications of Composites

On the basis of the various types of composites, there are numerous applications of composite materials that are used in various domains as discussed below:

- Bear high strength for developing hard materials like for weight ratio.
- Resist high level of toughness and improved fatigue.
- Highly stiff in nature.
- Better corrosion and resistance power.
- Components like shafts, rotors and brakes which are manufactured using aluminum matrix composites [28].
- Wings and the components used in airline industry are made from composites.
- Various components of materials are manufactured are marine industry [29].
- Has low maintenance cost.
- Highly durable for long terms.
- Have improved electrical insulation properties.
- Great anti-friction properties.
- Prone to corrosion.
- Highly resistant and reactive [30–32].
- Light weighting with composite materials [33].

8 Research Challenges

Nowadays, a lot of disadvantages are there for structuring any resistive material because of their different chemical and physical properties. Below given are some of the recent research challenges in the domain of composite materials [34].

- Since, the composite materials are somewhat very hard, so they are not broken easily, and hence, it becomes difficult to mold them easily in the required internal structure.
- Fiber defects may lead to poor structure and quality of various materials being used for the development purpose [27].
- Breakage of materials causes poor design and development of real composite materials rising difficult in managing the overall structure of the materials.
- Voids and interface defects may lead to incomplete and deteriorated structure of composite materials [35].
- Various failures in multi-axial fatigue.
- Advanced level of design concepts is required for sustainable composite designs like for recycling of materials and reassembling the materials.
- Cost factor is one of the major research challenges in the domain of composite

  Materials which need to be expanded to life cycle cost.

9 Future Scope

The recent tradeoffs between various parameters in the composite materials can lead to different domains in different sectors like manufacturing, cost analysis and performance analysis [36]. In Fig. 4, different future scopes of applicability of composite materials has been discussed. Below listed are some of the major future scopes of design and development of composite materials as:

- There are several current models which are needed to be improved for the better development of composite materials by incorporating other manufacturing defects.
- In future, the cost analysis parameter should be expanded to other forms life cycle cost methods.
- Other engineering domains like lightweight engineering should be incorporated with composite materials for improving performance, throughput, cost, etc.
- Design and development of composite materials should sustain effective conditions for good conditions for human and environment both.

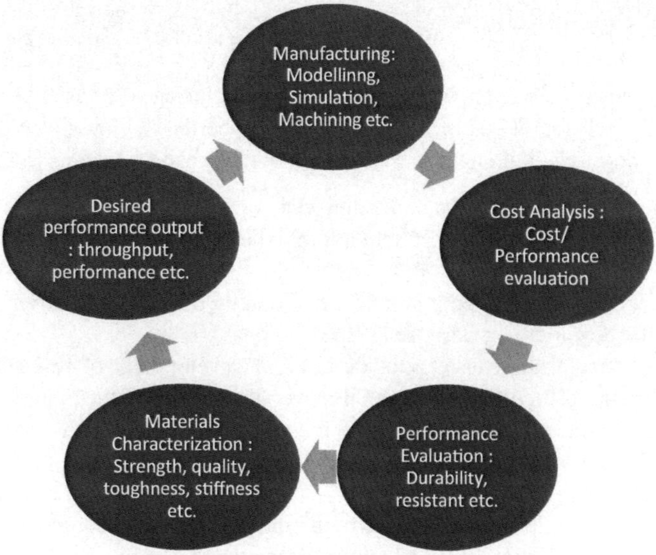

Fig. 4 Various future scopes of composite materials

10 Conclusion

Analysts have done a great deal of research on normal fiber strengthened polymer composites; however, SCB dust molecule has never been utilized in manufacture and utilization of epoxy-based composite requiring little to no effort. There is distinct degree on investigation of mechanical conduct under diverse testing states of SCB dust strengthened polymer composite. Cost and quality control of regular filler strengthened composite is the significant stone to use as elective material side effect creator and producers. Other than all these, the primary thought process is not to create a monetary normal fiber-based composite material for business utilization. Right now, use of composite materials and their effectiveness in airplane industry and different cases were considered. Several unmistakable highlights of composite materials make them to be basic materials in aviation and car and make them more viable than metals. The materials arrangement dependent on their properties and characters was clarified, and the effect conduct on various sorts of material has been explored. The methods of sway harm got to go from network cracking and delimitation through to fiber disappointment and infiltration. Harm modes association, sorts of speed and impact of different factors on sway conduct must be considered and comprehended to foresee any disappointment in composite materials.

References

1. Jayaramudu, J., Agwuncha, S. C., Ray, S. S., Sadiku, E. R., & Rajulu, A. V. (2015). Studies on the chemical resistance and mechanical properties of natural polyalthiacerasoides woven fabric/glass hybridized epoxy Composites. *Research Article Advanced Materials Letters, 6*(2), 114–119.
2. Chanda, B., Kumar, R., Kumar, K., & Bhowmik. S. (2015). Optimisation of mechanical properties of wood dust-reinforced epoxy composite using grey relational analysis. In K. N. Das et al. (Eds.), *Proceedings of Fourth International Conference on Soft Computing for Problem Solving, Advances in Intelligent Systems and Computing* 336. © Springer India. http://dx.doi.org/10.1007/978-81-322-2220-0_2.
3. Sarkia, J., Hassana, S. B., Aigbodiona, V. S., & Oghenevweta, J. E. (2011). Potential of using coconut shell particle fillers in eco-composite materials. *Journal of Alloys and Compounds, 509*, 2381–2385.
4. Ku, H., Wang, H., Pattarachaiyakoop, N., & Trada, M. (2011). A review on the tensile properties of natu- ral fiber reinforced polymer composites. *Composites: Part B,* 42, 856–874.
5. Qinglin, W., Chi, K., Yiqiang, W., & Lee, S. (2014). Mechanical, thermal expansion, and flammability properties of coextruded wood polymer composites with basalt fiber reinforced shells. *Materials and Design, 60,* 334–342

12 Karaduman, Y., Sayeed, M. M. A., Onal, L., & Rawal, A. (2014) Viscoelastic properties of surface modified jute fiber/ polypropylene nonwoven composites. *Composites: Part B.* http://dx.doi.org/10.1016/j.compositesb.2014.06.019.

7. Nayim, S. T. I., Hasan, M. Z., Seth, P. P., Gupta, P., Thakur, S., Kumar, D., Jamwal, A. (2020). Effect of CNT and TiC hybrid reinforcement on the micro-mechano-tribo behaviour of aluminium matrix composites. *Materials Today: Proceedings*, *21*, 1421–1424.
8. Verma, G., & Prakash, S. (2019). A study towards current trends, issues and challenges in internet of things (IoT) based System for intelligent energy management. In 4th *International Conference on Information Systems and Computer Systems (ISCON 2019)*, Venue IEEE Conference Record Number: #47742, GLA University, Mathura, Uttar Pradesh, India, pid 297, Nov. 21–22, 2019.
9. Verma, G., & Prakash, S. (2020) A comparative study based on different energy saving mechanisms based on green internet of things (GIoT). In *IEEE 8th International Conference on Reliability, Infocom Technology and Optimiza- tion (ICRITO-2020)*, IEEE Conference Record Number 48877, Amity University, Noida, India, Pid 631, June 4 05, 2020.
10. Gangwar, S., Payak, V., Pathak, V. K., Jamwal, A., & Gupta, P. (2020). Characterization of mechanical and tribological properties of graphite and alumina reinforced zinc alloy (ZA-27) hybrid metal matrix composites. *Journal of Composite Materials,* 0021998320938442.
11. Kumar, A., Arafath, M. Y., Gupta, P., Kumar, D., Hussain, C. M., Jamwal, A. (2020). Microstructural and mechano-tribological behavior of Al reinforced SiC-TiC hybrid metal matrix composite. *Materials Today: Proceedings, 21*, 1417–1420.
12. Garg, P., Jamwal, A., Kumar, D., Sadasivuni, K. K., Hussain. C. M., Gupta, P. (2019). Advance research progresses in aluminium matrix composites: Manufacturing & applications. *Journal of Materials Research and Technology, 8*(5) 4924–4939
13. Jamwal, A., Mittal, P., Agrawal, R., Gupta, S., Kumar, D., Sadasivuni, K. K., & Gupta, P. (2020). Towards sustainable copper matrix composites: Manufacturing routes with structural, mechanical, electrical and corrosion behaviour. *Journal of Composite Materials*, 0021998319900655.
14. Hossain, S., Rahman, M. M., Jamwal, A., Gupta, P., Thakur, S., & Gupta, S. (2019, September). Processing and characterization of pine epoxy based composites. In *AIP Conference Proceedings* (Vol. 2148, No. 1, p. 030017). AIP Publishing LLC.
15. Jamwal, A., Prakash, P., Kumar, D., Singh, N., Sadasivuni, K. K., Harshit, K., … Gupta, P. (2019). Microstructure, wear and corrosion characteristics of Cu matrix reinforced SiC–graphite hybrid composites. *Journal of Composite Materials, 53*(18), 2545–2553.

16. Jamwal, A., Seth, P. P., Kumar, D., Agrawal, R., Sadasivuni, K. K., & Gupta, P. (2020). Microstructural, tribological and compression behaviour of Copper matrix reinforced with Graphite-SiC hybrid composites. *Materials Chemistry and Physics*, 123090.
17. Gerstle, F. P. (2009). *Composites, Encyclopedia of Polymer Science and Engineering*. New York: Wiley.
18. Harris, B. S. (1991). A perspective view of composite materials. *Materials and Design, 12*(5), 259–271.
19. Raju, F. A., & Kumar, K. B. (2014) Design and analysis of horizontal tail of UAV using composite materials. *International Journal of Computer Trends and Technology (IJCTT) 4*(7).
20. Rout, A. K., & Sahoo, S. S. (2014). Study on erosion wear performance of jute-epoxy composites filled with industrial wastes using Taguchi methodology. In *Proceedings of Second IRF International Conference*. Mysore, India. ISBN: 978–93–84209–69–8.
21. Rout, L. K., & Sahoo, S. S. (2014) Study on erosion wear performance of jute-epoxy composites filled with industrial wastes using Taguchi methodology. In *Proceedings of Second IRF International Conference*. Mysore, India, ISBN: 978–93–84209–69–8.
22. Brien, D. J. O., Chin, W. K., Long, L. R., & Wetzel, E. D. (2014) Polymer matrix, polymer ribbon- reinforced transparent composite materials. *Composites: Part A, 56*, 161–171.
23. Kumar, R., Kumar, K., Sahoo, P., & Bhowmik, S. (2014). Study of mechanical properties of wood dust reinforced epoxy composite. *Procedia Materials Science, 6*, 551–556.
24. Brien, D. J. O., Chin, W. K., Long, L. R., & Wetzel, E. D. (2014). Polymermatrix, polymerribbon-reinforced transparent composite materials. *Composites: Part A, 56*, 161–171. Mahendra, G., Srividya, K., Reddy, C. K., & Kavitha, E. Hygrothermal degradation studies on e-glass woven rovings epoxy composite. *International Journal of Engineering Sciences & Research Technology*.
25. Verma, G., & Prakash, S. (2020) Design and implementation of modified unicode strategy for data security in IoT. *International Journal of Advanced Science and Technology (IJAST), 29*(06), 6271–6294. ISSN: 2005–4238.
26. Verma, G., & Prakash, S. (2020) Pneumonia classification using deep learning in healthcare. *International Journal of Innovative Technology and Exploring Engineering (IJITEE)*, 9(4), 1715–1723. ISSN: 2278–3075.
27. Chandra Rao, C. H., Madhusudan, S., Raghavendra, G., & Venkateswara Rao, E.) Investigation in to wear behavior of coir fiber reinforced epoxy composites with the Taguchi method. *International Journal of Engineering Research and Applications (IJERA)*. ISSN: 2248–9622
28. Jamwal, A., Agrawal, R., & Gupta, P. Application of multi-criteria decision-making techniques in the optimization of mechano-tribological properties of copper-SiC-Graphite hybrid metal Matrix composites. In Intelligent manufacturing (pp. 149–172). Springer.
29. Jamwal, A., Vates, U. K., Gupta, P., Aggarwal, A., & Sharma, B. P. (2019). Fabrication and characterization of Al_2O_3–TiC-reinforced aluminum matrix composites. In Advances in industrial and production engineering (pp. 349–356). Springer, Singapore.
30. Seth, P. P., Parkash, O., & Kumar, D. (2020). Studies on the effect of processing parameters on microstructure and properties of magnesium compacts prepared via powder metallurgy. *Transactions of the Indian Institute of Metals, 73*(11), 2715–2726.
31. Seth, P. P., Das, A., Bar, H. N., Sivaprasad, S., Basu, A., & Dutta, K. (2015). Evolution of dislocation density during tensile deformation of BH220 steel at different pre-strain conditions. *Journal of Materials Engineering and Performance, 24*(7), 2779–2783.
32. Bandil, K., Vashisth, H., Kumar, S., Verma, L., Jamwal, A., Kumar, D., … Gupta, P. (2019). Microstructural, mechanical and corrosion behaviour of Al–Si alloy reinforced with SiC metal matrix composite *Journal of Composite Materials, 53*(28–30), 4215–4223.
33. Sohag, M. A. Z., Gupta, P., Kondal, N., Kumar, D., Singh, N., & Jamwal, A. (2020). Effect of ceramic reinforcement on the microstructural, mechanical and tribological behavior of Al-Cu alloy metal matrix composite. *Materials Today: Proceedings, 21*, 1407–1411.
34. Hossain, S., Rahman, M. M., Chawla, D., Kumar, A., Seth, P. P., Gupta, P., … Jamwal, A. (2020). Fabrication, microstructural and mechanical behavior of Al-Al2O3–SiC hybrid metal matrix composites *Materials Today: Proceedings, 21*, 1458–1461.

35. Seth, P. P., Parkash, O., & Kumar, D. (2020). Structure and mechanical behavior of in situ developed Mg 2 Si phase in magnesium and aluminum alloys–a review. *RSC Advances, 10*(61), 37327–37345
36. Seth, P. P., Singh, N., Singh, M., Prakash, O., & Kumar, D. (2020). Formation of fine Mg2Si phase in Mg–Si alloy via solid-state sintering using high energy ball milling. *Journal of Alloys and Compounds, 821*, 153205.

An Overview of Proteus: The world's First Man-Made Non-cuttable Material

Dinbandhu, Ashish Thakur, E. Venugopal Goud, Kumar Abhishek, and Jay J. Vora

1 Introduction

Since the beginning, the developments in science and technology are focussed on two major fields. One is the advancements in effective and economical power units, and the second one is to attain the highest feasible driving force from the accessible power. The second advancement largely depends upon the attributes of materials used in engineering fields. For instance, the aviation sector needs materials with greater stiffness attributes besides lightweight. Similarly, there is a demand for corrosion-resistant materials with high strength attributes for the effective functioning of pressure vessels. In such conditions, there are no alternatives for engineers other than composite materials. Simply by modifying the arrangement, several attributes can be improved, hence, causing the composites to multipurpose and dependable and also an alternative for the traditional engineering materials [1–3].

Nature has its hierarchical framework to protect the ecological system from outrageous loads. For example, the freefall of grapefruit from a certain height does not harm the mash since pomelo peel comprises of vascular packs, and an open-pored

Dinbandhu (✉) · K. Abhishek
Department of Mechanical & Aero-Space Engineering, Institute of Infrastructure, Technology, Research and Management (IITRAM), Ahmedabad, Gujarat 380026, India

A. Thakur
Department of Metallurgical and Materials Engineering, Visvesvaraya National Institute of Technology, Nagpur, Maharashtra 440010, India

E. Venugopal Goud
Department of Mechanical Engineering, G. Pulla Reddy Engineering College, Kurnool, Andhra Pradesh 518007, India

J. J. Vora
Department of Mechanical Engineering, Pandit Deendayal Energy University (PDEU), Gandhinagar, Gujarat 382426, India

R. Agrawal et al. (eds.), *Recent Advances in Smart Manufacturing and Materials*, Lecture Notes in Mechanical Engineering,
https://doi.org/10.1007/978-981-16-3033-0_9

cell-like arrangement with the trusses made up of parenchymatic cells. Similarly, pirarucu, a species of arapaima gigas in the Amazon River, oppose the assault of piranhas' three-sided teeth arrangement through the hierarchical structure of their scales. Organic shells additionally obligate their hardness to a hierarchical framework, which developed over a decade. The formation of nacre in the abalone shell occurs due to the development of hard aragonite tiles with a natural, adaptable interlayer. It has extremely high fracture toughness, which is 3000 times greater than that of a single aragonite crystal [4–7]. Gao et al. [8] recommended a one-dimensional model to examine the interplay between a substantial volume portion of mineral slates and adaptable protein layers. Metals that need processing temperatures more than 300 °C are essentially missing in organic structures. It shows that nature has merely been proficient to fabricate materials and structural designs liable to transformations at terrestrial ecological conditions [9].

Multiscale material fabrication engineering has been profited by ongoing improvements, namely self-engendering photopolymer waveguides to accomplish exceptional attributes. Pham et al. [10] interpreted the noted advantages of local imperfections (dislocations) in metals into three-dimensional printed, polymeric crystal lattices with unit cells at mm extent. In this work, an emphasis has been given to the consideration of system-level interactions. But these interactions focus only on a monotonous unit cell which might be self-restricting. Shape reconfigurable materials (SRMs) were fabricated to sustain significant morphological changes upon little loading and retain back their ideal shape upon unloading. Novel designs were introduced to transform the power of the inbound mechanical waves into a local resonance. Examples of that concept vary from the atomistical level, through 3D printed frameworks up to infrastructural scale-like seismic meta barriers. There are certain materials patterns, produced by topology optimization algorithms, are fit for centering, turning, or scattering of impact waves [11–14].

In this context, a group of researchers and scholars from Germany's Fraunhofer Institute and the UK's Durham University have created the world's first mankind-manufactured uncuttable material. The nature-inspired material has been named "Proteus" after the evasive and shape-changing mythological Greek god. The tough cell flesh of the Citrus paradisi and the hard rupture-resistive aragonite carapace of the ormer (a marine creature) has inspired the team of researchers for this novel innovation. The shells of the abalone sea creatures are exceptionally flexible and tough, with a hard external lamina, and an internal lamina made up of fibres disposed over one another in a helical stairway structure. This distinctive arrangement permits for an amalgamation of characteristics that are rare in artificial materials: flexibility and strength [15–17].

These sea creatures are built from biological materials, namely aragonite stones, found in molluscs carapaces, which are interconnected with a bio-polymer material that fabricates them resilient to ruptures. Similarly, the team of researchers has replaced such combinations by implementing aluminium and ceramics materials to withstand the extremely powerful aggressive admission of cutting tools in the material. Thus, the Proteus is a structural material in which diminutive spheres of alumina ceramics are embedded within the laminae of aluminium-made flexible

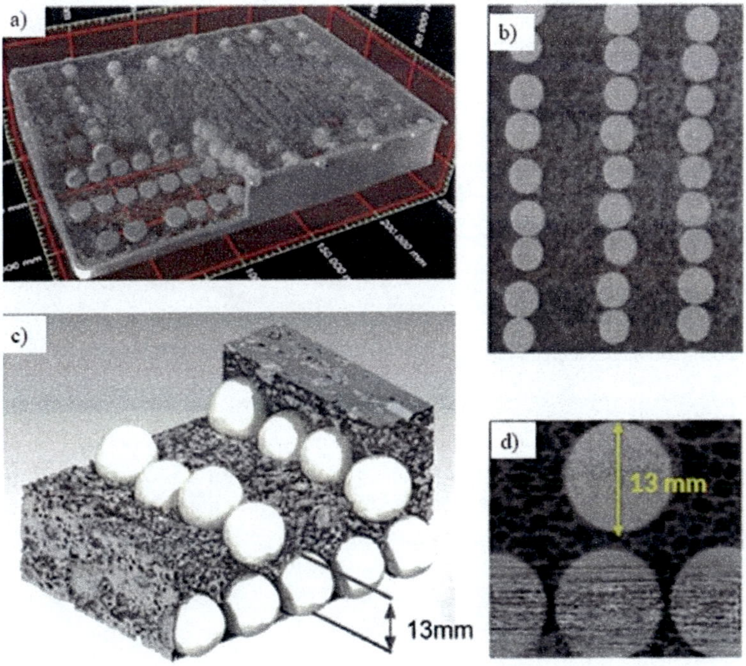

Fig. 1 Structural details of Proteus material: **a** sandwich panel specimen, **b** arrangement of ceramic spheres, **c** ceramic spheres not touching each other as well as parted by aluminium cells, **d** the lower magnitude of foam cells as compared to ceramic spheres [15]

cellular metallic foam at the inner side (core) and a coating of steel alloy on the outer surface as shown in Fig. 1. The ceramic materials reinforced in the flexible matrix structure are also made up of incredibly fine grains that toughen and withstand the cutting tools at a higher speed in a similar manner as a sandbag repels and stop a high-speed bullet fired from a gun. Due to this arrangement, the material becomes harder and denser but still 85% lighter than steel (just 15% the density of steel) [18–20].

2 Materials and Methods

The composition of Proteus includes ceramic and metallic ingredients, and hence, metallurgical processes were incorporated to obtain the final product. Figure 2 illustrates the steps involved in the making of Proteus. First, the aluminium powder and foaming agent (titanium dihydride) were blended properly through a rotating impeller to ensure uniform mixing. Then the blend was compressed (cold compacted) in a compressor and extruded in terms of dense rods. These rods were chopped off to size and thereafter placed with the tiny spheres of ceramic in a laminated grid, as one

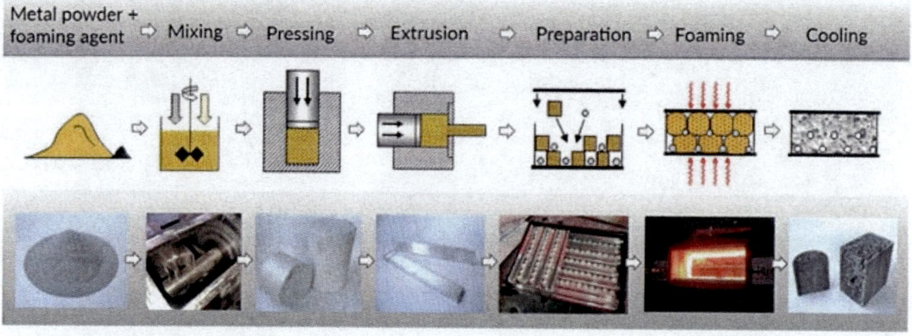

Fig. 2 Steps involved in the manufacturing of Proteus [15]

metallurgical lasagna. The laminated grid was spot-welded into a box-like structure made up of steel and then kept in a furnace at 760 °C for 15–20 min. Within the furnace, hydrogen gas discharges from the foaming agent that creates voids into the aluminium foam and offers it a cellular arrangement [15, 21].

3 Result and Discussion

This section discusses the significant mechanical characteristics of Proteus. It also reviews the effect of a variety of cutting tools employed for cutting this novel material.

3.1 Mechanical Characteristics

To evaluate the mechanical characteristics of the material, compression tests were performed on the cylindrical specimen (Fig. 3). From the loading–unloading test, the tensile stiffness (E) of the material was measured as 5.5 GPa, whereas the yield stress (σ_y) was measured according to ISO/DIS 13314 standards as 8.1 MPa. The Poisson's ratio (υ) was almost 0.0 initially upon loading and started rising towards 0.5 from 10% engineering strain onward. This novel material exhibited considerable deformability, surpassing 20% of engineering strain as compared to the existing cellular materials with the densification strain of $\varepsilon_d \approx 0.25$. Table 1 lists the important outcomes of the tests.

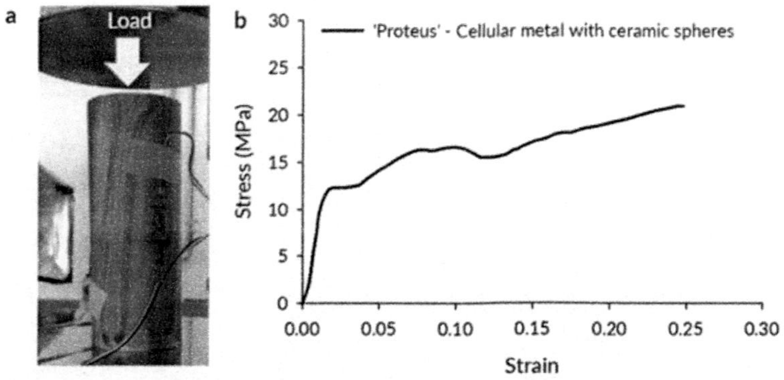

Fig. 3 Evaluation of mechanical characteristics of Proteus **a** quasistatic compression test of the cylindrical specimen, **b** the stress–strain curve showing deformability up to 25% [15]

Table 1 Important mechanical characteristics of Proteus [15]

Mechanical characteristics	Values
Young's modulus (E)	5.5 GPa
Yield stress (σ_y)	8.1 MPa
Poisson's ratio (v)	0 (initially), but ascends towards 0.5 from 10% engineering strain onward
Deformability	above 20% of engineering strain with the densification strain of $\varepsilon_d \approx 0.25$.

3.2 Resistance to the Cutting Tools

A few different cutting tools such as a power drill, waterjet cutter and angle grinder were employed to cut the newly developed material, but none of them were able to perform the thorough cutting. When a power drill or angle grinder impinges on the material, it cuts the outer layer. But once, the tool reaches the embedded ceramic spheres, their interaction causes high-frequency vibrations. These vibrations blunt the tool's sharp edges. As the cutting tool infiltrates the material, the spheres deliver fine ceramic powder particulates that behave like sand paper, dulling the cutting edges of the tool again. These fine particulates also refill the gaps in the aluminium foam, making the material ever harder and denser to cut-off. In other words, the larger the cutting force is implemented, the stronger and tougher the material becomes. Furthermore, an attempt was also made to cut the material using a high-pressure waterjet cutter, but no significant cutting effect was observed by the researchers. This happened due to the coiled surfaces of the ceramic spheres which broaden the waterjet, leading to a substantial reduction in the cutter speed yielding a poor cutting capacity. Figure 4 illustrates the impact of the angle grinder tool on the Proetus material. X-ray transmission images were also taken for better visualization and understanding of the cutting effect (Fig. 5) [15, 17, 19–21].

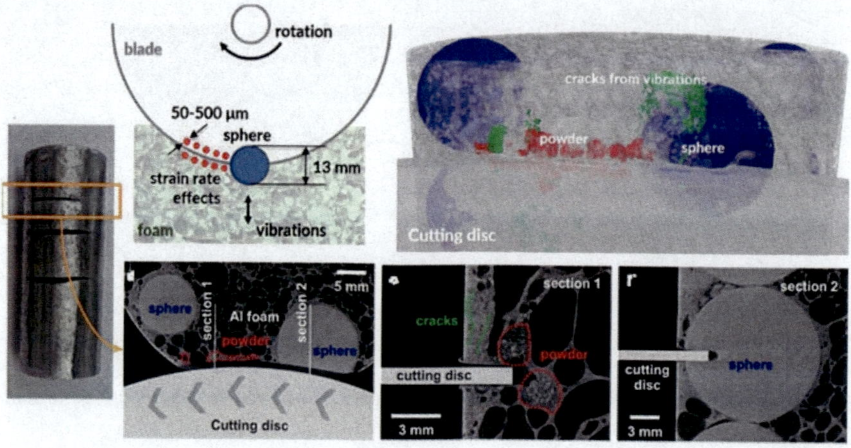

Fig. 4 Cutting effect of angle grinder tool on Proteus material [15]

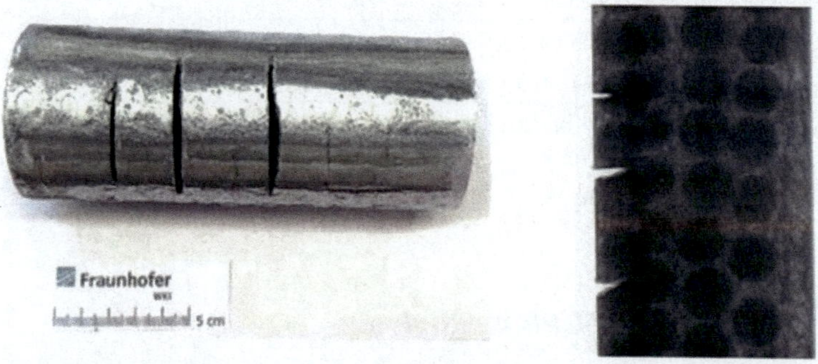

Fig. 5 X-ray transmission image for better visualization and understanding of cutting effect [15]

3.3 *Advantages and Applications*

Proteus is lightweight (just 15% the density of steel), strong and non-cuttable making it a well-suitable material for security and safety industries. As per the research team, this newly developed non-cuttable material will have exciting and useful applications in safety industries. The material can be useful in making lightweight armour, indestructible bike locks, high-security doors, protective equipment for the personnel working in cutting operations, work boots and elbow pads. It can be also used as a structural material for columns and beams. As per the researchers, the mass level production of the material is possible once it becomes commercially available as the raw materials used in it are abundant [15, 16, 19].

4 Conclusion

Proteus is a nature-inspired, newly developed uncuttable material which is strong and much lighter than steel. The ceramic spheres embedded in the aluminium foam matrix destroy the cutting tools and do not allow them for further penetration. Such arrangements enhance the mechanical characteristics of the Proteus making it suitable for structural applications. This material can have very effective and useful applications where it can be employed in making security equipment and items, namely lightweight armour, indestructible bike locks, high-security doors, protective equipment for the personnel working in cutting operations, work boots and elbow pads. Once the material becomes commercially available, it can be easily produced at a mass level because of the abundant availability of its raw materials.

References

1. Thakur, A., Purohit, R., Rana, R. S., Bandhu, D. (2018). Characterization and evaluation of mechanical behavior of epoxy-CNT-bamboo matrix hybrid composites. *Materials Today: Proceedings, 5*(2), 3971–3980. https://doi.org/10.1016/j.matpr.2017.11.655.
2. Bandhu, D., Thakur, A., Purohit, R., Verma, R. K., & Abhishek, K. (2018). Characterization and evaluation of Al7075 MMCs reinforced with ceramic particulates and influence of age hardening on their tensile behavior. *Journal of Mechanical Science and Technology, 32*, 3123–3128. https://doi.org/10.1007/s12206-018-0615-9
3. Sastry, M. N. P., Devi, K. D., & Bandhu, D. (2016). Characterization of Aegle Marmelos fiber reinforced composite. *International Journal of Engineering Research, 5*, 345–349.
4. Bührig-Polaczek, A., Fleck, C., Speck, T., Schüler, P., Fischer, S. F., Caliaro, M., & Thielen, M. (2016). Biomimetic cellular metals—Using hierarchical structuring for energy absorption. *Bioinspiration and Biomimetics., 11*, 045002. https://doi.org/10.1088/1748-3190/11/4/045002
5. Meyers, M. A., Chen, P. Y., Lopez, M. I., Seki, Y., & Lin, A. Y. M. (2011). Biological materials: A materials science approach. *Journal of the Mechanical Behavior of Biomedical Materials, 4*, 626–657. https://doi.org/10.1016/j.jmbbm.2010.08.005
6. Barthelat, F. (2015). *Architectured materials in engineering and biology: Fabrication, structure, mechanics and performance*. https://www.tandfonline.com/doi/abs/10.1179/1743280415Y.0000000008.
7. Li, X. W., Ji, H. M., Yang, W., Zhang, G. P., & Chen, D. L. (2017). Mechanical properties of crossed-lamellar structures in biological shells: A review. *Journal of the Mechanical Behavior of Biomedical Materials, 74*, 54–71. https://doi.org/10.1016/j.jmbbm.2017.05.022
8. Gao, H., Ji, B., Jäger, I. L., Arzt, E., & Fratzl, P. (2003). Materials become insensitive to flaws at nanoscale: Lessons from nature. *Proceedings of National Academay Science U. S. A., 100*, 5597–5600. https://doi.org/10.1073/pnas.0631609100
9. Garg, P., Jamwal, A., Kumar, D., Sadasivuni, K. K., Hussain, C. M., & Gupta, P. (2019). Advance research progresses in aluminium matrix composites: Manufacturing and applications. *Journal of Materials Research and Technology, 8*(5), 4924–4939.
10. Pham, M. S., Liu, C., Todd, I., & Lertthanasarn, J. (2019). Damage-tolerant architected materials inspired by crystal microstructure. *Nature, 565*, 305–311. https://doi.org/10.1038/s41586-018-0850-3
11. Haghpanah, B., Salari-Sharif, L., Pourrajab, P., Hopkins, J., & Valdevit, L. (2016). Multistable shape-reconfigurable architected materials. *Advanced Materials, 28*, 7915–7920. https://doi.org/10.1002/adma.201601650

12. Liu, Z., Zhang, X., Mao, Y., Zhu, Y. Y., Yang, Z., Chan, C. T., Sheng, P. (2000). Locally resonant sonic materials. *Science, 80, 289*, 1734–1736. https://doi.org/10.1126/science.289.5485.1734.
13. Matlack, K. H., Bauhofer, A., Krödel, S., Palermo, A., & Daraio, C. (2016). Composite 3D-printed metastructures for lowfrequency and broadband vibration absorption. *Proceedings of National Academy of Science U. S. A., 113*, 8386–8390. https://doi.org/10.1073/pnas.1600171113
14. Le, C., Bruns, T. E., & Tortorelli, D. A. (2012). Material microstructure optimization for linear elastodynamic energy wave management. *Journal of the Mechanics and Physics of Solids, 60*, 351–378. https://doi.org/10.1016/j.jmps.2011.09.002
15. Szyniszewski, S., Vogel, R., Bittner, F., Jakubczyk, E., Anderson, M., Pelacci, M., Chinedu, A., Endres, H. J., & Hipke, T. (2020). Non-cuttable material created through local resonance and strain rate effects. *Science and Reports, 10*, 1–24. https://doi.org/10.1038/s41598-020-65976-0
16. *Engineers just debuted the first "non-cuttable" material: A metal that turns tools against themselves.* Business Insider India. https://www.businessinsider.in/science/news/engineers-just-debuted-the-first-non-cuttable-material-a-metal-that-turns-tools-against-themselves/articleshow/77115090.cms, last accessed 2020/10/05.
17. *Proteus technology: New material is strong, light and non-cuttable.* https://scitechdaily.com/proteus-technology-new-material-is-strong-light-and-non-cuttable/, last accessed 2020/10/05.
18. *Non-cuttable material—Indestructible bike locks.* https://www.popularmechanics.com/science/a33368424/non-cuttable-material/, last accessed 2020/10/05.
19. *Engineers created Proteus, a new lightweight, non-cuttable material—Tech Explorist.* https://www.techexplorist.com/proteus-lightweight-non-cuttable-material/33922/, last accessed 2020/10/05.
20. *Proteus: The world's first manufactured non-cuttable material.* https://www.intelligentliving.co/proteus-non-cuttable-material/, last accessed 2020/10/05.
21. *Proteus, The shape-shifting and possibly non-cuttable material.* Hackaday. https://hackaday.com/2020/08/06/proteus-the-shape-shifting-and-possibly-non-cuttable-material/, last accessed 2020/10/05.

Multiresponse Regression Modeling in Face Milling of Al6061 Using Design of Experiments

Y. K. Mogal, S. K. Mahajan, S. B. Rane, and Rajeev Agrawal

1 Introduction

Aluminum and its alloys are utilized in an assortment of uses, for example, making vehicle and aviation segments, rocket parts, stockpiling compartments, marine applications, storage containers, etc., as a result of its low density, fantastic corrosion resistance, and better thermal and mechanical properties. These metals are easy to machine as compared to other metals. With the addition of various kinds of metals, properties of pure aluminum can be enhanced [1]. Nowadays, Al6061 is one of the most significant and widely used materials in the industries from Al 6xxx series alloys. It is created in 1935 and contains Mg and Si as its major alloying parts [2]. It is an easily available metal and has good mechanical properties, exhibits good weldability, commonly extruded. It is mostly used in aircraft, automobiles, boats, and packaging of food and beverage industries [3].

2 Literature Survey

Nowadays due to industrials rivalry, the utilization of reasonable improvement strategies for the right choice of process parameters is amazingly important to stay away from non-esteem included costs. The advancement of process parameters requires deliberate methodologies [4]. As indicated in literature, different conventional methods, for example, geometric programming, goal programming,

Y. K. Mogal (✉) · S. K. Mahajan · S. B. Rane
Department of Mechanical Engineering, Sardar Patel College of Engineering, Mumbai, India

R. Agrawal
Department of Mechanical Engineering, Malaviya National Institute of Technology Jaipur, Jaipur, Rajasthan, India

R. Agrawal et al. (eds.), *Recent Advances in Smart Manufacturing and Materials*, Lecture Notes in Mechanical Engineering,
https://doi.org/10.1007/978-981-16-3033-0_10

and dynamic programming have been viably applied to improve process parameters [5]. Among the different strategies, the Taguchi-based methodology has created exceptional and remarkable control that appears differently with customary practices. It has demonstrated a wide extent of present-day applications for making the item obtuse toward any wild factors [6].

Niranjan et al. [7]; Rajendra and Deepak [8] streamlined speed, feed rate, and depth of cut in the turning of Al 6061 for surface finish and material removal rate. Kishore et al. [9] considered the impact of cutting parameters on cutting force and surface quality performing CNC turning on Al6061-4 wt% TiC composite. Ugrasen et al. [10] determined the ideal process parameters concerning ultimate tensile strength (UTS) and hardness of the weld joint utilizing the Taguchi technique. Kandpal et al. [11] proposed a multiresponse improvement strategy utilizing the Taguchi approach and utility idea for electrical discharge machining (EDM) of Al6061/10%Al2O3 MMC. MisbahNiamat et al. [12] enhanced electrical discharge machining parameters for electrode wear rate (EWR) and material ejection rate of Aluminum 6061 T6 Alloy utilizing Taguchi plan of assessment. Adalarasan et al. [13] enhanced lesser cutting parameters using the Taguchi-based response surface method.

Shaik and Srinivas [14]; Kumar et al. [15], built up a coordinated framework to demonstrate and improve the processing parameters during end milling of Al 6061. Nghiep et al. [16] examined the mechanism of deflection error, and Tomadi et al. [17] introduced the forecast model in the end milling process. Kondayya and GopalaKrishna [18] introduced a non-dominated sorting genetic algorithm-II for modeling and multi-objective optimization of the CNC end milling process.

Niknam and Songmene [19] utilized Taguchi and ANOVA investigation to streamline burr size and surface finish parameters similarly Palanisamy et al. [20] utilized genetic algorithm by considering machining time as a target for slot milling activity. Shinge and Dabade [21] explored the impact of preparing factors in micro-milling of Al 6063 T6 utilizing Taguchi L16 orthogonal array. Hwan and Sang-Heon [22] endeavored to obtain ideal cutting conditions for burr minimization in face-processing tasks. Response surface methodology (RSM) has been utilized by Premnath [23] for finding the ideal machining parameters while milling Al6061/Al_2O_3/Gr. Baharudin et al. [24] utilized the Taguchi strategy to locate the ideal surface roughness for Al6061 face milling. Sukumar et al. [25] utilized an artificial neural network (ANN) model and Taguchi S/N ratio examination for getting the ideal blend of parameters to accomplish a decent surface finish in face milling activity. Nguyen [26] investigated the effects of milling type and various cutting conditions on the surface roughness using Taguchi and ANOVA analysis.

3 Research Gaps and Problem Definition

Al 6061-T6 is a typical alloy that is utilized in different modern applications for some reason since it has prevalent mechanical properties. At present, numerous endeavors are being attempted to improve different handling parameters utilizing the genetic algorithm, response surface methodology, regression analysis, neural network, grey relational analysis (GRA), Taguchi strategy, etc. In the present date, necessity of industry is to make items having high quality at low cost as well as high productivity, less machining, and production time. This paper will fulfill the industries need to deliver excellent items with minimal effort. Problem definition of underlying paper is to improve the processing parameters, for example, speed, feed rate, and cutting depth on surface roughness (Ra) and material removal rate (MRR) in the face milling process by utilizing Taguchi plan of experiment and multiresponse regression analysis.

4 Research Objectives and Research Methodology

The main objective of this paper is to investigate the optimum setting of machining parameters (speed, feed rate, and depth of cut) in face milling of Al6061-T6 alloy to accomplish the minimal surface roughness and high material removal rate utilizing the Taguchi plan of an experiment. Figure 1 shows the research methodology flowchart. Experimentation is done according to the L27 orthogonal array with 03 control factors and 03 levels for each factor. Further examination is completed using signal-to-noise (S/N) ratio investigation and analysis of variance (ANOVA), to figure out which process parameters are measurably noteworthy. At long last, a confirmation test is done to examine the improvement in optimization. Finally, a multiresponse regression examination is performed to study the combined impact of the two responses.

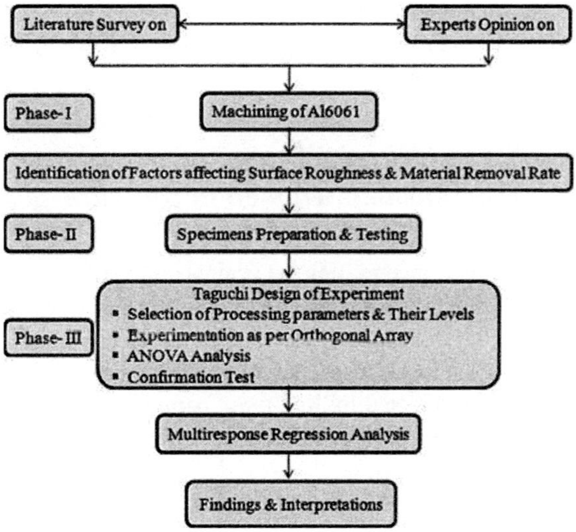

Fig. 1 Research methodology flowchart

Table 1 Chemical composition of Al6061 T6 alloy

Elements	Cr	Fe	Si	Mg	Mn	Cu	Zn	Ti	Al
Percentages	0.1	0.35	0.5	0.08	0.04	0.28	0.02	0.01	Balance

Table 2 Levels of input parameters used in this study

Input parameters	Level 1	Level 2	Level 3
Spindle speed (RPM)	1000	1500	2000
Feed rate (mm/min)	200	400	600
Depth of cut (mm)	0.4	0.8	1.2

5 Experimental Work

5.1 Material

In this paper, blocks of Al 6061 T6 alloy with a size 50 mm * 60 mm * 50 mm are used for experimentation. The chemical composition of the Al 6061 T6 alloy is given in Table 1.

5.2 Machining Parameters

Cutting speed (v), feed rate (f), and cutting depth (d) are chosen as input parameters similarly surface roughness and material removal rate are chosen as output factors for the examination. According to suggestions of the cutting instrument maker and the limit of the machine device, the levels of these input parameters were chosen as appeared in Table 2.

5.3 Experimental Setup and Procedure

For performing face milling, a Cosmos 05-axis vertical milling machine with a rotational speed of 8000 rpm and motor power of 11 kW was utilized as appeared in Fig. 2. A carbide face milling cutter having a diameter 63 mm, 05 inserts, and sixteen cutting edges for each insert manufactured by Tungaloy has been used as cutting tools. The impact of the selected milling input parameters on output factors and ideal settings of the parameters has been practiced utilizing Taguchi's L27 orthogonal array.

Fig. 2 Setup of vertical milling machine

5.4 *Measurement of Output Factors*

Surface roughness (Ra) is estimated by surface roughness analyzer SRT-6210. Three readings for surface roughness were taken for every surface, and its average is considered as a final value, to minimize the error. Material removal rate (MRR) is determined with the help of formula, [Material Removal Rate = Width of cut (mm) * Depth of cut (mm) * Feed Rate (mm/min)].

5.5 *Taguchi Signal to Noise Ratio Analysis*

Analysis of experimental data of surface finish (Ra) and material removal rate (MRR) is finished by utilizing the Taguchi plan in Minitab-19 software and the estimated S/N ratio values. From Fig. 3, it can presume that as rotational speed and cutting depth builds, surface roughness value diminishes, while feed rate diminishes at first up to 400 mm/min, over that it is incremented. Smaller the better trademark was utilized to decide the surface quality. Table 3 shows that speed at position 1, feed rate at position 3, and cutting depth at position 1 are the best estimations of parameters for surface finish. In Table 3, position 1, position 2, and position 3 are given to feed rate, cutting speed, and depth of cut, respectively, which demonstrates that the feed rate is having a most elevated effect on the S/N proportions of surface quality because of its delta worth and rank, and later, this is trailed by the cutting speed and depth of cut, respectively. Table 4 shows that speed at 1000 rpm, feed rate at 600 mm/min, and depth of cut at 0.4 mm ideal values of parameters for surface finish.

Larger the better trademark was utilized to determine the material removal rate. From Fig. 4, it can be seen that as feed rate and cutting depth build MRR increments,

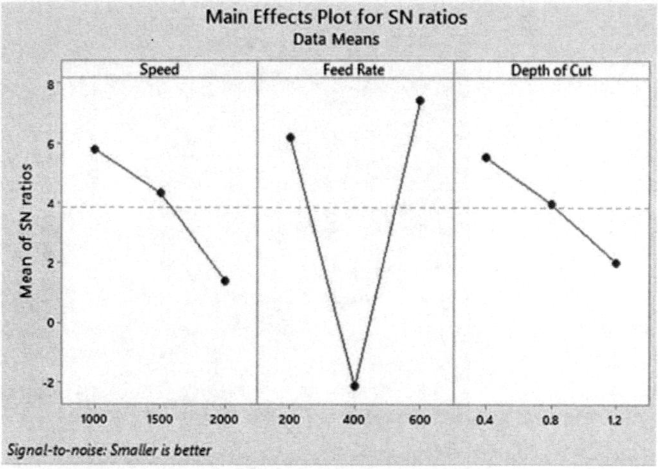

Fig. 3 Main effect plot of S/N ratios for surface roughness

Table 3 Surface roughness S/N ratio for each level of control parameters

Level	Speed (Rpm)	Feed rate (mm/min)	Depth of cut (mm)
1	**5.785**	6.186	**5.517**
2	4.312	-2.161	3.947
3	1.371	**7.443**	2.005
Delta	4.413	9.603	3.512
Rank	2	1	3

Bold values are the highest values, which shows the intensity of particular level on the effect of parameters

Table 4 Optimum control parameter values for surface roughness S/N ratio analysis

S. No.	Parameters	Optimum value
1	Speed (Rpm)	1000
2	Feed rate (mm/min)	600
3	Depth of cut (mm)	0.4

whereas MRR is consistent for all estimations of speed. Table 5 shows that feed rate and depth of cut at level 3 are the best values of cutting parameters for MRR. From Table 5, feed rate and depth of a cut have an equivalent impact on the S/N proportions of MRR because of its delta worth and rank. Cutting speed has no impact on the material removal rate because to calculate the MRR we have consider the formula, MRR = Depth of cut * Width of cut * Feed rate. Table 6 presumes that any value of speed, feed rate at 600 mm/min, and cutting depth at 1.2 mm is the ideal benefit of preparing parameters for MRR.

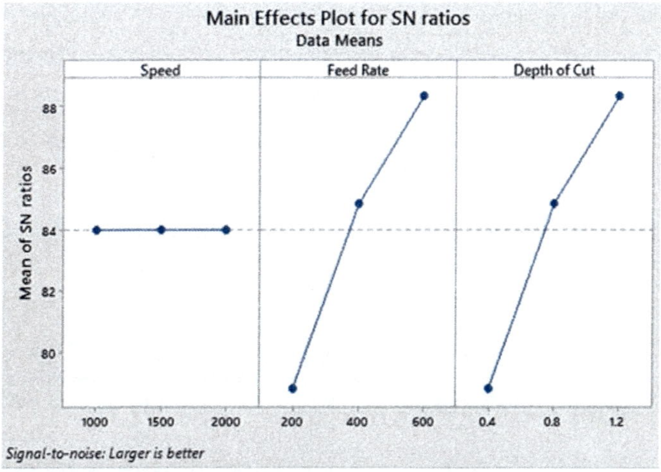

Fig. 4 Main effect plot of S/N ratios for material removal rate

Table 5 Material removal rate S/N ratio for each level of control parameters

Level	Speed (Rpm)	Feed rate (mm/min)	Depth of cut (mm)
1	84.00	78.81	78.81
2	84.00	84.83	84.83
3	84.00	**88.35**	**88.35**
Delta	0.00	9.54	9.54
Rank	3	1.5	1.5

Bold values are the highest values, which shows the intensity of particular level on the effect of parameters

Table 6 Optimum control parameter values for material removal rate S/N ratio analysis

S. No.	Parameters	Optimum value
1	Speed (Rpm)	Any value
2	Feed rate (mm/min)	600
3	Depth of cut (mm)	1.2

6 Analysis of Variance (ANOVA)

ANOVA is performed on the surface roughness and material removal rate to explore the impact of process parameters.

From the F-value shown in Table 7, it is clear that the commitment of feed rate is high for the surface finish, and later, this was trailed by cutting speed and cutting depth, respectively. P-estimation of feed rate is under 0.05, so this parameter is noteworthy to get the best quality surface, while the P-estimation of other parameters is more prominent than 0.05, so these are not critical.

Table 7 ANOVA table for surface roughness

Source	DF	Adj SS	Adj MS	*F*-value	*P*-value
Speed	2	1.5153	0.7576	1.58	0.230
Feed rate	2	3.5400	1.7700	3.70	0.043
Depth of cut	2	0.5951	0.2976	0.62	0.547
Error	20	9.5648	0.4782		
Total	26	15.2152			

Table 8 ANOVA table for material removal rate

Source	DF	Adj SS	Adj MS	*F*-value	*P*-value
Speed	2	0	0	0.00	1.000
Feed rate	2	1,658,880,000	829,440,000	60.00	0.000
Depth of cut	2	1,658,880,000	829,440,000	60.00	0.000
Error	20	276,480,000	13,824,000		
Total	26	3,594,240,000			

From Table 8, it is clear that for material removal rate, the involvement of feed rate and cutting depth is large, and cutting speed has no impact on it. *P*-value of feed rate and depth of cut is under 0.05, so these parameters are critical to getting the high material removal rate, while the *P*-value of cutting speed is more noteworthy than 0.05, so this is not huge.

7 Confirmation Test

A confirmation test is a critical and last piece of the Taguchi strategy. In this paper, optimal combination of parameters and their levels coincidently coordinate with one of the investigations in the orthogonal array (OA), in this manner confirmation test is not required.

8 Multiresponse Regression Analysis

After analyzing the effect of machining parameters on a single response individually, multiresponse regression analysis is carried out; for this purpose, response optimizer is utilized to recognize the combined impact of input variables on a single or a many output factors and draws an optimization plot. Response optimizer also permits the statistician to perform sensitivity investigation and enhancement in the previous solution. While performing the analysis with a response optimizer, make ensure that stored models should meet the assumptions of the main investigation and setting of variables should be inside the scope of the information that you used to fit the model else it gives erroneous outcomes.

Table 9 demonstrated that the objective for material removal rate is to expand it, and the objective for surface roughness is to limit it. Target value, upper and lower cutoff points, weight, and importance for both responses are also shown in Table 9. The weight decides the distribution of desirability on the interval between the lower (or upper) limit and the objective. The importance decides the impact of each response on the composite desirability. The values of weight and importance are considered from 0.1 to 10 which depict the shape of the desirability function and comparative significance of the response variable, respectively. As per the significance of one response over another, importance values are to be assigned. Higher values related to the most significant responses, lower values to less significant responses. Here both responses have the same importance value. Accordingly, both responses have an equivalent effect on the composite desirability.

Table 9 Responses and its boundary conditions for multiresponse regression analysis

Response	Goal	Lower	Target	Upper	Weight	Importance
Material removal rate	Maximum	4800	43,200	43,200	1	1
Surface roughness	Minimum	0.1	0.1	2.7791	1	1

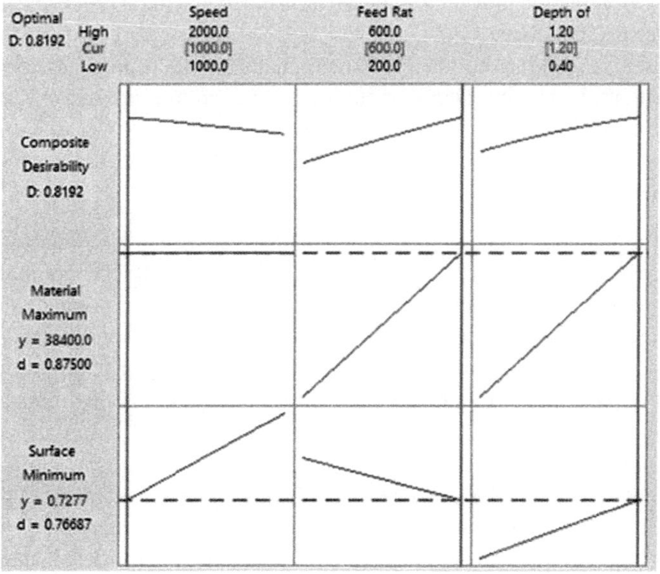

Fig. 5 Optimization plot of individual and composite desirability for surface roughness and material removal rate

Table 10 Multiple response prediction table

Solution	Speed	Feed rate	Depth of cut	Material removal rate fit	Surface roughness fit	Composite desirability
1	1000	600	1.2	38,400	0.727724	0.819154

The optimization plot (Fig. 5) shows the impact of all three input factors on the output parameters or composite desirability. The individual and composite desirability survey how well a blend of factors fulfills the objectives of the output factors. Optimized parameter settings of one and more than that response are assessed by individual and composite desirability, respectively. The value of desirability lies between 0 and 1. One indicates the perfect case; zero demonstrates that some responses are not lies within their adequate cutoff points. An engineer can adjust the values of parameters from the plot to accomplish bigger individual desirability for the response as per the requirement.

Table 10 shows the composite desirability (0.8192) is near to 1, which shows the settings appear to accomplish good outcomes for all responses in general. However, the individual desirability indicates that the settings are more effective at maximizing MRR (0.87500) than at minimizing surface roughness (0.76687). The present variable settings are speed = 1000 rpm, Feed rate = 600 mm/min, depth of cut = 1.2 mm.

Perceptions from the above optimization plot are as per the following:

1. Speed: Optimization plot shows that MRR is independent on speed and as speed increases Ra increases.
2. Feed rate: As feed rate builds, MRR increments and Ra diminishes.
3. Depth of cut: Increasing Depth of cut increments both responses.

9 Conclusion

This research is governed by the Taguchi method to identify the optimal set of parameters such as rotational speed, feed rate, and cutting depth on surface roughness and material removal rate in the face milling process. From Taguchi S/N proportion investigation, it can reason that speed at 1000 Rpm, feed rate at 600 mm/min, and depth of cut at 0.4 mm are the ideal qualities for better surface roughness and any value of speed, feed rate at 600 mm/min, and depth of cut at 1.2 mm are the ideal qualities for material removal rate.

ANOVA examination is recognized that the commitment of feed rate is high on the surface quality then by speed and depth of cut, respectively. On account of material removal rate, the feed rate and depth of cut both are similarly contributed, and cutting speed does not have any impact on MRR.

Multiresponse regression analysis concludes the speed at 1000 rpm, feed rate at 600 mm/min, and depth of cut at 1.2 mm is the ideal characteristics for multiresponse investigation.

10 Limitations and Future Scope

This research is restricted to optimize a couple of parameters including two response factors only. In the future, tool wear rate, tool life, cutting forces, energy consumption, etc., can be considered as a response factor, also tool material, type of workpiece material, type of coolants, coolant flow rate, tool geometry, machine condition, operator skill, environmental condition, costing, etc., can be considered as processing parameters. At the same time, validation and multiple objectives can likewise be accomplished by utilizing other techniques such as genetic algorithm, neural network, etc.

Acknowledgements The authors express sincere gratitude to all researchers who made literature available in this domain. The authors are earnestly appreciative of the reviewers for their important recommendations which assisted with improving the nature of this paper. At long last, the authors also grateful to those things, which are directly or indirectly, helped us for completing this research paper.

References

1. Deepak, D., & Rajendra, B. (2016). Optimization of machining parameters for turning of Al6061 using robust design principle to minimize the surface roughness. *Procedia Technology, 24*, 372–378.
2. Sanders, R. E. (2001). Technology innovation in aluminum products. *JOM, 53*, 21–25
3. Johnson, J., Bibin, K. T., & Anoop, S. (2018). Optimization of wire electric discharge machining parameters on Al 6061. *IJESRT, 7*, 447–455.
4. Gaitonde, V. N., Karnik, S. R., & Davim, J. P. (2009). Multi performance optimization in turning of free-machining steel using Taguchi method and utility concept. *Journal of Materials Engineering and Performance, 18*, 231–236.
5. Jamwal, A., Agrawal, R., Sharma, M., & Kumar, V. (2021). Review on multi-criteria decision analysis in sustainable manufacturing decision making. *International Journal of Sustainable Engineering*, 1–24.
6. Luo, M., Liu, G., & Chen, M. (2008). Mechanism of burr formation in slot milling Al-alloy. *International Journal of Materials and Product Technology, 31*, 63–71.
7. Niranjan, D. B., Shivashankar, G. S., SreenivasRao, K. V., & Praveen, R. (2017). Optimization of cutting process parameters on AL6061 using ANOVA and Taguchi method. *Materials Today: Proceedings, 4*, 10845–10849.
8. Rajendra, B., & Deepak, D. (2016). Optimization of process parameters for increasing material removal rate for turning Al6061 using S/N ratio. *Procedia Technology, 24*, 399–405.
9. SaiChaitanya Kishore, D., PrahladaRao, K., Ramesh, A. (2015). Optimization of machining parameters for improving cutting force and surface roughness in turning of Al6061-TiC in-situ metal matrix composites by using Taguchi method. *Materials Today: Proceedings, 2*, 3075–3083.
10. Ugrasen, G., Bharath, G., Kishor Kumar, G., Sagar, R., Shivu, P. R., & Keshavamurthy, R. (2018). Optimization of process parameters for Al6061-Al7075 alloys in friction stir welding using Taguchi technique. *Materials Today: Proceedings, 5*, 3027–3035.
11. Chandra Kandpal, B., Kumar, J., & Singh, H. (2017). Optimization and characterization of EDM of AA 6061/10%Al2O3 AMMC using Taguchi's approach and utility concept. *Production & Manufacturing Research, 5*, 351–370.

12. Misbah, N., Shoaib, S., Essam, S., Sikiru, O. I., & Qazi S. K. (2019). Experimental characterization of electrical discharge machining of Aluminum 6061 T6 alloy using different dielectrics. *Arabian Journal for Science and Engineering.*
13. Adalarasan, R., Santhanakumar, M., & Thileepan, S. (2016). Selection of optimal machining parameters in pulsed CO_2 laser cutting of Al6061/Al_2O_3 composite using Taguchi-based response surface methodology (TRSM). *International Journal of Advanced Manufacturing Technology.*
14. Shaik, J. H., & Srinivas, J. (2017). Optimal selection of operating parameters in end milling of Al-6061 work materials using multi-objective approach. *Mechanics of Advanced Materials and Modern Processes, 3*, 1–11.
15. Kumar, D., Chandna, P., & Pal, M. (2017). Efficient optimization of process parameters in 2.5 D end milling using neural network and genetic algorithm. *International Journal of System Assurance Engineering and Management.*
16. Nghiep, T. N., Sarhan, A. A. D., & Aoyama, H. (2018). Analysis of tool deflection errors in precision CNC end milling of aerospace Aluminum 6061-T6 alloy. *Measurement.*
17. Tomadi, S. H., Ghani, J. A., CheHaron, C. H., Mas, A. H., & Daud, R. (2017). Effect of cutting parameters on surface roughness in end milling of AlSi/AlN metal matrix composite. *Procedia Engineering, 184*, 58–69.
18. Kondayya, D., & Gopala Krishna, A. (2012). An integrated evolutionary approach for modelling and optimisation of CNC end milling process. *International Journal of Computer Integrated Manufacturing, 25*, 1069–1084.
19. Niknam, S. A., & Songmene, V. (2013). Simultaneous optimization of burrs size and surface finish when milling 6061–T6 Aluminium alloy. *International Journal of Precision Engineering and Manufacturing, 14*, 1311–1320.
20. Palanisamy, P., Rajendran, I., & Shanmugasundaram, S. (2007). Optimization of machining parameters using a genetic algorithm and experimental validation for end-milling operations. *International Journal of Advanced Manufacturing Technology, 32*, 644–655.
21. Shinge, A. R., & Dabade, U. A. (2018). The effect of process parameters on material removal rate and dimensional variation of channel width in micro-milling of Aluminium alloy 6063 T6. *Procedia Manufacturing, 20*, 168–173.
22. Hwan, L. S., & Sang-Heon, L. (2003). Optimization of cutting parameters for burr minimization in face-milling operations. *International Journal of Production Research, 41*, 497–511.
23. Premnath, A. A. (2015). Studies on machining parameters while milling particle reinforced hybrid (Al6061/Al_2O_3/Gr) MMC. *Particulate Science and Technology: An International Journal*. https://doi.org/10.1080/02726351.2015.1025457
24. Baharudin, B. T. H. T., Ibrahim, M. R., Ismail, N., Leman, Z., Ariffin, M. K. A., & Majid, D. L. (2012). Experimental investigation of HSS face milling to AL6061 using Taguchi method. *Procedia Engineering, 50*, 933–941.
25. Sukumar, M. S., VenkataRamaiah, P., & Nagarjuna, A. (2014). Optimization and prediction of parameters in face milling of Al-6061 using Taguchi and ANN approach. *Procedia Engineering, 97*, 365–371.
26. Nguyen, N.-T. (2020). A study on influence of milling types and cutting conditions on surface roughness in milling of Aluminum alloy Al6061 T6. *Universal Journal of Mechanical Engineering, 8*(4), 183–190.

Computational and Experimental Methods to Investigate Fracture Behavior of Functionally Graded Material Structures—A Critical Review

Manish Bhandari and Kamlesh Purohit

1 Introduction

Heterogeneous composite materials are very popular in engineering applications. The constituents of heterogeneous composite materials can be two or more different material phases and compositions. One of the recently developed heterogeneous materials is Functionally Graded Material (FGM). FGM is a typical composite material in which the material properties vary gradually in one or more directions in space. FGMs have extensively been used because of their attractive properties, reduced residual stress, enhanced thermal properties, etc. Functionally graded materials (FGM) are a class of composites that allow a gradual variation of material properties from one surface to another. These novel materials were proposed by the Japanese in 1984 and are projected as thermal barrier materials for applications in space planes, space structures, and nuclear reactors, to name only a few. As conceived and manufactured today, these materials are isotropic and non-homogeneous. The grades can be continuous on a microscopic level as shown in Fig. 1a, or they may be laminated composed of metals, ceramics, polymers, or variations of porosity and density as shown in Fig. 1b.

In a composite material when two dissimilar materials are bonded together, there is a very high chance that crack will occur at some extreme loading conditions, be it static, dynamic, or thermal loads. Cracks are likely to initiate at interfaces and grow into the weaker material section. Another problem is the presence of residual stresses due to the difference in coefficients of thermal expansion of two dissimilar materials. These problems can be resolved by gradually varying the volume fraction of the constituents. The gradation in properties of the material reduces thermal stresses, residual stresses, and stress concentration factors. Stress Intensity Factor (SIF) is a predominant fracture parameter in graded materials and subsequent prediction of

M. Bhandari (✉) · K. Purohit
MBM Engineering College, Jai Narain Vyas University, Jodhpur, Rajasthan 342003, India

R. Agrawal et al. (eds.), *Recent Advances in Smart Manufacturing and Materials*,
Lecture Notes in Mechanical Engineering,
https://doi.org/10.1007/978-981-16-3033-0_11

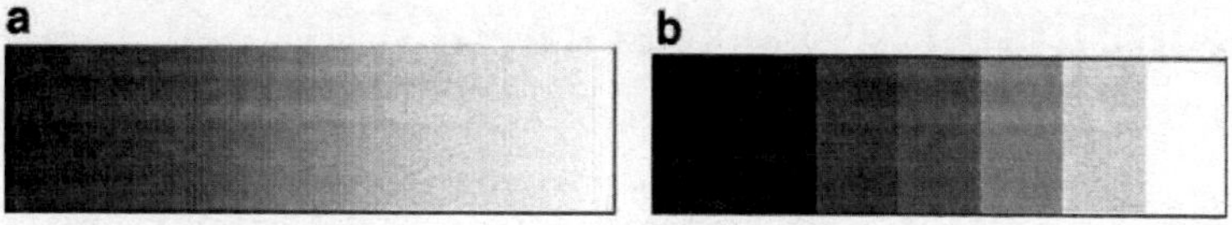

Fig. 1 **a** Continuously graded FGM. **b** Discretely layered FGMs

SIF values for various loading and gradient conditions has been the major area for research work regarding FGMs so far. Both experimental work and numerical work have been carried out to investigate fracture behavior of FGMs. The determination of SIFs expressions via numerical methods for various crack positions in FGMs has been carried out by various researchers. Typically the problem of crack is presented in a system of integral equations with appropriate boundary conditions. Integral equations are solved using transforms and numerical methods. The finite element method (FEM) has been used exhaustively in fracture modeling, including fracture in graded regions. Often experimental results are combined with FE models to yield further information.

The present review concentrates on the various computational and experimental methods which are important for the study of fracture behavior of functionally graded materials. An attempt has been made to classify various numerical methods used for the crack and fatigue analyses of FGMs. Finally, some vital suggestions for future scope of research in the area of FGMs are presented. It is hoped that this review paper will serve the interests of all the academicians, researchers, and engineers involved in the analysis and design of FGMs.

2 Finite Element Analysis of Cracks in FGM

The numerical methods exhibit better adaptability and application areas as compared with analytical techniques. Various numerical approaches such as the conventional finite element method (FEM), 2-D and 3-D FEM, Extended finite element method (XFEM), Element-free Galerkin method (EFGM), and Cohesive volumetric finite element (CVFE) have been used to study the fracture behavior of FGMs. The Finite element method (FEM) is the most popular method in computational mechanics because of its high accuracy and ability to deal with various complex shapes.

2.1 Finite Element Method (FEM)

Conventional FEM has been applied to calculate SIFs and design parameters in for various crack positions in FGM objects subject to different loading conditions. FEM has been used to simulate FGM for various locations and positions of crack to obtain fracture parameters such as complex SIFs and energy release rate [1], fracture toughness [2] and energy release rates of kinking of surface crack subjected to the initial strain resulting from stress relaxation [3]. FGM and homogeneous cracked beams are examined using FEM to evaluate crack tip responses to impact loading [4–6] and SIFs are obtained using regression analysis alongwith FEM. [7] FEM is used to simulate dynamic crack propagation in elastic–plastic graded double cantilever beam model [8]. Orthotropic FGMs have been presented by FEM, for mode I and mixed-mode crack problems considering plane elasticity and arbitrarily oriented crack [9, 10], for mode I and mixed-mode 2D problems [11] and for evaluating mixed-mode SIFs and T-stress [12]. Crack propagation is investigated in graded materials under monotonic and cyclic loading using finite element for linear elastic FGM [13]. Remeshing algorithm in conjunction with the FEM performed automatic simulation of crack propagation in FGMs [14–16]. The FE models are utilized in a parametric study of various elastic gradients in core on mixed-mode fracture performance of FG sandwich structures [17]. FGM specimen with local singularity and fracture characterization is studied with combination of optical caustics technology and FEM and SIF, material constants, graded index, and characteristic size are calculated [18]. FEM and the displacement correlation technique are used to evaluate the fracture parameters for orthotropic FGM coatings considering an embedded crack [19]. The concept of fast running crack in FGMs is investigated through the FEM under impact loading and obtained the DSIFs during the crack growing process [20]. Plane strain buckling problem is solved using FEM of an orthotropic graded coating with a crack embedded in FGM layer [21]. P-version of FEM is presented to estimate location and size of an open edge crack in a beam made of FGM [22]. The Muskhelishvili complex potential method with FEM is formulated to evaluate the model of an FGM beam [23]. Embedded crack problem accuracy is modeled using FEM in graded coating [24]. Capabilities and limitations of numerical modeling in fracture problems of FGMs are investigated by means of FEM and is recommended when the crack is not perpendicular to the direction of the material property variation [25]. Numerical modeling of crack propagation path in FGMs under mixed-mode loadings is presented to evaluate the direction angle as a function of SIFs at each increment of propagation [26]. A phase field formulation is presented for fracture in FGMs and a model is built using homogenization theory [27]. Two-dimensional mixed-mode crack propagation in FGMs is investigated using the finite element method (FEM) [28].

2.2 Two-Dimensional and Three-Dimensional FEM

2-D FEA is performed to calculate the stress field surrounding an edge crack in an elastic FG plate [29] Dynamic fracture of FGMs is modeled using an explicit cohesive volumetric finite element (CVFE) scheme that incorporates spatially varying constitutive and failure properties [30]. Quasistatic crack initiation under mixed-mode loading in planar (2-D plane stress) FGMs is considered to evaluate the applicability of the maximum tangential stress (MTS) criterion using FEA [31]. Fracture behavior of an arbitrarily oriented crack located in 2-D FGM is modeled under antiplane shear stress by discretizing the domain into finite elements [32]. 2-D plane strain FG finite element within the ABAQUS software environment is developed to model a crack growth under dynamic loading [33] and thermal loading [34] with virtual crack closure technique (VCCT).

3-D surface crack problems in FG coatings are examined using collapsed 20-node isoparametric 3-D finite elements to calculate SIFs in the thermal and structural problems [35]. Mixed-mode SIFs and the non-singular T-stress are evaluated in FGMs under steady thermal loads [36] and transient thermal loads [37] with the 2D and 3D finite element analyses. 3-D enriched FE methodology is applied to FGMs to compute fracture parameters, i.e., SIF, for elastic 3-D cracks in FGMs [38, 39]. 3D FEM is presented for mixed-mode fracture analysis of an FGM coating-bond under transient thermal loads [40], mechanical loads [41] and fatigue crack propagation [42] containing an inclined elliptical crack at the free surface. Strain singularity around the crack front is simulated by utilizing collapsed wedge-shaped singular elements. Mixed-mode SIFs for three-dimensional cracks in FGMs subject to mixed mode loading are computed using 3-D enriched finite elements [43].

2.3 Extended Finite Element Method (XFEM) and Element-Free Galerkin Method (EFGM)

Element-free Galerkin method (EFGM) is presented to evaluate the TSIFs 2-D thermo-elastic cracked functionally graded orthotropic medium [44]. Interfacial crack problems along with crack interactions are modeled by EFGM and XFEM under both mode-I and mixed-mode loadings. The extrinsic partition of unity enrichment technique has been successfully used to model both material and geometric discontinuities [45]. Fatigue crack growth is simulated using extended finite element method (XFEM). [46] and thermo-mechanical loads [47] of alloy/ceramic FGMs [48]. XFEM method is used to analyze and simulate thermal shock and transient heat conduction problems in a strip made of FGM for crack propagation [49]. Coupled dynamical system of equations are obtained from the extended finite element method (XFEM) and solved by the Newmark method in the time domain to calculate the DSIFs at each time step [50]. Transversely isotropic FGMs with discontinuity are modeled using XFEM and employed interaction integral method to evaluate the SIFs [51]. TSIF of an FGM (Al-Al_2O_3) plate under cyclic thermal load in presence of multiple discontinuities (cracks, holes, and inclusions) is simulated by employing XFEM formulation and domain-based interaction integral approach for calculating the TSIF at the crack tip [52, 53]. The fatigue crack growth problems in FGM are modeled using generalized Ramberg–Osgood material model and simulated using XFEM [54]. Both EFGM and XFEM are utilized to model and compare crack growth in fatigue loading. Partition of unity is used to track crack path in XFEM and a new enrichment criterion is proposed to track the crack path in EFGM [55]. Life of an FGM made of Al alloy and Alumina (ceramic) subject to cyclic/fatigue mixed mode loading is computed using both EFGM and XFEM [56]. The complex 3-D crack with delamination and transverse cracks pattern of FGM is simulated in direction of thickness and displacement discretization (in-plane) of XFEM [57]. An edge crack in 3D elastic FGM is solved by XFEM [58].

2.4 FEM in Thermal Loadings

Thermal crack propagation [59, 60] and thermomechanical loading on FGM cracked plate in FGM plate [61] is modeled using FEM when it is cooled from high temperature and curved or straight crack paths are observed on the ceramic surface. The axisymmetric crack problem for thermal barrier coatings is solved using FEM [36]. The effects of the thermal residual stresses and TSIFs are calculated by employing FEM for cracks developing from the surface [62]. Multiple surface cracking of thermal shock resistance [63] and transient temperature field in FGMs [64] are studied using FEM. Inclined cracks [65] and embedded cracks [66] located in FGMs subjected to thermal loading are investigated by developing a numerical procedure based on the FEM to evaluate J-integral components, I mode and II mode SIFs and the T-stresses at crack tips. Mixed-mode hygrothermal loading on orthotropic FGMs with cracks are analyzed by integrating the developed technique into a finite element analysis software [67, 68]. A set of analytical and numerical techniques combined with interaction energy integral method, a perturbation method, and FEM are developed for analyzing problems of cracked FG plate subject to mixed-mode thermal shock loading [69, 70]. A weight function method is introduced to analyze circumferentially cracked FG hollow cylinder under transient thermomechanical loading [71].

3 Experimental Studies of Cracks in FGM

Various experimental techniques have also been used to validate theoretical predictions of stress fields and SIFs in cracked graded materials. Often experimental results are combined with FE models to yield further information.

3.1 Static Loading

The 4-point bending test, for a short precrack at the machined notch in FGM beams allows accurate J-R measurements which is useful in fatigue [72, 73]. Complex SIF is determined by conducting static fracture experiments on epoxy-based FGM. Electrical strain gages are used to measure strains and homogeneous field equations are used to convert the strains to SIF [1]. Optical measurements near quasi-statically growing edge cracks in FGM beams made of glass-filled epoxy have been undertaken under static four-point bending [4, 6]. Single edge notch fracture tests are conducted on FG ethylene carbon monoxide copolymer (ECO) to find out the fracture parameters such as SIFs and fracture toughness [2]. The distribution of fracture toughness or R-curve behavior with stable crack growth in a three-point-bending FGM specimen is evaluated. The FGM specimen is fabricated by powder metallurgy using partially stabilized zirconia (PSZ) and stainless steel (SUS 304) [74]. Coherent gradient sensing (CGS) to interpret experimentally obtained fringes has also been utilized and developed specific relations for out-of-plane deformation and stress field near crack tips with the optical method [75]. The influences of material gradient variation on crack initiation, singularity field, and SIFs at the crack tip were analyzed using three-point-bending of FGM beams with crack location on both the stiff and the compliant sides [18].

3.2 *Dynamic Loading*

Various experimental research has been carried out on cracked FGM structure under static and dynamic loading. These are tabulated and shown in Table 1.

3.3 *Thermal Loading*

Bending test is done for delamination crack propagation taking into account TBC shrinkage [91]. Thermal cracking has been depicted experimentally in a ceramic/metal FG plate. Curved or straight crack paths are observed on the ceramic surface when it is cooled from high temperature [59]. An experimental study is conducted for the thermal fracture behavior of plasma-sprayed yttria stabilized zirconia-NiCoCrAlY bond coat alloy compositionally graded thermal barrier coatings (TBC) when subjected to a thermal shock loading using a continuous CO_2 laser [92]. Thermal shock fracture behavior of metal/ceramic FGMs is evaluated by a controlled burner heating method using a H_2/O_2 combustion flame [93]. Resistance to crack growth is investigated using Laser thermal shock experiments in conjunction with crack arrest principles to estimate [94] and record response of multiple surface cracking [95] in a FG yttria stabilized zirconia (YSZ) -bond coat (BC) alloy (NiCoCrAlY) TBC. Failure of graded TBCs is modeled under cyclic surface heating by laser irradiation on the basis of fracture mechanics. Performance of heat insulation and fracture behavior of thermal barrier-type FGM coatings is studied [96]. The experimental and analytical investigation are presented on the thermal fatigue

Table 1 Cracked FGM structure under dynamic loading

Experimental techniques	Result parameters
One-point impact loading	Crack growth resistance behavior [7, 76]
Digital image correlation technique	Real-time displacement field [77], R-curves for crack growth [78] SIF KI, KII, and T-stress are obtained [79]
Four-point bending test	Effect of grading on increase of interface fracture toughness and a decrease due to aging modified by a stiffening layer [80]
Compact tension specimen test	Maximum circumferential stress theory is justified [81]
Low-velocity impact loading with reflection coherent gradient sensing technique	Mode-I crack initiation and growth behaviors are studied [82]
Constant load amplitude conditions by a Shimadzu servo-hydraulic testing machine	The crack growth paths and fracture surfaces [83]
Impact loading	Feasibility of introducing compositional gradients to the core [17]
Dynamic photoelasticity coupled with high-speed photography	Crack tip velocities and dynamic stress fields [84]
Mixed-mode dynamic fracture experiments under the conditions of low-velocity impact	Transient crack tip deformations for pre- and post-crack initiation periods [85]
Monotonic and cyclic bending loading	Propagation trajectory [86]
Multistep infiltration technique under monotonic and cyclic four-point bend loading	Crack initiation and propagation [13] and fatigue crack propagation [87]
Four points bending tests on notched beams	The fracture processes in quasi-brittle FGMs [88]
Dynamic fracture experiments	Transient mode I crack growth behavior [89]
Optical experiments reflection-mode CGS and high-speed photography	Mixed-mode dynamic crack growth behavior [20]
Photoelastic visualization	DSIFs [23]
Tensile cyclic load	Fatigue life increased for central notched specimens [90]

fracture mechanism of three types of FG composites which were fabricated using squeeze infiltration process [97].

4 Conclusion

FGMs have revolutionized the field of material science by involving graded microstructure and have signified to be a multi-functional material in handling extreme conditions effectively. The flexibility in design and its performance under thermomechanical environments are the main reasons for its potential applications in various areas. The recent progress in the numerical analysis and experimental studies of fracture behavior of FGM have been reviewed in this paper. In spite of the variety of methods used to date of analysis of fracture behavior of FGMs, several common themes have emerged. Many of these works provide a fundamental understanding of the basic fracture behavior of the material. Therefore, modeling and analysis of FGMs are vital for the development of this emerging new material.

The work which may require more exploration and future work is as follows:

1. It has been observed that the work has been mainly focused on modeling, both analytical and FEM, and has mainly focused on estimation of SIF for cracks in graded regions.
2. Experimental studies have been limited because of processing difficulties, though some experiments have provided validation to theory.
3. Three-dimensional elasticity solutions were found to be the most accurate solutions for crack analysis under thermal environment of FGM.
4. Although significant research has been made recently in the investigation of fracture in graded materials, a consolidated understanding may require more work.
5. 3D solutions for geometric nonlinear effects require more exploration.
6. The computational effort for 3D analysis is very high and hence approximate methods/2D theories is gaining more attention.
7. Among several numerical methods, EFGM, and XFEM were found to consider the effect of inclusions which may be further extended.

An effort has been made to include most of the important contributions in the current area of interest highlighting the most pertinent literature available to research engineers studying FG plate structures.

References

1. Marur, P. R., & Tippur, H. V. (2000). Numerical analysis of crack-tip fields in functionally graded materials with a crack normal to the elastic gradient. *International Journal of Solids and Structures, 37*, 5353–5370.
2. Li, H., Lambros, J., Cheeseman, B. A., & Santare, M. H. (2000) Experimental investigation of the quasi-static fracture of functionally graded materials. *International Journal of Solids and Structures, 37*, 3715–3732.
3. Ueda, S., & Shindo, Y. (2000). Crack kinking in functionally graded materials due to an initial strain resulting from stress relaxation. *Journal of Thermal Stresses, 23*, 285–290.
4. Rousseau, C.-E., & Tippur, H. V. (2000) Compositionally graded materials with cracks normal to the elastic gradient. *Acta Materialia, 48*, 4021–4033.
5. Rousseau, C.-E., & Tippur, H. V. (2001) Influence of elastic gradient profiles on dynamically loaded functionally graded materials: cracks along the gradient. *International Journal of Solids and Structures, 38*, 7839–7856.
6. Rousseau, C.-E., & Tippur, H. V. (2001) Dynamic fracture of compositionally graded materials with cracks along the elastic gradient: Experiment and analysis. *Mechanics of materials, 33*, 403–421.

7. Rousseau, C.-E., & Tippur, H. V. (2002) Evaluation of crack tip fields and stress intensity factors in functionally graded elastic materials: Cracks parallel to elastic gradient. *International Journal of Fracture, 114*, 87–112.
8. Wang, Z., & Nakamura, T. (2004). Simulations of crack propagation in elastic–plastic graded materials. *Mechanics of Materials, 36*, 601–622.
9. Kim, J.-H., & Paulino, G. H. (2003). An accurate scheme for mixed-mode fracture analysis of functionally graded materials using the interaction integral and micromechanics models. *International Journal of Numerical Methods in Engineering, 58*, 1457–1497.
10. Kim, J.-H., & Paulino, G. H. (2003). Mixed-mode J-integral formulation and implementation using graded elements for fracture analysis of nonhomogeneous orthotropic materials. *Mechanics of Materials, 35*, 107–128.
11. Kim, J.-H., & Paulino, G. H. (2003). The interaction integral for fracture of orthotropic functionally graded materials: evaluation of stress intensity factors. *International Journal of Solids and Structures, 40*, 3967–4001.
12. Kim, J.-H., & Paulino, G. H. (2003). T-stress, mixed-mode stress intensity factors, and crack initiation angles in functionally graded materials: a unified approach using the interaction integral method. *Computer Methods Application Mechanics and Engineering, 92*, 1463–1494.
13. Tilbrook, M., Rutgers, L., Moon, R., & Hoffman, M. (2005). Fracture and fatigue crack propagation in graded composites. Functionally graded materials VIII. In *Proceedings of the eight international symposium on multifunctional and functionally graded materials*, Materials Science Forum Trans Technology Publications Ltd., Uetikon-Zuerich, Switzerland, pp. 492–493, 573–580.
14. Jamwal, A., Vates, U. K., Gupta, P., Aggarwal, A., & Sharma, B. P. (2019). Fabrication and characterization of Al_2O_3–TiC-reinforced aluminum matrix composites. In *Advances in industrial and production engineering* (pp. 349–356). Singapore: Springer
15. Kim, J.-H., & Paulino, G. H. (2005). Mixed-mode crack propagation in functionally graded materials. *Materials Science Forum, 492*, 409–414.
16. Kim, J.-H., & Paulino, G. H. (2007). On fracture criteria for mixed-mode crack propagation in functionally graded materials. *Mechanics of Advanced Materials and Structures, 14*, 227–244.
17. Kirugulige, M. S., Kitey, R., & Tippur, H. V. (2005). Dynamic fracture behavior of model sandwich structures with functionally graded core: A feasibility study. *Composites Science and Technology, 65*, 1052–1068
18. Kakkar, K., Rawat, N., Jamwal, A., & Aggarwal, A. (2018). Optimization of surface roughness, material removal rate and tool wear rate in EDM using Taguchi method. *International Journal of Advanced Research Ideas Innovations Technology, 4*(2), 16–24.
19. Dag, S., & Ayse Ilhan, K. (2008). Mixed-mode fracture analysis of orthotropic functionally graded material coatings using analytical and computational methods. *Journal of Applied Mechanics, 75*, 1–9.
20. Kirugulige, M., & Tippur, H. V. (2008). Mixed-mode dynamic crack growth in a functionally graded particulate composite: experimental measurements and finite element simulations. *Journal of Applied Mechanics, 75*, 1–14.
21. Aloulou, W., Yildirim, B., El-Borgi, S., & Zghal, A. (2009). Buckling of an orthotropic graded coating with an embedded crack bonded to a homogeneous substrate. *International Journal of Solids and Structures, 46*, 1890–1900.
22. Yu, Z., & Chu, F. (2009). Identification of crack in functionally graded material beams using the p-version of finite element method. *Journal of Sound and Vibration, 325*, 69–84.
23. Jaroniek, M. (2010). Experimental model of fracture of functionally graded materials. *Journal of Theoretical and Applied Mechanics, 48*, 71–86.
24. Rekik, M., Neifar, M., & El-Borgi, S. (2010). An axisymmetric problem of an embedded crack in a graded layer bonded to a homogeneous half-space. *International Journal of Solids and Structures, 47*, 2043–2055.
25. Martínez-Pañeda, E., & Gallego, R. (2015) Numerical analysis of quasi-static fracture in functionally graded materials. International Journal of Mechanics in Materials Design, 1, 405–424.

26. Benamara, N., Boulenouar, A., & Aminallah, M. (2017). Strain Energy density prediction of mixed-mode crack propagation in functionally graded materials. *Periodica Polytechnica Mechanical Engineering, 61*(1), 60–67.
27. Hirshikesh, S. N., Annabattula, R. K., & Martínez-Pañeda, E. (2019). Phase field modelling of crack propagation in functionally graded materials. *Composites Part B, 69*, 239–248.
28. Chafia, M., Boulenouara, A. (2019). A numerical modelling of mixed mode crack initiation and growth in functionally graded materials. *Materials Research, 22*(3), e20180701.
29. Jamwal, A., Prakash, P., Kumar, D., Singh, N., Sadasivuni, K. K., Harshit, K., Gupta, S., & Gupta, P. (2019). Microstructure, wear and corrosion characteristics of Cu matrix reinforced SiC–graphite hybrid composites. *Journal of Composite Materials, 53*(18), 2545–2553.
30. Kandula, S. S. V., Abanto-Bueno, J., Geubelle, P. H., & Lambros, J. (2005). Cohesive modeling of dynamic fracture in functionally graded materials. *International Journal of Fracture, 132*, 275–296.
31. Kumar, A., Arafath, M. Y., Gupta, P., Kumar, D., Hussain, C. M., & Jamwal, A. (2020). Microstructural and mechano-tribological behavior of Al reinforced SiC-TiC hybrid metal matrix composite. *Materials Today: Proceedings, 21*, 1417–1420.
32. Torshizian, M. R., Kargarnovin, M. H., & Nasirai, C. (2011). Mode III fracture of an arbitrary oriented crack in two dimensional functionally graded material. *Mechanics Research Communications, 38*, 164–169.
33. Zhou, L. M., Meng, G. W., Li, X. L., & Li, F. (2016). Analysis of dynamic fracture parameters in functionally graded material plates with cracks by graded finite element method and virtual crack closure technique. *Advances in Materials Science and Engineering, 8085107*, 1–14.
34. Burlayenko, V. N. (2016). Modelling thermal shock in functionally graded plates with finite element method. *Advances in Materials Science and Engineering, 7514638*, 1–12.
35. Yildirim, B., Dag, S., & Erdogan, F. (2005). Three dimensional fracture analysis of FGM coatings under thermomechanical loading. *International Journal of Fracture, 132*, 369–395.
36. Yildirim, B., & Erdogan, F. (2004). Edge crack problems in homogenous and functionally graded material thermal barrier coatings under uniform thermal loading. *Journal of Thermal Stresses, 27*, 311–329.
37. Dag, S., Yildirim, B., & Erdogan, F. (2006). Three dimensional analysis of periodic cracking in FGM coatings under thermal stresses. In *CP973, multiscale and functionally graded materials*, pp. 676–681.
38. Ayhan, A. O. (2007). Stress intensity factors for three-dimensional cracks in functionally graded materials using enriched finite elements. *International Journal of Solids and Structures, 44*, 8579–8599.
39. Ayhan, A. O. (2009). Three-dimensional mixed-mode stress intensity factors for cracks in functionally graded materials using enriched finite elements. *International Journal of Solids and Structures, 46*, 796–810.
40. Kosker, S., Dag, S., & Yildirim, B. (2010). Three dimensional modeling of inclined surface cracks in FGM coatings. *Materials Science Forum, 631*, 109–114.
41. Kheirikhah, M. M., & Khalili, S. M. R. (2011). Fracture analysis of semielliptical cracks at the interface of two functionally gradient materials using three-dimensional finite-element method. *Proceedings of IMechE 2011, 225*, 103–110.
42. Sabuncuoglu, B., Dag, S., & Yildirim, B. (2012). Three dimensional computational analysis of fatigue crack propagation in functionally graded materials. *Computational Materials Science, 52*, 246–252.
43. Sheikhi, J., Poorjamshidian, M., & Peyman, S. (2015). Mixed-mode stress intensity factors for surface cracks in functionally graded materials using enriched finite elements. *Journal of Solid Mechanics, 7*, 1–12.
44. Chen, J. (2005). Determination of thermal stress intensity factors for an interface crack in a graded orthotropic coating-substrate structure. *International Journal of Fracture, 133*, 303–328.
45. Pathak, H., Singh, A., & Singh, I. V. (2012). Numerical simulation of bi-material interfacial cracks using EFGM and XFEM. *International Journal of Mechanics and Materials in Design, 8*, 9–36.

46. Bhattacharya, S., Singh, I. V., & Mishra, B. K. (2013). Fatigue-life estimation of functionally graded materials using XFEM. *Engineering with Computers, 29*, 427–448.
47. Gangwar, S., Payak, V., Pathak, V. K., Jamwal, A., & Gupta, P. (2020). Characterization of mechanical and tribological properties of graphite and alumina reinforced zinc alloy (ZA-27) hybrid metal matrix composites. *Journal of Composite Materials, 54*(30), 4889–4901.
48. Hosseini, S. S., Bayesteh, H., & Mohammadin, S. (20133). Thermo-mechanical XFEM crack propagation analysis of functionally graded materials. *Materials Science and Engineering A, 561*, 285–302.
49. Ivanov, I. V., Sadowski, T., & Pietras, D. (2013). Crack propagation in functionally graded strip under thermal shock. *The European Physical Journal of Special Topics, 222*(7), 1587–1595.
50. Rokhi, M. M., & Shariati, M. (2013). Coupled thermoelasticity of a functionally graded cracked layer under thermomechanical shocks. *Archive Mechanics, 65*(2), 71–96.
51. Golia, E., & Kazemib, M. T. (2014). XFEM modeling of fracture mechanics in transversely isotropic FGMs via interaction integral method. *Procedia Materials Science (20th European Conference on Fracture (ECF2014), 3*, 1257–1262.
52. Bhattacharya, S., Singh, I. V., & Mishr, B. K. (2014). Fatigue life simulation of functionally graded materials under cyclic thermal load using XFEM. *International Journal of Mechanical Sciences, 82*, 41–59.
53. Sharma, K., Bhattacharya, S., Sonkar, V. (2016). XFEM simulation on mixed-mode fatigue crack growth of functionally graded materials. *Journal of Mechanical Engineering and Biomechanics, 1*(1), 46–55.
54. Bhattacharya, S., Sharma, K., & Sonkar, V. (2017). Numerical simulation of elastic plastic fatigue crack growth in functionally graded material using the extended finite element method. *Mechanics of Advanced Materials and Structures, 24*(16), 1367–1380.
55. Pant, M., Bhattacharya, S. (2017). Fatigue crack growth analysis of functionally graded materials by EFGM and XFEM. *International Journal of Computational Methods, 14*(1750004), 1–33.
56. Pant, M., & Bhattacharya, S. (2019). Fatigue crack growth analysis of functionally graded materials by EFGM and XFEM. *International Journal of Computational Methods, 14*(1), 1–16.
57. Li, D. H., Yang, X., Qian, R. L., & Xu, D. (2018). Static and dynamic response analysis of functionally graded material plates with damage. *Mechanics of Advanced Materials and Structures, 14*, 1–14.
58. Sonkar, V., Bhattacharya, S., & Sharma, K. (2019). A three dimensional fracture analysis of an edge crack using XFEM. *Material Science Forum, 969*, 315–320.
59. Fujimoto, T., & Noda, N. (2000). Crack propagation in a functionally graded plate under thermal shock. *Archive of Applied Mechanics, 70*, 377–386.
60. Fujimoto, T., & Noda, N. (2001). Two crack growths in a functionally graded plate under thermal shock. *Journal of Thermal Stresses, 24*, 847–862.
61. Jamwal, A., Aggarwal, A., Gautam, N., & Devarapalli, A. (2018). Electro-discharge machining: recent developments and trends. *International Research Journal of Engineering Technology, 5*, 433–448.
62. Vena, P., Gastaldi, D., & Contro, R. (2005). Effects of the thermal residual stress field on the crack propagation in graded alumina/zirconia ceramics. Functionally graded materials VIII (FGM2004). In *Proceedings of the eighth international symposium on multifunctional and functionally graded materials*, Materials Science Forum 2005, pp. 492–493, 177–182.
63. Jamwal, A., Mittal, P., Agrawal, R., Gupta, S., Kumar, D., Sadasivuni, K. K., & Gupta, P. (2020). Towards sustainable copper matrix composites: Manufacturing routes with structural, mechanical, electrical and corrosion behaviour. *Journal of Composite Materials, 54*(19), 2635–2649.
64. Wang, B.-L., & Mai, Y.-W. (2007). On thermal shock behavior of functionally graded materials. *Journal of Thermal Stresses, 30*, 523–528.
65. Sohag, M. A. Z., Gupta, P., Kondal, N., Kumar, D., Singh, N., & Jamwal, A. (2020). Effect of ceramic reinforcement on the microstructural, mechanical and tribological behavior of Al-Cu alloy metal matrix composite. *Materials Today: Proceedings, 21*, 1407–1411.

66. Hossain, S., Rahman, M. M., Chawla, D., Kumar, A., Seth, P. P., Gupta, P., Kumar, D., Agrawal, R., & Jamwal, A. (2020). Fabrication, microstructural and mechanical behavior of Al-Al_2O_3-SiC hybrid metal matrix composites. *Materials Today: Proceedings, 21*, 1458–1461.
67. Nayim, S. T. I., Hasan, M. Z., Jamwal, A., Thakur, S., & Gupta, S. (2019, September). Recent trends and developments in optimization and modelling of electro-discharge machining using modern techniques: A review. In *AIP Conference Proceedings, 2148*(1), 030051.
68. Jamwal, A., Agrawal, R., & Gupta, P. Application of multi-criteria decision-making techniques in the optimization of mechano-tribological properties of copper-SiC-graphite hybrid metal matrix composites. In *Intelligent manufacturing* (pp. 149–172). Cham: Springer.
69. Hossain, S., Rahman, M. M., Jamwal, A., Gupta, P., Thakur, S., & Gupta, S. (2019, September). Processing and characterization of pine epoxy based composites. In *AIP Conference Proceedings, 2148*(1), 030017.
70. Zhang, Y., Guo, L., & Noda, N. (2014). Investigation Methods for thermal shock crack problems of functionally graded materials–Part II: Combined analytical-numerical method. *Journal of Thermal Stresses, 37*, 325–339.
71. Eshraghi, I., Soltani, N., & Dag, S. (2016). Weight function method for transient thermomechanical fracture analysis of a functionally graded hollow cylinder possessing a circumferential crack. *Journal of Thermal Stresses, 39*, 1182–1199.
72. Carpenter, R. D., Paulino, G. H., Munir, Z. A., & Gibeling, J. C. (2000). A novel technique to generate sharp cracks in metallic/ceramic functionally graded materials by reverse 4-point bending. *Scripta Material, 43*, 547–552.
73. Moon, R. J., Hoffman, M., Hilden, J., Bowman, K. J., Trumble, K. P., & Rödel, J. (2002). R-curve behavior in alumina–zirconia composites with repeating graded layers. *Engineering Fracture Mechanics, 69*(14–16), 1647–1665.
74. Garg, P., Jamwal, A., Kumar, D., Sadasivuni, K. K., Hussain, C. M., & Gupta, P. (2019). Advance research progresses in aluminium matrix composites: manufacturing and applications. *Journal of Materials Research and Technology, 8*(5), 4924–4939.
75. Rousseau, C.-E. (2006). Critical examination of the use of coherent gradient sensing in measuring fracture parameters in functionally graded materials. *Journal of Composite Materials, 40*, 1763–1782.
76. Rousseau, C.-E., & Tippur, H. V. (2002) Influence of elastic variations on crack initiation in functionally graded glass-filled epoxy. *Engineering Fracture Mechanics, 69*, 1679–1693.
77. Abanto-Bueno, J., & Lambros, J. (2002). Investigation of crack growth in functionally graded materials using digital image correlation. *Engineering Fracture Mechanics, 69*, 1695–1711.
78. Abanto-Bueno, J., & Lambros, J. (2006). Parameters controlling fracture resistance in functionally graded materials under mode I loading. *International Journal of Solids and Structures, 43*, 3920–3939.
79. Abanto-Bueno, J., & Lambros, J. (2006). An experimental study of mixed mode crack initiation and growth in functionally graded materials. *Experimental Mechanics, 46*, 179–196.
80. Bahr, H.-A., Balke, H., Fett, T., Hofinger, I., Kirchhoff, G., Munz, D., Neubrand, A., Semenov, A. S., Weiss, H.-J., & Yang, Y. Y. (2003). Cracks in functionally graded materials. *Materials Science and Engineering, A362*, 2–16.
81. Forth, S. C., Favrow, L. H., Keat, W. D., & Newman, J. A. (2003). Three-dimensional mixed-mode fatigue crack growth in a functionally graded titanium alloy. *Engineering Fracture Mechanics, 70*, 2175–2185.
82. El-Hadek, M.A., & Tippur, H. V. (2003). Dynamic fracture parameters and constraint effects in functionally graded syntactic epoxy foams. *International Journal of Solids and Structures, 40*, 1885–1906.
83. Xu, F. M., Zhu, S. J., Zhao, J., Qi, M., Wang, F. G., Li, S. X., & Wang, Z. G. (2004). Effect of stress ratio on fatigue crack propagation in a functionally graded metal matrix composite. *Composites Science and Technology, 64*, 1795–1803.
84. Jain, N., & Shukla, A. (2006). Mixed mode dynamic fracture in particulate reinforced functionally graded materials. *Experimental Mechanics, 46*, 137–154.

85. Kirugulige, M. S., & Tippur, H. V. (2006). Mixed-mode dynamic crack growth in functionally graded glass-filled epoxy. *Experimental Mechanics, 46*, 269–281.
86. Nayim, S. T. I., Hasan, M. Z., Seth, P. P., Gupta, P., Thakur, S., Kumar, D., & Jamwal, A. (2020). Effect of CNT and TiC hybrid reinforcement on the micro-mechano-tribo behaviour of aluminium matrix composites. *Materials Today: Proceedings, 21*, 1421–1424.
87. Jamwal, A., Seth, P. P., Kumar, D., Agrawal, R., Sadasivuni, K. K., & Gupta, P. (2020). Microstructural, tribological and compression behaviour of copper matrix reinforced with Graphite-SiC hybrid composites. *Materials Chemistry and Physics, 251*, 123090.
88. Comi, C., & Mariani, S. (2007). Extended finite element simulation of quasi-brittle fracture in functionally graded materials. *Computer Methods Applications Mechanics and Engineering, 196*, 4013–4026.
89. Jain, N., & Shukla, A. (2007). Asymptotic analysis and reflection photoelasticity to study transient crack propagation in graded materials. *Journal of Mechanics of Materials and Structures, 2*, 595–612.
90. Ulukoy, A., Topcu, M., & Tasgetiren, S. (2016). Experimental investigation of aluminum matrix functionally graded material: Microstructural and hardness analyses, fretting, fatigue, and mechanical properties. *Journal of Engineering Tribology, 230*(2), 143–155.
91. Balke, H, Bahr, H.-A., Semenov, A. S., Hofinger, I., Häusler, C., Kirchhoff, G., & Weiss, H.-J. (2000) Graded thermal barrier coatings. Cracking due to laser irradiation and determining of interface toughness. *Proceedings of 6th International Symposium on Functionally Graded Materials. Ceramic Transactions, 114*, 205–212.
92. Bandil, K., Vashisth, H., Kumar, S., Verma, L., Jamwal, A., Kumar, D., Singh, N., Sadasivuni, K. K., & Gupta, P. (2019). Microstructural, mechanical and corrosion behaviour of Al–Si alloy reinforced with SiC metal matrix composite. *Journal of Composite Materials, 53*(28–30), 4215–4223.
93. Kawasaki, A., & Watanabe, R. (2002) Thermal fracture behavior of metal/ceramic functionally graded materials. *Engineering Fracture Mechanics, 69*, 1713–1728.
94. Rangaraj, S., & Kokini, K. (2003). Estimating the fracture resistance of functionally graded thermal barrier coatings from thermal shock tests. *Surface and Coatings Technology, 173*, 201–212.
95. Kokini, K., & Rangaraj, S. V. (2005). Time-dependent behavior and fracture of functionally graded thermal barrier coatings under thermal shock. *Materials Science Forum, 2005*(492–493), 379–384.
96. Xiong, H.-P., Kawasaki, A., Kang, Y.-S., & Watanabe, R. (2005). Experimental study on heat insulation performance of functionally graded metal/ceramic coatings and their fracture behavior at high surface temperatures. *Surface and Coatings Technology, 194*, 203–214.
97. Lee, W. J., Park, B. G., Park, I. M., Park, Y. H., Oak, J. J., & Kimura, H. (2009). Thermal fatigue cracking and extension behaviors of squeeze infiltrated Al18B4O33/Mg functionally graded materials. *Materials Transactions, 50*, 864–871.

Investigation on the Machining of Inconel-718 Using EDM

Niteen Patil, M. R. Patil, Rakesh Chaudhari, and Praveen Kumar Loharkar

1 Introduction

In recent years, applications of materials with inferior machinability characteristics are gaining wide importance in industries such as aviation, aerospace, and several other engineering industries. This is primarily due to the material's high strength for a given magnitude of weight, superior heat resistant capability and large hardness. Inconel 718 is one such alloy based on nickel that is characterized by superior yield strength, ultimate tensile strength, and creep strength even at high temperatures. It is an important alloy in various engineering applications owing to its slow-paced response to age –hardening. This allows the application of heat in processes such as annealing and joining without "spontaneous-hardening". In addition, the weldability of Inconel-718 material is better than other nickel-based super-alloys which are hardened using aluminum and titanium. However, it possesses poor machinability and requires non-conventional machining to get the required shape and geometry.

Non-conventional machining overcomes the limitations of conventional machine tools to machine the advanced materials. Electrical discharge (ED) machining is one of those non-conventional methods which has proven merits such as lesser stresses during processing, reduced work-hardening effects, and very little damage at microstructural level [1]. Therefore, EDM process was chosen to perform the parametric study on the machinability of Inconel 718.

The electrical discharge (ED) machining process is a significant non-conventional material removal process. It is quite precise in terms of the quality of the product. Quite a few optimization studies are published which have discussed the role of

N. Patil · M. R. Patil
Department of Mechanical Engineering, SES's RCPIT, Shirpur, Dhule, Maharashtra 425405, India

R. Chaudhari (✉) · P. K. Loharkar
Department of Mechanical Engineering, SVKM's NMIMS, MPSTME, Shirpur Campus, Dhule, Maharashtra 425405, India

R. Agrawal et al. (eds.), *Recent Advances in Smart Manufacturing and Materials*,
Lecture Notes in Mechanical Engineering,
https://doi.org/10.1007/978-981-16-3033-0_12

"machining parameters" that significantly affect the response parameters. Choudhary and Jadoun [2] carried out a review on various approaches used by researchers to augment the efficiency of ED machining and enlisted the relevant process parameters which are either electrical or non-electrical based and electrode-based and powder-based. The outcome of a few related research studies is presented subsequently.

In one of the earlier parametric studies on EDM, an investigation on precision drilling of "micro holes" in the borosilicate glass was performed by Yan et al. [3]. In this research, micro EDM was combined with micro-ultrasonic vibration machine. The experiments demonstrated the criticality of choosing machining parameters. In a similar context, Diver et al. [4] produced tapered holes using micro EDM based on newly proposed technique. The work presented in this article showed that high-quality tapered holes (reverse) can be produced exhibiting superior geometrical features with cycle times same as that for machining of straight holes using the same process. Further studies on "deep micro-holes" drilling were done by Tai et al. [5]. The authors investigated drilling to the depth of 320 μm in micro-electrical discharge machined tool steel. The holes were characterized using microstructural imaging using "optical microscope", "scanning-electron microscope (SEM)" and "co-focal laser-scanning microscope" to evaluate the correlation between multiple process and output parameters including enlargement of the hole, wear rate of the electrodes used, MRR, wear-ratio. Consequently, the settings of most suited parameters were revealed, which could improve the quality of micro-ED machining. Reduction in the wear rate can be attributed to grain refinement due to localized heat treatment of the material caused due to friction [6].

As far as machining of metals is concerned, Singh et al. [7] studied the variable "pulsed current" and found its effect on MRR, diametral overcut, wear of the electrode, and surface finish in ED machining of EN-31 steel. The output parameters were found to be synergistically affected by the "pulsed-current" while copper and aluminum electrodes led to superior machining rates.

Machining of ceramics is also reported in the published literature. Puertas et al. [8] determined the effect of ceramic machining on the "surface roughness", "electrode wear (EW)" and MRR. The ceramic used in this study was a cemented carbide. The work confirmed the efficacy of the using a combination of techniques such as design of factorial experiments and "multiple linear-regression", in developing the functions affected by multiple variables. The advantage of combining different techniques was further established through the research performed by Luis et al. [9]. In a similar study, the effect of "electrical pulse duration" on the micro-EDM parameters was investigated by Seong-Min et al. [10]. The primary focus of the study was MRR.

Kanlayasiri et al. [11] studied the machining variables' effect on the "surface roughness" of wire-ED machined die-steel. The developed model for surface roughness was validated using experimental values. The model had very less prediction error (7%). It was identified using the ANOVA technique.

An approach based on utilizing hybrid combination of techniques such as "Taguchi method (TM)" and "PCA- principal component analysis" was employed for optimization based on multiple objectives in pulsed Nd: YAG laser beam cutting (LBC)

Table 1 ED machining factors and their levels

S. No.	Factor	Unit	Levels		
			1	2	3
1	Current (*A*)	Ampere	20	30	40
2	Pulse-on duration (*B*)	microseconds	200	300	400
3	Pulse-off duration (*C*)	microseconds	100	150	200

[12]. The material used as a workpiece was nickel-based alloy. It resulted in superior cut qualities. Similarly, Rajmohan et al. [13] explored the effect of ED machine parameters on the MRR in the machining of SS 304 using L9 orthogonal array-based experimental design.

In this paper, optimization study on the parameters, which have a critical influence on the machining of Inconel 718 using ED is presented. The objective of the work was to evaluate the process efficacy in terms of material removal rate (MRR). "Taguchi method" was used as a technique for optimization. The values of parameters that lead to signal-to-noise ratio of larger the better type have been used to find the optimum value of response parameter, that is, MRR.

2 Materials and Method

Electrode is the most critical component of Electrical discharge machine. Therefore, the selection of a suitable electrode for the EDM process is important. Based on the literature reviewed [1], copper electrode was used while Inconel-718 was selected as the work-piece material for the present study.

The experimental trials were carried out using an experimental design based on "Taguchi method". In this experiment, the three input parameters that were used are "current", "pulse- on duration" and "pulse-off duration". The corresponding levels of the factors used in the investigation are shown in Table 1.These are based on the study carried out by Ahmad and Lajis [14].

5 mm diameter holes were drilled on Inconel-718 plates of sizes 5 × 76 × 240 mm and 5 × 53 × 220 mm based on the randomized experimental trials. The MRR is evaluated and measured in mm^3/min. The experiments were performed on ECOLINE 200 EDM (C.P. SERIES) is shown in Fig. 1. The dielectric fluid used in the study was commercially available EDM oil.

3 Results and Discussion

The drilled holes on plate of Inconel are as shown in Fig. 2. Table 2 shows the experimental results for EDM of Inconel Alloy 718 by using copper electrode.

Fig. 1 ECOLINE 200 EDM

Fig. 2 ED machined Inconel 718 plate

Table 2 Experimental results of ED machined Inconel-718

S. No.	Current (C), A	Pulse-on duration (T_{on}), μs	Pulse-off duration (T_{off}), μs	MRR, mm^3/min
1	20	200	100	19.44
2	20	300	150	16.82
3	20	400	200	15.11
4	30	200	150	33.01
5	30	300	200	31.67
6	30	400	100	30.41
7	40	200	200	35.08
8	40	300	100	32.16
9	40	400	150	30.74

3.1 Signal to Noise (S/N) Ratio for ED Machining

S/N was calculated using "Minitab Software". The response parameter material removal rate should have a larger value. Hence, the study corresponds to "Larger is better". In graphical representation of the main effects plot for means, "*A*" is taken as "current", "*B*" is taken as "pulse-on duration" and "*C*" is taken as "pulse-off duration". The corresponding tables of S/N calculation (Table 3), table for "means" (Table 4) and table for "S/N" (Table 5) are as follows. It is clearly evident from the Table 3 that the maximum S/N ratio was obtained for the levels 3, 1, 3 for the factors *A*, *B*, *C*, respectively. Refer to Table 1 for the coded levels. Tables 2 and 3, that is, response tables for means and signal-to-noise ratio, respectively, signify the ranking of the factors with "current" being the highest-ranked factor and the "pulse-off duration" being the least ranked.

Figures 3 and 4 show the plots of main effects for means and S/N ratio, respectively.

From these two plots, it is clearly evident that the larger magnitude of MRR corresponds to third level of factor *A* that is "current", first level of factor *B* ("pulse-on duration") and first level of factor *C* ("pulse-off duration"), respectively. The results are comparable to the findings of earlier researchers which have revealed the dominance of "current" on the EDM process.

Table 3 S/N ratio calculation

A	*B*	*C*	MRR	S/N ratio	Mean
1	1	1	19.44	25.7739	19.44
1	2	2	16.82	24.5165	16.82
1	3	3	15.11	23.5853	15.11
2	1	2	33.01	30.3729	33.01
2	2	3	31.67	30.0130	31.67
2	3	1	30.41	29.6603	30.41
3	1	3	35.08	30.9012	35.08
3	2	1	32.16	30.1463	32.16
3	3	2	30.74	29.7541	30.74

Table 4 Response table for means

Level	*A*	*B*	*C*
1	17.12	29.18	27.34
2	31.70	26.88	26.86
3	32.66	25.42	27.29
Delta	15.54	3.76	0.48
Rank	1	2	3

Table 5 Response table for signal-to-noise ratios: larger is better

Level	*A*	*B*	*C*
1	24.63	29.02	28.53
2	30.02	28.23	28.21
3	30.27	27.67	28.17
Delta	5.64	1.35	0.36
Rank	1	2	3

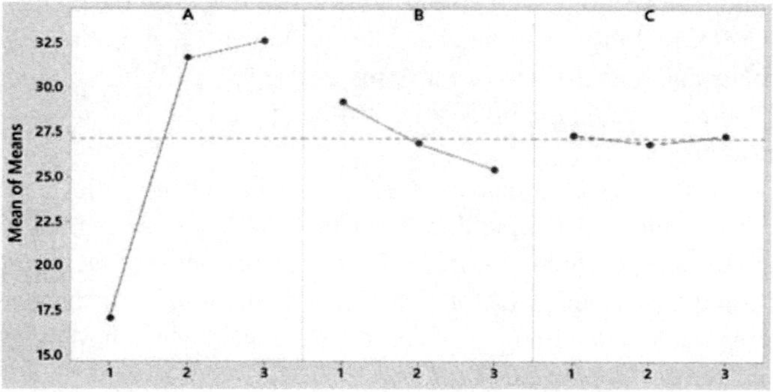

Fig. 3 Main effect plot for means

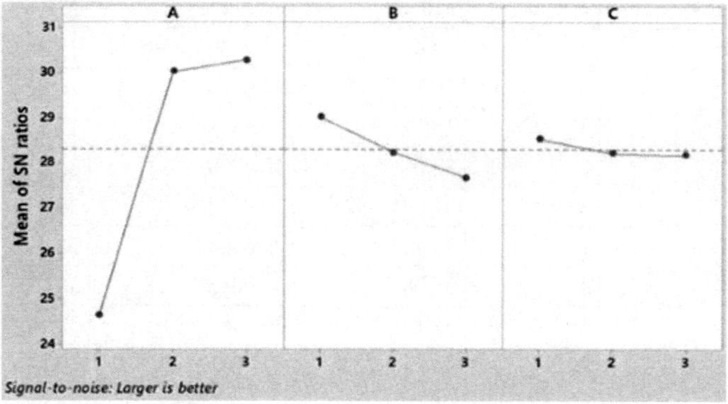

Fig. 4 Main effect plot for S/N ratios

4 Conclusion

Drilling experiments were successfully performed on a 5 mm thick Inconel-718 plate using copper as the electrode for evaluating MRR at different combination of the process parameters. For optimization of parameters of ED machined Inconel-718, relatively high magnitudes of "current", i.e., 20, 30, and 40 A were used with varying "pulse-on duration" and "pulse-off duration". Taguchi L9 orthogonal array was used for designing the experiments. Following conclusions have been made based on the study.

- It has been found that at higher "current", MRR had increased and maximum material removal was obtained at 40 A.
- In case of "pulse-on duration" and "pulse-off duration", the material removal rate had increased initially. However, it was found to be decreasing at higher values of corresponding parameters.
- It is quite evident from the study that "current" is the most critical parameter for MRR as compared to the other two parameters considered for this experimental investigation.
- Based on the results obtained, the maximum material removal rate corresponded to the experimental combination of 40 A "current", 200 microseconds "pulse-on duration" and 200 microseconds "pulse-off duration".

References

1. Kale, S. M., & Khedekar, D. S. (2016). Optimization of process parameters in electric discharge machining of Inconel 718 by using copper electrode. *IOSR Journal of Mechanical and Civil Engineering, 13*(3), 51–55.
2. Choudhary, S. K., & Jadoun, R. S. (2014). Current advanced research development of electric discharge machining: A review. *International Journal of Research in Advent Technology, 2*(3), 2321–9637.
3. Yan, B. H., Wang, A. C., Huang, C. Y., & Huang, F. Y. (2002). Study of precision micro holes on borosilicate glass using micro EDM combined with micro ultrasonic vibration machining. *International Journal of Machine Tools & Manufacture*, 1105–1111.
4. Diver, C., Atkinson, J., Helml, H. J., & Li, L. (2004). Micro-EDM drilling of tapered holes for industrial applications. *Journal of Materials Processing Technology, 149*(1–3), 296–303.
5. Tai, T. Y., Masusawa, T., & Lee, H. T. (2007). Drilling microholes in hot tool steel by using micro-electro discharge machining. *Materials Transactions, 48*(2), 205–210.
6. Chaudhari, R., Ingle, A., & Kalita, K. (2017). Tribological investigation of effect of grain size in 304 austenitic stainless steel. *Transactions of the Indian Institute of Metals, 70*(9), 2399–2405.
7. Singh, S., Maheshwari, S., & Pandey, P. C. (2004). Some investigations into the electric discharge machining of hardened tool steel using different electrode materials. *Journal of Materials Processing Technology, 149*(1), 272–277.
8. Puertas, I., Luis, C. J., & Alvarez, L. (2004). Analysis of the influence of EDM parameters on surface quality, MRR and EW of WC–Co. *Journal of Materials Processing Technology, 153*, 1026–1032.
9. Luis, C. J., & Puertas, I. (2007). Methodology for developing technological tables used in EDM processes of conductive ceramics. *Journals of Material Processing Technology, 153*, 189–197.
10. Son, S., Lim, H., Kumar, A. S., & Rahman, M. (2007). Influences of pulsed power condition on the machining properties in micro EDM. *Journal of Materials Processing Technology, 190*(1–3), 73–76.
11. Kanlayasiri, K., & Boonmung, S. (2007). Effects of wire-EDM machining variables on surface roughness of newly developed DC 53 die steel: Design of experiments and regression model. *Journal of Materials Processing Technology, 192*, 459–464.
12. Nayim, S. T. I., Hasan, M. Z., Jamwal, A., Thakur, S., & Gupta, S. (2019, September). Recent trends and developments in optimization and modelling of electro-discharge machining using modern techniques: A review. *AIP Conference Proceedings, 2148*(1), 030051.

13. Rajmohan, T., Prabhu, R., Rao, G. S., & Palanikumar, K. (2012). Optimization of machining parameters in electrical discharge machining (EDM) of 304 stainless steel. *Procedia Engineering, 38*, 1030–1036.
14. Ahmad, S., & Lajis, M. A. (2013). Electrical discharge machining (EDM) of Inconel 718 by using copper electrode at higher peak current and pulse duration. *IOP Conference Series: Materials Science and Engineering, 50*(1), 012062.

Multi-objective Structural Analysis of Kevlar Fiber Reinforced Polymer Composite

K. Kesavan, P. Kiran, M. Sivaguru, S. Indira Prasanth, R. Sudharsan, G. Raj Kumar, and R. Vijayanandh

1 Introduction

Structural failures occur because of forces acting on the structure. These can be weight or the load that it is carrying or dynamic forces by the external moving fluids [1]. Structural failure is initiated when the material in a structure is stressed to its strength limit thus causing fracture or excessive deformation [2]. The major problem in structural engineering is sustainability. The sustainability of structures is affected by environmental issues [3] such as high aerodynamic loadings, external artificial loads, etc. Structurally, these issues are been affected the material's strength and load-carrying capability [5]. Composites are mainly used in structural engineering for the construction of bridge structures, aircraft structures, automotive structures, etc. [6]. Also, they are popularly used in construction because of high stiffness to weight ratio and strength to weight ratio when compared to reinforced concrete and steel [7]. But comparatively, KFRP (Kevlar Fiber Reinforced Polymer) is fit to handle peak load than other composites therefore the same material is selected as major platform of this work [8].

1.1 Kevlar Fiber Reinforced Polymer [KFRP]

Kevlar fiber comes in the Aramid family and it is a type of synthetic fiber. It is widely used in all the material makings as the replacement of alloys and composites [9]. The fibers of Kevlar are made up of long molecular chains, which are highly linked

K. Kesavan · P. Kiran · M. Sivaguru · S. Indira Prasanth · R. Sudharsan · G. Raj Kumar · R. Vijayanandh (✉)
Department of Aeronautical Engineering, Kumaraguru College of Technology, Coimbatore, Tamil Nadu, India

R. Agrawal et al. (eds.), *Recent Advances in Smart Manufacturing and Materials*, Lecture Notes in Mechanical Engineering,
https://doi.org/10.1007/978-981-16-3033-0_13

with strong bonding. Kevlar fiber having the major notable structural advantages such as low elongation, high structural rigidity, low conductivity, high resistance, low shrinkage, high toughness, high dimensional stability, and high flame resistance [10]. Kevlar fiber has 0.60 of fiber volume expansion, 1.38 of specific gravity, its Young's modulus is about 87 GPa and its shear modulus is about 2.2 GPa. Kevlar fiber has a high tensile strength of nearly about 1280 MPa. Kevlar has a compressive strength of 335 MPa, shear strength is 49 MPa [11]. Nowadays, the real-time problems due to the complexity are arising. To reduce it, learning about the integrational effect of the respective fibers should be made. This work deals with the systematic analysis of Kevlar fibers' optimization by applying tensile and bending loads. The primary properties of Kevlar fiber are not affected anywhere and undergo the aforesaid tests with the arrangement of fibers from one another [12].

1.2 Problem Identification and Its Solution Technique

Material science is a kind of physics field, which can able to provide a solution for any kind of engineering problems [12]. Comparatively, in structural engineering, the material need is quite high to fulfill the requirements such as high mechanical loading conditions, ultra thermal loading conditions, high electrical loading conditions, etc., In this work deals with a similar kind of problem, which is nothing but, provide the suitable suggestion for high complicated loading environments through composite material [13]. Relatively, Kevlar can withstand abnormal loading conditions so it is analyzed for the purpose to understand its structural performance. In this regard, the main elemental study is conducted in Kevlar composites to get the best fiber orientational angle, which can have the capacity to withstand high loading conditions under tensile environments. The advanced numerical tool, i.e., ANSYS Workbench has solved the structural conditions of various Kevlar's test models [14–19]. The targeted applications of this work are aerospace-relevant structural domains, so the handling of aerodynamic loading conditions itself generates a major problem. Generally, the working environment of the aerospace application is affecting the components in a non-linear manner. The non-linearity of the working environments are reacting in an object in two ways, which are non-linear in aerodynamic pressure loading and non-linear structural loads such as fluttering, 'g' force variations, etc.

1.3 Literature Survey

This paper [1] dealt about the material's performance to stimulate the honeycomb Deployable Energy Absorber [DEA] by nonlinear dynamic finite element code. The prototype was manufactured by Kevlar/epoxy composite with the orientation of + or − 45° with respect to the loading axis. Once the structure was expanded, DEA becomes a more efficient structure, stiffness, and high strength. The cell wall was 0.01

thick Kevlar-129 fabrics that were impregnated with reinfusion 8601 Epoxy resin. To get accurate results of Kevlar/epoxy, the LS-DYNA computational tool was used, in which primarily a three-point bending test was conducted. High fine structural meshes were implemented in this computation. For MAT 24, Kevlar/epoxy composite was an isotropic elastic–plastic material with an orientation of + or −45°. Therefore, two LS-DYNA models were used for the prediction of the behavior of material Kevlar/epoxy [1]. Through this work [1], the computational procedures were learned. This paper [2] provided the mechanical properties of the Kevlar KM2 single fiber. Three types of experiments such as longitudinal tensile tests, transverse compressive tests, and torsional tests were conducted. All experiments were performed with the standard test method described in ASTM D3379-75 tested on MTS 810 servo-hydraulic material testing machine. This model proposed various kinds of properties of a single fiber of a Kevlar composite like ultimate tensile strength, failure strain, etc., From this paper [2], design inputs, mechanical properties were attained. In this paper [3], the author used kevlar aluminum alloy as the fiber and the polymer as the resin and thereby formed polymer composite. The author modeled the polymeric fiber composites and finite element simulation and mechanical properties were examined. FEA was computed and the displacement field is suited to identify and quantifying stress intensities in the local region of the composite to determine parameters critical to the performance of the composite. By creating a fiber-matrix interface the given stress was uniformly distributed inefficient manner. In this simulation, the surface region and inner fiber matrix identified as zones that support more stress during the deformation by bending since one surface was extended and while the other surface is compressed. The interior was partially compressed to create low-density impact-resistant composites [3]. In the study of FEA of fiberglass uni-directional E-type was analyzed in the Framework of ABAQUS software. The simulation was run twice, for ply-1 with an angle of 45°, ply-2 with an angle of − 45°, and ply-3 with an angle of 90°. In the second simulation, the orientation angle is 0° for all the plies. Both simulation and experimental analyses were made. The composite plate was made of 6 layers of uni-direction E-glass and each layer had the thickness of 0.25 mm. The dimension of the specimen was 250 mm × 25 mm × 1.5 mm, which was one side fixed and the other side was given concentrated force and 20 kN. In experimental the tensile test was undertaken using a material test system (MTS) machine. The maximum stress obtained from the experimental result was 0.23 and the maximum stress obtained from the analysis method was 0.46 [4].

2 Structural Optimization—Methodology Used

2.1 The Geometry of the Model

In this paper, entire analyses of tensile tests are completed with a simple design by ASTM D3039 Model. Kevlar fiber is an emerging one and a simple change in orientation may give high property variations. Thus, the ASTM standard gives accurate dimensions of 10 mm thickness, 25 mm breadth, and 175 mm length. Aside from tensile test, the bending test also incorporated in this optimization. In bending test, D790 based ASTM test specimens are used for the construction of all the 38 models. The composite numerical tools, i.e., ANSYS ACP pre/post processors are unique facilities, which are used for composite constructions. The physical model of dog-bone shape specimen is drawn in ANSYS Design Modeler, in which the standard dimensions are used. The different models and their angle details are provided as follows:

In all the 38 models, the ten layers used, in which one layer is given at the thickness of 1 mm and a 10 mm thickness is constructed for the test specimen. The structural analyses are executed for all the 38 models and the results are listed in Table 1.

Table 1 Comprehensive report of model numbers versus orientational angles

Model No.	Orientational angles	Model No.	Orientational angles
Model-1	0-15-30-15-0-0-15-30-15-0	Model-20	90-0-75-0-90-90-0-75-0-90
Model-2	15-0-30-0-15-15-0-30-0-15	Model-21	0-45-90-45-0-0-45-90-45-0
Model-3	30-15-0-15-30-30-15-0-15-30	Model-22	45-0-90-0-45-45-0-90-0-45
Model-4	30-0-15-0-30-30-0-15-0-30	Model-23	45-90-0-90-45-45-90-0-90-45
Model-5	0-30-45-30-0-0-30-45-30-0	Model-24	90-45-0-45-90-90-45-0-45-90
Model-6	30-0-45-0-30-30-0-45-0-30	Model-25	90-0-45-0-90-90-0-45-0-90
Model-7	45-30-0-30-45-45-30-0-30-45	Model-26	0-90-45-90-45-0-90-45-90-45
Model-8	45-0-30-0-45-45-0-30-0-45	Model-27	0-0-0-0-0-0-0-0-0-0
Model-9	0-45-60-45-0-0-45-60-45-0)	Model-28	15-15-15-15-15-15-15-15-15-15
Model-10	45-0-60-0-45-45-0-60-0-45	Model-29	30-30-30-30-30-30-30-30-30-30
Model-11	60-45-0-45-60-60-45-0-45-60	Model-30	45-45-45-45-45-45-45-45-45-45
Model-12	60-0-45-0-60-60-0-45-0-60	Model-31	60-60-60-60-60-60-60-60-60-60
Model-13	0-60-75-60-0-0-60-75-60-0	Model-32	75-75-75-75-75-75-75-75-75-75
Model-14	60-0-75-0-60-60-0-75-0-60	Model-33	90-90-90-90-90-90-90-90-90-90
Model-15	75-60-0-60-75-75-60-0-60-75	Model-34	15-30-0-30-15-15-30-0-30-15
Model-16	75-0-60-0-75-75-0-60-0-75	Model-35	30-45-0-45-30-30-45-0-45-30
Model-17	0-75-90-75-0-0-75-90-75-0	Model-36	45-60-0-60-45-45-60-0-60-45
Model-18	75-0-90-0-75-75-0-90-0-75	Model-37	60-75-0-75-60-60-75-0-75-60
Model-19	90-75-0-75-90-90-75-0-75-90	Model-38	75-90-0-90-75-75-90-0-90-75

2.2 *Discretization*

The conversion of the physical model into the Finite Element Model (FEM) is executed in the ANSYS Mesh Tool, in which the sub-components of FEM are overloaded with the character of reactions with respect to the applied load. The fine structural mesh facilities are implemented in entire analysis. The overall maximum nodes are recorded as 258,749 and the maximum elements are noted as 745,821. The maximum quality is obtained as 0.9745. In addition to that, the grid convergence study also organized, in which five different mesh models are implemented. The coarse mesh with proximity and curvature, the medium mesh set-up with proximity and curvature, the fine mesh set-up with proximity and curvature, face mesh set-up and overall refinements are the five different mesh facilities implemented in this grid convergence investigation. Finally, the fine with proximity and curvature is mesh case fulfilled the criteria of grid convergence test so the same one is used for all other 76 tests.

2.3 *Construction of Composite Computational Model*

After the discretization, the furthermore construction of a composite test specimen is needed, which must be constructed from the base physical model. So, the composite numerical tool, ANSYS ACP is used in this work for the construction of Kevlar composite. In which, epoxy resin is used as a matrix because of its low delamination property with Kevlar fibers. Also, the composite test specimens are constructed in an organized manner with the help of unique facilities such as rosette, solid model, and orientational set-up. In addition to these facilities, this composite numerical tool has the facility to change the orientation of the fibers without affecting other facilities. Therefore, through this fiber orientational flexibility, the angles of the Kevlar fibers are modified and the test specimens are prepared for comparative analysis. Based on the angle modification, a total of 38 different models are generated and the comparative structural analyses are executed.

2.4 *Boundary Conditions and Discussions*

The FEA tool, ANSYS Structural Analysis platform is used for the solution and results compilation. The standard experimental tests were organized to estimate the ultimate load, which are used in this comparative analysis as boundary conditions. The common tensile load of 1000 N is applied at one end of all the 38 models. The other ends of 38 models are arrested with fixed support. The structural results of best model (M-27) are revealed in Figs. 1, 2 and 3, in which three different structural outputs such as deformation, normal stress, and equivalent stress are computed. The

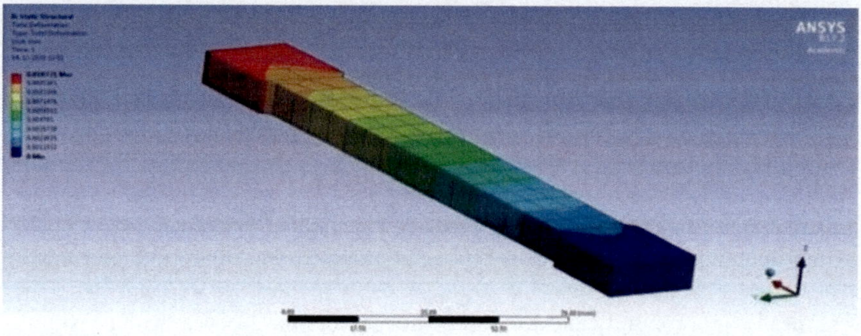

Fig. 1 Displacement variations in the test specimen

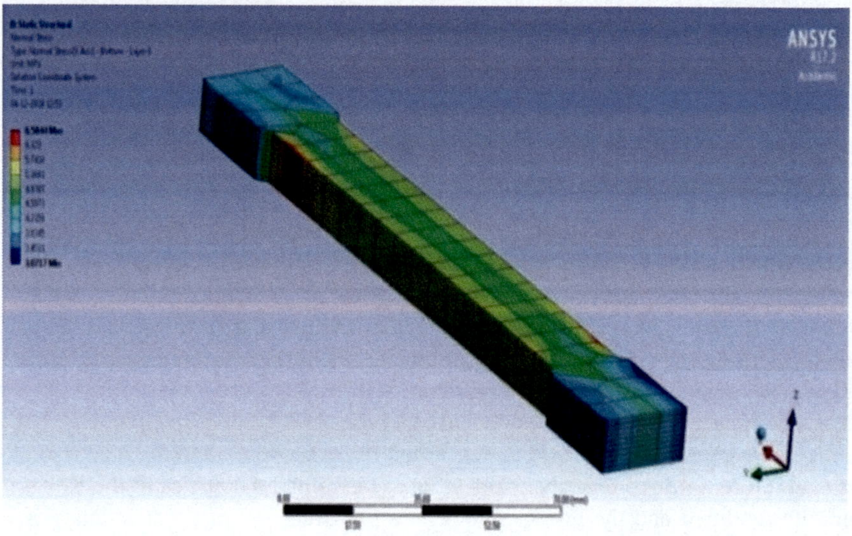

Fig. 2 Normal stress variations in the test specimen

comprehensive test results of 38 models are revealed in Figs. 4, 5, 6 and 7. From the comprehensive analysis, it is observed that M-27–M-33 models are reacted low stress forces than all remaining models. Also, M-27 model is reacted very lower deformation than others so M-27 is fit to handle tensile test in an effective manner and M-27–M-31 models are capable to withstand tensile load.

Fig. 3 Equivalent stress variations in the test specimen

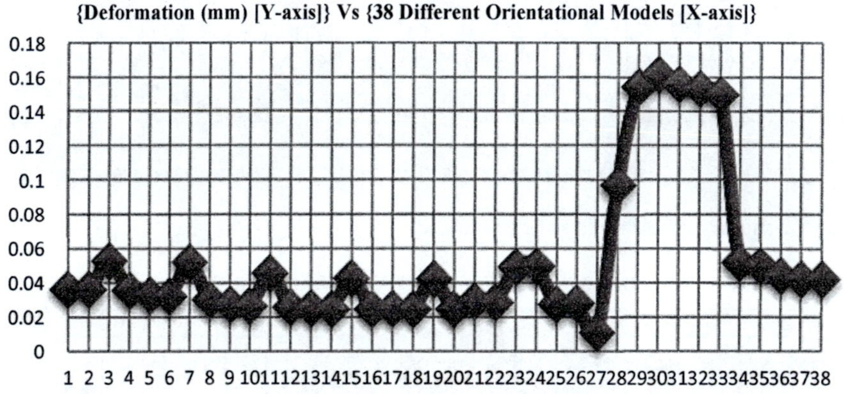

Fig. 4 Variations of total deformation under tensile load

3 Computational Structural Results and Its Discussion

Common external loads are given to all the models for comparative purpose. The orientational angles are picked from the literature survey and field works. From the comparative results, it is understood that low reacted geometrical setup can able to withstand high amount external loads. Thus this work precisely concentrated on low deformed orientational setup, low stresses setup, and low strain energy setup. Under this perspective M-27 and M-31 orientational angles are shortlisted to handle

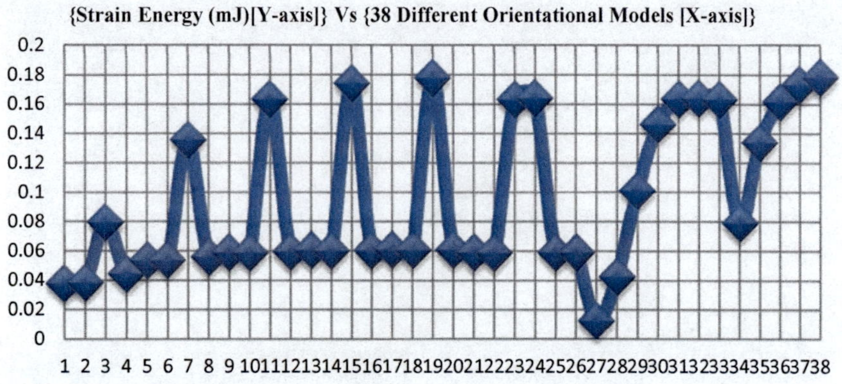

Fig. 5 Variations of strain energy under tensile load

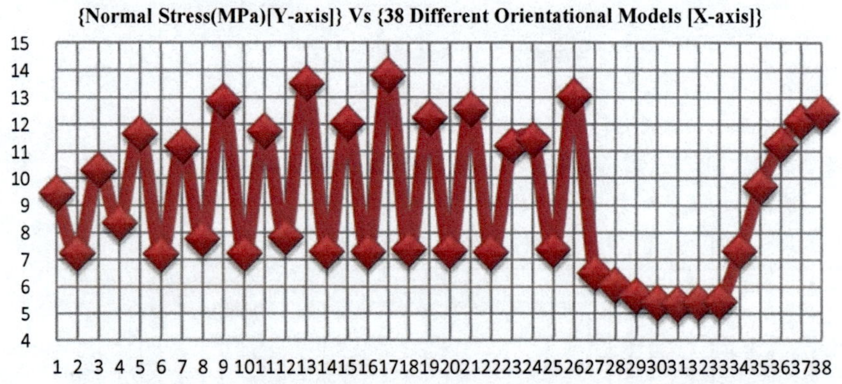

Fig. 6 Variations of normal stress under tensile load

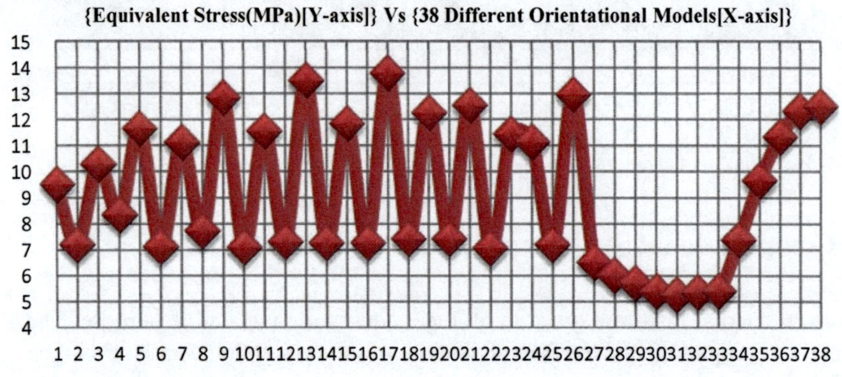

Fig. 7 Variations of equivalent stress under tensile load

high amount real-time tensile loads. The model finalizations are clearly confirmed in Figs. 4, 5, 6 and 7. The same perspective is extended for bending load and its optimization too. Under the bending loads, the Model 27 performed well in low deformation and strain energy, Models 13 and 20 are induced low stresses than other models. Thus these three Models are shortlisted as best orientational angles to withstand high amount of flexural loads. The model finalizations are clearly confirmed in Figs. 12, 13, 14, 15, 16 and 17.

3.1 Tensile Test Results—M-27 Model

3.2 Tensile Test—Comparative Results

3.3 Bending Test Results—M-20 Model

These optimizational studies are also incorporated the bending load-based behavioral effects on KFRP. The bending load of 537.09 N is given at mid of all the models. The comparative flexural results are revealed in Figs. 12, 13, 14, 15, 16 and 17. From the comparison, it has been strongly observed that M-20 and M-27 models and its corresponding orientational angles are performed well. Especially, M-20 model is reacted lower equivalent stress than all other models. Figures 8, 9, 10 and 11 are revealed the structural outcomes of M-20, in which displacement, normal stress, shear stress, and equivalent stress are projected. Apart from that, M-27 model is reacted lower deformation than all other models. Thus M-27 and M-20 are selected as best bending test withstanding models.

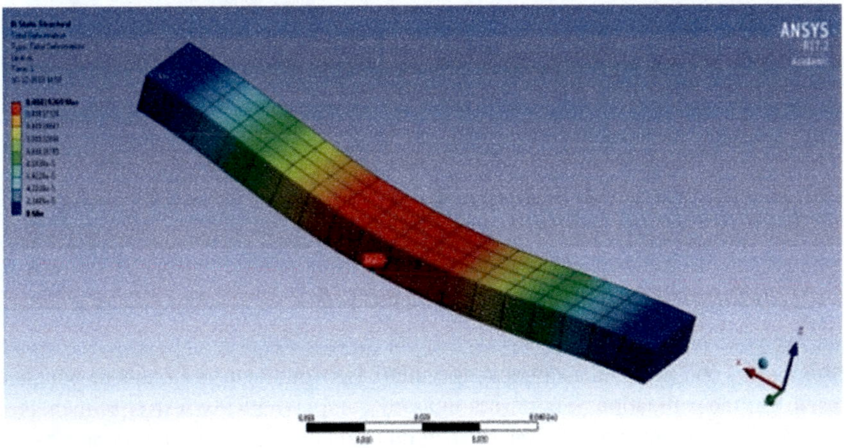

Fig. 8 Displacement variations in the test specimen

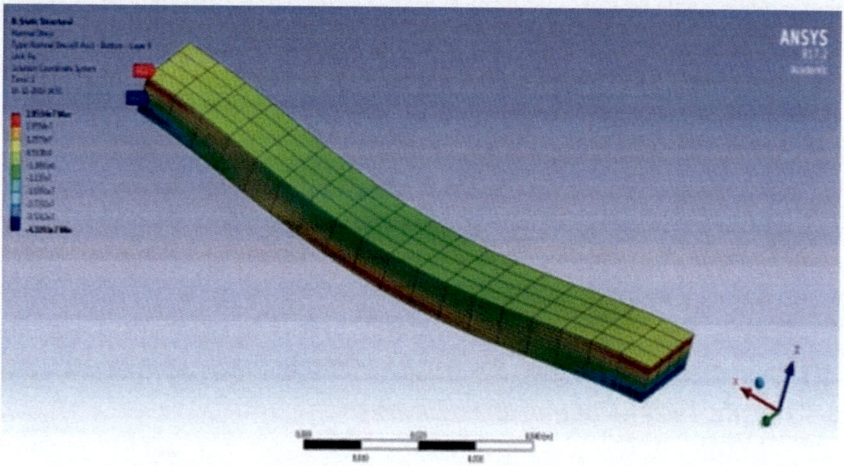

Fig. 9 Normal stress variations in the test specimen

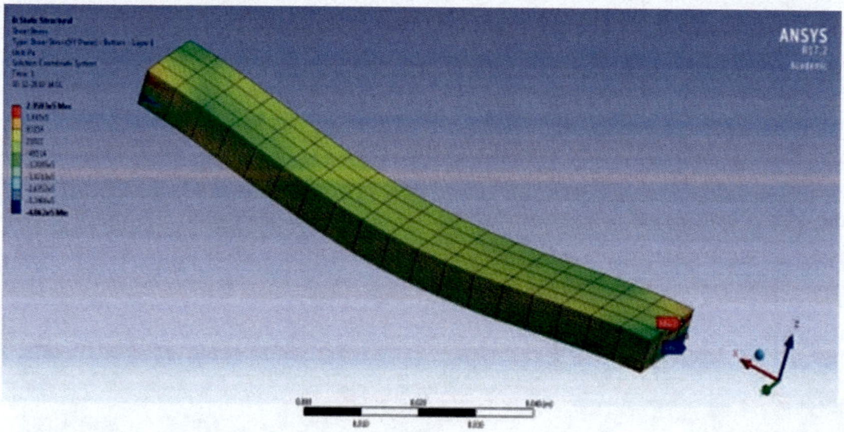

Fig. 10 Shear stress variations in the test specimen

3.4 Bending Test—Comparative Results

3.5 Validation Study of Methodology Used

In this work, conventional analytical formula-based outcome is obtained, which is used for the validation. The stress-induced sample KFRP's test specimen is 120.41 MPa under the computational structural analysis. Thus the same sample test results are compared with Eq. (1) outcome.

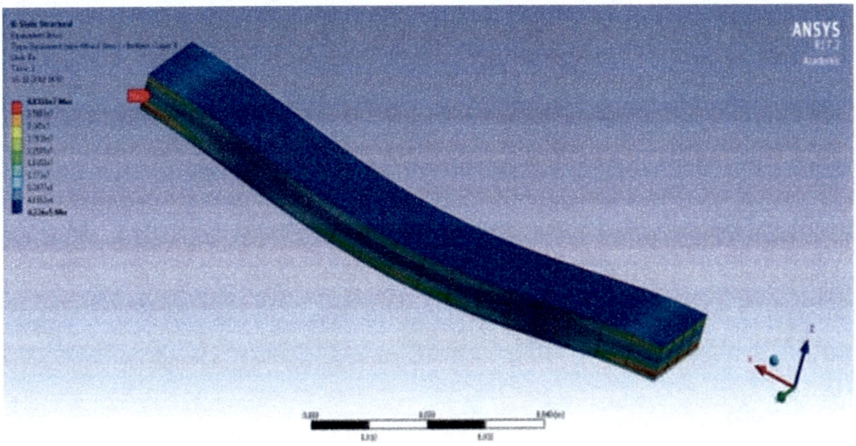

Fig. 11 Equivalent stress variations in the test specimen

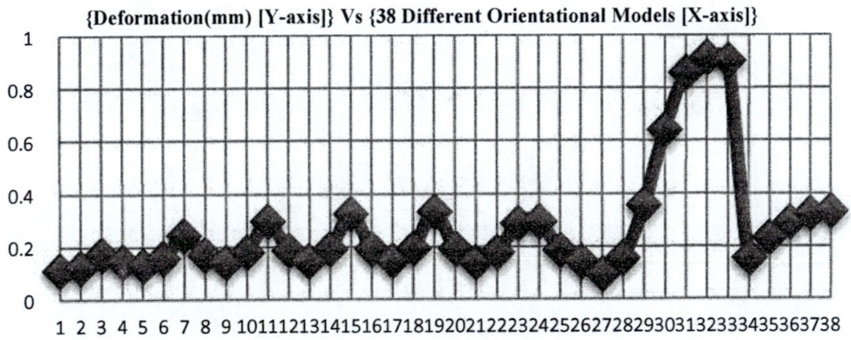

Fig. 12 Variations of total deformation under bending load

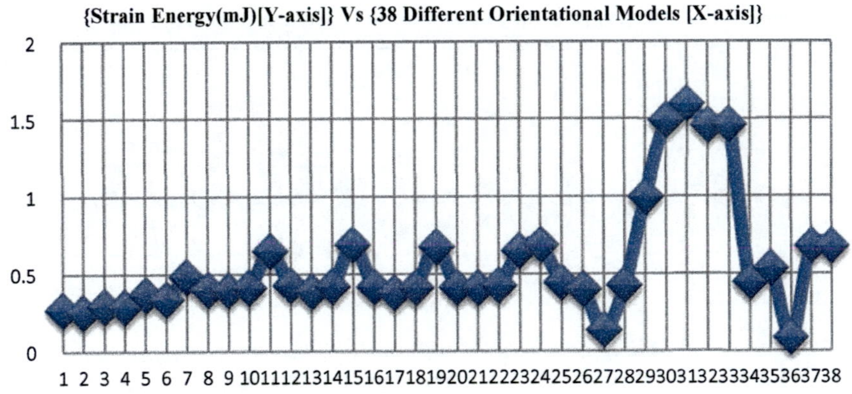

Fig. 13 Variations of strain energy under bending load

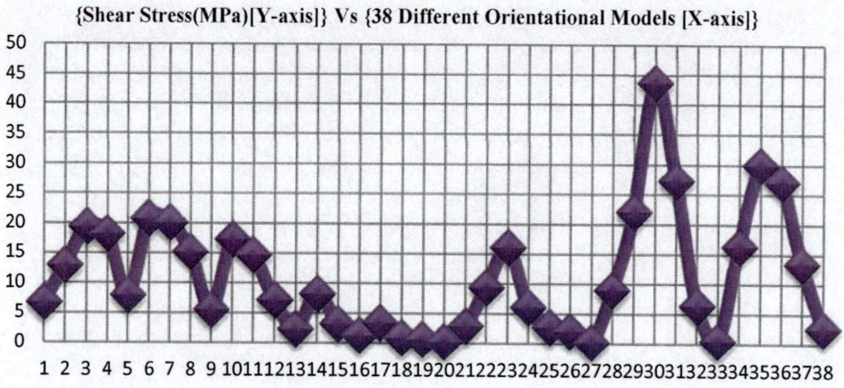

Fig. 14 Variations of shear stress under bending load

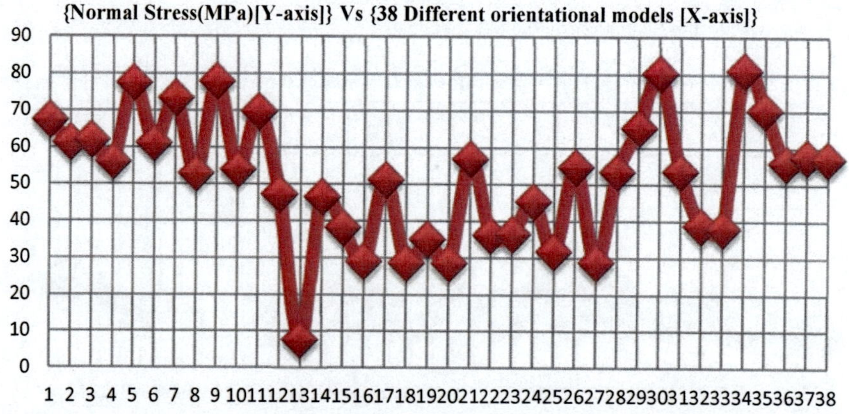

Fig. 15 Variations of normal stress under bending load

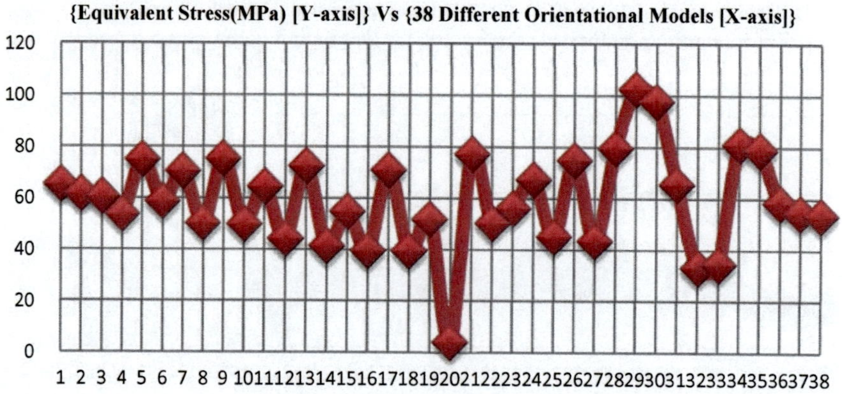

Fig. 16 Variations of equivalent stress under bending load

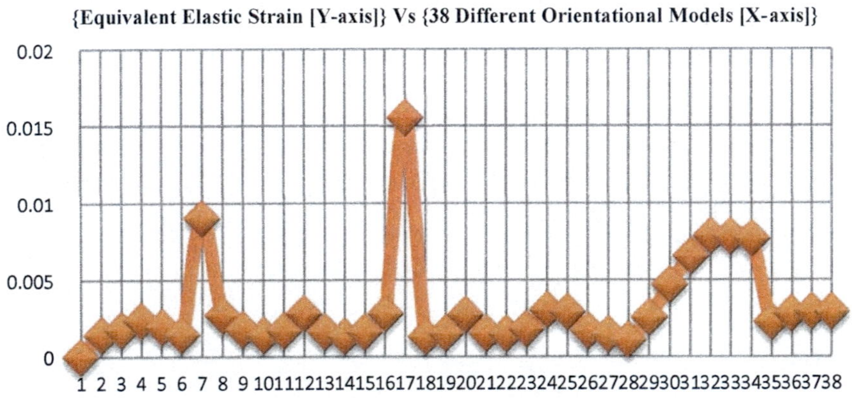

Fig. 17 Variations of equivalent elastic strain under bending load

$$\text{Flexural Strength}\,\sigma = \frac{3 * P * L}{2 * b * (t)^2} \tag{1}$$

where 'P' is the maximum applied load, 'L' is the length between two supports, 'b' is the width of the specimen and 't' is the thickness of the test specimen.

$$\sigma = \frac{3 * 537.09 * 0.1}{2 * 0.02 * (0.00555)^2} \Rightarrow \frac{161.127}{0.0000012321} \Rightarrow 130.77\,\text{MPa}$$

The error percentage is obtained as 7.9, which is comes under the allowable limit therefore it is clearly confirmed that the proposed methodology is capable to provide reliable outcomes. Finally, through the verified computational structural analysis outcomes are validated, which will be suggested to implement in real-time applications.

4 Conclusions

Owing to the great need for high strength to weight ratio material in structural engineering, the Kevlar-based composite material is proposed. So in this work, the internal elemental study is conducted and the structural results are noted for the given boundary conditions under tensile and bending loads. The internal element selected for this optimization is fiber, in which the angle orientations are modified and thereby 38 different test models are created. An advanced numerical tool, i.e., ANSYS Workbench has been used to solve the structural analyses on all the models. In which, the total deformations, strain energy, shear stress, equivalent stress, and normal stress are played the top role in the selection process of best orientational angles. To enhance the reliability of the structural outcomes, the grid convergence study and conventional analytical formula-based validations are executed. Finally, the fiber optimizations are executed with the help of predicted structural parameters. The model 27 is performed excellent in both the loading conditions. The model 20 is short listed to withstand high flexural load than other models. Also, it is observed that the model 31 is performed well by reacted low structural outputs than others under tensile load. Therefore the orientational angles of 60-60-60-60-60-60-60-60-60-60, 90-0-75-0-90-90-0-75-0-90, and 0-0-0-0-0-0-0-0-0-0 are more suitable to withstand high structural loads.

References

1. Polanco, M. A., Kellas, S., & Jackson, K. E. (2009). Evaluation of Material Models within LS-DYNA for a Kevlar/Epoxy Composite Honeycomb, Corpus ID: 53621356, Document ID-20090022372, Conference Paper, pp. 1–16.
2. Cheng, M., Chen, W., & Weerasooriya, T. (2005). Mechanical properties of Kevlar KM2 single fiber. *Journal of Engineering Materials and Technology, Transactions of the ASME, 127*, 197–203. https://doi.org/10.1115/1.1857937
3. Thiyagarajan, K., Martin, L., Elayaraja, N., & Sivaprakasam, P. (2014). Modelling of Kevlar—Al alloys and finite element simulation of mechanical properties. *ARPN Journal of Engineering and Applied Sciences, 9*(5), 599–607. ISSN 1819-6608.
4. Nurhaniza, M., Ariffin, M. K. A., Ali, A., Mustapha, F., & Noraini, A. W. (2010) Finite element analysis of composites materials for aerospace applications. *IOP Conference on Series: Materials Science and Engineering, 11*, 012010, 1–7. https://doi.org/10.1088/1757-899X/11/1/012010.
5. Raj Kumar, G., Vijayanandh, R., Senthil Kumar, M. (2017). Experimental testing and numerical simulation on natural composite for aerospace applications. *AIP Conference Proceedings, 1953*, 090045-1–090045-5. https://doi.org/10.1063/1.5032892.
6. Vijayanandh, R., Naveenkumar, K., Senthil Kumar, M., & Raj Kumar, G. et al. (2018). Material optimization of high-speed micro aerial vehicle using FSI simulation. *Procedia Computer Science, 133*, 2–9. ISSN 1877-0509. https://doi.org/10.1016/j.procs.2018.07.002.
7. Rajagurunathan, M., Raj Kumar, G., Vijayanandh, R., Vishnu, V., Rakesh Kumar, C., & Mohamed Bak, K. (2018). The design optimization of the circular piezoelectric bimorph actuators using FEA. *International Journal of Mechanical and Production Engineering Research and Development, 8*(7), 410–422. ISSN(E): 2249-8001
8. Raj Kumar, G., Senthil Kumar, M., Vijayanandh, R., Raja Sekar, K., Mohamed Bak, K., & Varun, S. (2019). The mechanical characterization of carbon fiber reinforced epoxy with carbon nanotubes. *International Journal of Mechanical and Production Engineering Research and Development, 9*(1), 243–255. ISSN(E): 2249-8001.
9. Vijayanandh, R., Senthil Kumar, M., Naveenkumar, K., & Raj Kumar, G. (2018). Design optimization of advanced multi-rotor unmanned aircraft system using FSI. *Lecture Notes in Mechanical Engineering, 28*, 299–310. https://doi.org/10.1007/978-981-13-2718-6.
10. Vijayanandh, R., Venkatesan, K., Ramesh, M., Raj Kumar, G., & Senthil Kumar, M. (2019). Optimization of orientation of carbon fiber reinforced polymer based on structural analysis. *International Journal of Scientific & Technology Research, 8*(11), 3020–3029. ISSN 2277-8616.
11. Raj Kumar, G., Balasubramaniyam, S., Senthil Kumar, M., Vijayanandh, R., Raj Kumar, R., & Varun, S. (2019). Crash analysis on the automotive vehicle bumper. *International Journal of Engineering and Advanced Technology, 8*(6S3), 1602–1607. ISSN: 2249–8958. https://doi.org/10.35940/ijeat.F1296.0986S319.

12. Naveen Kumar, K., Vijayanandh, R., Bruce Ralphin Rose, J., Swathi, V., & Narmatha, R. (2019). Research on structural behavior of composite materials on different cantilever structures using FSI. *International Journal of Engineering and Advanced Technology, 8*(6S3), 1075–1086. ISSN: 2249-8958. https://doi.org/10.35940/ijeat.F1178.0986S319.
13. Naveen Kumar, K., Vijayanandh, R., Raj Kumar, G., Sanjeev, B., Balachander, H., & Guru Prasad, S. Comparative approaches for fatigue life estimation of aluminium alloy for aerospace applications. *International Journal of Vehicle Structures & Systems, 10*(4), 282–286. ISSN: 0975-3540. https://doi.org/10.4273/ijvss.10.4.11.
14. Raj Kumar, G., Vijayanandh, R., Mohammad Bak, K., Shyam Chander, R., & Arawinth, R. (2018). Experimental testing on mechanical properties effect of aluminum foam. *International Journal of Mechanical and Production Engineering Research and Development, 8*(7), 1047–1059. ISSN (E): 2249-8001.
15. Venkatesan, K., Ramanathan, K., Vijayanandh, R., et al. (2020). Comparative structural analysis of advanced multi-layer composite materials. *Materials Today: Proceedings, 27*(Part 3), 2673–2687. https://doi.org/10.1016/j.matpr.2019.11.247
16. Venkatesan, K., Geetha, S., Vijayanandh, R., Raj Kumar, G., Jagadeeshwaran, P., & Raj Kumar, R. (2020). Advanced structural analysis of various composite materials with carbon nano-tubes for property enhancement. *AIP Conference Proceedings, 2270*, 030005-1–030005-6. https://doi.org/10.1063/5.0019367.
17. Mirrudula, P., Kaviya Priya, P., Malavika, M., Raj Kumar, G., Vijayanandh, R., & Senthil Kumar, M. (2020). Comparative structural analysis of the sandwich composite using advanced numerical simulation. *AIP Conference Proceedings, 2270*, 040005-1–040005-5. https://doi.org/10.1063/5.0019370.
18. Indira Prasanth, S., Kesavan, K., Kiran, P., Sivaguru, M., Sudharsan, R., & Vijayanandh, R. (2020). Advanced structural analysis on E-glass fiber reinforced with polymer for enhancing the mechanical properties by optimizing the orientation of fiber. *AIP Conference Proceedings, 2270*, 040006-1–040006-5. https://doi.org/10.1063/5.0019378.
19. Bhagavathiyappan, S., Balamurugan, M., Rajamanickam, M., Vijayanandh, R., Raj Kumar, G., & Senthil Kumar, M. (2020). Comparative computational impact analysis of multi-layer composite materials. *AIP Conference Proceedings, 2270*, 040007-1–040007-5. https://doi.org/10.1063/5.0019380.

Generation of Tool Path in Fused Filament Fabrication

Krishnanand, Shivam Soni, Ankit Nayak, and Mohammad Taufik

1 Introduction

Additive manufacturing is the method of manufacturing in which product is made by layer over layer addition of material using the digital data in form of CAD model of the part to be made. As normal printers print the letters on a page by depositing ink in two dimensions, say x and y only, but in additive manufacturing, including x and y dimension, the material is also deposited in z dimension to form a physical three-dimensional product. That is why it is also termed 3D printing. There are various methods of 3D printing, all these methods could be categorized into three categories based on the form of build material used—liquid-based, powder-based and solid-based [1]. In liquid-based, build material is in liquid form like stereolithography (SLA) [2]. If build material is in powdered form then it is powdered-based 3D printing like selective laser sintering (SLS), selective laser melting (SLM), etc. Similarly, laminated object manufacturing (LOM) and fused filament fabrications (FFF) are examples of solid-based 3D printing, because build material used is in solid form [3].

3D printing uses a CAD model as raw information for movement of nozzle/laser or material deposition tool to deposit the material in layers. The additive manufacturing process can be divided into three phases [4]. Process of 3D printing starts with making the CAD model of the part to be manufactured, then this design data in form of CAD model is transported to 3D printing process planning tool where it is converted into manufacturing data and finally, this manufacturing data is used by

Krishnanand · S. Soni · M. Taufik (✉)
Maulana Azad National Institute of Technology (MANIT) Bhopal, Bhopal, Madhya Pradesh 462003, India
e-mail: mohammad.taufik@manit.ac.in

A. Nayak
Banasthali Vidyapith, Vanasthali, Rajasthan 304022, India

R. Agrawal et al. (eds.), *Recent Advances in Smart Manufacturing and Materials*, Lecture Notes in Mechanical Engineering,
https://doi.org/10.1007/978-981-16-3033-0_14

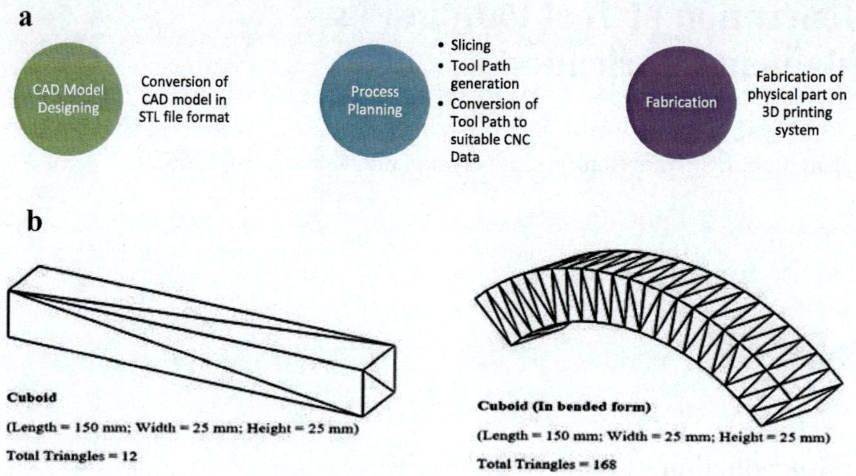

Fig. 1 Three phases in additive manufacturing process (**a**) and Tessellated CAD models in STL file format (**b**)

additive manufacturing or 3D printing system to manufacture the part in physical form (see Fig. 1a). Kulkarni et al. [5] divided process planning into two domains—model domain and layered domain. In model domain build orientation of part and support generation is decided, while in layered domain slicing, tool path planning is performed [6, 7].

STL (Stereolithography) file format is the bridge between CAD modeling tools and additive manufacturing process planning tools. STL file is the data of CAD model in form of tessellated (triangulated) surface model (see Fig. 1b) [6, 8]. It is de facto industry standard as the raw data for any 3D printing system. However, there are various other file formats like—IGES, STEP, etc. but STL file is simple to generate and having capability of defining any shape with any number of edges [9]. During tessellation, CAD model is converted into triangulated surface model. These triangles are called facets. Each facet is represented by its coordinates of vertices and unit normal vector (see Fig. 2).

In process planning of additive manufacturing build orientation, support structure generation, slicing, and tool path generation are performed. Build orientation affect the support generation, mechanical properties, certain shrinkage tolerances, and time for manufacturing [5]. Suitable tool path planning improves the precision, surface quality, strength and also reduce build time and material wastage [10]. In this study, tool path generation process and factor affecting the amount of material extrusion for fused filament fabrication (FFF) process have been presented.

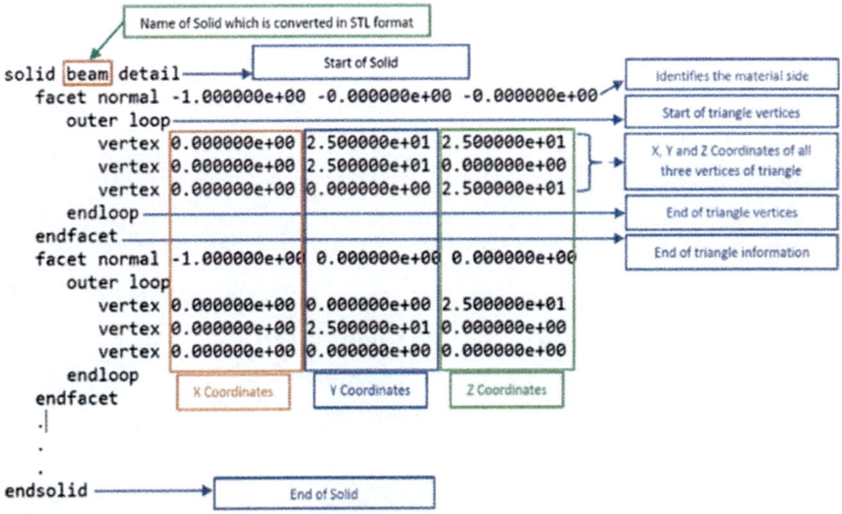

Fig. 2 Example of facet representation in STL file in ASCII format

2 Tool Path Generation in Fused Filament Fabrication (FFF)

STL file is transported from CAD modeling tool to some additive manufacturing process planning tools like—Cura, Slicer, etc. In these tools build orientation of the tessellated surface model is decided such that it would be optimum for the build time, part quality, and support material. Then tool path generation starts with the slicing process.

2.1 Slicing

Slicing is the process in which a plane parallel to build surface, called slicing plane or cutting plane, will cut the STL model at distances equal to layer height along the build direction (z-axis) to form a 2D contour. Slicing may be uniform or adaptive. If intersection planes are at equal distance, then they will slice the part at uniform intervals and hance this named as uniform slicing. If the distance between planes varies according to model curvature then it will result in adaptive slicing. [5]. STL file is used as input by slicing tool and follows an algorithm (see Fig. 3a [12]) to output a list of line segments in random order. Since STL model containing distinct facets, orientation of these facets is used to differentiate between the interior and exterior of the object to be created.

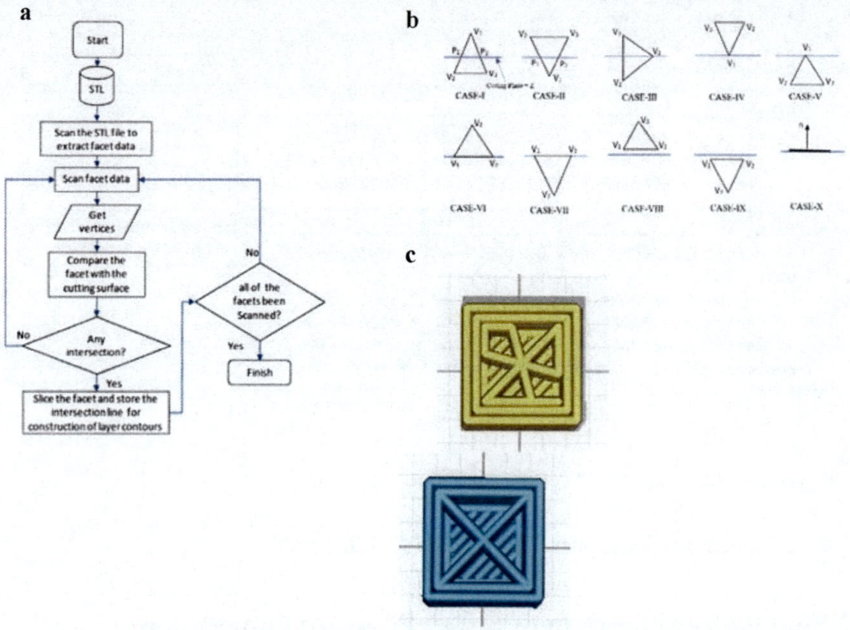

Fig. 3 Slicing algorithm in FFF (**a**) [12]; cases of the intersection of slicing plane with facets (**b**) [11] and different in-fill pattern affects the amount of material extrusion (**c**)

Finding Intersection Points of Facets with Cutting Plane. The algorithm finds the points where facet vertices and cutting planes are intersecting. These facets can be represented by F and as mentioned in Sect. 1, facet is represented by coordinates of its vertices and normal vector. There is total of 10 cases of the intersection of facets with cutting plane (see Fig. 3b [11]). As in case I, facet intersect the cutting plane at point P_1 and $P_{2,}$ these are the point which will be joined to form a line segment to form a closed polygon as a contour of the slice (2D layer). In cases IV and V, only one vertex of the facet is touching the cutting plane, these two cases result in redundant information. In case VI and VII, two vertices of facet are on cutting plane, so information obtained during this case, is value of these two vertices along with *Z* coordinate value of third vertex of the facet. This *Z* coordinate value is used to decide whether to keep or discard this information. In case VIII and IX, there is no intersection between facet and slicing plane, so these facets are ignored by algorithm and move to the next value of slicing plane. In case *X*, facet is parallel to slicing plane, all three vertices are the intersection point with cutting plane.

Generation of line segments. Output of this slicing algorithm are the points of intersection of a particular facet with slicing plane. Two points of intersection from same facet are linked together to form a line segment. Once all the facets have been checked by the algorithm, then output will be number of line segments in random order; so next step is to arrange these line segments to form a closed contour.

2.2 Contour Construction

Line segments generated by the slicing algorithm are in random order. To make a closed contour out of those line segments, these need to be sorted and arranged in appropriate order. An algorithm is followed to arrange these line segments in sequence (see Fig. 4a [9]). A line segment has two ends, one is called head and another one tail. When two lines segment joined together then joining point is head for one line and tail for the second line. This head-to-tail approach is used to connect the line segments by comparing the head of one line segment to the tail of another line segment. Slicing file is opened and first line segment is transferred to contour file. Then next line segment is read and compared with line segment in contour file. If it is neighboring to line segment in contour file then it is also transferred to contour file otherwise next line segment is checked from slice file. This process repeated for all the line segments in slicing file to check them. This comparison process to join line segments results in a closed contour. Closed contour will define the outer boundary for the material deposition for respective layer. After obtaining contour of particular sliced layer, next step is to generate tool path to fill that contour.

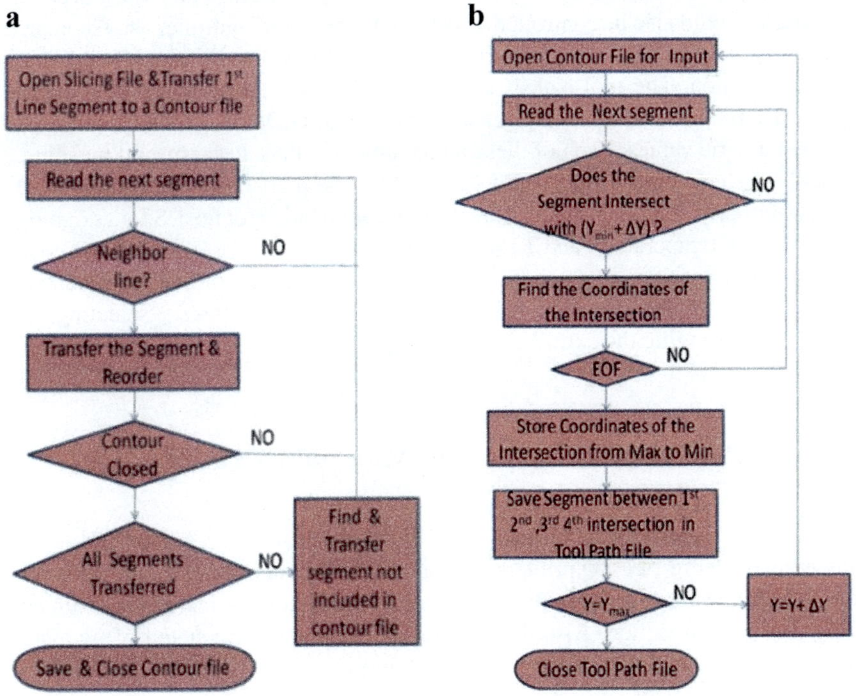

Fig. 4 Algorithm for contour construction (**a**) and Contour filling algorithm (**b**) [9]

2.3 Contour Filling

After getting closed contour next goal is to obtain a path that can provide suitable fill pattern for it. The product fabrication quality and time depend on the tool path and type of fill [13].

The raster angle, their overlapping, and distance between the two consequent lines are user-dependent. Raster angle and density will affect the build quality and time of the product. Party check algorithm is used when the contour is defined in polygon and interior region of the contour has to be found (see Fig. 4b [9]). This algorithm is based on the fact that a straight line intersects any closed curve at even number of times.

2.4 G-Code Generation

After generating the tool path in fused filament fabrication, the next step is to send the signals to motors of FFF machine to move the nozzle on that path for depositing the material as per the part geometry to fabricate a 3D part. Rate of material extrusion from nozzle should also be controlled as per need. The microcontroller uses standard G-Code instructions. G-code are commands used in CNC code; each G-code has a special function. The tool path is converted into G-code by slicing tools of 3D printers. There are statements in G-code starting with G, M, F, S, T, etc., followed by some numeric values. Each of these alphanumeric values has a special meaning. Generally, a statement starting with M is used for the preparation and setup of 3D printers, similarly, statements with G are motion command, F for feed, S for assigning target values of temperature, and T for choosing the extruder. An example of G-code used in FFF 3D printers is given below (see Fig. 5a). In case of dual extruder 3D printers there is some extra code to set the temperature of the nozzle, selecting the extruder, and changing the extruder during the printing process (see Fig. 5b).

3 Factor Affecting the Amount of Material Extrusion in FFF

A microcontroller of FFF 3D printers calculates the amount of filament required for printing. FFF printers used calculations based on the volume of material consumed during the part fabrication. Also, for printing the same object, amount of filament required could differ as there are several parameters on which it depends i.e. (i) Layer height, (ii) Infill density, (iii) Infill pattern (see Fig. 3c).

Layer height is defined as the thickness or height of the single printed layer. Using smaller layer height, the print quality increases, the amount of material extruded per layer decreases, increases the resolution of print and the time required to print the

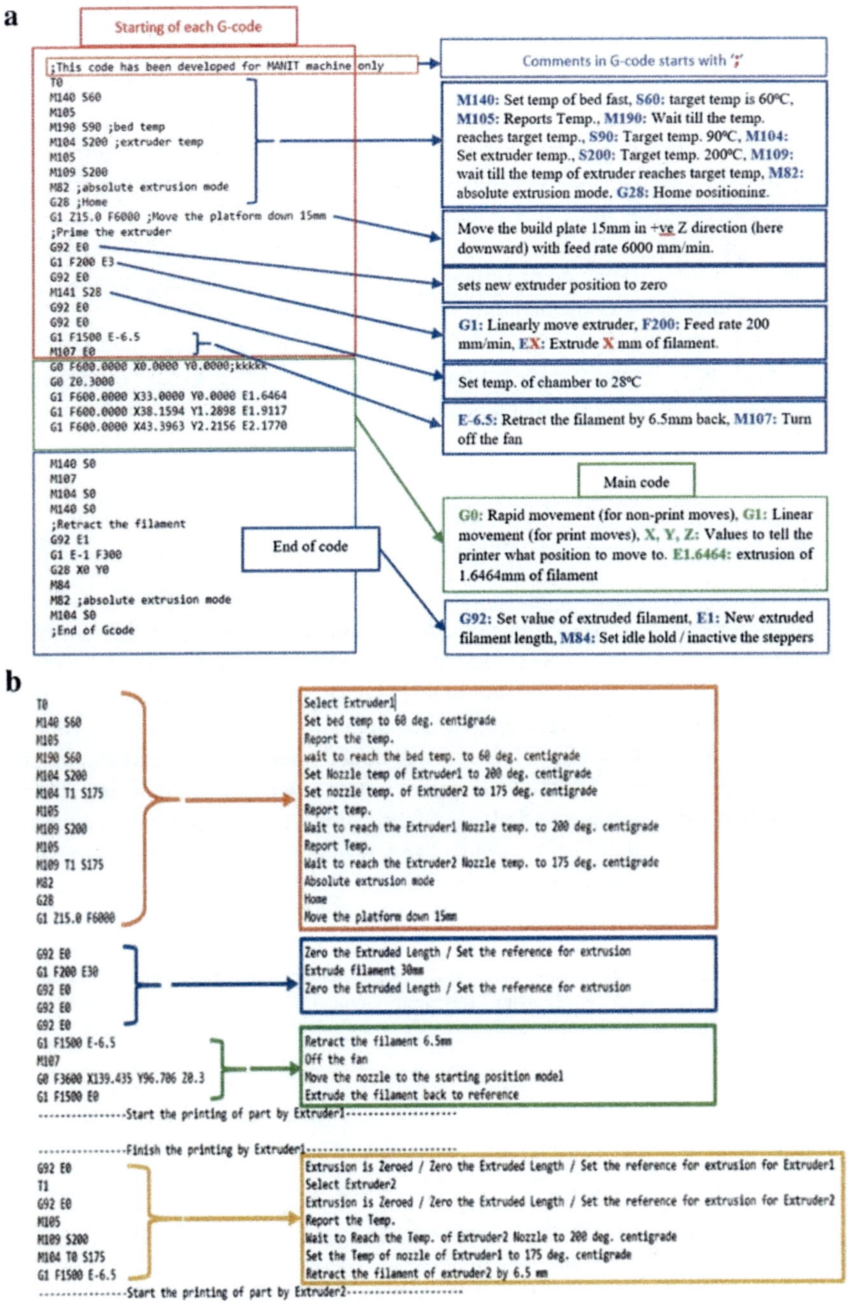

Fig. 5 Example of G-code for single extruder (**a**) and dual extruder (**b**) 3D printers

object increases. Infill density is the amount of material or filament required to print the internal structure of the object. Increase in the value of infill density increases the amount of material extruded per layer and this will increase the strength of the object. Usually, 20% infill density is preferred. Infill is the internal structure of the object and it provides strength and stability to the object. Some infill types require less material and are used for objects having less strength. For higher-strength object, the material required will be more. The diameter of the nozzle, the diameter of filament, and layer height are the factors used in the calculation of extruded filament. For example, one has to print a line of 100 mm length with 0.3 mm layer height using a nozzle of 0.4 mm diameter then amount of extruded material can be calculated. The volume of the line, V is 100 × 0.3 × 0.4 cubic mm or 12 cubic mm. The diameter of filament, D is 1.75 mm. Cross-section of filament, A will be $\pi\left(\frac{1.75}{2}\right)^2$ sq.mm or 2.40528 mm^2. So, the length of filament required for printing 100 mm line = volume of print/cross-section of filament = 12/2.40528 or 4.9890 mm.

4 Conclusion

Tool path generation is a critical step in the fused filament fabrication process. In this study, a detailed discussion on STL files and slicing along with basic algorithms for slicing, contour construction, and contour filling have been presented. Suitable tool path planning improves the precision, surface quality, strength and help to reduce build time and material wastage. Hence, it will directly lead the productivity. Layer height, infill density, infill pattern, and number of layers are the factors which affect the material extrusion during fused filament fabrication. A pellet extruder could also be integrated with existing FFF 3D printers and as future scope, a suitable tool path planning could be performed for the best performance of this pellet as well as filament extrusion system.

Acknowledgements This work was supported by the Science and Engineering Research Board (SERB)—DST under its Start-up Research Grant (SRG) scheme [Grant number: SRG/2019/000943].

References

1. Taufik, M., & Jain, P. K. (2013). Role of build orientation in layered manufacturing : A review. *International Journal of Manufacturing Technology Management, 27*, 47–73.
2. Gibson, I., Rosen, D. W., & Stucker, B. (2010). Additive manufacturing technologies: Rapid prototyping to direct digital manufacturing.
3. Krishnanand, Soni, S., & Taufik, M. (2020). Design and assembly of fused filament fabrication (FFF) 3D printers. *Materials Today Proceedings.*
4. Taufik, M., & Jain, P. K. (2016). CNC-assisted selective melting for improved surface finish of FDM parts. *Virtual Physical Prototype, 11*, 319–341.

5. Kulkarni, P., Marsan, A., & Dutta, D. (2000). Review of process planning techniques in layered manufacturing. *Rapid Prototype of Journal, 6*, 18–35.
6. Taufik, M., & Jain, P. K. (2017). Laser assisted finishing process for improved surface finish of fused deposition modelled parts. *Journal of Manufacturing Processes, 30*, 161–177.
7. Taufik, M., & Jain, P. K. (2020). Part surface quality improvement studies in fused deposition modelling process: A review.
8. Ding, D. H., Pan, Z. X., Dominic, C., & Li, H. J. (2015). Process planning strategy for wire and ARC additive manufacturing. *Advances in Intelligent Systems and Computing, 363*, 437–450.
9. Eragubi, M. (2013). Slicing 3D CAD model in STL format and laser path generation *4*, 410–413.
10. Jin, Y., He, Y., Fu, J. Z., Gan, W. F., & Lin, Z. W. (2014). Optimization of tool-path generation for material extrusion-based additive manufacturing technology. *Additive Manufacturing, 1*, 32–47.
11. Adnan, F. A., Romlay, F. R. M., & Shafiq, M. (2018). Real-time slicing algorithm for stereolithography (STL) CAD model applied in additive manufacturing industry. *IOP Conference on Series Materials Science Engineering, 342.*
12. Vatani, M., Rahimi, A. R., Brazandeh, F., & Sanati, A. An enhanced slicing algorithm using nearest distance analysis for layer manufacturing, pp. 669–674
13. Brown, A. C., & Brown, A. (2014). Development of a stereolithography (STL) slicing and G-code output algorithm for an entry level 3D printer/Andrew, C., Conradie, P., & Beer, D. (2014). Vaal Univeristy of Technology.

A Study on Mechanical Attributes of Epoxy-Carbon Fiber-*Terminalia bellirica* Embedded Hybrid Composites

Dinbandhu, Kumar Abhishek, Ashish Thakur, M. Nagaphani Sastry, K. Devaki Devi, and Anshumali Nishant

1 Introduction

The advancements in engineering applications have been centered around two significant fields, since inception, i.e., developments in efficient and cost-effective power units and attaining the maximum possible driving force from the available power. The latter essentially relies on the attributes of materials utilized in engineering fields. For example, an aeronautical engineer wants materials with superior stiffness qualities as well as lightweight. Likewise, an engineer with thermal engineering background needs corrosion-resistant materials with high strength features for the successful functioning of pressure vessels. Such situations are creating an ambiguity in the material's selection and letting the engineers think beyond the regular materials. Recent advancements in composites are providing a solution to engineers in terms of quality as well as cost. Essentially by changing the structure, multiple qualities can be enhanced henceforth, making the composites multipurpose and trustworthy and a substitute for the conventional designing materials [1–4].

Dinbandhu (✉) · K. Abhishek
Department of Mechanical & Aero-Space Engineering, Institute of Infrastructure, Technology, Research and Management (IITRAM), Ahmedabad, Gujarat 380026, India

A. Thakur
Department of Metallurgical and Materials Engineering, Visvesvaraya National Institute Of Technology, Nagpur, Maharashtra 440010, India

M. Nagaphani Sastry · K. Devaki Devi
Department of Mechanical Engineering, G. Pulla Reddy Engineering College, Kurnool, Andhra Pradesh 518007, India

A. Nishant
Department of Mechanical Engineering, Vidya Vihar Institute of Technology, Purnea, Bihar 854301, India

R. Agrawal et al. (eds.), *Recent Advances in Smart Manufacturing and Materials*, Lecture Notes in Mechanical Engineering,
https://doi.org/10.1007/978-981-16-3033-0_15

Machining of composite materials plays a significant role in their possible applications in engineering fields. During the initial days, composites were machined through conventional machining processes due to simple adaptability as well as to minimize the capital costs. But later, as substantial innovations in tool materials and tooling design took place, the conventional manufacturing techniques became obsolete in terms of productivity and quality. These days, several nonconventional machining processes namely abrasive water jet (AWJ) machining, laser machining, spark erosion machining, ultrasonic-assisted machining, wire electrical discharge machining, and electrochemical spark machining are recommended by scholars and fabricators for the machining of composite materials to meet the ongoing global demand. But still, there is a need for a high level of automation for the mass manufacturing of composite components to cut down the processing costs and compete with other materials. Developments in the nonconventional machining processes propose a prospect to deal with these materials cost-effectively, thus recognizing the broad potential of the composites [5–9].

Owing to their several advantageous attributes namely lightweight, corrosion resistance, and high fatigue strength, composites have become an essential part of today's developments. Their widespread application areas include aircraft, automobiles, sports items, electronic packaging, medical equipment, space vehicles, and homebuilding. Composites are structural materials that consist of two and/or several constituent elements. These elements, knowns as the matrix phase and reinforcing phase, are amalgamated at the macroscopic level and are mutually insoluble. The reinforcing materials, such as flakes/fibers, or particles, are embedded in the matrix phase, which is usually continuous. The matrix phase, which is more in terms of volume, holds the entire system of composite material and maintains its integrity. The most common composite materials include polymer matrix composites (PMC), metal matrix composites (MMC), ceramic matrix composites (CMC), and carbon–carbon composites (CCC)s. PMCs are further categorized into three sub-categories namely aramid fiber-reinforced polymer (AFRP) composites, carbon fiber reinforced polymer (CFRP) composites, and glass fiber reinforced polymer (GFRP) composites [10–13].

In recent times, more attention is given to natural fiber composites (i.e., biocomposites) by several researchers and academicians. Biocomposites comprise natural/synthetic resins with the reinforcement of natural fibers. These fibers demonstrate lots of beneficial attributes. As these fibers are low dense, comparatively lightweight composite materials are produced that reveal applications specific attributes. They are also cost-effective and their easy processing, besides being an extremely renewable source, reduces the reliance on overseas and domestic energy products. The latest developments in using natural fibers (for instance coir, jute, wax, straw, cellulose, switchgrass, hemp, bamboo, and kenaf) in composites are studied by several researchers. Effortless accessibility and economical existence caused the natural fiber an ideal option. In the meantime, a new trend has been observed among the researchers where hybrid natural fiber composites are developed by incorporating two or more than two reinforcements. These reinforcements can be either natural or synthetic or both [14–16].

Ashok et al. [17] evaluated the mechanical attributes of luffa/carbon fiber hybrid composites. For the preparation of composites, the epoxy matrix was used with varying percentages of luffa and carbon fibers. The outcomes of the study reveal enhancement in the mechanical performance of the composites. Failure analysis was also done through a scanning electron microscope. Bakshi et al. [18] employed an injection molding technique for the development of a sustainable composite material. Calcium-rich marble waste with varying concentrations (by weight) has been utilized as reinforcement in polypropylene resin. Mechanical and thermal attributes of the samples were examined, and it was observed that the developed composite is technically feasible and can be used in civil structural works. Oliveira et al. [19] utilized the fibers of sugarcane bagasse and particulates of disposed rubber for the preparation of a new hybrid composite. They have also evaluated the impact of aforesaid reinforcement by implementing the full factorial design of 2^5. An observation was drawn from the work that these particles have a severe effect on the mechanical performance of the materials. The size and quantity of the rubber particles affect the physical and mechanical attributes whereas the presence of sugarcane fibers influences the toughness, strength, and stiffness. Prabhuram et al. [20] studied the chemical and mechanical attributes of the developed hybrid composite made up of wood powder particles along with the fibers of coconut coir. The technique employed for preparing the material was hand layup. Water absorption capability along with impact, tensile, and flexural strength were evaluated for the said composites and it was found that the reinforced particles highly influence chemical and mechanical performances. Akash et al. [21] evaluated the water-absorbing capability and several other mechanical attributes for hemp/sisal fiber hybrid composites. The samples were prepared in the varying composition of the resin and reinforcements with the cold pressing method. Fracture analysis of the composites was also performed through a scanning electron microscope. The outcomes of the work recommend using the aforesaid fibers in preparation for inexpensive biocomposites. Chowdhary et al. [22] conducted a review on the degradation of mechanical attributes of the hybrid composites. The work also explores the features of failure and wear analysis done by several researchers. Sinha et al. [23] reported the recent advancements in the mechanical performance of hybrid composites. The authors suggested using fly ash and red mud along with the fibers of bamboo, ramie, pineapple, coir, and abaca, as the reinforcement in the preparation of polymer composites. An emphasis is also given on the optimization of mechanical performances of the hybrid polymer composites.

The current work is an endeavor to make a novel composite material by dispersing *Terminalia bellirica* + carbon fibers in the epoxy resin with their varying weight percentage. The work also focuses on assessing the mechanical attributes namely hardness, flexural, and tensile strength of the developed composite models.

2 Materials and Methods

Preparation of this novel composite was done using the following raw materials: epoxy resin (LY556), fibers of *T. bellirica*, carbon fibers, and hardener (HY951). The bark of *T. bellirica* was extracted from a village, Sohagpur, Gopalganj, Bihar. Procurement of hardener and epoxy material was done from Chemicote Engineers, Bengaluru, Karnataka. Carbon fibers were collected from Bhaskar Exports, Kakinada, Andhra Pradesh. The properties of epoxy resin and hardener have been illustrated in Tables 1 and 2, respectively [24].

The epoxy-hardener mixture is taken at the ratio of 1:10 (1 g of hardener is mixed in every 10 g of epoxy). Then the mixture is completely mixed for some time and is utilized for preparing laminates. Initially, the fibers were extracted from the bark and treated with NaOH solution (10gms of NaOH with 1 L of water). These extracted fibers were soaked in the solution for about 1 h (Fig. 1). After that, they were taken

Table 1 Properties of epoxy resin [24]

S. No.	Features	Description
1	Density @ 25 °C	11.15–1.20 g/cm^3
2	Viscosity @ 25 °C	10,000–12,000 mPa s
3	Chemical composition	Bisphenol-a-based epoxy resin
4	Product type	Epoxies—Bisphenol-A based
5	Visual aspect	Clear, pale yellow liquid

Table 2 Properties of hardener [24]

S. No.	Features	Description
1	Viscosity @ 25 °C	10–20 mPa s
2	Specific gravity @ 25 °C	0.95–1.05 g/cm^3

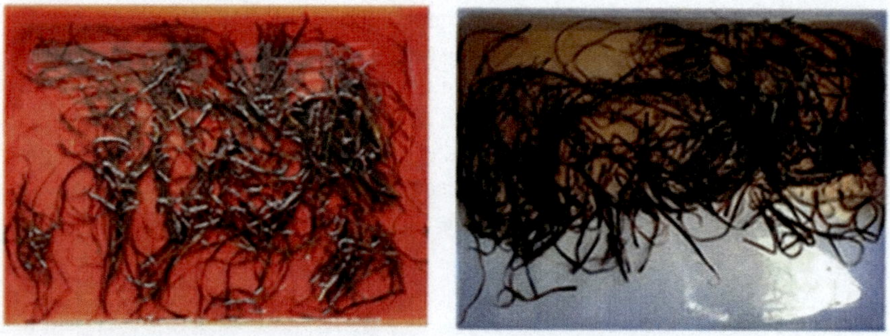

a) Fiber soaked in NaOH solution

b) After 1 hr soaking in NaOH solution

Fig. 1 Processing of *T. bellirica*'s fibers before composites preparation

Table 3 Composition of the prepared composite specimens

S. No.	Name assigned to the samples (%)	Composition of the samples
1	0	00 gm CF + 20 gm Terminalia
2	5	05 gm CF + 15 gm Terminalia
3	10	10 gm CF + 10 gm Terminalia
4	15	15 gm CF + 05 gm Terminalia
5	20	20 gm CF + 00 gm Terminalia

out, cleaned with the distilled water two to three times, and placed under the sun for getting dried. This treatment was done to avoid pull out of the fibers during composite preparation. Using hand-layup techniques, the composite specimens were prepared. To do this, a glass mold of $200 \times 200 \times 03\ mm^3$ was utilized where the treated fibers were dispersed into the epoxy matrix. Carbon fibers with varying percentages (by weight) were also dispersed with the purpose of enhancing the tensile strength, stiffness, temperature tolerance, and minimizing the thermal expansion of the prepared composites. A solidification time of 24 h was given and after that, composites were extracted from the mold and cut down for further evaluation. By keeping the epoxy percentage fixed (80% by weight) along with the varying fiber percentage of carbon and *T. bellirica*, a total of five samples were prepared. Table 3 depicts the composition of prepared composite samples. One of the prepared specimens is shown in Fig. 2.

Fig. 2 Prepared composite specimen

Table 4 Tensile strength of the composite samples

S. No.	Composition of the samples	Tensile strength (MPa)
1	00 gm CF + 20 gm Terminalia	13.96
2	05 gm CF + 15 gm Terminalia	26.77
3	10 gm CF + 10 gm Terminalia	35.48
4	15 gm CF + 05 gm Terminalia	79.35
5	20 gm CF + 00 gm Terminalia	95.50

3 Result and Discussion

Evaluation of mechanical properties specifically the tensile test, flexural test, and hardness test was accomplished for the prepared specimens. For that, the samples were ready based on the ASTM standards. Flexural strength and tensile strength were inspected on a universal testing machine (UTM) for the prepared samples whereas, for hardness tests, a Rockwell hardness tester was employed. The outcomes of these testing have been discussed in the subsequent section.

3.1 Tensile Test

The tensile test evaluates the potential of a material to endure applied forces before rupture. The experiments have been carried out on a UTM and the outcomes were recorded as shown in Table 4. The outcomes are evident in increasing tensile strength with the varying percentage of carbon and *T. bellirica*'s fibers. The highest value of the tensile strength (i.e. 95 MPa) has been obtained for sample 5 (having 20 gm CF + 00 gm Terminalia). This is due to the addition of carbon fibers in reinforcement. The graphical conclusion of the obtained outcomes has been exhibited in Fig. 3.

3.2 Flexural Test

When a material is having some structural applications then it becomes compulsory for an engineer to examine its flexural strength and for that flexural test is supposed to be conducted. The flexural test evaluates the performance of a material subjected to simple beam loading before rupture. The tests were performed on a universal testing machine with a special setup of three-point loading and the outcomes were recorded as shown in Table 5. Sample 3 (having 10 gm CF + 10 gm Terminalia) yields the best result (i.e. 41,230.23 MPa) among all the samples being tested. This is due to the increasing percentage of carbon fibers in reinforcement which makes the material

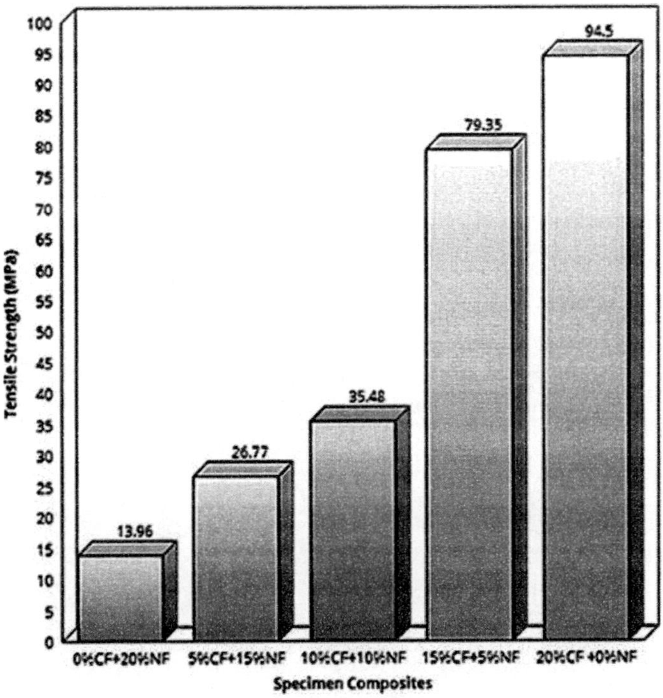

Fig. 3 The graph depicts the tensile strength of the composite samples

Table 5 Flexural strength of the composite samples

S. No.	Composition of the samples	Flexural modulus (MPa)
1	00 gm CF + 20 gm Terminalia	8577.87
2	05 gm CF + 15 gm Terminalia	16,342.71
3	10 gm CF + 10 gm Terminalia	41,230.23
4	15 gm CF + 05 gm Terminalia	37,276.62
5	20 gm CF + 00 gm Terminalia	19,829.41

brittle. The graphical conclusion of the obtained outcomes has been exhibited in Fig. 4.

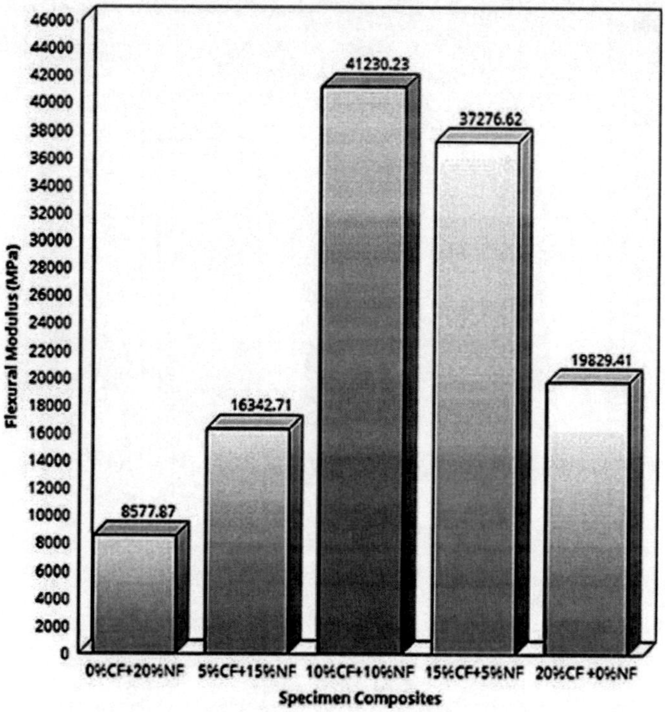

Fig. 4 The graph depicts the flexural strength of the composite samples

3.3 Hardness Test

The hardness test evaluates the capability of a material to resist wear and tear against the applied forces. The tests were performed on a Rockwell hardness tester and the outcomes were recorded as shown in Table 6. The outcomes depict a descending trend in Rockwell hardness number. The highest hardness value (i.e. 99) has been obtained for sample 1 (having 0 gm CF + 20 gm Terminalia). This happened because the carbon fibers were not continuous, and the matrix phase was not able to hold them

Table 6 Harness of the composite samples

S. No.	Composition of the samples	Rockwell hardness number (HRC)
1	00 gm CF + 20 gm Terminalia	99
2	05 gm CF + 15 gm Terminalia	95
3	10 gm CF + 10 gm Terminalia	90
4	15 gm CF + 05 gm Terminalia	86
5	20 gm CF + 00 gm Terminalia	80

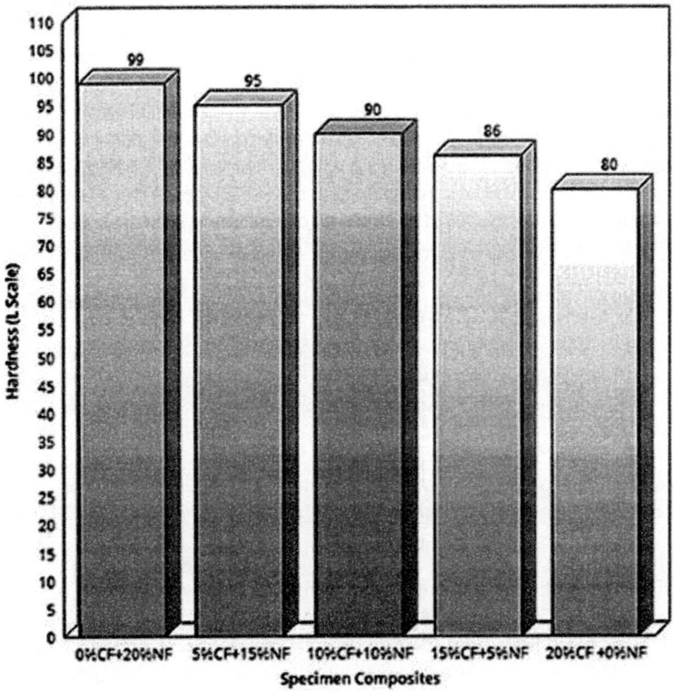

Fig. 5 The graph depicts the hardness of the composite samples

properly. The graphical conclusion of the obtained outcomes has been exhibited in Fig. 5.

4 Conclusions

In this work, hybrid polymer matrix composites have been prepared and their mechanical properties have been evaluated. The study uncovers the following conclusive findings:

i. An epoxy-based composite reinforced with natural fiber and carbon fiber is effectively made using the hand lay-up technique.
ii. The tensile strength of the composites increases with an increase in the weight percentage of the reinforcements.
iii. The increasing weight percentage of carbon fiber makes the composite brittle which is not recommended for structural applications.
iv. The descending trend of Rockwell hardness number shows that there is a lack of interaction between the reinforced materials and the epoxy resin. Hence, the material is not recommended for applications subjected to wear and tear.

References

1. Thakur, A., Purohit, R., Rana, R. S., & Bandhu, D. (2018). Characterization and evaluation of mechanical behavior of epoxy-CNT-bamboo matrix hybrid composites. *Materials Today Proceedings, 5*, 3971–3980. https://doi.org/10.1016/j.matpr.2017.11.655
2. Bandhu, D., Mohan, M. M., Praveen, G., Yadav, K., Reddy, K. J., & Shavali, G. S. (2016). Mechanical properties of Makhana (Euryale Ferox Salisbury) reinforced composite. *IJSRD—International Journal of Science Research Development, 4*, 2321–2613.
3. Hull, D., & Clyne, T. W. (1996). An introduction to composite materials. *Cambridge University Press*. https://doi.org/10.1017/cbo9781139170130
4. Ngo, T.-D. (2020) Introduction to composite materials. In *Fiber Composites*. IntechOpen. https://doi.org/10.5772/intechopen.91285.
5. Komanduri, R. (1997). Machining of fiber-reinforced composites. https://www.tandfonline.com/doi/abs/10.1080/10940349708945641.
6. Kumar, D., & Singh, K. K. (2015). An approach towards damage free machining of CFRP and GFRP composite material: A review. *Advanced Composite Materials, 24*, 49–63. https://doi.org/10.1080/09243046.2014.928966
7. Kuruc, M. (2020). Machining of composite materials by ultrasonic assistance. *Advanced Science Technology Research of Journal, 14*, 140–144. https://doi.org/10.12913/22998624/118943.
8. Ablyaz, T. R., Muratov, K. R., Shlykov, E. S., Shipunov, G. S., & Shakirzyanov, T. V. (2019). Electric-discharge machining of polymer composites. *Russian Engineering Research, 39*, 898–900. https://doi.org/10.3103/S1068798X19100058
9. Mohamed, S. B., Rashid, R. A., Muhamad, M., & Ismail, J. (2019). Composite materials and types of machining. In *SpringerBriefs in applied sciences and technology* (pp. 1–14). Springer Verlag. https://doi.org/10.1007/978-981-13-1804-7_1.
10. Bandhu, D., Thakur, A., Purohit, R., Verma, R. K., & Abhishek, K. (2018). Characterization & evaluation of Al7075 MMCs reinforced with ceramic particulates and influence of age hardening on their tensile behavior. *Journal of Mechanical Science and Technology, 32*, 3123–3128. https://doi.org/10.1007/s12206-018-0615-9
11. Naga Phani Sastry, M., Devaki Devi, K., & Bandhu, D. (2016). Characterization of Aegle Marmelos fiber reinforced composite. *International Journal of Engineering Research, 5*, 345–349. https://doi.org/10.17950/ijer/v5i2/009.
12. Shavali, G. S., Venugopal Goud, E., Praveen, G., Yadav, K., & Bandhu, D. (2016). Tensile and flexural characterization of Nomex and E-glass fibre reinforced epoxy composites. *IJSRD—International Journal of Science Research Development, 4*, 2321–2613.
13. Tripathi, D. R., Vachhani, K. H., Kumari, S., Dinbandhu, & Abhishek, K. (2020). Experimental investigation on material removal rate during abrasive water jet machining of GFRP composites. *Materials Today Proceedings, 26*, 1389–1392. https://doi.org/10.1016/j.matpr.2020.02.280.
14. Shesan, O. J., Stephen, A. C., Chioma, A. G., Neerish, R., & Rotimi, S. E. (2019). Improving the mechanical properties of natural fiber composites for structural and biomedical applications. In *Renewable and sustainable composites*. IntechOpen. https://doi.org/10.5772/intechopen.85252.
15. Jeyapragash, R., Srinivasan, V., & Sathiyamurthy, S. (2020). Mechanical properties of natural fiber/particulate reinforced epoxy composites—A review of the literature. *Materials Today: Proceedings*, 1223–1227. https://doi.org/10.1016/j.matpr.2019.12.146.
16. Vinayagamoorthy, R. (2019). Trends and challenges on the development of hybridized natural fiber composites. *Journal of Natural Fibers*, 1–18. https://doi.org/10.1080/15440478.2019.1598916.
17. Ashok, K. G., & Damodaran, A. (2020). Mechanical and morphological properties of luffa/carbon fiber reinforced hybrid composites. *Materials Today Proceedings*. https://doi.org/10.1016/j.matpr.2020.05.716

18. Bakshi, P., Pappu, A., Patidar, R., Gupta, M. K., & Thakur, V. K. (2020). Transforming marble waste into high-performance, water-resistant, and thermally insulative hybrid polymer composites for environmental sustainability. *Polymers (Basel), 12*, 1781. https://doi.org/10.3390/polym12081781.
19. Oliveira, P. R., Ribeiro Filho, S. L. M., Panzera, T. H., Christoforo, A. L., Durão, L. M. P., & Scarpa, F. (2020). Hybrid polymer composites made of sugarcane bagasse fibres and disposed rubber particles. *Polymers and Polymer Composites*, 096739112094345. https://doi.org/10.1177/0967391120943459.
20. Prabhuram, T., Elilraja, D., Prathap Singh, S., & Durairaj, I. (2021) Investigation of mechanical and chemical properties of the coir fiber and wood powder reinforced hybrid polymer composite. In Lecture Notes in Mechanical Engineering, pp. 285–292. Springer. https://doi.org/10.1007/978-981-15-4745-4_26.
21. Akash, S. R., Patil, V., & Girisha, K. G. (2021) An experimental study on hemp/sisal fiber embedded hybrid polymer composites. In Lecture Notes in Mechanical Engineering. pp. 293–301. Springer. https://doi.org/10.1007/978-981-15-4745-4_27.
22. Somaiah Chowdary, M., Raghavendra, G., Niranjan Kumar, M. S. R., Ojha, S., & Om Prakash, M. (2020) A review on the degradation of properties under the influence of liquid medium of hybrid polymer composites. *SN Application Science, 2*, 1708. https://doi.org/10.1007/s42452-020-03502-7.
23. Sinha, A. K., Narang, H. K., Bhattacharya, S. (2020). Mechanical properties of hybrid polymer composites: A review. https://doi.org/10.1007/s40430-020-02517-w.
24. Priya, I. I. M., & Vinayagam, B. (2018). Enhancement of bi-axial glass fibre reinforced polymer composite with graphene platelet nanopowder modifies epoxy resin. *Advances in Mechanical Engineering, 10*, 168781401879326. https://doi.org/10.1177/1687814018793261

Development and Analysis of Sustainable and Innovative Surface Finishing Process Through Combined Effects of Ball and Roller Burnishing

C. S. Jawalkar

1 Introduction

Burnishing is a ***sustainable*** (no-chips are produced) machining process, widely used in improving the surface roughness and surface hardness of metallic surfaces [1, 2]. It is an effective technique as there is no material wastage, and it improves the performance of ferrous and non-ferrous metals [3]. It can be done on the same work setting such as on lathes, without work-changeovers. Burnished surfaces have higher wear resistance and improved fatigue behaviors [4]. Surface finish and surface micro hardness of components play vital roles in improving the product quality since it introduces compressive residual stresses, produces accurate finely finished and densely compact surfaces which can resists wear [5]. In burnishing process, the tool used is a hard ball or a roller; which deforms the surface layers by means of point contact, which concentrates and deforms the peaks into adjacent valleys plastically using varied tool materials [6]. The surface roughness in burnishing process can be varied through optimization of the process parameters [7]. The roller burnishing process uses hardened rollers as the tool; thereby producing accurate, well-finished and compact surfaces. Figure 1 schematically illustrates burnishing phenomenon, resembling the road-rolling process.

The conventional burnishing process has over the last few years, been further innovated and hybridized using various combinations for different applications. Some of these processes are illustrated along with the major findings in Table 1.

C. S. Jawalkar (✉)
Department of Production and Industrial Engineering, Punjab Engineering College (Deemed University), Chandigarh 160012, India
e-mail: csjawalkar@pec.edu.in

R. Agrawal et al. (eds.), *Recent Advances in Smart Manufacturing and Materials*,
Lecture Notes in Mechanical Engineering,
https://doi.org/10.1007/978-981-16-3033-0_16

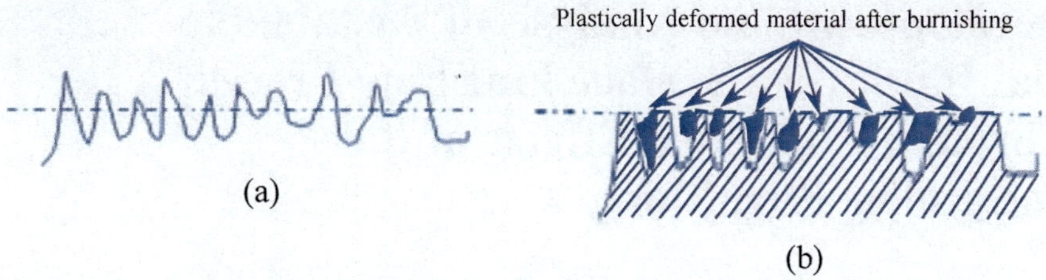

Fig. 1 Schematic sketch of a surface. **a** Before burnishing. **b** After burnishing

1.1 Burnishing Using Combined Tool

Through the conducted literature review it was found that combining "ball and roller tools" in a single set-up had not been effectively tried; which would make the study interesting as well as enhance the utility; hence the objective was to develop such economic and innovative tooling. It can be used on conventional machines like lathes. Figure 2 illustrates schematic sketch of the developed tool, which had the options to carry out individual burnishing either by ball, by roller or with combined effects.

The tool consisted of two rollers, two ball elements, square metal casing and lock nuts. The design of tool was made and fabricated in-house for experimentation and trial production purpose. The multiple options in tool increased the flexibility. Mechanism of combined tool is illustrated as follows the subsequent roller and ball tools generate more work hardening effect into the surface of such hard ductile materials (low-carbon steel) and also help in deforming material plastically, rather more evenly (due to multiple action) into adjacent valleys which reduces peaks and valleys.

2 Experimentation

The process parameters and their range which affect the machining characteristics were carefully selected based on pilot experimental studies. Pilot experiments confirmed that cutting speeds (from 300 to 550 RPM), tool feeds (from 0.1 to 0.5 mm/rev) and tool passes (from 1 to 3) were ideal for such studies and also provided useful results. Considering the practical constraints and available cutting parameters on lathe, final experimentation was designed on low C steels (0.05–0.15% carbon) which is commonly used in making beams, piston rods, vehicle parts and in making structural shapes for bridges, buildings etc. At each experimental run, depth of cut (around 10–20 microns) was given. In this study, surface roughness and micro hardness were response variables. Table 2 illustrates process parameters and their range.

Table 1 Recent findings and application in burnishing

Important findings	Reported by
Ball burnishing was discussed as a promising method in initiating stress induced phase transformation from austenite to martensite on Nitinol (Ni 50.8, Ti 49.2; a shape memory alloy)	Fu et al. [8]
Analytical modeling studies were reported using ball burnishing, which could predict surface roughness of thermally sprayed coatings through post treatments	Hiegemann et al. [9]
Hardened and highly polished steel rollers were brought into pressure contact using softer work piece. As pressure exceeded the yield point of work material, surfaces were plastically deformed through cold forming	Murthy and Kotiveerachary [10], Hassan [11]
Roller burnishing, as a cold forming process, was used on previously machined or ground external and internal surfaces. It removed minute surface irregularities produced during machining or shearing operations and proved to be a superior process, yielding better results than grinding	Korzynskia and Pacanab [12], El Tayeb et al. [13]
In studies on roller burnishing of En-8 specimens, applying DOE, micro hardness increased, without appreciable change in its surface micro structures	Walia and Kumar [14]
A study was reported on enhancing the tribological properties of TaC material at elevated temperature (≈2501 °C); through laser surface texturing and burnished WS_2 additions, remarkable increasing the wear life	Jussi et al. [15]
In cryogenic burnishing of Co–Cr–Mo alloys, through proper selection of burnishing conditions, improved surface integrity, refinement in micro structures and high-hardness helped in getting favorable structures	Yang et al. [16]
Comprehensive analysis of surface integrity in ball and roller burnishing were extensively studied and reported in aluminim and brass with different compositions	Khaberry and Axir [17]
Ball grinding, ball burnishing and ball polishing were sequentially used and roughness improvement was reported from 1.83 to 0.035 mm on STAVAX plastic mold SS (hardness ≈ 50 H_{Rc})	Shiou and Chang [18]

(continued)

Table 1 (continued)

Important findings	Reported by
Dynamic burnishing process increased the polishing effect on machined surfaces by an average of 17% using the non contact type imagining and analysis LASER methods on experimented INCONEL pieces	Kapłonek and Nadolny [19]
In studies on optimization of surface roughness and micro hardness of AZ91D Mg alloys through ball burnishing; the best surface roughness of 0.336 μm and best micro hardness of 102.7 H_V was achieved	Buldum and Cagan [20]
The studies have presented an input factor-based optimization for simultaneously enhancing power factor (PFB) considering the speed, feed and penetration depth as its main factors, while burnishing SKD61 steels	Nguyen et al. [21]
An analytical model based on contact mechanic theorem was developed and proposed; in this, surface plastic strain increased by an increase in static force and vibration amplitude; while it decreased with an increase in ball radius and feed rate	Reza Amini [22]
The 3-D explicit nonlinear finite element analysis model (FEM) of the ultrasonic assisted burnishing process was established and calibrated	Liu et al. [23]

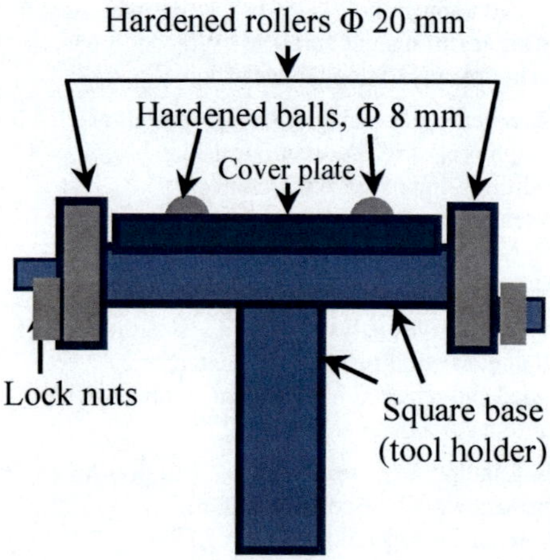

Fig. 2 Schematic sketch of the innovative combined burnishing tool

Table 2 Process parameters used and their range

No	Process parameter	Range/levels	Unit
1	Spindle speed	325, 420, 550	RPM
2	Feed rate	0.1, 0.2, 0.4	mm/rev
3	Passes	1, 2, 3	Numbers

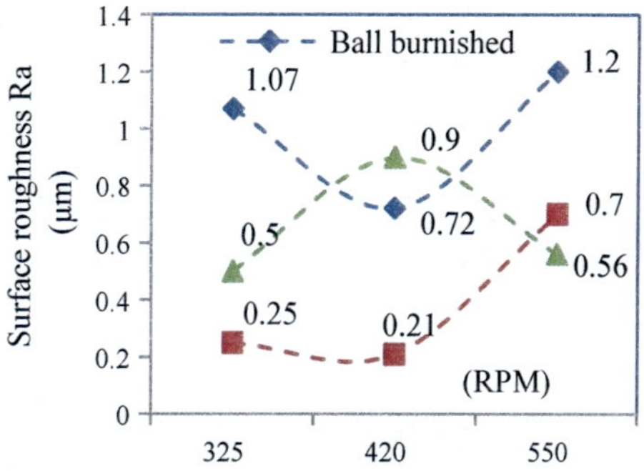

Fig. 3 Effect of speed on R_a

2.1 *Effect of Cutting Speed and Feed on Surface Roughness*

In surface roughness studies, experiments were conducted using ball, roller and combined burnishing tool on low-carbon steel specimens to analyze the effect of variation in spindle speeds and feeds. Two readings were taken at each levels and its average was taken. The variation of surface roughness with spindle speed is shown in Fig. 3. Optimum value of surface finish was found to be 0.21 ($R_a$in μm, in case of roller burnishing at second parametric level, 420 RPM). In case of ball burnishing, optimum value of surface finish (0.72, $R_a$in μm) was found at second level (420 RPM) and in case of combined burnished tool, optimum surface roughness value (0.5, $R_a$in μm) was found at first speed level (325 RPM). Roller burnishing process provided optimum surface roughness values (0.21, $R_a$in μm), owing to better line contact and regularization of deformed peaks into adjacent valleys. Variations in feed rate also affected surface roughness (Fig. 4). Surface finish improved with an increase in feed rate for ball and roller burnishing. In feed variation studies, optimum surface finish was achieved in roller burnishing ($R_a = 0.25$ μm).

2.2 *Effect of Cutting Speed and Feed on Surface Hardness*

In surface hardness studies; ball, roller and combined burnishing tools were used on low-carbon steel to analyze effects of variation in spindle speeds and feeds. Two readings were taken at each levels and its average was taken. Initial surface hardness before burnishing was in range of 50–55 $H_{Rc.}$ Increase in plastic deformation due to

Fig. 4 Effect of feed on R_a

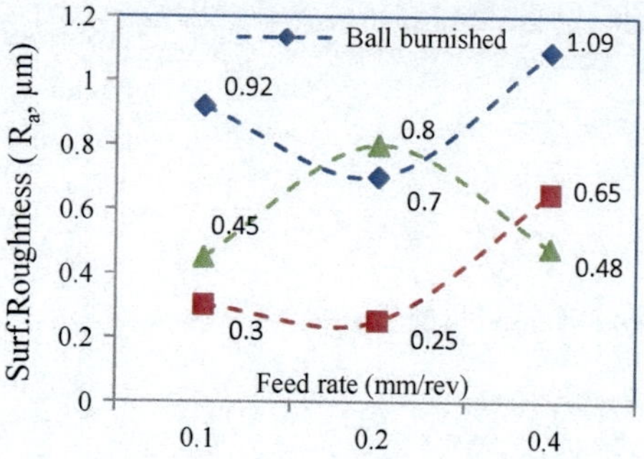

increased surface contact from roller tools resulted in more work hardening effect (compared to ball tool) thereby increasing surface hardness. As seen in Fig. 5, hardness decreased with an increase in spindle speed for ball and combined burnishing tools. In roller burnishing, hardness decreased from 63 to 61 H_{Rc} from 325 to 420 RPM value and then marginally increased upto 66 H_{RC} at 550 RPM (Fig. 6).

The best results were obtained with a feed rate of 0.1 mm/rev in case of combined and ball burnishing tools. Roller burnishing tools gave an increasing trend with respect to the increase in feed rate and the reverse was observed in case of ball and combined burnishing process. The combined burnishing tool gave best performance in surface micro hardness (77 H_{Rc}), since there are both the elements (roller and ball tools) which move subsequently and compress the material evenly.

Fig. 5 Effect of speed on H_{RC}

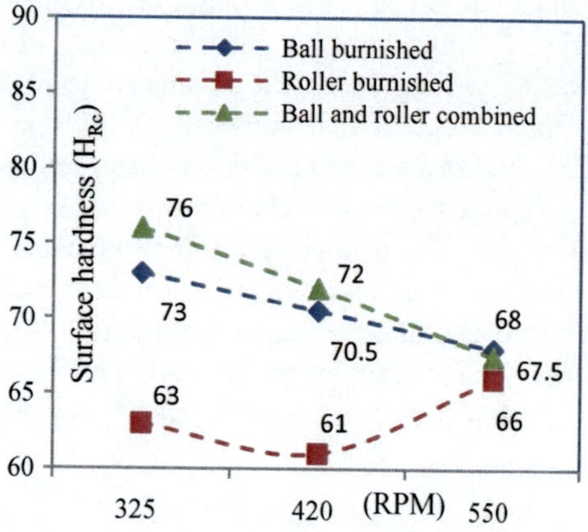

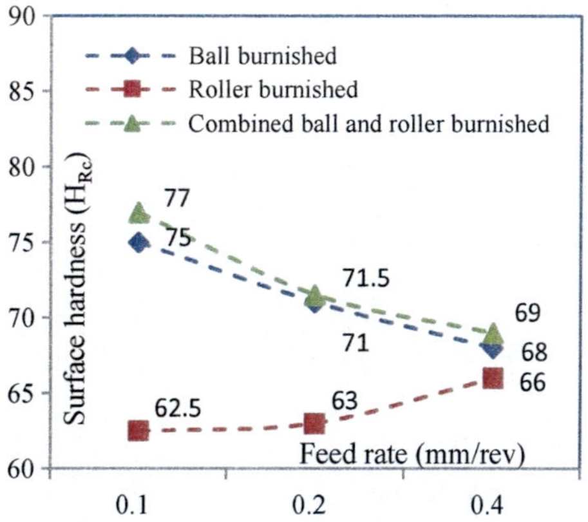

Fig. 6 Effect of feed on H_{RC}

3 Parametric Studies on Surface Roughness Using Design of Experiments

In surface roughness measurements, for precision engineering applications, fine surface finishes are desired. In order to investigate the effect of surface roughness using combined roller and ball burnishing tool, further experiments were designed and conducted on low-carbon steel material. The response characteristics were surface finish and surface micro hardness. Taguchi parametric design methodology was adopted in this phase and the experiments were conducted using the standard orthogonal array L_9 [24]. The results are indicated in Table 3 and the results of ANOVA studies is shown in Table 4, which indicates that feed rate had the highest contribution (51.94%), followed by number of passes (31.67%) and speed (2.34%).

Table 3 Results on roughness (R_a) studies

Trial No.	Parameter trail condition				S/N ratio dB
	(A)	(B)	(C)	(D)	
	Speed RPM	Feed mm/rev	Pass	Mean	
1	325	0.1	1	0.46	6.72
2	325	0.2	2	0.85	14.10
3	325	0.4	3	1.41	29.93
4	420	0.1	2	0.99	0.08
5	420	0.2	3	0.89	9.75
6	420	0.4	1	0.9	9.08
7	550	0.1	3	0.28	10.83
8	550	0.2	1	0.32	9.86
9	550	0.4	2	0.42	7.48

Table 4 ANOVA studies in R_a

Source	SS	DOF	*V*	*P* (%)	*F*-Ratio
Speed	0.052	2	0.0258	2.335	0.166
Feed	1.147	2	0.573	51.935	3.693
Passes	0.699	2	0.349	31.669	2.252
Error	0.311	2	0.155	14.060	
T	2.208	8		100	

SS Sum of squares, *DOF* Degree of freedom, *V* Variance, *P* Percentage effect

In low-carbon steels, it was seen that at 550 RPM (Fig. 7) the surface roughness value was the best (= 0.3 μR_a). At intermediate speed (420 RPM), the value of surface roughness became slightly higher ($\approx$ 0.92 μR_a); a possible reason for this could be that once the higher peaks get cut-off, a group of smaller peaks adjacent to one another could generate some deformities, these peaks get can further get lowered through subsequent burnishing action. As the feed rate increased from 0.2 to 0.4 (mm/rev), the surface roughness increased (from 0.6 to 0.9 μR_a). An almost linear tread was evidenced, the likely reason behind this phenomenon was due to the continuous rubbing action (compression), the peaks deformed plastically and the surface gradually got leveled. As the number of passes increased, initially the surface roughness reduced (due to material flow into valleys); with further increase in passes (three passes), the surface became relatively coarse. Increases in passes from two

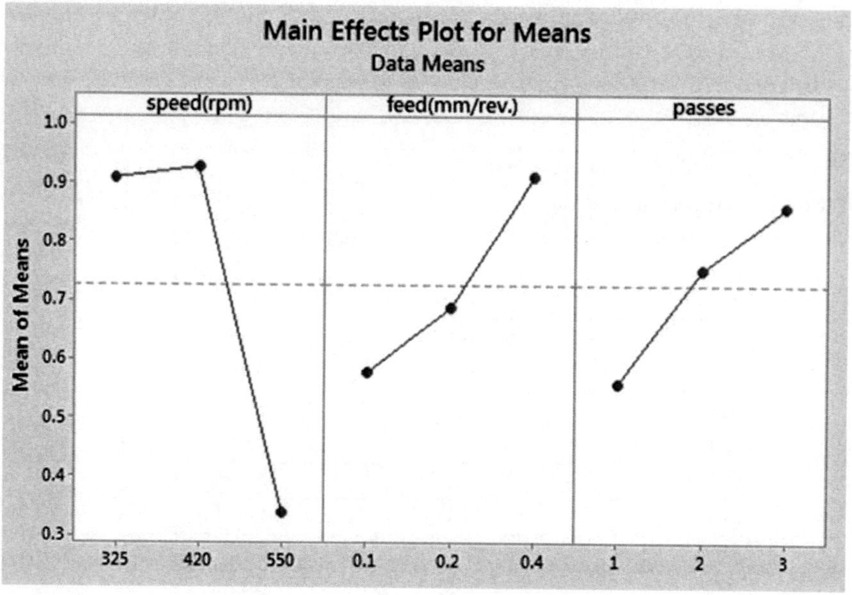

Fig. 7 Main effects on surface

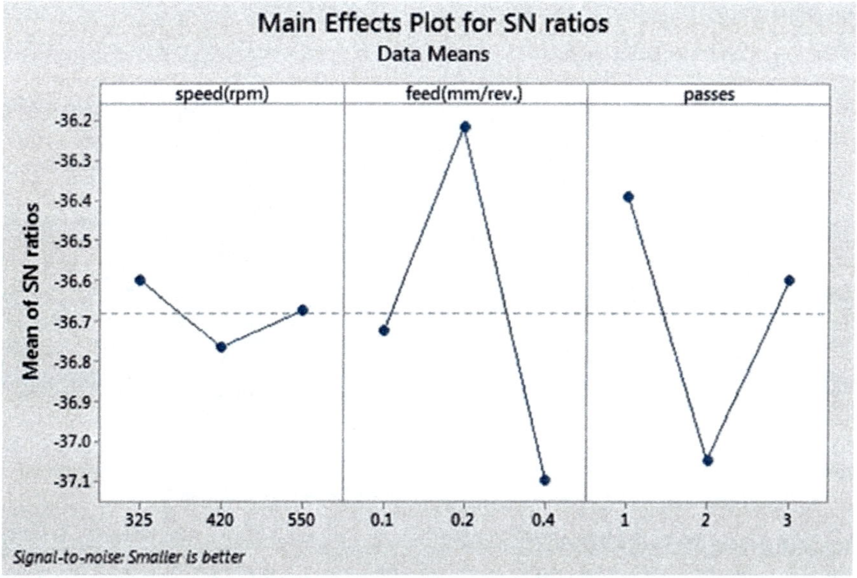

Fig. 8 Main effect plot for S/N ratio

to three or more can cause more compression, resulting in more plastic flow of the material and sometimes it may also cause folding, thereby making the surface rough. The graphs of signal to noise (S/N) ratio are illustrated in Fig. 8.

3.1 Parametric Study on Surface Micro Hardness Using Design of Experiments

Using the same procedure as illustrated for surface roughness, a detailed study using Taguchi's orthogonal array L_9 was conducted for micro hardness as it is illustrated in Tables 5 and 6 illustrates the results of ANOVA studies, which shows that spindle speed had the highest contribution (34.1%) in surface hardness, followed by number of passes 31.06% and feed rate as 26.59%. In case of micro hardness studies, in low-carbon steel specimens; the optimum hardness value ($\approx 69\ H_{Rc}$) was seen using spindle speed of 420 RPM. As the feed rate increased, the micro hardness first slightly decreased and it then increased ($\approx 71.5\ H_{Rc}$) at the third level (feed = 0.4 mm/rev). The phenomenon can primarily be attributed to the work hardening effect, which makes the surface hard due to atomic level entanglements. The continuous subsequent pressure of the ball and roller on job increased the surface hardness. As the number of passes increased, the micro hardness first increased and then decreased (Fig. 9); the optimum value was found to be $\approx 71\ H_{Rc}$. Figures 9 and 10 illustrate the said phenomenon along with corresponding signal to noise ratios.

Table 5 Results on hardness studies

Trial no.	Parameter trail condition				S/N ratio (dB)
	(A)	(B)	(C)	(D)	
	Speed RPM	Feed mm/r	Passes	Mean	
1	325	0.1	1	66	34.577
2	325	0.2	2	65	34.489
3	325	0.4	3	72	35.364
4	420	0.1	2	74	35.619
5	420	0.2	3	65	34.491
6	420	0.4	1	68	34.876
7	550	0.1	3	66	34.616
8	550	0.2	1	64	34.356
9	550	0.4	2	75	35.712

Table 6 ANOVA studies in hardness

Source	SS	DOF	*V*	*P* (%)	*F*-Ratio
Speed	178.82	2	89.410	34.104	4.137
Feed	139.42	2	69.711	26.590	3.226
Passes	162.87	2	81.436	31.063	3.768
Error	43.22	2	21.611	8.243	
T	1063.4	8		100	

SS Sum of squares, *DOF* degree of freedom, *V* Variance, *P* Percentage effect

3.2 Micro Structural Studies

In order to gain some insights into the surface micro structure, some studies were carried out through material testing of the burnished surfaces. The micro structural results obtained through these tests are shown in Fig. 11. The investigated surfaces were seen to be uniformly dispersed and they contained fine-equiaxed grains of pearlite and ferrite. Here were no segregations (Test reference- IS: 7739 (PART V)—1976 (RA 1996); grain size number was 8) without any noticeable changes in surface micro-structures.

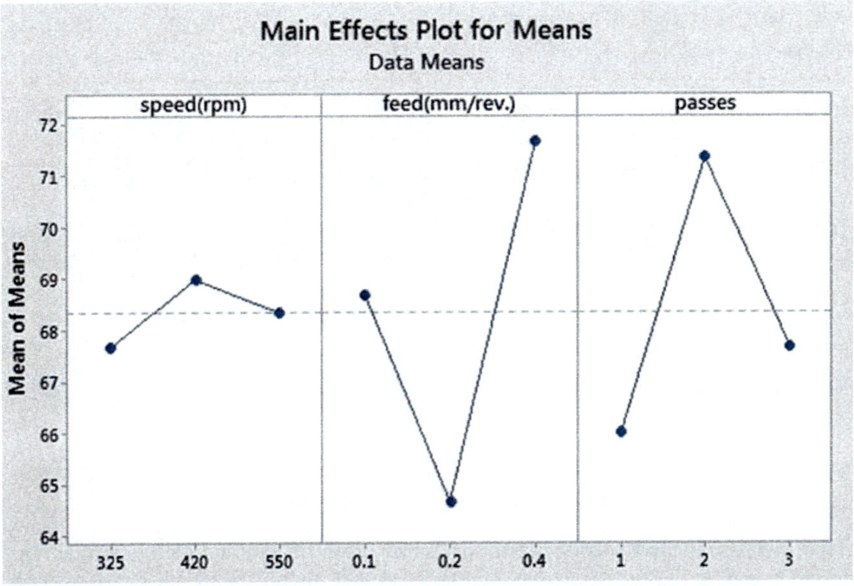

Fig. 9 Main effects on μ-hardness

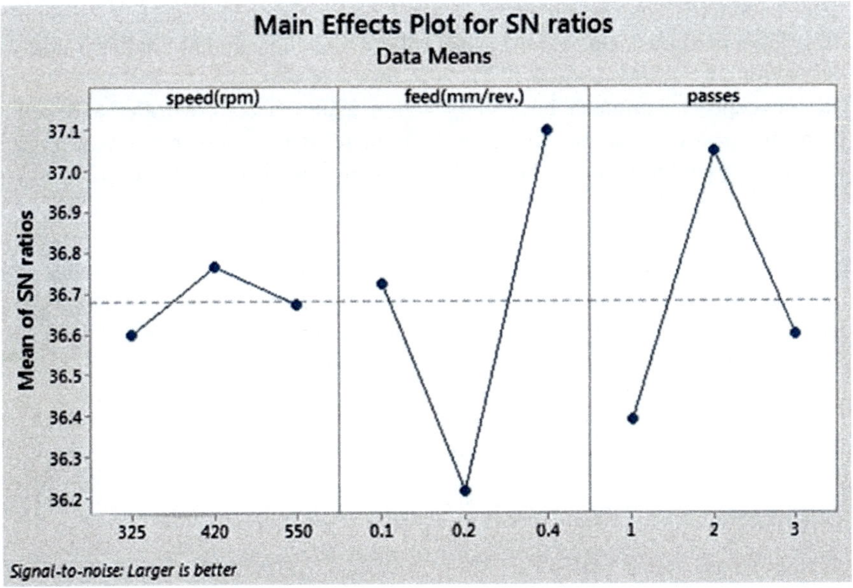

Fig. 10 Main effect plot for S/N ratio

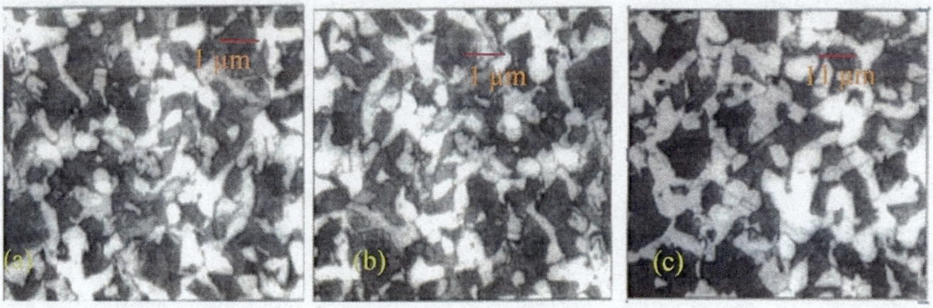

Fig. 11 Micro structures of **a** Ball, **b** Roller and **c** Combined burnished (400 X)

4 Conclusions

Through the conducted experiments on low-carbon steel specimens, the following key contributions in the form of conclusions have be drawn:

- The novel-innovated burnishing tool (combination of ball and roller burnishing tool) was an economic and sustainable approach; being a chip-less (non waste producing) process, it increased tool interchangeability provisions and provided improved results.
- In the conducted experiments on low C steel specimens, the best surface finish of 0.21 (R_a in μm) was achieved in case of roller burnishing, at the spindle speed of 420 RPM.
- The concept of combined burnishing action using successive roller and ball burnishing tools in a single set-up led to an increase in surface micro hardness, which was found to be maximum (value = 76 H_{Rc} at spindle speed of 325 rpm) in comparison with individually used tools (ball, roller and combination).
- As illustrated through experimental studies, surface finish obtained through combined tool had a marginal edge of improvement over ball burnished tools.
- The spindle speed was found to be the most significant variable (among speed, feed and passes) affecting the surface roughness and micro hardness.
- Surface micro structural studies have illustrated valuable insights into texture and integrity, thereby opening new avenues in the specific field of research.

References

1. Deepak, M., & Ravindra, T. (2013). A review on ball burnishing process. *Journal of Scientific and Research Publications, 3*(4), 1–2.
2. Pavan, K., & Purohit, G. K. (2013). Design and development of ball burnishing tool. *International Journal of Engineering Research and Technology, 6*(2), 733–738.
3. Luo, L., Lijiyang, W., Jianying, L., & Qunpeng, Z. (2005). Investigations of the burnishing process with PCD tool on non-ferrous metals. *International Journal of Advanced Manufacturing Technology, 25*, 454–459.

4. Jawalkar, C. S., & Walia, R. S. (2009). Study of Roller Burnishing process on En-8 specimens using Design of Experiments. *Journal of Mechanical Engineering and Research, 1*(1), 39–46.
5. Shneider, Yu. G. (1967). Characteristics of the burnished components. *Mech. Tooling Journal, 38*, 19–22.
6. Degarmo Paul, E., Black, J., & Kohser (2005). *Materials and processes in engineering*, 8th ed. (pp. 1096–1112). Prentice Hall of India.
7. Loh, N. H., Tam, S. C., & Miyazawa, S. (1991). Investigations on the surface roughness produced by ball burnishing. *International Journal of Machine Tools Manufacturing, 31*(1), 75–81.
8. Fu, C. H., Sealya, M. P., Guo, Y. B., & Weib, X. T. (2014). Austenite–martensite phase transformation of biomedical Nitinol by ball burnishing. *Journal of Materials Processing Technology, 214*, 3122–3130.
9. Hiegemann, L., Weddeling, C., Ben, K. N., & Tekkaya, A. E. (2015). Prediction of roughness after ball burnishing of thermally coated surfaces. *Journal of Materials Processing Technology, 217*, 193–201.
10. Murthy, R. L., & Kotiveerachary, B. (1981). Burnishing of metallic surfaces—A review. *Precision Engineering Journal, 3*, 172–179.
11. Hassan, A. M. (1997). The effects of ball and roller burnishing on the surface roughness and hardness of some nonferrous metals. *Journal of Materials Processing Technology, 72*, 385–391.
12. Mieczyslaw, K., & Andrzej, P. (2010). Center- less burnishing and influence of its parameters on machining effects. *Journal of Materials Processing Technology*, 1217–23.
13. El-Tayeb, N. S. M., Low, K. O., & Brevern, P. V. (2009). On the surface and tribological characteristics of burnished cylindrical Al-6061. *Tribology International, 42*, 320–326.
14. Walia, R. S., & Sandeep, K. (2007). *Optimization of surface finishing conditions in roller burnishing*. Master's Thesis, Punjab Engineering College, Chandigarh.
15. Oksanen, J., Timo, H. J., Tervakangas, S., Laakso, P., Kilpi, L., Ronkainen, H., & Koskinen, J. (2014). Tribological properties of Ta-C at elevated temperature. *Tribology International, 70*, 94–103.
16. Shu, Y., Dillon, O., Puleo, D., & Jawahir, I. S. (2013). Effect of cryogenic burnishing on surface integrity modifications of Co-Cr-Mo biomedical alloy. *Journal of Biomedical Materials Research Part B Applied Biomaterials, 101B*(1). https://doi.org/10.1002/jbm.b.32827
17. El-Khabeery, M. M., & El-Axir, M. H. (2003). Experimental techniques for studying the effects of milling roller burnishing parameters on surface integrity. *International Journal of Machine Tools and Manufacture, 42*, 1705–1719.
18. Shiou, F. J., & Chang, C. H. (2008). Ultra precision surface finish of NAK 80 mould tool steel using sequential ball burnishing and ball polishing. *Journal of Materials Processing Technology, 205*(1–3, 26), 249–258.
19. Wojciech, K., & Krzysztof, N. (2016), Laser method based on imaging and analysis of scattered light used for assessment of cylindrical surfaces after dynamic burnishing process. *International Journal of Surface science and Engineering. 10*(1), 55.
20. Buldum, B., & Cagan, S. (2018). Study of ball burnishing process on the surface roughness and microhardness of AZ91D alloy. *Experimental Techniques, 42*, 233–241.
21. Trung-Thanh, N., Cao, L.-H., Dang, X.-P., Nguyen, T.-A., & Trinh, Q.-H. (2019). Multi-objective optimization of the flat burnishing process for energy efficiency and surface characteristics. *Materials and Manufacturing Processes, 34*(16), 1888–1901.
22. Amini, R. T. S. (2019). Analytical modeling of ultrasonic surface burnishing process: Evaluation of through depth localized strain. *International Journal of Mechanical Sciences, 151*, 118–132.
23. Liu, Z., Yang, M., & Deng, J. (2020). A predictive approach to investigating effects of ultrasonic-assisted burnishing process on surface performances of shaft. *International Journal of Advanced Manufacturing Technology, 106*, 4203–4219.
24. Walpole Ronald. (2005). Statistics and probability for engineers and scientists. Pearson Publications.

Predictive Soft Modeling of Turning Parameters Using Artificial Neural Network

N. Gupta and R. S. Walia

1 Introduction

Difficult to machine materials as titanium and its alloys and glass-reinforced composites are widely being used in various industries as aerospace, nuclear, and other sectors. Their high melting point, electrochemical capability [1], favors the usage. Titanium-based alloys are very hard having very high melting point in range of 3000–4000 °C. Similarly, many hard composites have been developed for commercial usage as glass fiber-reinforced plastics [2]. Their machining too is a challenging task. Major difficulties encountered are enhanced surface damage, rapid tool wear; bouncing back phenomenon [3]. Previously, grinding was the only available manufacturing process to machine them; however, many alternates are available today. Their machining too is very difficult. Hard turning has emerged as a potential machining process of such hard materials due to its versatility as efficiency, less power consumption, maintenance of complex shapes, and tolerances. Surface integrity is one of desirable outcome of any hard turning process. It is indication of surface finish.

The surface integrity is often characterized by surface roughness values. Prediction of surface roughness values is an demanding area. The prediction of surface roughness values helps in selecting optimum process parameters as well as improvement in process capability.

Conventional optimization techniques such as Taguchi-ANOVA, MANOVA, response surface methodology, full factorial, partial factorial, partial factorial, and CCD have been widely employed by researchers in optimizing surface roughness values.

However, nowadays, researchers are often employing Newer and unconventional modeling techniques based on machine learning as artificial neural modeling,

N. Gupta (✉) · R. S. Walia
Mechanical Engineering Department, DTU, Delhi, India

R. Agrawal et al. (eds.), *Recent Advances in Smart Manufacturing and Materials*, Lecture Notes in Mechanical Engineering,
https://doi.org/10.1007/978-981-16-3033-0_17

particle swarm optimization, ant colony optimization, and bootstrap are some of soft modeling techniques used in prediction of outcomes.

In this analytical study, ANN modeling in optimizing the surface roughness has been done.

2 ANN Technique in Optimization of Output Responses in Turning

In research [4], artificial neural network-based feed-forward model was used for predicting roughness of machined surface. In another research [5], ANN was used in determination of tool wear. In a separate research [1], ANN was used for predictive modeling of surface roughness, force, and temperature. Dave et al. [6] in their research work applied ANN for prediction of MRR. In yet another research work, ANN was used in optimization of turning parameters during turning of polymers [7].

In yet another research, prediction of roughness of machined surface and forces developed during metal cutting was done using ANN [8].

In a research work, ANN predictive modeling was used to model surface roughness alongside using conventional optimization methods [9].

Machinability investigation while turning AISI 52,100 steel was done using ANN in empirical research [10]. In a separate research work, cutting forces were predicted using ANN-based models and traditional methods of regression analysis while the turning of red brass (C23000) was done [11].

In a research work [12], AISI 316 stainless steel was machined with PVD-coated carbide tools. The output responses of Surface roughness were optimized using ANN.

In a separate research work, hard machining of AISI4140 was done using ANN [13]. The statistically determined results were in agreement to the ANN results.

In another experimental study, Inconel 718 was hard turned. The input machining variables selected were metal machining velocity, feed rate, and cutting depth/depth of cut.

The output parameters chosen were roughness of machined surface and cutting force. Among different models selected for modeling, ANN was the most accurate one [14].

3 Experimental Dataset

The dataset for the present analysis has been taken from research work by Gupta and Kumar [3]. In their research work, glass fiber-reinforced plastics rods were taken as raw material. The six different input parameters taken were as shown below in Tables 1 and 2 [2].

Table 1 Input parameters and their levels [2]

Input parameters	Level 1	Level 2	Level 3
Tool nose Radius/mm	0.4	0.8	Nil
Tool rake Angle/°	−6	0	6
Feed rate (mm/rev)	0.05	0.1	0.2
Cutting rate (rpm)	420	840	1210
Cutting Environment	Dry	Wet	Cooled
Depth of cut (/mm)	0.2	0.8	1.4

Table 2 Experiments using Taguchi L18 O.A

S. no.	A	B	C	D	E	F
1	1	1	1	1	1	1
2	1	1	2	2	2	1
3	1	1	3	3	3	3
4	1	2	1	1	2	2
5	1	2	2	2	3	3
6	**1**	**2**	**3**	**3**	**1**	**1**[a]
7	1	3	1	2	1	3
8	1	3	2	3	2	1
9	1	3	3	1	3	2
10	2	1	1	3	3	2
11	2	1	2	1	1	3
12	**2**	**1**	**3**	**2**	**2**	**1**[a]
13	2	2	1	2	3	1
14	2	2	2	3	1	2
15	2	2	3	1	2	3
16	2	3	1	3	2	3
17	2	3	2	1	3	1
18	**2**	**3**	**3**	**2**	**1**	**2**[a]

[a]Sample data for testing

Total readings were divided into training and testing data sets. As per 80:20 rule, out of total 18 data sets, 3 data sets were used for validation and rest 15 for training of ANN network in MATLAB.

The 80:20 rule is well-known rule. It means that 80% of data set will be used to train the ANN feed-forward back-propagation model, while the testing will be done on 20% of data sets.

4 Artificial Neural Network

In this present work, a feed-forward neural network is used. The output parameter to be predicted was roughness of machined surface. The feed forward propagation method is used. It has two hidden layer apart from one input and one output layer. The neural network has six input.

The ANN network selected was comprising two hidden layers, as, one layer structure has considerable errors [15]. MATLAB software was used for training this network, and the ANN was trained with the feed-forward back-propagation algorithm.

The weights of the network connections were decided on the basis of input machining variables and corresponding output machining variable. It is inbuilt determined by the software using the training and testing data on application of regression method.

A code was developed in MATLAB, and the results were obtained. The six inputs were taken, and one output of surface roughness was analyzed. Feed-forward back-propagation model was used. Numbers of epochs selected were 100. 80–20 training to testing rule was followed. No. of hidden layers selected were two, having 10 neurons each.

Levenberg–Marquardt algorithm was used for training of data (Fig. 1).

The selection of epochs was kept at 100 as increasing epochs stopped affecting output after certain value. In the literature, it had been established that increasing no. of epochs will not increase accuracy of system. Often too many epochs resulted in complexity of system. It ultimately leads to slowing of the system.

In Fig. 2, the ANN output graphs from MATLAB were shown. The R value for all training sets is coming as 0.950, and testing was coming as 0.987. The overall R value was coming as 0.959, which signifies a very high similarity between the experimental values of surface roughness and ANN predicted values.

5 Observations

As per Table 3, the ANN predicted values were within $+-0.015\%$ of experimental values. It proved the accuracy of ANN predictive modeling.

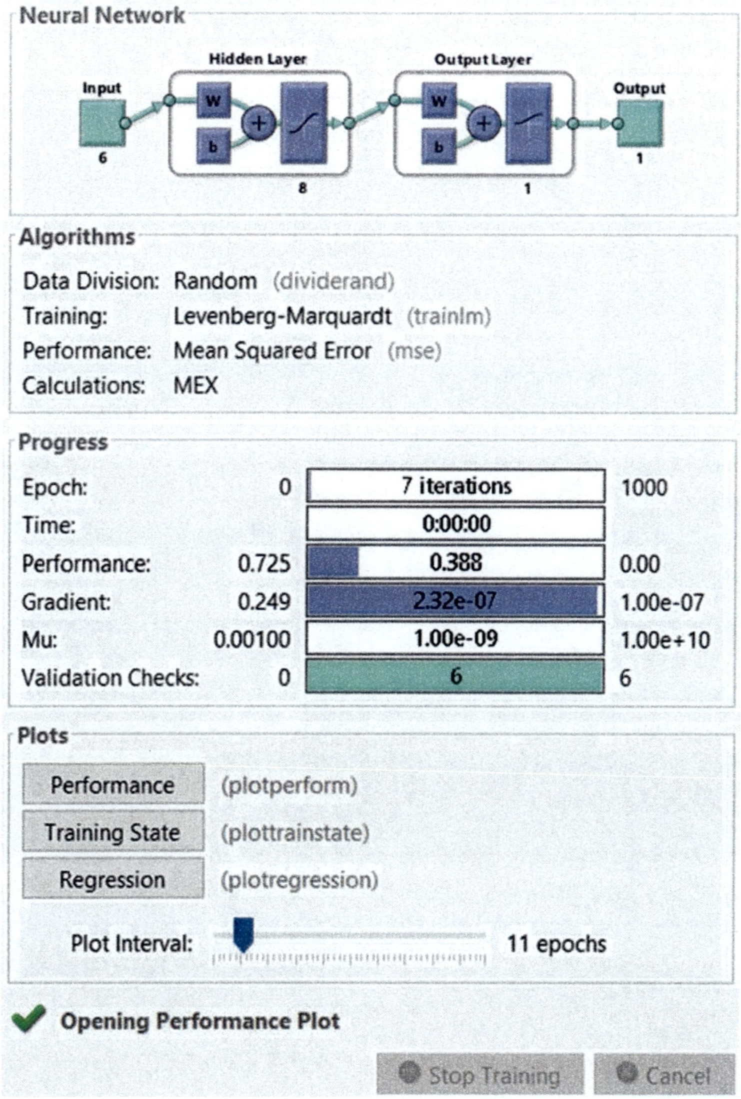

Fig. 1 MATLAB derived ANN architecture

In Table 3, column 2 is experimentally determined values of surface roughness, whereas column 3 gives ANN predicted values of surface roughness. The column 4 gives MSE corresponding to each of the experimental run. As per the table above, the MSE was coming out to be 0.051, which signifies agreement between the experimental results and ANN predicted values.

The MSE is fairly accurate parameter to determine accuracy of neural networks.

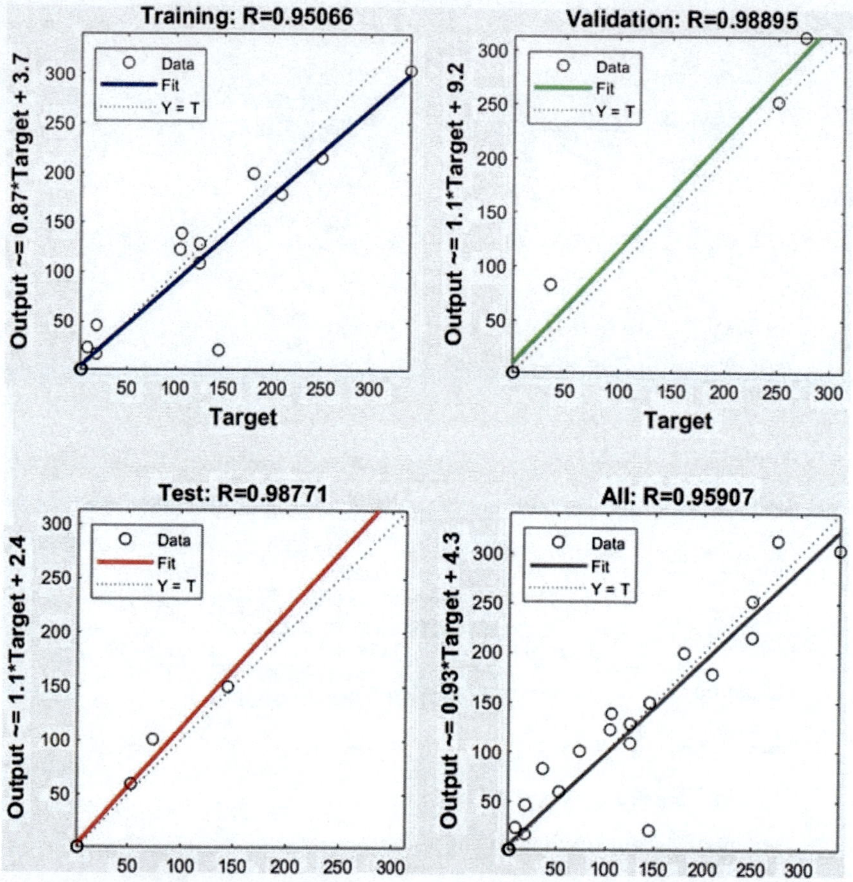

Fig. 2 ANN output graphs

6 Conclusions

The major conclusions are as follows:

1. The foremost conclusion from this research is the utility of predictive modeling before start of actual experimentation processes. Based on previous researches, a database could be maintained to explore possibilities of improvement in processes and machine tool based on predicted outcomes by ANN.
2. Combination of ANN with traditional techniques as Taguchi, etc.
3. ANN can be used as accurate predictive tool in machining and manufacturing activities.

The utility of machine learning in optimization of manufacturing process is a real-time application. The benefit is higher levels of accuracy coupled with reliable results over researches conducted in different countries. People in mechanical and allied

Table 3 Experiments output versus ANN output

S. no.	Exp-SR	ANN-SR	MSError
1	1.397	1.377	0.014316
2	1.453	1.431	0.015141
3	3.076	3.081	−0.00163
4	1.366	1.371	−0.00366
5	1.53	1.541	−0.00719
6*	**2.4**	2.3381	0.025792
7	1.513	1.5473	−0.02267
8	1.636	1.6362	−0.00012
9	2.263	2.26	0.001326
10	1.547	1.532	0.009696
11	1.606	1.611	−0.00311
12*	**1.966**	1.947	0.009664
13	1.597	1.582	0.009393
14	1.63	1.621	0.005521
15	2.237	2.221	0.007152
16	1.94	1.951	−0.00567
17	1.486	1.494	−0.00538
18*	**1.973**	1.968	0.002534
		MSE	0.051101

fields are now more receptive toward unconventional techniques of data optimization. The main benefit lies in predicting the optimal set of input conditions in order to optimize the output responses.

References

1. Gupta, N., Agrawal, A. K., & Walia, R. S. (2019, May). Soft modeling approach in predicting surface roughness, temperature, cutting forces in hard turning process using artificial neural network: An empirical study. In *International Conference on Information, Communication and Computing Technology* (pp. 206–215). Singapore: Springer.
2. Davim, J. P. (2009). *Machining composite materials*. Wiley-ISTE.
3. Gupta, M., & Kumar, S. (2015). Investigation of surface roughness and MRR for turning of UD-GFRP using PCA and Taguchi method. *Engineering Science and Technology, an International Journal, 18*(1), 70–81.
4. Sangwan, K. S., Saxena, S., & Kant, G. (2015). Optimization of machining parameters to minimize surface roughness using integrated ANN-GA approach. *Procedia Cirp, 29*, 305–310.
5. Karabulut, Ş. (2015). Optimization of surface roughness and cutting force during AA7039/Al2O3 metal matrix composites milling using neural networks and Taguchi meth od. *Measurement, 66*, 139–149.
6. Dave, H. K., Desai, K. P., & Raval, H. K. (2008, October). Investigations on prediction of MRR and surface roughness on electro-discharge machine using regression analysis and artificial

neural network programming. In *Proceedings of the World Congress on Engineering and Computer Science* (pp. 123–128)

7. Chabbi, A., Yallese, M. A., Nouioua, M., Meddour, I., Mabrouki, T., & Girardin, F. (2017). Modeling and optimization of turning process parameters during the cutting of polymer (POM C) based on RSM, ANN, and DF methods. *The International Journal of Advanced Manufacturing Technology, 91*(5–8), 2267–2290.
8. Labidi, A., Tebassi, H., Belhadi, S., et al. (2018). Cutting conditions modeling and optimization in hard turning using RSM, ANN and desirability function. *Journal of Failure Analysis and Prevention, 18*, 1017–1033. https://doi.org/10.1007/s11668-018-0501-x
9. Mia, M., Khan, M. A., & Dhar, N. R. (2017). *International Journal of Advanced Manufacturing Technology, 93*, 975. https://doi.org/10.1007/s00170-017-0566-9
10. Panda, A., Sahoo, A. K., Panigrahi, I., & Rout, A. K. (2018). Investigating machinability in hard turning of AISI 52100 bearing steel through performance measurement: QR, ANN and GRA Study. *International Journal of Automotive and Mechanical Engineering, 15*, 4935–4961.
11. Hanief, M., Wani, M. F., & Charoo, M. S. (2017). Modeling and prediction of cutting forces during the turning of red brass (C23000) using ANN and regression analysis. *Engineering Science and Technology, An International Journal, 20*(3), 1220–1226.
12. Rao, K. V., & Murthy, P. B. G. S. N. (2018). Modeling and optimization of tool vibration and surface roughness in boring of steel using RSM, ANN and SVM. *Journal of Intelligent Manufacturing, 29*(7), 1533–1543.
13. Meddour, I., Yallese, M. A., Bensouilah, H., Khellaf, A., & Elbah, M. (2018). Prediction of surface roughness and cutting forces using RSM, ANN, and NSGA-II in finish turning of AISI 4140 hardened steel with mixed ceramic tool. *The International Journal of Advanced Manufacturing Technology, 97*(5–8), 1931–1949.
14. Subhash, N., Sambedana, S., Raj, P. N., & Jagadeesha, T. (2019). Experimental study on tool wear and optimization of process parameters using ANN-GA in turning of super-duplex stainless steel under dry and wet conditions. In *Advances in Manufacturing Technology* (pp. 411–420). Singapore: Springer.
15. Das, B., Roy, S., Rai, R. N., & Saha, S. C. (2015). Studies on effect of cutting parameters on surface roughness of al-cu-TiC MMCs: An artificial neural network approach. *Procedia Computer Science, 45*, 745–752.

Welding Performance of Dissimilar AZ91 and AZ31 Mg Alloys Using Developed Friction Stir Welding Set-Up

Umesh Kumar Singh and Avanish Kumar Dubey

1 Introduction

The lightweight materials reduce the amount of energy consumption and carbon emission by automobiles which will help in current exigency of energy and sustainable development of automotive industries. Magnesium (Mg) alloys are structural materials which have lightweight, high strength and good vibration damping capacity [1, 2] and fulfil the requirement of automobile industries. In automobile industries, many times the fabricated components require their joining to fulfil their functionality. But weld characteristics of fusion welded Mg alloys are not very good due to the presence of pores and cracks in welded zone [3]. These defects are formed due to the presence of very active chemical elements in their compositions [4] and restrict their application in automobile industries. These difficulties can be removed by the applications of solid-state joining processes.

Friction stir welding (FSW) is a solid-state joining process developed in the year 1991 by The Welding Institute, UK [5, 6]. In FSW, heat utilised for welding the materials is generated by friction caused by high pressure, rotation and translation of the tool relative to the work material and also plastic deformation of work material. The main part of the tool is pin which is rotating and flow the plasticised material from front side of pin to back side of pin along the weld line and causes turbulence, and after cooling, the material is solidified and formed joint [1, 6].

Its initial application was limited to aluminium (Al) alloys but researchers have extended its application to other similar [1, 3] and dissimilar alloys such as Al-Al [7], Mg-Mg [8], Al-Mg [9], steel [10], Al-cupper [11] and Al-titanium [12]. Researchers have utilised forfeited milling machine for FSW but Singh and Dubey [13] have used radial drilling machine for dissimilar FSW of AZ91-AZ31 alloys. They observed the joint characteristics and found 85.09% joint efficiency at 850 rpm

U. K. Singh (✉) · A. K. Dubey
Mechanical Engineering Department, MNNIT Allahabad, Prayagraj, UP 211004, India

R. Agrawal et al. (eds.), *Recent Advances in Smart Manufacturing and Materials*, Lecture Notes in Mechanical Engineering,
https://doi.org/10.1007/978-981-16-3033-0_18

rotational speed (RotS) and 55 mm/min welding speed (WeldS). Microstructure of welded joint shows mainly two types of grains and joint efficiency decreases with increasing heat input. Sunil et al. [8] investigated the impact of process parameters on crack formation in FSW of AZ91-AZ31 alloys. They obtained a sound joint at 1400 rpm RotS and 25 mm/min WeldS with joint strength of 183 $\pm$ 15 MPa. The nugget zone contains finer grains and distributed $Al_{12}Mg_{17}$ intermetallic compound. Zhang et al. [14] analysed the impact of two different RotS on microstructure, textural variation and mechanical properties of AM60-AZ31 alloys by FSW. They found relatively larger grains at higher RotS as compared to lower RotS but RotS did not have any obvious effects on textural variation. Larger grains support for basal slip activation and twinning in the region away from the interface of nugget and thermo-mechanically affected zone.

Available literature survey suggests that the Mg alloys are advance as well as superior quality materials for automobile industries but there is a need for optimising the process to obtain the better joint property. In current research work, authors have tried to investigate the joint strength of dissimilar (AZ91-AZ31) Mg alloys generated by using developed FSW set-up on drilling (radial) machine. Microstructures just below the shoulder, i.e. in crown zone of welded joint, have also been observed.

2 Materials and Methods

2.1 *Experimental Method*

In this research work, two different compositions of AZ series Mg alloys (AZ91 and AZ31) have been chosen for joining. The compositions of samples are listed in Table 1. The 110 mm length, 55 mm wide and 5 mm thick plates have been taken for experimentation. The tool having 18 mm shoulder diameter and 5 mm cylindrical pin diameter was made of H13 tool steel. AZ91 and AZ31 materials workpieces were clamped on advancing side (AS) and retreating side (RS), respectively. The tool RotS and WeldS were fixed at 580 rpm and 55 mm/min, respectively. For producing the joint, a developed FSW set-up on drilling (radial) machine [13] was utilised. The working set-up with tool is shown in Fig. 1.

Table 1 Compositions of used parent metals [13]

Alloy(s)	Mg	Al	Zn	Mn	Si (max.)	Fe (max.)	Cu (max.)	Ni (max.)	Other(s) (max.)
AZ91	Bal	8.5–9.5	0.5–0.9	0.17–0.4	0.05	0.005	0.03	0.002	0.02
AZ31	Bal	2.5–3.5	0.6–1.4	0.2 min.	0.1	0.005	0.05	0.005	0.30

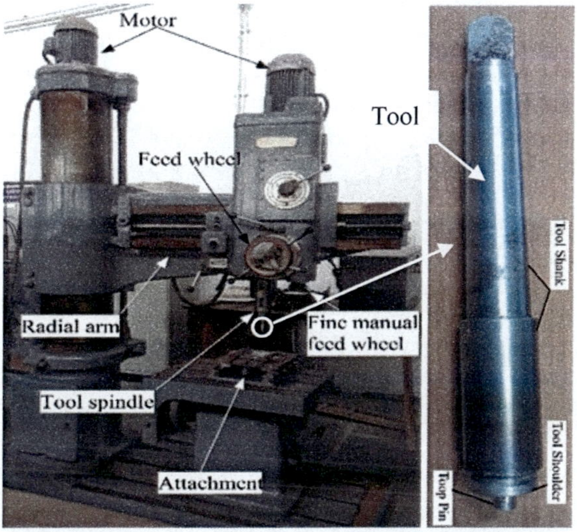

Fig. 1 Working set-up with tool

2.2 Testing Method

Samples for tensile test (TT), microhardness (MH) and microstructural observations were cut using wire cut-EDM. All samples were cut across the weld line. Milling operation has been used for providing the smoothness (finishing) on TT samples. TT was carried out at 1 mm/min strain rate using 25 kN capacity BISS, India tensile testing machine. Microstructural samples were polished by following standard set methods and then etched with picral reagent for taking SEM images to observe the joint characteristics. MH was measured in 1 mm regular interval at 1.5 mm depth (from top surface) across the weld line by using 200 g load and 15 s dwell time on RADICAL Vickers hardness tester.

3 Results and Discussion

3.1 Microstructure

Materials in FSW process experiences intense plastic deformation and turbulence in welded zone due to which grains have different features at different locations. The observation of the types of grains formed after FSW process is very important because it decides the joint property. The SEM images shown in Fig. 2 were taken at different locations of AS and RS in crown zone of welded joint. Very fine equiaxed grains were observed in SEM images just below the shoulder as compared to size of grains obtained at some distance away from the shoulder bottom surface as shown in Fig. 2 in marked areas. Since the material which is in direct contact with shoulder

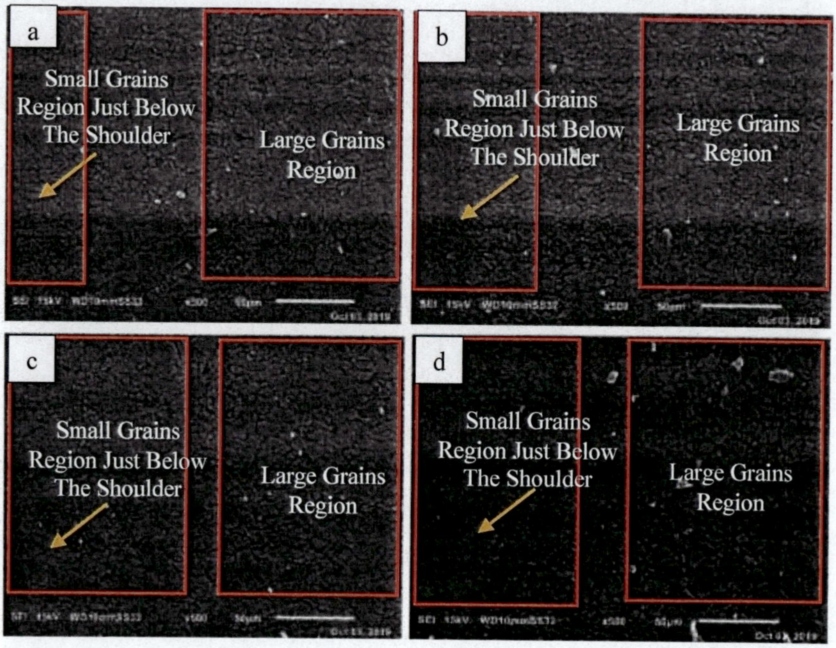

Fig. 2 SEM images just below the shoulder of crown zone, **a**, **b** advancing (AZ91) side; **c**, **d** retreating (AZ31) side

experiences more deformation as compared to the material at some distance away from the shoulder bottom surface, and also heat losses are more because of convective heat transfer from the top surface of welded zone which is at just below the shoulder.

3.2 *Microhardness*

MH of welded sample is shown in Fig. 3. From Fig. 3, it can be seen that the values of MH are fluctuating when measured across the weld line. The average measured value of MH of base materials AZ91 and AZ31 was 71.0 and 65.0 Hv, respectively. The maximum and minimum value of MH 85.50 and 62.03 Hv, respectively, were obtained on retreating (AZ31) side. The fluctuation in MH may be caused by distribution of secondary strengthening phase ($Al_{12}Mg_{17}$) and variation in texture of material in welded zone due to intense deformation (plastic). Also, the metals/alloys generally contain anisotropic hardness.

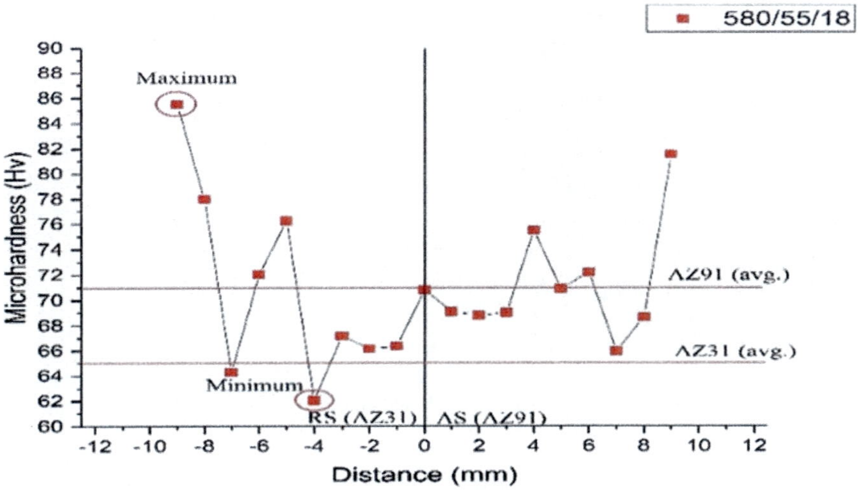

Fig. 3 Microhardness of welded zone across the weld line

3.3 *Tensile Strength*

TT of welded sample has been done for evaluating the efficiency and strength of welded joint as compare to AZ91 base metal, and for observing the fracture behaviour, fractographic image of fractured surface has also been taken as shown in Figs. 4 and 5. Tensile test sample breaks from AS (AZ91) near the weld centre line with very small (negligible) quantity of reduction in cross-sectional area. The maximum strength (tensile) of AZ91 was 246.34 MPa, and AZ31 was 261.79 MPa, respectively, but the joint has 186.74 MPa (75.80% of AZ91) which means reduction in tensile strength. The reduction in tensile strength mainly due to the variation in grain size, variation in texture and distribution of different phases (strengthening). These phenomena

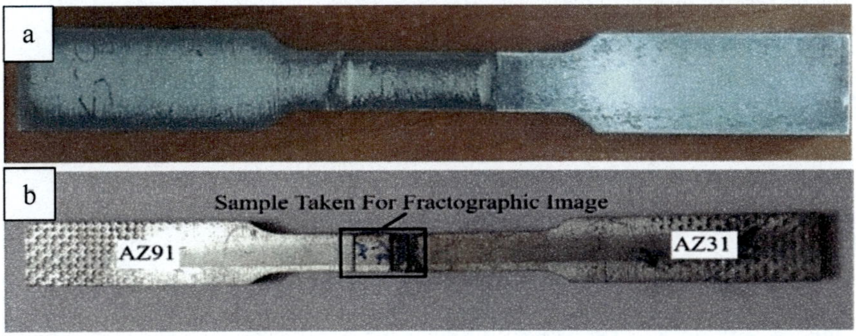

Fig. 4 Tensile test sample, **a** initial (before finishing); **b** after test

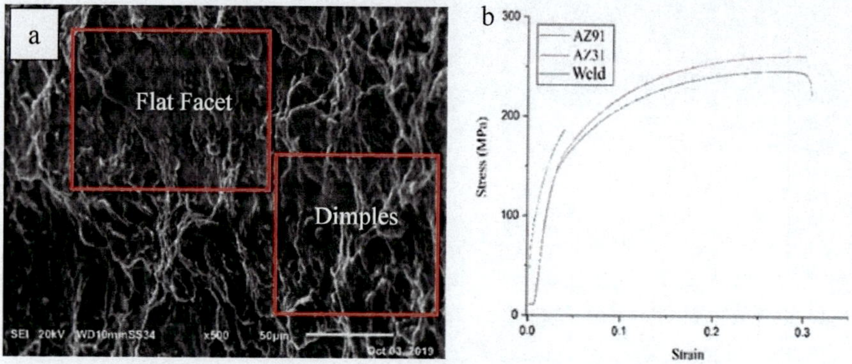

Fig. 5 **a** Fractograph fractured surface; **b** stress–strain graph

are caused by intense deformation (plastic) in welded zone [15]. The fractured surface has flat facet as well as dimple-like structure which indicate mixed type fracture.

4 Conclusions

AZ91 and AZ31 dissimilar magnesium alloys can be welded using developed FSW set-up on drilling (Radial) machine at selected process parameters. From the results of welded joint, following conclusion can be drawn:

- The welded joint is free from defects which are generally presented in fusion welding of magnesium alloys.
- The joint has maximum 85.5 Hv microhardness within the welded zone on retreating side.
- The maximum joint strength obtained is 75.80% of AZ91, and welded sample breaks from AZ91 (i.e. advancing) side. The fracture surface has mixed type fracture behaviour.

Acknowledgements Authors wants to acknowledge UGC New Delhi-110002, India and MNNIT Allahabad, Prayagraj-211004, India through TEQIP-III program for giving research assistantship.

References

1. Suhuddin, U. F. H. R., Mironov, S., Sato, Y. S., Kokawa, H., & Lee, C. W. (2009). Grain structure evolution during friction-stir welding of AZ31 magnesium alloy. *Acta Materialia, 57*, 5406–5418.

2. Asadi, P., Mahdavinejad, R. A., & Tutunchilar, S. (2011). Simulation and experimental investigation of FSP of AZ91 magnesium alloy. *Materials Science and Engineering A, 528*(21), 6469–6477.
3. Park, S. H. C., Sato, Y. S., & Kokawa, H. (2003). Effect of micro-texture on fracture location in friction stir weld of Mg alloy AZ61 during tensile test. *Scripta Materialia, 49*, 161–166.
4. Zhou, M., Morisada, Y., & Fujii, H. (2020). Effect of Ca addition on the microstructure and the mechanical properties of asymmetric double-sided friction stir welded AZ61 magnesium alloy. *Journal of Magnesium and Alloys, 8*, 91–102.
5. Mishra, R. S., & Ma, Z. Y. (2005). Friction stir welding and processing. *Materials Science and Engineering Reports, 50*, 1–78.
6. Fattah-alhosseini, A., & Taheri, A. H. (2015). Effect of friction stir welding on corrosion behavior of pure copper in 3.5 wt.% NaCl solution. *Journal of Manufacturing Processes, 20*, 98–103.
7. Gupta, S. K., Pandey, K. N., & Kumar, R. (2018). Multi-objective optimization of friction stir welding process parameters for joining of dissimilar AA5083/AA6063 aluminum alloys using hybrid approach. *Proceedings of IMechE Part L: Journal of Materials: Design and Applications, 232*(4), 343–353.
8. Sunil, B. R., Reddy, G. P. K., Mounika, A. S. N., Sree, P. N., Pinneswari, P. R., Ambica, I., Babu, R. A., & Amarnadh, P. (2015). Joining of AZ31 and AZ91 Mg alloys by friction stir welding. *Journal of Magnesium and Alloys, 3*, 330–334.
9. Li, P., You, G., Wen, H., Guo, W., Tong, X., & Li, S. (2019). Friction stir welding between the high-pressure die casting of AZ91 magnesium alloy and A383 aluminum alloy. *Journal of Materials Processing Technology, 264*, 55–63.
10. Sharma, G., & Dwivedi, D. K. (2017). Study on microstructure and mechanical properties of dissimilar steel joint developed using friction stir welding. *International Journal of Advanced Manufacturing Technology, 88*, 1299–1307.
11. Patel, N. P., Parlikar, P., Dhari, R. S., Mehta, K., & Pandya, M. (2019). Numerical modelling on cooling assisted friction stir welding of dissimilar Al-Cu joint. *Journal of Manufacturing Processes, 47*, 98–109.
12. Kar, A., Choudhury, S. K., Suwas, S., & Kailas, S. V. (2018). Effect of niobium interlayer in dissimilar friction stir welding of aluminium to titanium. *Materials Characterization, 145*, 402–412.
13. Singh, U. K., & Dubey, A. K. (2020). Study of joining performance of dissimilar Mg alloys in friction stir welding. *Journal of Mechanical Engineering Science.* https://doi.org/10.1177/0954406220959096
14. Zhang, J., Liu, K., Huang, G., Chen, X., Xia, D., Jiang, B., Tang, A., & Pan, F. (2020). Optimizing the mechanical properties of friction stir welded dissimilar joint of AM60 and AZ31 alloys by controlling deformation behaviour. *Materials Science and Engineering A, 773*, 138839.
15. Singh, U. K., & Dubey, A. K. (2021). Study of weld characteristics in friction stir welding of dissimilar Mg-Al-Zn magnesium alloys under varying welding conditions. *Journal of Materials Engineering and Performance.* https://doi.org/10.1007/s11665-021-05893-z

A Note on Preparation of Electroless Nickel Coating on Alumina Micro-particulates as the Forerunner to Reinforce Al-MMCs

D. Vijay Praveen, D. Ranga Raju, M. V. J. Raju, and T. Nancharaiah

1 Introduction

In Al-based MMCs, the addition of hard ceramic particulates into a base alloy, which enhances the mechanical properties, becomes more prevalent in the synthesis and characterization of novel composites [1]. Variation of reinforcement in matrix material, mechanical and thermal properties of the matrix, as well as particulates and degree of micro-structural integrity, will influence the strength of the MMCs [2, 3]. However, there exist some challenges viz. wettability, interfacial reactions and bonding between base material and reinforcements of composites which may lead to the failure of the composites [4]. Metal coating on ceramic particles can improve the wettability and bonding between matrix and reinforcements. In addition to that, the coated surface can minimize the interfacial reaction between the base material and hard reinforcements, which may improve the mechanical properties [5–7]. Several routines are available in surface coating processes like chemical vapor deposition (CVD), physical vapor deposition (PVD), chemical and electro-chemical techniques. Among all, electroless coating route becomes very famous as the deposit is not dependent on current distribution, and it is almost uniform in thickness, regardless of the size or shape of the plated surface. Using this technique, the coating can be done for both metallic and non-metallic particles and surface [8]. This process finds wide applications because of its excellent characteristics like abrasion, corrosion and

D. Vijay Praveen (✉)
Andhra University, Vishakhapatnam 530003, India

D. Ranga Raju
Srinivasa Institute of Engineering and Technology, Amalapuram 533201, India

M. V. J. Raju
College of Engineering (A), Andhra Unviersity, Vishakhapatnam 530003, India

D. Vijay Praveen · T. Nancharaiah
Department of Mechanical Engineering, Bapatla Engineering College, Bapatla 522102, India

R. Agrawal et al. (eds.), *Recent Advances in Smart Manufacturing and Materials*,
Lecture Notes in Mechanical Engineering,
https://doi.org/10.1007/978-981-16-3033-0_19

wear resistance, lubricity, solderability and hardness. This method has gained good recognition nowadays in preparing composite coatings [7–9].

Leon et al. have investigated the nickel plating on micrometer-sized alumina particulates with different in sizes and observed consistent nickel films on the surface of the alumina particles [4]. DS Kumar et al. studied the effect of sensitization and activation under different conditions and analyzed micro-structural characteristics on electroless nickel coating on nano-sized alumina particles [7]. Chintada and Koona used EN route for the preparation of nanocomposite coatings of ZnO particles on mild steel surface further analyzed the effect of composite coating on microhardness and corrosion resistance [10].

Sudhakar et al. have reported fundamental aspects, characteristics and various properties of EN composite coatings. They concluded that the incorporation of EN coated ceramic particles in the various alloy could improve physical and mechanical properties [11]. Vijay et al. have studied hardness and compressive strength of Ni-coated boron carbide (B_4C) particle reinforced 601AC/201AC selective layered functionally graded materials. Specimens were prepared at three different numbers of layers at the standard powder metallurgy route. From the experimental results, they observed the highest compressive strength and hardness in Ni-coated three-layered FGM with 201AC as base material [12].

In light of the above literature, it was observed that most of the authors had explored the optimal procedure and techniques involved in EN coating process on various reinforcements [13–20]. The main objective of the present work is to prepare the nickel plating on Al_2O_3 powder by standard electroless plating procedures and micro-structural studies, which further can be reinforced into various novel alloying elements to study the physical, mechanical and machining properties.

2 Materials and Methodology of Electroless Plating

2.1 Materials

Alumina (α-type) particles of 2–20 μm (avg.) size of 99% purity, supplied by M/s Aarshadhaatu Green Nanotechnologies India Private Limited, Guntur, used in the present work.

2.2 Electroless Coating Process

In electroless coating route, metal ions are reduced to metal by the action of chemical reducing agents. The metal ions are electron acceptors, which react with electron donors. This electroless coating method is an autocatalytic process, which accelerates

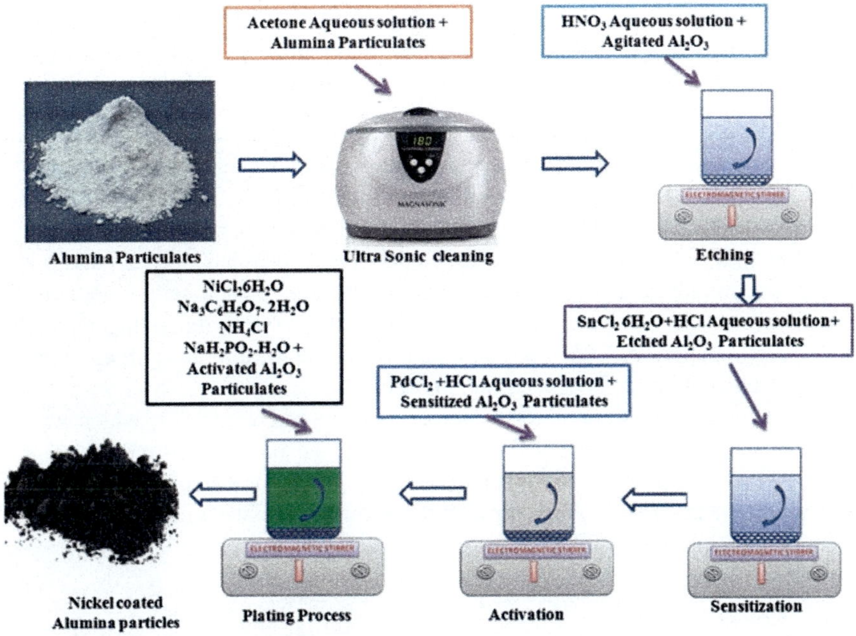

Fig. 1 Basic steps in nickel plating

the electroless chemical reaction allowing oxidation of the reducing agent used. The basic steps involved in the standard coating process are shown in Fig. 1.

Ultrasonic Cleaning A weighed quantity of 40 g of alumina particles was cleaned by mixing with distilled water with 20 ml of acetone for a period of 20 min ultrasonicator [7]. In this process, volatile compounds like oil and dust particles are eliminated, which may present on the surface of the alumina particles. Further, the particles are to be washed with distilled water followed by drying. Now, the surface of the particles is very smooth and clean.

Etching Upon ultrasonic cleaning, the etching process is to be carried out, for obtaining microscopic roughing on the alumina particles surface. 10 ml/l of HNO_3 is to be taken along with one liter distilled water in a beaker, and stirring is to be done for a period 15 min using an electromagnetic stirrer. Afterward, particles are washed with distilled water for 5 min and then dried out.

Sensitization In the sensitization process, the surface of the reinforcements will be crumbled, which prepares the surface for activation. Stannus chloride hexa hydride ($SNCl_2.6H_2O$) 10 g/l, hydrochloric acid (HCL) 30 ml/l, stirred with the particles for a period of 20 min followed by rinsing with distilled water and then dried.

Activation In the activation process, the pores on the surface are activated for plating on the particles. Palladium chloride ($PdCl_2$) 0.2 gm/l along with HCL 3 ml/l were stirred for a period of 20 min. Furthermore, the particles were washed and dried.

Plating Process The plating process is carried by stirring the palladium-activated alumina particles in a freshly prepared coating bath. Nickel chloride, hexahydrate ($NiCl_2.6H_2O$) 30 g/l, ammonium chloride (NH_4Cl) 50 g/l, trisodium citrate ($Na_3C_6H_5O_7$. $2H_2O$) 40 g/l and sodium hypophosphite monohydrate ($NaH_2PO_2.H_2O$) 25 g/l were incorporated in one liter of distilled water. In addition to this, sodium hydroxide was used to maintain pH value 8. The bath temperature was maintained 85 ± 2 °C. Pretreated alumina particles were stirred in the prepared coating bath for a period of 20 min. Subsequently, the coated alumina particles were cleaned once again in distilled water and dried up.

After the coating process, it was observed that the microparticles, which were initially in white, were turned to black color. It is the preliminary indication of the successful coating on the alumina particles. Microstructural studies, using XRD analysis, SEM and EDAX, were carried out for further confirmation of the nickel coating on the alumina particles.

2.3 Microstructural Studies of the Alumina Powders

Surface morphology of alumina particles, both uncoated and coated, was studied using Scanned electron microscope (FEI-Quanta FEG 200, USA) with energy dispersive spectroscopy attachment. Partical phase analysis was done using X-ray diffraction (XRD) method (D8 Discover, Bruker, US), at step size of 0.1° per step, scanning rate one step per second with CuKα radiation ($\lambda = 1.54$ A°).

3 Results and Discussion

3.1 Densities of the Particles

Density of the nickel-coated and uncoated alumina particles was determined using standard Archimedean principle. Averages of three readings were taken for measuring the density. From the results, density of the un coated particles was observed to be 3.94 g/cc and for the coated particles 4.12 g/cc. The deposition of the nickel on the alumina particles has increased the density of the particles [4, 7].

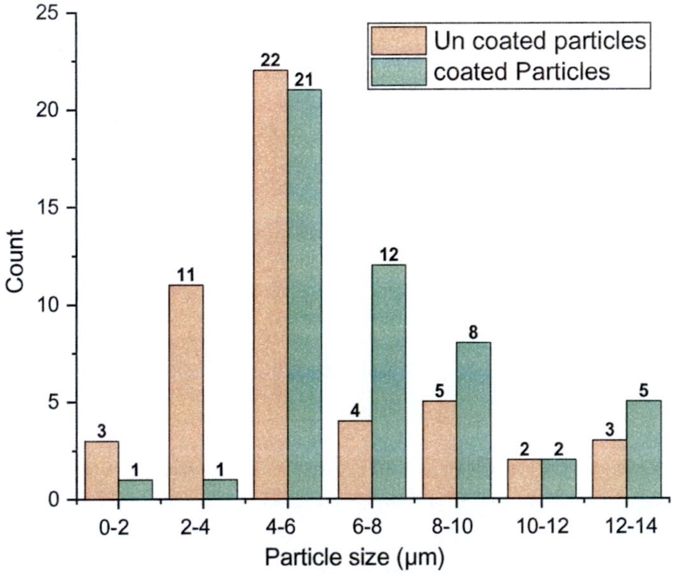

Fig. 2 Particle size variation

3.2 *Particle Size Measurement*

Particle size was measured using ImageJ (version 1.48) software by considering 50 readings of the uncoated and coated alumina particles SEM images. The sizes of the particles were observed to be 4.74 ± 1.7 μm for uncoated particles and 6.89 ± 2.65 μm for the nickel-coated alumina particles. Variation in size of the fifty particulates was depicted in Fig. 2.

3.3 *XRD Method*

The X-ray diffraction patterns of the alumina particles were shown in Fig. 3. The peaks were identified as per the reported patterns (JCPDS 75–787 and 04-850 for Al_2O_3 and Ni). New peaks were observed in the XRD pattern of Ni/Al_2O_3 at 2 theta values of 44.505, 51.844 and 76.366, which were related to nickel. The presence of nickel on micrometer-sized alumina particles was confirmed in XRD analysis.

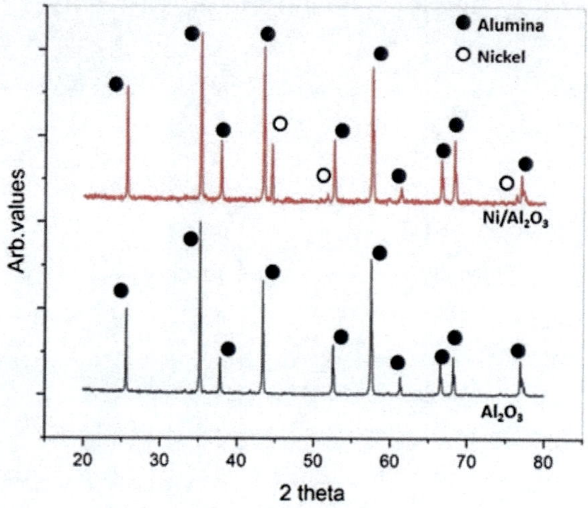

Fig. 3 XRD of alumina particles (uncoated and coated)

3.4 *SEM and EDAX Analysis*

As the surface of Al_2O_3 particles was plated with EN coating process, the existence of nickel on the surface can be confirmed by observing SEM and EDAX of both particulates. Figure 4 shows the scanned electron microscope images of both the particles (uncoated and coated).

Very smooth and clean surfaces were observed on pure alumina particles. After nickel plating on the Al_2O_3 particles, the surface appeared to be rough. The corresponding EDAX confirms the presence of the nickel on the alumina particles.

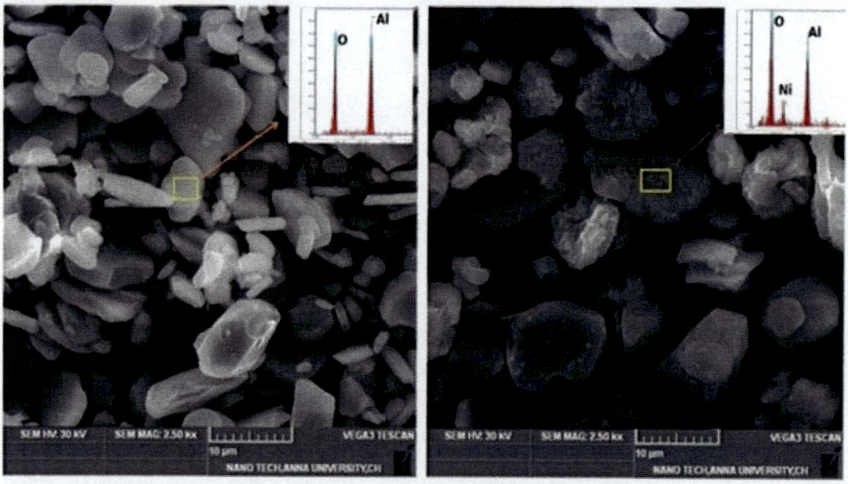

Fig. 4 SEM images of uncoated and coated Al_2O_3 particles with EDAX

4 Conclusion

Micrometer-sized alumina particles were successfully coated by nickel using standard electroless plating technique. SEM, EDAX and XRD analysis has confirmed the presence of nickel coating on the surface of alumina particles.

Further, these nickel-coated alumina particulates can be reinforced into different aluminum alloys to fabricate novel Al-MMCs to assess the role of nickel-coated alumina particulates on mechanical, tribological and machining properties of the composites.

References

1. Bhoi, N. K., Singh, H., & Pratap, S. (2020). Developments in the aluminum metal matrix composites reinforced by micro/nano particles—A review. *Journal of Composite Materials, 54*(6), 813–833.
2. Garg, P., Jamwal, A., Kumar, D., Sadasivuni, K. K., Hussain, C. M., & Gupta, P. (2019). Advance research progresses in aluminium matrix composites: Manufacturing & applications. *Journal of Materials Research and Technology, 8*(5), 4924–4939.
3. Jamwal, A., Mittal, P., Agrawal, R., Gupta, S., Kumar, D., Sadasivuni, K. K., & Gupta, P. (2020). Towards sustainable copper matrix composites: Manufacturing routes with structural, mechanical, electrical and corrosion behaviour. Journal of Composite Materials, 0021998319900655
4. Leon, C. A., & Drew, R. A. L. (2000). Preparation of nickel-coated powders as precursors to reinforce MMCs. *Journal of Materials Science, 35*(19), 4763–4768.
5. Ramesh, C. S., Keshavamurthy, R., Channabasappa, B. H., & Ahmed, A. (2009). Microstructure and mechanical properties of Ni–P coated Si3N4 reinforced Al6061 composites. *Materials Science and Engineering: A, 502*(1–2), 99–106.
6. Ramesh, C. S., Keshavamurthy, R., Channabasappa, B. H., & Pramod, S. (2010). Friction and wear behavior of Ni–P coated Si3N4 reinforced Al6061 composites. *Tribology International, 43*(3), 623–634.
7. Kumar, D. S., Suman, K. N. S., & Kumar, P. R. (2016). A study on electroless deposition of nickel on nano alumina powder under different sensitization conditions. *International Journal of Advanced Science and Technology, 97*, 59–68.
8. Loto, C. A. (2016). Electroless nickel plating—A review. *SILICON, 8*(2), 177–186.
9. Rajan, T. P. D., Pillai, R. M., & Pai, B. C. (1998). Reinforcement coatings and interfaces in aluminium metal matrix composites. *Journal of Materials Science, 33*(14), 3491–3503.
10. Chintada, V. B., & Koona, R. (2020). Preparation and properties of composite electroless Ni-P-ZnO coatings. *Materials Research Innovations, 24*(2), 67–74.
11. Sudagar, J., Lian, J., & Sha, W. (2013). Electroless nickel, alloy, composite and nano coatings–A critical review. *Journal of alloys and compounds, 571*, 183–204.
12. Vijay, S., & Srinivasa Rao, C. H. (2020). A comparative investigation of hardness and compression strength of Nickel coated B4C reinforced 601AC/201AC selective layered functionally graded materials. *Materials Research Express, 7*(1), 016527.
13. Agarwala, R. C., & Agarwala, V. (2003). Electroless alloy/composite coatings: A review. *Sadhana, 28*(3–4), 475–493.
14. Brenner, A., & Riddell, G. E. (1946). Electroless plating by a process of controlled self continuing reduction. *Proc Amer Eletropl Soc, 33*, 16.
15. Deuis, R. L., Subramanian, C., Yellup, J. M., Strafford, K. N., & Arora, P. (1995). Study of electroless nickel plating of ceramic particles. *Scripta Metallurgica et Materialia, 33*(8), 1217–1224.

16. Sahoo, P., & Das, S. K. (2011). Tribology of electroless nickel coatings—A review. *Materials & Design, 32*(4), 1760–1775.
17. Mousavian, R. T., Damadi, S. R., Khosroshahi, R. A., Brabazon, D., & Mohammadpour, M. (2015). A comparison study of applying metallic coating on SiC particles for manufacturing of cast aluminum matrix composites. *The International Journal of Advanced Manufacturing Technology, 81*(1–4), 433–444.
18. Islam, M., Azhar, M. R., Khalid, Y., Khan, R., Abdo, H. S., Dar, M. A., ... & Burleigh, T. D. (2015). Electroless Ni-P/SiC nanocomposite coatings with small amounts of SiC nanoparticles for superior corrosion resistance and hardness. *Journal of Materials Engineering and Performance, 24*(12), 4835–4843.
19. Devarakonda, S. K. (2020). Preparation and characterization of electroless Ni coated nano alumina powder under different sensitization-activation conditions. *Metallurgical and Materials Engineering.*
20. Praveen, D. V., Raju, D. R., & Raju, M. J. (2019). Influence of nickel coating of Al2O3 (P) reinforced AA-7075 metal matrix composites on hardness, impact strength and tensile properties. *Materials Research Express, 6*(12), 126537.

Investigation on Mechanical Properties of Aluminum-Boron Carbide Metal Matrix Composites

Anantha Krishnan Nair, Arun Kalmadi, Nitin Kumar, and B. P. Dileep

1 Introduction

Mixing of different materials, such as ceramics, metals and polymers, opens up possibilities of an infinite number of variations of the properties in the material [1]. These combinations have been made use of for thousands of years by mankind since the beginning of civilization. Concrete is a good example of an ancient compound [2]. It is an aggregate mixture of cement and sand. Resistance to squashing provided good compressive strength [3] to concrete. Addition of metal rods or wires to concrete in recent times has increased its tensile [4] and bending strength and is now referred to as reinforced concrete [5]. In simple terms, a composite is a combination of components [6]. Combining two or more natural or artificial elements having different physical or chemical properties [7] results in a composite material, wherein the resulting composite is stronger than its components taken individually [8]. The identity of each individual component is not lost; the components combine together contributing their most useful traits, thereby improving the final product.

The study conducted by Mohanty et al. [9] from IIT Madras was focused on fabrication of aluminum 1100 matrix composite with boron carbide reinforcement and its properties. Up to 25 wt.% of B_4C in Al was prepared at 873 K for the study. Previous studies provided details regarding the temperature range at which aluminum reacted with B_4C on heating them. It is said that at temperatures above 973 K and below 1173 K, they reacted together to form phases like Al2BC, Al3BC and a few other complicated phases. Based on such previously established facts, temperature was controlled to limit it at 873 K for a period of 90 min. It was observed that there was decrease in density and electrical conductivity while there was an

A. K. Nair · A. Kalmadi · N. Kumar · B. P. Dileep (✉)
Department of Mechanical Engineering, Amrita School of Engineering, Amrita Vishwa Vidyapeetham, Bangalore 560035, India
e-mail: bp_dileep@blr.amrita.edu

R. Agrawal et al. (eds.), *Recent Advances in Smart Manufacturing and Materials*, Lecture Notes in Mechanical Engineering,
https://doi.org/10.1007/978-981-16-3033-0_20

increase in hardness by up to 11 times, which was quite significant. But increasing the percentage of reinforcement resulted in decrease of strength of the composite [10]. Judiciously heat treating the composite in solid state was recommended for reduction in embrittlement of metal matrix composite because a tendency of brittle fracture was observed at higher boron carbide percentages.

Vasudevan et al. [11] worked on aluminum alloy AA5083-based composites with B_4C (7, 9 and 10% composition) ceramic particles as reinforcement. Fabrication of composite was done via stir casting method. Salt fog test (ASTM B117) was used for investigation of the friction stir welds' and base metal's corrosion behavior, and estimation of corrosion rate was done using weight loss measurement. The test results indicated that increasing content of B_4C particles resulted decrease in corrosion resistance of the composite. The study by Mehta and Badheka [12–15] aimed at investigating wear behavior of surface Al-B_4C composites fabricated via friction stir processing. Base metal used was Al-6061-T6, and the reinforcement media used was boron carbide particles (800 mesh). The number of capping and stirring passes was restricted to 1 pass and 3 passes, respectively, [16, 17]. The direction of consecutive passes for processing was same in specimen-1, and the direction of passes for processing specimen-2 was reversed by rotating the sample between consecutive by 180°. Pin-on-Disk (ASTM G99 standard) testing was used to conduct wear analysis on an as received base metal sample and the fabricated specimen supported by scanning electron microscopy imaging, microstructural characterization and micro-hardness data.

1.1 Metal Matrix Composites (MMC)

In MMCs, the matrix material is metal, and the reinforcement can be either a metal, a ceramic compound or an organic compound. The reinforcement added may not always have structural contribution to the matrix. Depending on the requirement, some may enhance properties such as wear or thermal resistance. The matrix is monolithic and continuous material. The matrix material could be metals and alloys based on aluminum, copper, magnesium or alloys with high heat resistance, or even alloys based on rare earth metals as per requirement.

Demand from defense and aerospace industries for materials with increased toughness and strength in comparison to currently available materials in the market has motivated studies in aluminum alloys and composites. Aluminum (Al) metal and its alloys have a major drawback in the form of relatively low modulus of elasticity when compared to more widely available structural materials such as steel. Aluminum-based MMC is among those commonly used for automotive and aerospace components. They provide a distinct advantage due to lower density and improved thermal properties and resistance to wear and corrosion. The matrix is aluminum or its alloy like aluminum-silicon, aluminum-magnesium or aluminum-copper. Reinforcement particles are chosen based favorable properties they contribute to the matrix such as high stiffness and low density. Popular reinforcement materials for aluminum are ceramics such as silicon carbide, aluminum oxide, boron carbide, boron nitride or carbon. Boron carbide (B_4C), with a density 2.52 g/cm^3 and elastic modulus 460 GPa, is one of the particles that falls in the category of positive reinforcement for an aluminum MMC.

1.2 Aluminum-Boron Carbide (Al-B_4C)

Al-B_4C composites retain their low density while combining properties of Al such as ductility with those of B_4C such as hardness resulting in a relatively stiff material which could find its use in defense industry as an armor plate material and in nuclear technology as a structural neutron absorber.

2 Materials and Methods

2.1 Powder Mixing

The mass of each material to be taken for a specimen of a particular composition was calculated using rule of mixture. The volume of compacted specimen was to be obtained which was taken as a constant with the dimensions of the expected specimen to be 30 mm diameter and 10 mm thickness. Using the density of Al and B_4C powders individually, the mass of the powder required for each specimen with the composition of 98% Al + 2% B_4C, 96% Al + 4% B_4C, 94% Al + 6% B_4C and 92% Al + 8% B_4C was calculated as in Table 1.

The mixing of powders can be done by various methods like ball milling or centrifugal mixing. Powders for each specimen were mixed in a mortar and pestle setup making an "8" patterned motion for even distribution particles in the mix.

Table 1 Theoretically calculated mass of respective elements

Composition (%)		Total density (g/cc)	Total mass (g)	Mass of Al (g)	Mass of B_4C (g)
98	2	2.6964	19.059	18.678	0.381
96	4	2.6928	19.034	18.272	0.761
94	6	2.6892	19.008	17.868	1.140
92	8	2.6856	18.983	17.464	1.518
90	10	2.682	18.957	17.062	1.895

2.2 Compaction

The compaction of samples was done using a die and plunger setup with a circular cross section for compacting the samples. An initial load of 80 kN was applied on pure Al powder specimen, and the loads were increased progressively for the next samples up to 180 kN. The loading force was fixed at 180 kN for compacting the composite specimen after observing that the load produced the closest density to the theoretically values.

2.3 Sintering

Sintering of samples is essential to ensure proper binding of the reinforcement particles in the matrix by diffusing across the boundaries creating a single solid piece. Sintering of all the compacted specimen was done in a muffle furnace with thermocouples to control the temperature. The temperature was fixed at 793 K for a period of 1 h.

3 Results and Discussion

The composite casting with Al as base material and B_4C reinforcements with varying particles from 2 to 10% was successfully prepared by compaction process. From Table 2, the hardness values of the developed composite are higher as compared to the base material.

To ascertain the optimal load value, pure Al samples were compacted under the following load until desired density was achieved as shown in Fig. 1.

After optimizing the load, the compacted samples' dimensions were measured so as to generate a graph between the composition of specimen and the compaction density as shown in Fig. 2.

Scanning electron microscopy imaging was done to study the microstructure of the pure Al, B_4C powder and the compacted specimen as follows (Figs. 3, 4, 5, 6, 7 and 8).

Table 2 Hardness number with respect to composition

Composition of specimen	Hardness Value (BHN)
100% Al	100
98% Al + 2% B_4C	108
96% Al + 4% B_4C	116
94% Al + 6% B_4C	125
92% Al + 8% B_4C	126

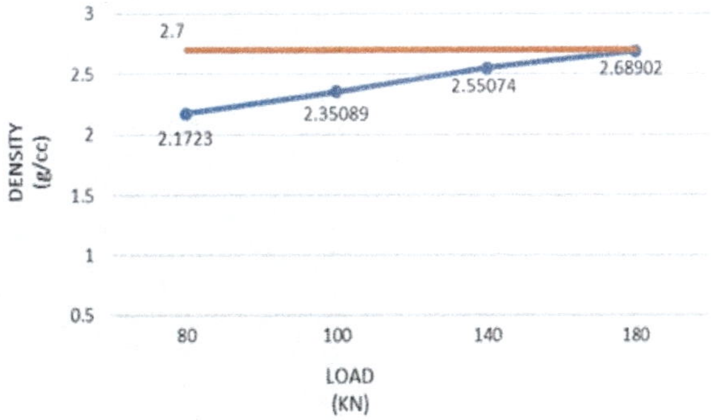

Fig. 1 Compacted density with respect to load on pure Al samples

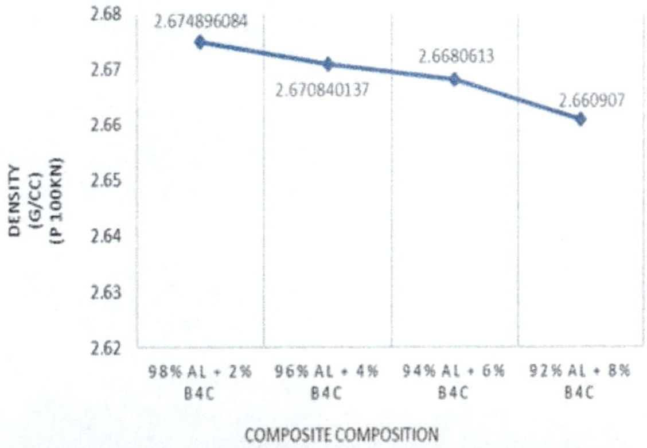

Fig. 2 Compacted density with respect to composites

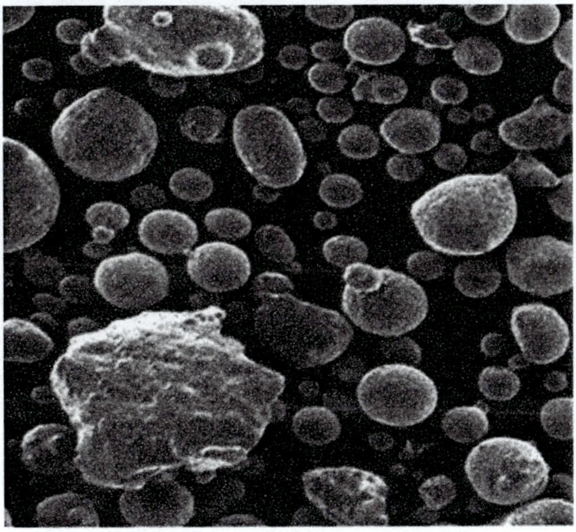

Fig. 3 Spherical structure of Al particles

Fig. 4 Flake-like structure of B_4C particles

Fig. 5 SEM image (100X) of 98% Al + 2% B_4C

4 Conclusion

Al and B_4C metal matrix composites were successfully fabricated by using powder metallurgy. The compaction load was optimized to achieve the dimension, and the required density of specimen was recorded to be 180 kN. Increasing the wt.% of

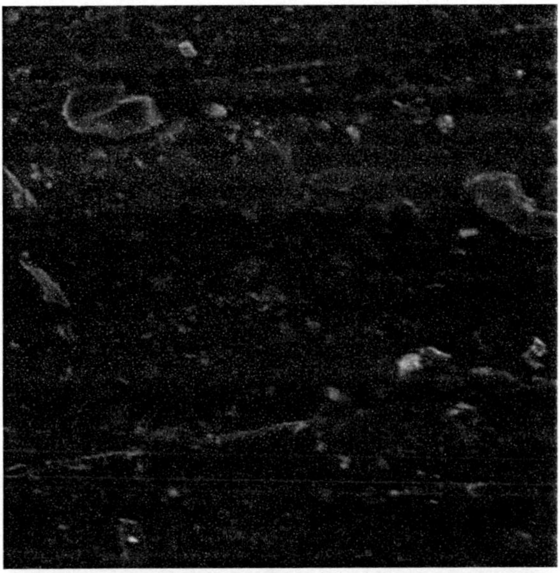

Fig. 6 SEM image (100X) of 96% Al + 4% B_4C

Fig. 7 SEM image (100X) of 94% Al + 6% B_4C

B_4C reinforcement particles in the composite resulted in the marked increase of hardness of the composite by 20.6%, then the base aluminum standalone specimen. The SEM imaging further strengthens the above claims with there being no visible cracks or voids in the bonding of B_4C reinforcement particles with the Al matrix and also shows that there were no unwanted particle inclusions nor the existence of any unnaturally large voids present in the composite.

Fig. 8 SEM image (100X) of 92% Al + 8% B_4C

References

1. Vitala, H. R., Dileep, B. P., & Ravi Kumar, V. (2018). IOP conference series: Materials science and engineering. *IOP Publishing, 377*(1), 012032.
2. Harris, B. (2001). *Engineering composite materials* (2nd ed.). Cambridge: Woodhead Publishing and Maney Publishing. 030026-6.
3. Mohanty, R. M., Balasubramanian, K., & Seshadri, S.K. (2007). *Boron carbide-reinforced aluminium 1100 matrix composites: Fabrication and properties*. Received 12 June 2007, Received in revised from 31
4. Ravishankar, M. K., Vitala, H. R., & Dileep, B. P. (2020). *Investigation on mechanical properties of equal channel angular formed Al6061 and graphite metal matrix composite* (pp. 471–480). Lecture Notes in Mechanical Engineering.
5. Vasudevan, N., Bhaskar, G. B., Rajendra Prasad, A. C., & Suresh, S. M. (2017). *Corrosion study on AA5083 aluminum alloy-boron carbide composite*. ICAMMAS17
6. Mehta, K. M., & Badheka, V. J. (2011). *Wear behavior of boron-carbide reinforced aluminum surface composites fabricated by Friction Stir Processing*. Gandhinagar, Gujarat, India: Department of Mechanical Engineering, School of Technology, Pandit Deendayal Petroleum University.
7. Dileep, B. P., & Ravishankar, M. K. (2019). Weight optimization of gimbal joint for aircraft application. *AIP Conference Proceedings, 2200*, 020083.
8. Shinohara, A. H., et.al. (1999). UNICAM-University of Campions, Faculty of Mechanical Engineering, Department of Material Engineering, Campions-SP, Brazil.
9. Sellers, R. D., & Carr, Jr., V. J. (2007). Applied Research Associates, Air Force Research Laboratory.
10. Jensen, R. E. (2013). Ph.D. Thesis, Department of Chemistry, Virginia Polytechnic Institute and State.
11. Wojcik, A. B., et al. (2007). *Hybrid glass technologies*. NJ, March: Princeton.
12. Ravi Kumar, V., & Prakash, K. R. (2018). Mathematical modeling of unsaturated isophthalic resin's curing cycle. *International Journal of Mechanical and Production Engineering Research and Development, 8*, 805–810.
13. Townsend Brown, T. (1999). *Levitation on earth's E-field technology*. Brauda.

14. Herold, G. G. (2006). *Investigation of oil insulation strength in power transformer using prototype model*. Erlangen Germany: Institute of Electrical Power Engineering, Erlangen – Nurnberg University.
15. Dileep, B. P., & Sridhar, B. R. (2018). *International Journal of Mechanical and Production Engineering Research and Development, 8*(2), 189–194.
16. Ravikimar, V., Dileep, B. P., & Vital, H. R. (2017). *AIP Conference Proceedings, 1859*, 020020.
17. Ravikumar, V., Dileep, B. P., Mohan Kumar, S., & Phanibhushana M. V. (2017). *AIP Conference Proceedings, 1859*, 020037-1–020037-6.

Research Opportunities in Industry 4.0: A Literature Review

Shivam and Manish Gupta

1 Introduction

In the year 2011, the term industry 4.0 was coined after an association in Germany, a European country, made up of corporate, political, and academic representatives. The term is new to everyone, but very important for the whole world to compete in the world market and a shift from the Industrial Revolution 3.0, which was the automation in the industries. Industry 4.0 refers to a new stage in the industrial revolution based on interconnectivity, machine learning and access to data in real time.

In the middle of the eighteenth century, the entire journey of revolution began in the world economy, i.e., in the industry. England is known for being the birthplace of the Industrial Revolution; in the latter half of the eighteenth century, it represented a phase of growth that changed largely rural, agrarian societies in Europe and America into industrialized, urban ones. The Industrial Revolution started in Britain and spread to the rest of the world, including the United States, in the 1830s and '40s, and that was only possible by the game-changing use of steam power. More often this period of the revolution referred as the first industrial revolution to set it apart; from the late nineteenth century to the early twentieth century, the second industrial revolution took place and saw rapid advances in the steel, electric, and automotive industries. After the second revolution with the use of automation in the industry with the help of computer and robots, the third industrial revolution came into the picture and with the use of technology-like big data, artificial intelligence, Internet of things in the cyber-physical systems; that forms the concept of industrial revolution 4.0. Industry 4.0 or smart industry refers to the advancement of technology from embedded systems to cyber-physical systems (CPS) [1].

Shivam (✉) · M. Gupta
Department of Mechanical Engineering, Motilal Nehru National Institute of Technology Allahabad, Prayagraj, India

R. Agrawal et al. (eds.), *Recent Advances in Smart Manufacturing and Materials*, Lecture Notes in Mechanical Engineering,
https://doi.org/10.1007/978-981-16-3033-0_21

Many authors worked on the topic industry 4.0 and took one of the technologies of industry 4.0 for their study purpose-like Internet of things, robotics, 3D printing; the basic definition of these technologies by different authors, IoT includes platforms that connects multiple sensors and data devices in order to create a full vision of the behavior of an organization, a system, a business processes or a phenomenon [1] Big data analytics is characterized as the ability to process extremely large data sets to identify relationship patterns (correlation, causality) between data to be used in detecting market trends, customer behavior and preferences [1], 3D printing is a technology that enables artifacts to be produced by subsequent printing of adhesive materials, such as polymers, [1] and when technologies associated with the part of the supply chain-like logistics then it is called logistics 4.0 [2], when associated with procurement it is redefined and called procurement 4.0 [3] and with human resource it is called human resource management 4.0 (HRM 4.0) [4].

Smart supply chain management is defined as the practice of the Internet of Things, advanced robotics, advanced big data analytics in supply chain management: Position sensors in everything, construct networks everywhere, automate everything, and analyse each and everything to significantly improve productivity and customer satisfaction and digital supply chain can be defined as DSC is made up of systems (e.g., software, hardware, communication networks) that facilitate connections between organizations that are globally distributed and orchestrate the activities of supply chain partners. These activities include the procurement, processing, storage, movement, and sale of a commodity.

This paper aims to give some conclusive points and future scope study. What study lags in industry 4.0 from the journal's papers submitted between 2010 and 2019 in reputed journals and conferences? Industry 4.0 and its innovations aligned with it, like the Internet of Things (IoT), radio frequency identification (RFID), big data, 3D printing, artificial intelligence, and machine learning.

2 Literature Review

Research work was carried out on the journals, conference papers, and white papers submitted and published by the various authors from 2011 to 2019 in reputed journals. Those open access papers were used in the review process. This review paper has reviewed 119 articles from 68 different journals and white paper institutions. The chart showing the number of articles based on the year and also a pie chart showing the various topics which were used by the authors to show their work are as follows (Figs. 1 and 2).

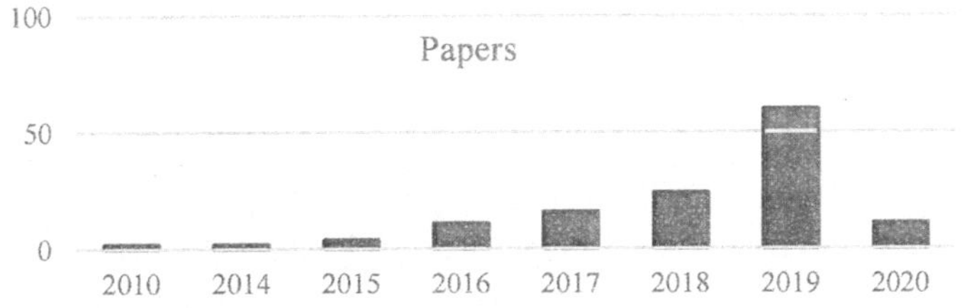

Fig. 1 Bar graph showing the year wise distribution of papers

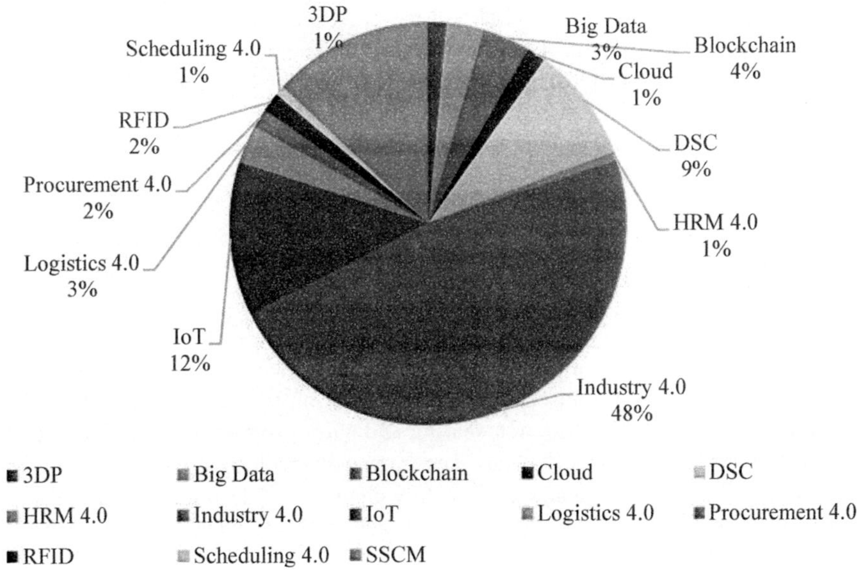

Fig. 2 Pie Chart showing the topic wise distribution of papers

2.1 Internet of Things (IoT)

Many authors discuss the management of the supply chain using the technology Internet of Things (IoT). IoT works as the connection between computer to the physical world by transferring real-time data. The author, Yuvraaj et al., proposed managing the supply chain using IoT and low-power wireless communication systems. They use RFID tag with the Arduino board for indoor tracking of the goods and outdoor pursuit using coding. The author creates a platform system based on IoT technology to provide a full range of customer search services through the Internet [5].

The authors Mohamed Abdel et al. proposed a framework that systems automate products' identification process, globally, trace and monitor goods, achieve transparency, and lowering time and costs for customer satisfaction [6]. Author Bendaya et al. discuss the latest developments in the application of IoT and technological enablers of IoT. He identified the literature gaps concerning the potential of IoT's role in addressing supply chain management [7]. In the context of industry, David Henriques et al. investigated which are the most suitable information technology enablers that can assist organizations in IoT implications [8].

Luis Hernan Contreras Pinochet attempted to deepen the research on the influence of the IoT products' attributes on the functional and emotional experiences and consumer's purchase intention. His paper shows the importance of IoT products' characteristics at the time of purchase [9]. One author worked on how a conventional refrigerator can be changed into IoT-enabled smart refrigerators without changing long-standing production lines in the industry. Suvi Nenonen worked on finding the enablers and hindrances in developing smart services in smart campuses [10]. Authors Zhi Li et al. are proposing an IoT-based monitoring and tracing framework for the pre-packaged food supply chain using RFID and extensible markup language (XML) to enable the exchange of information between applications and stakeholders. [11].

2.2 *Big Data*

A new intelligent supply chain integration and management framework based on the cloud of things was proposed by Junwei Yan et al. to provide versatile and agile approaches to promote resource sharing and participant communication over the entire life cycle of the supply chain. The framework built by the authors would make it easier for users to manage each supply chain connection through the real-time data collected and analyzed from the intelligent perception and network access convergence subsystems [2].

A hybrid cloud that essentially incorporates the supply chain network with versatility and productivity has been suggested by some authors. To enzure customer loyalty, a supply chain network must add value, which can be best accomplished by partnership with the hybrid cloud [3].

2.3 *Additive Manufacturing*

Cheryl Druehl et al. attempted to determine the technological implications and its impacts on the supply chain. They took the technologies like 3DP, virtual reality, and driverless vehicles and tried to find out its impacts on the supply chain and implications on the industry's managerial and IT policy [4].

The effect of 3D printing technology on the supply chain is being studied by Lukáš Kubáč et al. To highlight the effect, he compared the traditional supply chain using the AM supply chain and describes the benefits of the AM model over the conventional model over cost savings, speed response, quality assurance, and environmental impact [5].

2.4 Blockchain

Yassine Issaoui et al. worked on applying blockchain technology to determine its applications in smart logistics. The application was classified into four clusters and the four clusters were information, transport, finance, and management [6].

Francesco Lango et al. designed and developed a software connector to connect an Ethereum-like blockchain with the information systems of enterprises, enabling companies to share information with their partners with different levels of visibility and through the blockchain to check data authenticity, integrity, and invariability over time and thus build confidence with their partners [7].

2.5 Industry 4.0

Technology is the key to growth in today's world of development, as technology plays an enormous role in everyday lives today. Nevertheless, the increasingly connected culture in which we live also affects the world of industry. The name given to the growing mix of legacy manufacturing and industrial platforms and the newest smart technology is Industry 4.0. Many of the authors worked with using various technologies in their paper-like Lorenzo Ardito et al. provide insightful information on, which digital technologies can enable the SCM-M integration of supply chain management-marketing. In particular, by relying on real illustrative examples, the authors highlighted the part in which these solutions play crucial role in information acquisition, storage, and elaboration for SCM-M integration [8]. Three paradigms and six Industry 4.0 principles were identified by some editorials, and also they discussed five technologies frequently in Industry 4.0 [9].

3 Methodology

3.1 Systematic Literature Review Approach

The steps incorporated for the systematic literature review process are

1. Identification of research
2. Selection of studies
3. Quality assessment of studies

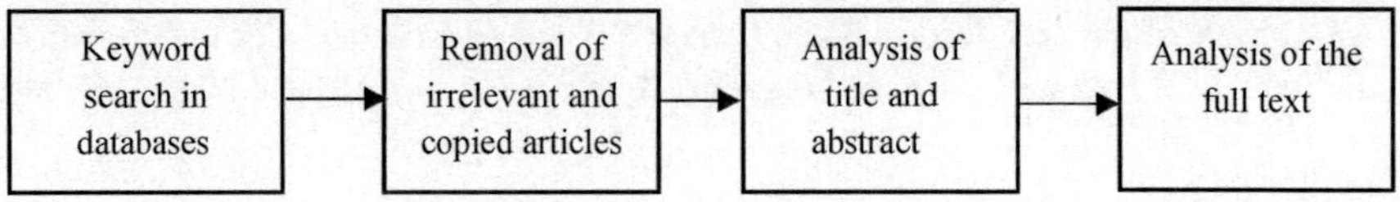

Fig. 3 Systematic Literature Review Process

4. Extraction of data and examination of process.

Upon searching the databases, too many results evolved from the first search strings, i.e., by identifying relevant literature on Industry 4.0 by the assessment of Industrial Reports, academics articles, conference proceedings, and guest editorial published in the right journals, conference proceedings, and white paper websites. Results from the surveys, books, books chapter were removed from the search results after that analyzed the search results and rejected the copied ones. In the second step, papers were analyzed to ensure that the subject under study was addressed in the selected paper. Those papers were identified based on the keyword identification associated with Industry 4.0, i.e., smart supply chain, IoT, digital supply chain, big data, etc. By scrutinizing the titles, keywords, and abstracts of 280 Articles. The articles with the abstract which do not comply with inclusion criteria were excluded, with 119 articles remaining. At the third stage of the literature review, analysis of the articles' done based on the quality that underwent a peer review (Fig. 3).

3.2 *Analysis of the Literature*

Literature was analyzed on the basis of latest technologies inherent with industry 4.0 and technology used by the industry. Technology, with the function of the supply chain and technologies, included making a supply chain smart enough. Table 1 shows the topic and references used in this literature review to give conclusions of the work carried out by different researchers and points on future scope of study in industry 4.0 and its related technologies and Fig. 4 is a histogram showing the name of the top 10 journals in which papers related to these topics have been published more.

4 Conclusion

After reviewing the papers, we concluded these points which can be used by the researchers for further research in the relevant areas of Industry 4.0.

- Several studies conducted by the authors tried to find out which technologies will benefit in which part of the supply chain, a supply chain, consists of many functions starting from the procurement to the fulfillment; the author studies individual function with the association of industry 4.0 technologies.

Table 1 Table showing topic and reference of literature review

Topic	References
3DP	[5, 10]
Augmented Reality, RFID	[11]
Big Data	[12–14]
Blockchain	[7, 15–18]
Cloud	[2, 3]
DSC	[4, 19–27]
HRM 4.0	[28]
Industry 4.0	[9, 18, 29–85]
IoT	[86–99]
Logistics 4.0	[6, 42, 100, 101]
Procurement 4.0	[76, 102]
RFID	[103]
Scheduling 4.0	[104]
SSCM	[105–118]

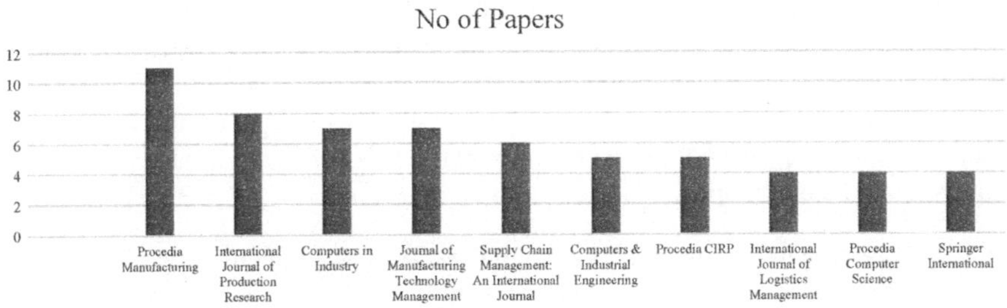

Fig. 4 Papers published in the top 10 journals

- Some authors focus on industry-specific studies, i.e., in the area of food and beverages supply chain, agriculture supply chain, and pharmaceutical supply chain, to know how the technologies will impact the industry on implementing the technologies.
- Studies focus on creating the methodologies to implement a smart supply chain in small enterprise among micro, small and medium enterprises (MSME). These studies were done to analyze the cost-effectiveness, where investing to return in long terms is unacceptable.
- Some studies focused on the economic aspects, i.e., cost-effectiveness on using industry 4.0 technologies, i.e., radio frequency identification RFID concerning micro, medium, and small enterprises.
- Some of the authors studied how to use the industry 4.0 technologies-like RFID and embedded systems in eliminating the quality defect.

- Several studies proposed an SCM model considering various factors: Business, technology, sustainable development, collaboration, and management strategy.
- Several studies were conducted on the supply chain's particular function and named them procurement 4.0, logistics 4.0, and scheduling 4.0. Like using the industry 4.0 technologies in the procurement process and made it smart, the name coined for this 'Procurement 4.0' likewise for using it with the logistics part; it is named 'Logistics 4.0'. In the case of the scheduling of machines, it is termed as 'scheduling 4.0'.
- Studies conducted by authors focus on reviewing the various aspects of the smart supply chain (SCM) on Enterprise resource planning (ERP) and explored the potential opportunities to redirect the methods with the use of industry 4.0 technologies.

5 Limitation and Future Scope

Several papers have been studied to analyze the work carried out by different researchers in the field of industry 4.0 and its technologies, the objective of this study to give an overview on the work lagging in this field and future work which can be done on this topic, the limitation of this paper is that there were too many papers which are not open access, but the papers may be very useful to the research purpose those papers are not included in writing this review paper, the following points are describing the future scope study can be done on the industry 4.0 and using its related technologies;

- The authors researched the effects of digitalization and automation on the supply chain, but the study of organizational and cultural factors affecting the application of Industry 4.0 in the supply chain was scarce to the best of my knowledge.
- Authors discussed the business, technology, sustainable development, collaboration, and management strategy perspective to study SCM's preparedness, but risks on a human operator, customer perspective not reviewed to date to the best of knowledge.
- Authors studied quality management through RFID, but many technologies are still there to be used concerning quality management.
- Industry-specific empirical and qualitative case studies can be established in the context of Industry 4.0 aided supply chain.
- The economic viability of Industry 4.0 application and its technologies adoption can be considered for the research purpose.
- None of the authors studied the impact of mobile technologies in working conditions considering the factors (e.g., safety, ergonomics) to the best of my knowledge.
- Papers discussed the implementation of Industry 4.0 in small industries, but no one considered micro-level enterprises, which is equally crucial for our country like India.

- Many authors studied the impact of Industry 4.0 on procurement, scheduling, and logistics part of the supply chain. Still, some other functions can be studied and not discussed by the other authors to the best of my knowledge.
- Business management processes like ERP discussed by authors considering it for smart supply chain but, none of the authors thought BPR (Business Process Reengineering) to the best of my knowledge.

References

1. World Economic Forum. (2019). WEF Supply Chain 4.0 2019 Report. White Paper. Available: http://www3.weforum.org/docs/WEF_Supply_Chain_4.0_2019_Report.pdf.
2. Yan, J., et al. (2014). Intelligent supply chain integration and management based on cloud of things. *International Journal of Distributed Sensor Networks, 2014.*
3. Sundarakani, B., Kamran, R., Maheshwari, P., & Jain, V. (2019). Designing a hybrid cloud for a supply chain network of Industry 4.0: A theoretical framework. *Benchmarking.*
4. Druehl, C., Nikkola, J. H., & Carrillo, J. (2018). *Technological innovations: Impacts on supply chains* (p. 1). Springer International. Online resource, 353pp.
5. Kubáč, L., & Kodym, O. (2017). The impact of 3D printing technology on supply chain. *MATEC Web Conference, 134,* 1–8.
6. Issaoui, Y., Khiat, A., Bahnasse, A., & Ouajji, H. (2019). Smart logistics: Study of the application of blockchain technology. *Procedia Computer Science, 160*(2018), 266–271.
7. Longo, F., Nicoletti, L., Padovano, A., d'Atri, G., & Forte, M. (2019). Blockchain-enabled supply chain: An experimental study. *Computer and Industrial Engineering, 136*(July), 57–69.
8. Ardito, L., Petruzzelli, A. M., Panniello, U., & Garavelli, A. C. (2019). Towards Industry 4.0: Mapping digital technologies for supply chain management-marketing integration. *Business Process Management Journal, 25*(2), 323–346.
9. Koh, L., Orzes, G., & Jia, F. (2019). The fourth industrial revolution (Industry 4.0): Technologies disruption on operations and supply chain management. *International Journal of Operations & Production Management, 39*(6), 817–828.
10. Ryan, M. J., Eyers, D. R., Potter, A. T., Purvis, L., & Gosling, J. (2017). 3D printing the future: Scenarios for supply chains reviewed. *International Journal of Physical Distribution & Logistics Management, 47*(10), 992–1014.
11. Edirisinghe, R. (2019). Digital skin of the construction site: Smart sensor technologies towards the future smart construction site. *Engineering, Construction and Architectural Management, 26*(2), 184–223.
12. Li, Q., & Liu, A. (2019). Big data driven supply chain management. *Procedia CIRP, 81,* 1089–1094.
13. Fosso Wamba, S., Gunasekaran, A., Papadopoulos, T., & Ngai, E. (2018). Big data analytics in logistics and supply chain management. *International Journal of Logistics Management, 29*(2), 478–484.
14. Sanders, N. R., Boone, T., Ganeshan, R., & Wood, J. D. (2019). Sustainable supply chains in the age of AI and digitization: Research challenges and opportunities. *Journal of Business Logistics, 40*(3), 229–240.
15. Munir, M. S., Bajwa, I. S., & Cheema, S. M. (2019). An intelligent and secure smart watering system using fuzzy logic and blockchain. *Computers & Electrical Engineering, 77*(July), 109–119.
16. Dolgui, A., Ivanov, D., Potryasaev, S., Sokolov, B., Ivanova, M., & Werner, F. (2019). Blockchain-oriented dynamic modelling of smart contract design and execution in the supply chain. *International Journal of Production Research,* 1–16.

17. Pournader, M., Shi, Y., Seuring, S., & Koh, S. C. L. (2019). Blockchain applications in supply chains, transport and logistics: A systematic review of the literature. *International Journal of Production Research*, 1–19.
18. Treiblmaier, H. (2018). The impact of the blockchain on the supply chain: A theory-based research framework and a call for action. *Supply Chain Management, 23*(6), 545–559.
19. Butner, K. (2010). The smarter supply chain of the future. *Strategic Leadership, 38*(1), 22–31.
20. Büyüközkan, G., & Göçer, F. (2018). Digital supply chain: Literature review and a proposed framework for future research. *Computers & Industrial Engineering, 97*.
21. Horváth, D., & Szabó, R. Z. (2019). Driving forces and barriers of Industry 4.0: Do multinational and small and medium-sized companies have equal opportunities? *Technological Forecasting and Social Change, 146*(May), 119–132.
22. Schrauf, S., & Berttram, P. (2016). Industry 4.0: How digitization makes the supply chain more efficient, agile, and customer-focused. *Strateg. &Technology*, 1–32. Available: https://www.strategyand.pwc.com/media/file/Industry4.0.pdf.
23. Garay-Rondero, C. L., Martinez-Flores, J. L., Smith, N. R., Caballero Morales, S. O., & Aldrette-Malacara, A. (2019). Digital supply chain model in Industry 4.0. *Journal of Manufacturing Technology Management.*
24. Schniederjans, D. G., Curado, C., & Khalajhedayati, M. (2020). Supply chain digitisation trends: An integration of knowledge management. *The International Journal of Production Economics, 220*, 0–1.
25. Ghobakhloo, M., & Ching, N. T. (2019). Adoption of digital technologies of smart manufacturing in SMEs. *Journal of Industrial Information Integration, 16*,.
26. de Giovanni, P. (2019). Digital supply chain through dynamic inventory and smart contracts. *Mathematics, 7*(12).
27. Seepma, A. P., de Blok, C., & Van Donk, D. P. (2020). Designing digital public service supply chains: four country-based cases in criminal justice. *Supply Chain Management, 2019.*
28. Liboni, L. B., Cezarino, L. O., Jabbour, C. J. C., Oliveira, B. G., & Stefanelli, N. O. (2019). Smart industry and the pathways to HRM 4.0: Implications for SCM. *Supply Chain Management, 24*(1), 124–146.
29. Rauch, E., Linder, C., & Dallasega, P. (2020). Anthropocentric perspective of production before and within Industry 4.0. *Computers & Industrial Engineering, 139*, 105644.
30. Chauhan, C., & Singh, A. (2019). A review of Industry 4.0 in supply chain management studies. *Journal of Manufacturing Technology Management*.
31. Fatorachian, H., & Kazemi, H. (2020). Impact of Industry 4.0 on supply chain performance. *Production Planning & Control*, 1–19.
32. Barata, J., Rupino Da Cunha, P., & Stal, J. (2018). Mobile supply chain management in the Industry 4.0 era: An annotated bibliography and guide for future research. *Journal of Enterprise Information Management, 31*(1), 173–192.
33. Strange, R., & Zucchella, A. (2017). Industry 4.0, global value chains and international business. *Multinational Business Review, 25*(3), 174–184.
34. Pérez-Lara, M., Saucedo-Martínez, J. A., Marmolejo-Saucedo, J. A., Salais-Fierro, T. E., & Vasant, P. (2018). Vertical and horizontal integration systems in Industry 4.0. *Wireless Networks, 2.*
35. Büchi, G., Cugno, M., & Castagnoli, R. (2020). Smart factory performance and Industry 4.0. *Technological Forecasting and Social Change, 150*, 119790.
36. Ghobakhloo, M., & Fathi, M. (2020). Corporate survival in Industry 4.0 era: the enabling role of lean-digitized manufacturing. *Journal of Manufacturing Technology Management, 31*(1), 1–30.
37. Krykavskyy, Y., Pokhylchenko, O., & Hayvanovych, N. (2019). Supply chain development drivers in industry 4.0 in Ukrainian enterprises. *Oeconomia Copernicana, 10*(2), 273–290.
38. Dallasega, P. (2018). Industry 4.0 fostering construction supply chain management: Lessons learned from engineer-to-order suppliers. *IEEE Engineering Management Review, 46*(3), 49–55.

39. Ding, B. (2018). Pharma Industry 4.0: Literature review and research opportunities in sustainable pharmaceutical supply chains. *Process Safety and Environment Protection, 119,* 115–130.
40. Hahn, G. J. (2020). Industry 4.0: A supply chain innovation perspective. *International Journal of Production Research, 58*(5), 1425–1441.
41. Hidayatno, A., Destyanto, A. R., & Hulu, C. A. (2019). Industry 4.0 technology implementation impact to industrial sustainable energy in Indonesia: A model conceptualization. *Energy Procedia, 156,* 227–233. https://doi.org/10.1016/j.egypro.2018.11.133.
42. Tjahjono, G. P. B., Esplugues, C., & Ares, E. (2017). What does Industry 4.0 mean to Supply Chain? *Procedia Manufacturing, 13,* 1245–1252.
43. Jose Ignacio Rodriguez Molano, S. J. G. M., & Parra, K. D. L. (2018). Impact of implementing Industry 4.0 in Colombia's supply chains. *Springer International, 6,* 12–22.
44. Nagy, J., Oláh, J., Erdei, E., Máté, D., & Popp, J. (2018). The role and impact of industry 4.0 and the internet of things on the business strategy of the value chain-the case of Hungary. *Sustainability, 10*(10).
45. Ivanov, D., Dolgui, A., & Sokolov, B. (2019). The impact of digital technology and Industry 4.0 on the ripple effect and supply chain risk analytics. *International Journal of Production Research, 57*(3), 829–846.
46. Winkelhaus, S., & Grosse, E. H. (2020). Logistics 4.0: A systematic review towards a new logistics system. *International Journal of Production Research, 58*(1), 18–43.
47. Da Xu, L., Xu, E. L., & Li, L. (2018). Industry 4.0: State of the art and future trends. *International Journal* of *Production Research, 56*(8).
48. Müller, J. M., & Voigt, K. I. (2018). The impact of Industry 4.0 on supply chains in engineer-to-order industries—An exploratory case study. *IFAC-Papers, 51*(11), 122–127.
49. Hofmann, E., & Rüsch, M. (2017) Industry 4.0 and the current status as well as future prospects on logistics. *Computers & Industrial Engineering, 89.*
50. Roblek, V., Meško, M., & Krapež, A. (2016). A complex view of Industry 4.0. *SAGE Open, 6*(2).
51. Ghobakhloo, M. (2018). The future of manufacturing industry: A strategic roadmap toward Industry 4.0. *Journal of Manufacturing Technology Management*, 29(6).
52. Müller, F., Jaeger, D., & Hanewinkel, M. (2019). Digitization in wood supply—A review on how Industry 4.0 will change the forest value chain. *Computers and Electronics in Agriculture, 162.*
53. Bag, S., Telukdarie, A., Pretorius, J. H. C., & Gupta, S. (2018). Industry 4.0 and supply chain sustainability: Framework and future research directions. *Benchmarking.*
54. Sanders, A., Elangeswaran, C., & Wulfsberg, J. (2016). Industry 4.0 implies lean manufacturing: Research activities in industry 4.0 function as enablers for lean manufacturing. *Journal of Industrial Engineering and Management, 9*(3), 811–833.
55. Buasuwan, P. (2018). Rethinking Thai higher education for Thailand 4.0. *Asian Education and Development Studies, 7*(2), 157–173.
56. Amaral, A., Jorge, D., & Peças, P. (2019). Small medium enterprises and Industry 4.0: Current models' ineptitude and the proposal of a methodology to successfully implement Industry 4.0 in small medium enterprises. *Procedia Manufacturing, 41,* 1103–1110.
57. Villa, A., & Taurino, T. (2019). SME innovation and development in the context of Industry 4.0. *Procedia Manufacturing, 39,* 1415–1420.
58. Li, D., Landström, A., Fast-Berglund, Å., & Almström, P. (2019). Human-centred dissemination of data, information and knowledge in industry 4.0. *Procedia CIRP, 84,* 380–386.
59. Bigliardi, B., Bottani, E., & Casella, G. (2020). ScienceDirect ScienceDirect Enabling technologies, application areas and impact of industry 4. 0: a bibliographic analysis. *Procedia Manufacturing, 42*(2019), 322–326.
60. Müller, J. M. (2019). Business model innovation in small- and medium-sized enterprises: Strategies for industry 4.0 providers and users. *Journal of Manufacturing Technology Management, 30*(8), 1127–1142.

61. Pacchini, A. P. T., Lucato, W. C., Facchini, F., & Mummolo, G. (2019). The degree of readiness for the implementation of Industry 4.0. *Computers in Industry, 113*,.
62. García, S. G., & García, M. G. (2019). Industry 4.0 implications in production and maintenance management: An overview. *Procedia Manufacturing, 41,* 415–422.
63. Albers, A., Stürmlinger, T., Mandel, C., Wang, J., de Frutos, M. B., & Behrendt, M. (2019). Identification of potentials in the context of design for industry 4.0 and modelling of interdependencies between product and production processes. *Procedia CIRP, 84,* 100–105.
64. Da Costa, M. B., Dos Santos, L. M. A. L., Schaefer, J. L., Baierle, I. C., & Nara, E. O. B. (2019). Industry 4.0 technologies basic network identification. *Scientometrics, 121*(2), 977–994.
65. Bălan, C. (2018). The disruptive impact of future advanced ICTs on maritime transport: a systematic review. *Supply Chain Management, 25*(2), 157–175.
66. Hennig, M., Reisinger, G., Trautner, T., Hold, P., Gerhard, D., & Mazak, A. (2019). TU Wien Pilot Factory Industry 4.0. *Procedia Manufacturing, 31,* 200–205.
67. Dossou, P.-E. (2019). Using industry 4.0 concepts and theory of systems for improving company supply chain: The example of a joinery. *Procedia Manufacturing, 38*(2019), 1750–1757.
68. Kolla, S., Minufekr, M., & Plapper, P. (2019). Deriving essential components of lean and industry 4.0 assessment model for manufacturing SMEs. *Procedia CIRP, 81,* 753–758.
69. Veile, J. W., Kiel, D., Müller, J. M., & Voigt, K. I. (2019). Lessons learned from Industry 4.0 implementation in the German manufacturing industry. *Journal of Manufacturing Technology Management*.
70. Atik, H., & Ünlü, F. (2019). The measurement of Industry 4.0 performance through Industry 4.0 Index: An empirical investigation for Turkey and European countries. *Procedia Computer Science, 158,* 852–860.
71. Lin, D., Lee, C. K. M., Lau, H., & Yang, Y. (2018). Strategic response to Industry 4.0: an empirical investigation on the Chinese automotive industry. *Industrial Management & Data Systems, 118*(3), 589–605.
72. Mogos, M. F., Eleftheriadis, R. J., & Myklebust, O. (2019). Enablers and inhibitors of industry 4.0: Results from a survey of industrial companies in Norway. *Procedia CIRP, 81*(2013), 624–629.
73. Magadán, L., Suárez, F. J., Granda, J. C., & García, D. F. (2020). Low-cost real-time monitoring of electric motors for the Industry 4.0. *Procedia Manufacturing, 42*(2019), 393–398.
74. Fareri, S., Fantoni, G., Chiarello, F., Coli, E., & Binda, A. (2020). Estimating Industry 4.0 impact on job profiles and skills using text mining. *Computers in Industry, 118*,.
75. Bal, H. Ç., & Erkan, Ç. (2019). Industry 4.0 and competitiveness. *Procedia Computer Science, 158,* 625–631.
76. Glas, A. H., & Kleemann, F. C. (2016). The impact of Industry 4. 0 on procurement and supply management: A conceptual and qualitative analysis. *International Journal of Business and Management Invention, 5*(6), 55–66.
77. Kuo, C.-C., Shyu, J. Z., & Ding, K. (2019). Industrial revitalization via industry 4.0—A comparative policy analysis among China, Germany and the USA. *Global Transitions, 1,* 3–14.
78. Hubert Backhaus, S. K., & Nadarajah, D. (2019). Investigating the relationship between industry 4.0 and productivity: A conceptual framework for Malaysian manufacturing firms. *Procedia Computer Science, 161*, 696–706.
79. Javaid, M., & Haleem, A. (2020). Impact of industry 4.0 to create advancements in orthopaedics. *Journal of Clinical Orthopaedics and Trauma (JCOT)*.
80. Valeria, E. (2020). Characteristics and skills of leadership in the context of Industry 4.0. *Procedia Manufacturing, 43,* 543–550.
81. Lass, S., & Gronau, N. (2020). A factory operating system for extending existing factories to Industry 4.0. *Computers in Industry, 115*,.
82. Zimmermann, M., Rosca, E., Antons, O., & Bendul, J. C. (2019). Supply chain risks in times of Industry 4.0: Insights from German cases. *IFAC-PapersOnLine, 52*(13), 1755–1760.

83. Torbacki, W., & Kijewska, K. (2019). Identifying key performance indicators to be used in logistics 4.0 and industry 4.0 for the needs of sustainable municipal logistics by means of the DEMATEL method. *Transportation Research Procedia, 39*(2018), 534–543.
84. Ebrahimi, M., Baboli, A., & Rother, E. (2019). The evolution of world class manufacturing toward Industry 4.0: A case study in the automotive industry. *IFAC-PapersOnLine, 52*(10), 188–194.
85. Wang, M., Asian, S., Wood, L. C., & Wang, B. (2020). Logistics innovation capability and its impacts on the supply chain risks in the Industry 4.0 era. Modern Supply Chain Research and Applications.
86. Yuvaraj, S., & Sangeetha, M. (2016). Smart supply chain management using internet of things(IoT) and low power wireless communication systems. In *Proceedings of 2016 IEEE International Conference on Wireless Communication and Signal Process. Networking, WiSPNET 2016* (pp. 555–558).
87. Witkowski, K. (2017). Internet of things, big data, Industry 4.0—Innovative solutions in logistics and supply chains management. *Procedia Engineering, 182,* 763–769.
88. Ben-Daya, M., Hassini, E., & Bahroun, Z. (2019). Internet of things and supply chain management: A literature review. *International Journal of Production Research, 57*(15–16), 4719–4742.
89. Shenkoya, T., & Dae-Woo, C. (2019). Impact of IoT on social innovation in Japan. *Asia Pacific Journal of Innovation and Entrepreneurship, 13*(3), 341–353.
90. Henriques, D., Pereira, R. F., Almeida, R., & Mira da Silva, M. (2019). IT governance enablers in relation to IoT implementation: A systematic literature review. *Digit. Policy, Regul. Gov., 22*(1), 32–49.
91. Pinochet, L. H. C., Lopes, E. L., Srulzon, C. H. F., & Onusic, L. M. (2018). The influence of the attributes of 'Internet of Things' products on functional and emotional experiences of purchase intention. *Innovation and Management Review, 15*(3), 303–320.
92. Hoerlsberger, M. (2019). Innovation management in a digital world. *Journal of Manufacturing Technology Management, 30*(8), 1117–1126.
93. Ghanbari, Z., Jafari Navimipour, N., Hosseinzadeh, M., & Darwesh, A. (2019). Resource allocation mechanisms and approaches on the Internet of Things. *Cluster Computing, 22*(4), 1253–1282.
94. Aheleroff, S., et al. (2020). IoT-enabled smart appliances under industry 4.0: A case study. *Advanced Engineering Informatics, 43*, 101043.
95. Tu, M. (2018). An exploratory study of internet of things (IoT) adoption intention in logistics and supply chain management a mixed research approach. *International Journal of Logistics Management, 29*(1), 131–151.
96. Nenonen, S., van Wezel, R., & Niemi, O. (2019). Developing smart services to smart campus. *Emerald Reach Proceedings Series, 2,* 289–295.
97. Li, Z., Liu, G., Liu, L., Lai, X., & Xu, G. (2017). IoT-based tracking and tracing platform for prepackaged food supply chain. *Industrial Management & Data Systems, 117*(9), 1906–1916.
98. Zhang, Y., Zhao, L., & Qian, C. (2017). Modeling of an IoT-enabled supply chain for perishable food with two-echelon supply hubs. *Industrial Management & Data Systems, 117*(9), 1890–1905.
99. Parry, G. C., Brax, S. A., Maull, R. S., & Ng, I. C. L. (2016). Operationalising IoT for reverse supply: The development of use-visibility measures. *Supply Chain Management, 21*(2), 228–244.
100. McFarlane, D., Giannikas, V., & Lu, W. (2016). Intelligent logistics: Involving the customer. *Computer and Industrial Engineering, 81*, 105–115.
101. Wang, Y., Rodrigues, V. S., & Evans, L. (2015). The use of ICT in road freight transport for CO2 reduction—An exploratory study of UK's grocery retail industry. *International Journal of Logistics Management, 26*(1), 2–29.
102. Bienhaus, F., & Haddud, A. (2018). Procurement 4.0: Factors influencing the digitisation of procurement and supply chains. *Business Process Management Journal, 24*(4), 965–984.

103. Zhong, R. Y., & Huang, G. Q. (2014). RFID-enabled learning supply chain: A smart pedagogical environment for TELD. *International Journal of Engineering Education, 30*(2), 471–482.
104. Bányai, T., Illés, B., & Bányai, Á. (2018). Smart scheduling: An integrated first mile and last mile supply approach. *Complexity, 2,* 2018.
105. Yen, L. W. X. Y. A. J. D. C. (2016). *International Journal of Logistics Management, 27*(2).
106. Frazzon, E. M., Rodriguez, C. M. T., Pereira, M. M., Pires, M. C., & Uhlmann, I. (2019). Towards supply chain management 4.0. *Brazilian Journal of Operations & Production Management, 16*(2), 180–191.
107. Khan, M. Z., Al-Mushayt, O., Alam, J., & Ahmad, J. (2010). Intelligent supply chain management. *Journal of Software Engineering and Applications, 03*(04), 404–408.
108. Ivanov, D., Dolgui, A., Sokolov, B., Werner, F., & Ivanova, M. (2016). A dynamic model and an algorithm for short-term supply chain scheduling in the smart factory industry 4.0. *International Journal of Production Research, 54*(2), 386–402.
109. Frank, A. G., Dalenogare, L. S., & Ayala, N. F. (2019). Industry 4.0 technologies: Implementation patterns in manufacturing companies. *International Journal* of *Production Economics, 210,* 15–26.
110. Seyedghorban, Z., Tahernejad, H., Meriton, R., & Graham, G. (2020). Supply chain digitalization: Past, present and future. *Production Planning & Control, 31*(2–3), 96–114.
111. Lima-Junior, F. R., & Carpinetti, L. C. R. (2020). An adaptive network-based fuzzy inference system to supply chain performance evaluation based on SCOR® metrics. *Computer and Industrial Engineering, 139,*.
112. Hofmann, E., Sternberg, H., Chen, H., Pflaum, A., & Prockl, G. (2019). Supply chain management and Industry 4.0: Conducting research in the digital age. *International Journal of Physical Distribution & Logistics Management, 49*(10), 945–955.
113. Crawford, C. (2019). The smart supply chain: A digital revolution. *AATCC Review, 19*(3), 38–45.
114. Oh, J., & Jeong, B. (2019). Tactical supply planning in smart manufacturing supply chain. *Robotics* and *Computer-Integrated Manufacturing, 55,* 217–233.
115. Makris, D., Hansen, Z. N. L., & Khan, O. (2019). Adapting to supply chain 4.0: An explorative study of multinational companies. *Supply Chain Forum, 20*(2).
116. Wilding, R., & Wagner, B. (2019). Guest editorial. *Supply Chain Management, 24*(1), 1–4.
117. Staniek, M., & Sierpiński, G. (2016). Smart platform for support issues at the first and last mile in the supply chain—The concept of the S-Mile project. *Scientific Journal of Silesian University of Technology, 92,* 141–148.
118. Scholz-Reiter, B., Teucke, M., Özsahin, M.-E., & Sowade, S. (2015). Smart label-supported autonomous supply chain control in the apparel industry. In *International Congress* on *Logistics* and *SCM Systems* (ICLS 20A09).

Effect of Temperature and Humidity on Tribological Properties of Rail and Wheel Using Pin-On-Disc

Ajeet Yadav, Sachin, Vineet Dubey, Rabesh Kumar Singh, and Anuj Kumar Sharma

1 Introduction

Railway transportation is very economic, environmentally friendly relative to another transportation mode. Friction and wear between railway track and wheel are very crucial key challenges for the smooth functioning and operation of the railway network. There are basically three types of friction mechanism, in which first is elasto-plastic deformation which mostly occurs in lubricated interfacing surface (very thin layer), second is adhesion mechanism of friction which generally occurs in cleaned interfacing surface, that arises due to the transformation of electrons between two interfacing surfaces. The third category belongs to viscous drag which generally arises in thick layered interfacing surfaces [1]. Olofsson et al. [2] performed a sliding test on the pin on disc set-up (wheel as disc and rail as a pin) to show how the wear rate was influenced by velocity (sliding) of rail and wheel on room temperature. As per the result, the wear rate is maximized at a higher sliding velocity in comparison with lower sliding velocity. Zhu et al. [3] observed that at a lower temperature, an increment in absolute humidity reduces the value of friction coefficient and investigated the variation friction coefficient with respect to temperature, but temperature limits within 20 °C. Olofsson et al. [4] has surveyed on the tribology of rail and wheel in an open environment and showed a variation of friction coefficient with varying season (on monthly basis) and also explained the variation of friction coefficient with respect to temperature in different contact pressure. Hardwick et al. [5] described that under wet condition wear increases when slip increase between rail and wheel in comparison of dry condition. To achieve transition between wear for wet and grease contact, more energy is required and also suggested that the application of grease in controlled quantity proved to be beneficial for preventing surface damages. Yang et al.

A. Yadav · Sachin · V. Dubey · R. K. Singh (✉) · A. K. Sharma
Centre for Advanced Studies, Dr. A.P.J. Abdul Kalam Technical University, Lucknow, Uttar Pradesh 226031, India

R. Agrawal et al. (eds.), *Recent Advances in Smart Manufacturing and Materials*, Lecture Notes in Mechanical Engineering,
https://doi.org/10.1007/978-981-16-3033-0_22

[6] explain the effect of temperature on the self-lubrication coating of rail and wheel. The lubricant coatings can be categorized as lower temperature coatings (temperature range −200 to 20 °C) ex-soft material coatings like Au, Ag, etc and moderate temperature coatings for temperature range 20–500 °C such as Mo, Tungsten, etc. At lower temperatures, conventional lubricants are become ineffective due to changes in the properties of lubricants. Lubricants used at a higher temperature also become ineffective due to properties change. Deters et al. [7] experimented on two roller method of rail and wheel and gave different conclusions. With increasing creep and acting pressure, the wear increased (by volume), when circumferential speed (test rollers) increased, and then, there is a reduction in the volume of wear. High temperature favours the formation of wear protection layer (oxide layer) which prevents the wear rate [8]. Chen et al. [9] mentioned the importance of relative humidity /water vapour on two interfacing surfaces and also proved that tribological properties cannot be predicted on the behalf of mechanical properties of two interfacing surfaces of metals. Dearden et al. [10] explain the wear of rail and wheel separately and how wear can be controlled. For better wear resistant, hard steel is suitable but using low ductile steel is not suitable. Vo et al. [11] used a tool which works on finite element method (FEM) to examine the temperature variation of wheel-rail contact in dry, wet and lubrication condition and concluded that there is a minor effect of environment on the contact pressure and area of contact, but shear stress is influenced. The increase in friction leads to the increase in traction forces, which further causes the generation of shear stress. Maximum surface damage occurs in dry conditions (higher slip causes the higher temperature led to an increase in friction coefficient). Khalladi et al. [12] tested these properties having four environmental contaminants, i.e. sulphur, sand, cement, and phosphate, and described the effect of these contaminants on the wear and friction mechanism of the wheel and rail. They conclude that sand causes more adhesion with respect to the other three. The presence of silica in these contaminants penetrates the contact surface enforces the surface damage and increases the abrasive mechanism. Lyu et al. [13] experimented in different environmental conditions, with and without oxidation and suggested that adhesive wear is dominant with low relative humidity (relative) and become a challenging issue, i.e. rate of wear increases at higher relative humidity (85%) adhesive wear transform into oxidative wear mechanism overall less wear with higher relative humidity within the temperature range (3–20 °C) and relative humidity range (40–85%). It was revealed that variation of wear with respect to relative humidity in both conditions, clean and with oxidation. Zhu et al. [14] experimented about friction between wheel and rail on different temperature, relative humidity, and oxides of iron and conclude that coefficient of friction become stable on higher relative humidity and decreases with increasing relative humidity. Knothe et al. [15] argued that the contact temperature can not exceed 500 °C in normal running conditions. It happens only in high-speed trains that cross 500 °C and can achieve 600 °C (critical point). Stoff et al. [16] adhesive wear exists within the relative humidity range of 10–70% and then after corrosive wear prevails, abrasive wear highly prevails in dry environment condition. Olofsson et al. [17] proved that by using leaves as lubricant is more beneficial than higher relative relative humidity. Odabas [18] and Singh et al. [19] have

proved that the frictional coefficient decreases with increase in acting pressure and loading condition due to formation of surface oxides. This paper contains the study of friction and wear behaviour of wheel and track interfacing surfaces on different atmospheric condition like temperature and humidity.

2 Experiment

2.1 Experimental Set-up

For the testing of wear and friction behaviour of rail wheel and track interfacing surfaces, a pin–disc tribometer has been used. A circular disc of wheel and pin of the track loaded (controllable) with dead load has been assembled with the machine as shown in Fig. 1. The experimental set-up is installed in a highly ventilated room, and Fig. 1a shows the whole set-up including the controller, monitor for data storage, and analysis. In Figure 1b, the side view of the pin–disc set-up comprises of LVDT sensor, closed chamber, and thermocouple. Figure 1c shows the pin holder with pin the top view of the closed chamber in which all the experiments are performed. Figure 1d shows the top view of the closed chamber in which all the experiments are performed. Figure 1e shows the top view of the pin holder and disc holder, and Fig. 1f shows the disc and pin after the experiments are performed. The experimental set-up has been installed with the proximity sensor to measure the speed/RPM of the rotating disc, load cells for the frictional force, LVDT for wear measurement, and

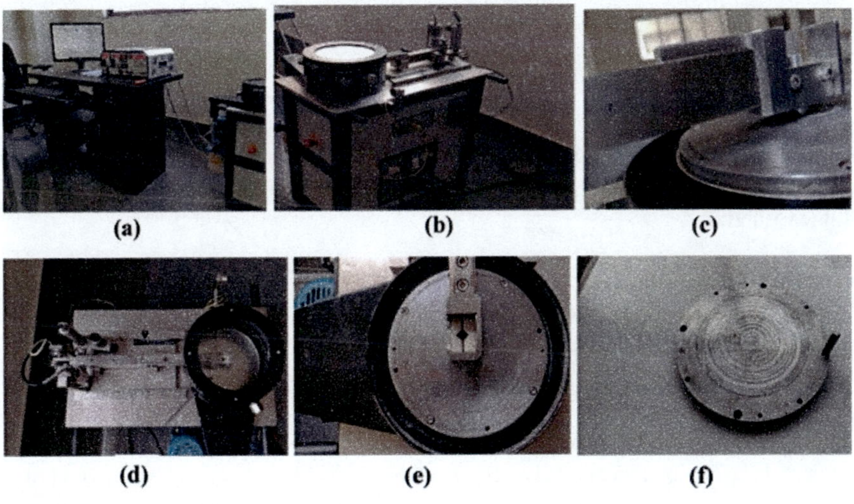

Fig. 1 **a** Front view of pin–disc tribometer set-up, **b** side view of pin–disc tribometer set-up, **c** side view of pin and pin holder, **d** top view of pin-disc closed chamber, **e** top view of disc and pin holder, **f** disc and pin after getting wear

Table 1 Properties of test specimen [20]

Specification	Grade	C % Max	Mn % Max	P % Max	S % Max	Si % Max	Al % Max
Chemical properties of the wheel	IRS: R-34/03	0.57–0.67	0.60–0.85	0.03	0.03	0.15	–
Chemical properties railway track	880	0.60–0.80	0.80–1.30	0.03	0.03	0.10–0.50	0.015

Table 2 Mechanical properties of test specimen [20]

S. no.	Specification	Yield strength (Mpa) Min	Tensile strength (Mpa) Min
1	Track (Grade 880)	460	880
2	Wheel (IRS: R-34/03)	50% of ultimate tensile strength	820–940

K-type of thermocouple for the temperature measurement. All the sensors installed in the machine have an accuracy of $0.1 \pm 1\%$. A separate humidity sensor and the humidifier have been used to measure and control the closed chamber humidity. The accuracy of the humidity sensor is $\pm$ 5%. A closed chamber as shown in Fig. 1d has been used for all the experiments in which the temperature and humidity can be controlled and a temporary environment can be created for experimentation.

2.2 *Material of Test Specimen*

The test specimen used in the tribological experiment is arranged from Research Design and Standard Organization (RDSO), Lucknow. The wheel material has been used as a disc and track material as a pin. The properties of these materials are given in Tables 1 and 2.

2.3 *Test Procedure*

Experiments are performed in two parts, i.e. case 1 and case 2, as mentioned in Table 3; in first case, it is tried to investigate the effect of various temperature on wear and friction by keeping load and speed as constant. Second case is to absolve the effect of humidity on wear and friction at constant temperature. The variation of parameters is tabled in Table 3. Each experiment has been performed thrice to get the accurate variation of the wear and friction, and after each experiment, it was taken care that no

Table 3 Variation of parameters for experimentation

S. no.	Time (min)	Temperature	Loading (N)	Humidity (%)	Speed of disc (RPM)
Case 1					
1	5	25	5	60	200
2	5	35	5	60	200
3	5	45	5	60	200
Case 2					
1	5	25	10	45	300
2	5	25	10	60	300
3	5	25	10	80	300

dust particle or wear particles are available on the test specimen to be experimented with by using solvent 0.1 M HCl.

3 Result and Discussion

3.1 Effect of Temperature on Wear and Frictional Force

All the experiments are performed as per the variation of parameters in Table 3, and the speed is constant at 200 RPM. The behaviour of the wear and friction is observed by keeping constant acting pressure, speed, and humidity on different temperatures, i.e. 25, 35 and 45 °C.

At temperature 25 °C, Fig. 2a, b shows that when relative motion starts the wear and friction rate increases uniformly, and when the contact temperature (sliding

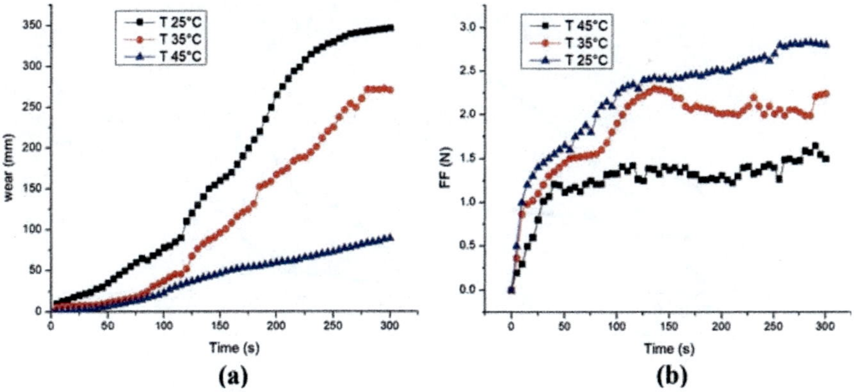

Fig. 2 **a** Variation of wear with time on different humidity, **b** variation of frictional force with time on different humidity

temperature) rises it results in the decrement of wear and friction rate and reaches nearby constant rate. When the temperature increases, then the slope of the time–wear and time–frictional force is lower as compare to at 25 °C. As the temperature increased, the contact temperature consequently increases at a higher rate than at lower temperature; hence, the wear and friction both rates become stable sooner. From these investigations, it is clear that at constant acting pressure, speed, and humidity, the wear and friction rates are the function of the temperature. The variation of the frictional force with respect to time is very fluctuating in nature as shown in Fig. 2b.

3.2 *Effect of Humidity on Wear and Frictional Force*

In this stage, all the experiments are performed by keeping speed and the acting pressure constant on different humidity, i.e. 45, 60 and 80%, to investigate the effect of humidity on wear and friction rate.

Figure 3a shows the variation of wear depth at different relative humidity (45, 60 and 80%) at a fixed temperature of 25 °C. For lower humidity, the slope of the wear-time is high and uniform; and when the sliding temperature increases, it became constant. At higher humidity, the slope is lower than as at lower humidity. Figure 3b shows the variation of the frictional force with the variation of humidity at a constant temperature of 25 °C; from here too, it has been observed that at lower humidity the frictional force–time slope is higher at lower humidity; and as humidity increases, the slope also get decreased and lowest at higher humidity though the nature of the variation of the frictional force is fluctuating. It can also be analysed that at higher humidity the wear and frictional rates are mostly independent of the other dynamic condition.

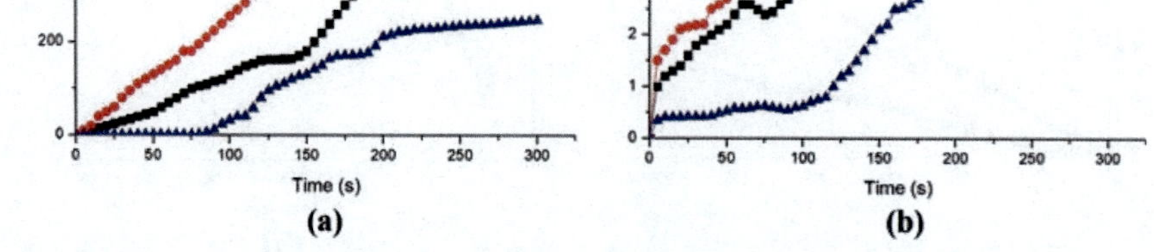

Fig. 3 Wear variation with respect to time **a** at different relative humidity and temperature 25 °C, **b** frictional force variation on different relative humidity and temperature 25 °C

From these experiments, it is investigated that lower relative humidity and temperature results in the maximum rate of wear and friction rate, and at higher relative humidity, the wear and friction rate decreases up to saturation level. The rate of wear and friction is inversely proportional to relative humidity. It is also clear from these experiments that lower temperatures and relative humidity are not suitable conditions for the functioning of the railway system and summers with high humidity are the favourable condition for the functioning.

4 Conclusion

The overall investigations from the above experiments can be concluded as:

- In general condition, the wear and friction in railways are the functions of temperature and relative humidity of the environment; both decrease with increasing the temperature and relative humidity.
- At higher relative humidity water vapour act as lubricants and hence results in the decrement of wear and friction phenomenon up to saturation level.
- Wear and friction are mostly dependent on acting pressure and the speed of the wheel of the railways system at lower temperatures and relative humidity. The sliding temperature is very essential to the curtailment of the wear and friction.

Further, investigation needs to be carried to find the relation of different types of wear and friction mechanism under varying environmental conditions and how they influence the functioning of the railway system.

Acknowledgements The authors want to thank Mr. Anurag Chaubey, Junior Research Engineer CAS, AKTU, who provided his kind support and helping in whole experimentation and especially Research and Design Standard Organization (RDSO), Lucknow, which provided us the railway scrap material for test specimen.

References

1. Batchelor, A. W., & Stachowiak, G. W. (1995). Tribology in materials processing. *Journal of Materials Processing Technology, 48*(1–4), 503–515. https://doi.org/10.1016/0924-0136(94)01689-X
2. Olofsson, U., & Telliskivi, T. (2003). Wear, plastic deformation and friction of two rail steels—a full-scale test and a laboratory study. *Wear, 254*(1–2), 80–93. https://doi.org/10.1016/S0043-1648(02)00291-0
3. Zhu, Y., Olofsson, U., & Chen, H. (2013). Friction between wheel and rail: a pin-on-disc study of environmental conditions and iron oxides. *Tribology Letters, 52*(2), 327–339. https://doi.org/10.1007/s11249-013-0220-0
4. Olofsson, U., & Lyu, Y. (2017). Open system tribology in the wheel-rail contact—A literature review. *Applied Mechanics Reviews, 69*(6). https://doi.org/10.1115/1.4038229

5. Hardwick, C., Lewis, R., & Eadie, D. T. (2014). Wheel and rail wear—Understanding the effects of water and grease. *Wear, 314*(1–2), 198–204. https://doi.org/10.1016/j.wear.2013.11.020
6. Yang, J. F., Jiang, Y., Hardell, J., Prakash, B., & Fang, Q. F. (2013). Influence of service temperature on tribological characteristics of self-lubricant coatings: A review. *Frontiers of Materials Science, 7*(1), 28–39. https://doi.org/10.1007/s11706-013-0190-z
7. Deters, L., & Proksch, M. (2005). Friction and wear testing of rail and wheel material. *Wear, 258*(7–8), 981–991. https://doi.org/10.1016/j.wear.2004.03.045
8. Stott, F. H. (2002). High-temperature sliding wear of metals. *Tribology International, 35*(8), 489–495. https://doi.org/10.1016/S0301-679X(02)00041-5
9. Chen, Z., He, X., Xiao, C., & Kim, S. H. (2018). Effect of humidity on friction and wear—A critical review. *Lubricants, 6*(3), 1–26. https://doi.org/10.3390/lubricants6030074
10. Dearden, J. (1960). The wear of steel rails and tyres in railway service. *Wear, 3*(1), 43–59. https://doi.org/10.1016/0043-1648(60)90174-5
11. Vo, K. D., Tieu, A. K., Zhu, H. T., & Kosasih, P. B. (2014). A tool to estimate the wheel/rail contact and temperature rising under dry, wet and oily conditions. *WIT Transactions on The Built Environment, 135*, 191–201. https://doi.org/10.2495/CR140151
12. Khalladi, A., & Elleuch, K. (2017). Tribological behavior of wheel-rail contact under different contaminants using pin-on-disk methodology. *Journal of Tribology, 139*(1), 1–9. https://doi.org/10.1115/1.4033051
13. Lyu, Y., Zhu, Y., & Olofsson, U. (2015). Wear between wheel and rail: A pin-on-disc study of environmental conditions and iron oxides. *Wear, 328–329*, 277–285. https://doi.org/10.1016/j.wear.2015.02.057
14. Zhu, Y., Lyu, Y., & Olofsson, U. (2015). Mapping the friction between railway wheels and rails focusing on environmental conditions. *Wear, 324–325*, 122–128. https://doi.org/10.1016/j.wear.2014.12.028
15. Yevtushenko, A., & Ivanyk, E. (1997). Determination of temperatures for sliding contact with applications for braking systems. *Wear, 206*(1–2), 53–59. https://doi.org/10.1016/S0043-1648(96)07318-8
16. Barthel, A. J., Gregory, M. D., & Kim, S. H. (2012). Humidity effects on friction and wear between dissimilar metals. *Tribology Letters, 48*(3), 305–313. https://doi.org/10.1007/s11249-012-0026-5
17. Olofsson, U., & Sundvall, K. (2004). Influence of leaf, humidity and applied lubrication on friction in the wheel-rail contact: Pin-on-disc experiments. In *Proceedings of the Institution of Mechanical Engineers, Part F, 218*(3), 235–242. https://doi.org/10.1243/0954409042389364
18. Odabas, D. (2018). The influence of the temperature on dry friction of AISI 3315 steel sliding against AISI 3150 steel. *IOP Conference Series: Materials Science and Engineering, 295*(1). https://doi.org/10.1088/1757-899X/295/1/012021
19. Singh, R. K., Dixit, A. R., Sharma, A. K., Tiwari, A. K., Mandal, V., & Pramanik, A. (2018). Influence of graphene and multi-walled carbon nanotube additives on tribological behaviour of lubricants. *International Journal of Surface Science and Engineering, 12*(3), 207–227. https://doi.org/10.1504/IJSURFSE.2018.094773
20. Steel authority of India Ltd. https://www.sail.co.in/sites/default/files/product-brochure/2020-04/Railway-Products.pdf

State of the Art in Dry Electric Discharge Machining Process: A Critical Review

Mrityunjaya Chauhan and Gangadharudu Talla

1 Introduction

EDM is one of the non-traditional manufacturing processes used widely in small hole drilling, mold making, die making, manufacturing complex shapes, etc. EDM process is preferred over many manufacturing processes because this process helps to obtain good surface finish, creating complex shapes, low tool wear rate, and also less heat-affected zone.

The machining principle of EDM is based on the melting and vaporization of the material due to energy produced during sparking between the electrode tool and the workpiece. In conventional EDM, dielectric fluid (like kerosene, paraffin oil, transformer oil, lubricating oil, etc.) is used which makes this manufacturing process hazardous for the environment.

From the development of the EDM technique in the late 1940s [1], oil has been used as the dielectric fluid, which is hazardous for our environment. To safeguard the environment, some researchers have been regularly working on the advancement of the EDM process by exploring different dielectric fluids and by varying machining parameters.

In Dry-EDM, oil-based dielectric fluids have been replaced by dielectrics like air, oxygen, nitrogen, argon, etc. This gaseous dielectric does not produce any toxic fumes while machining, and hence we call Dry-EDM an environmental-friendly machining process. In this, high-velocity gases are allowed to flow through the tool electrode. The high-velocity gases pass through the inter-electrode gap which helps to remove debris particles from the machining zone. Also, the gaseous dielectric helps to prevent excessive heating of the tool and the workpiece which in turn improves the life of the electrode tool as well as the workpiece. Heat-affected zone (HAZ) is less in Dry-EDM as competed to HAZ in conventional EDM.

M. Chauhan (✉) · G. Talla
Department of Mechanical Engineering, National Institute of Technology, Warangal, India

R. Agrawal et al. (eds.), *Recent Advances in Smart Manufacturing and Materials*,
Lecture Notes in Mechanical Engineering,
https://doi.org/10.1007/978-981-16-3033-0_23

2 Dry-EDM Process

2.1 *History of Dry-EDM Development*

NASA is the first organization to coin the term 'Dry-EDM' and published a technical report on Dry-EDM in 1985 [2]. Kunieda et al. [3] in 1991 manifested their interest and introduced oxygen gas (O_2) into the discharge gap with water (H_2O) as a dielectric fluid and observed a great improvement in MRR. He revealed the viability of using air as a dielectric medium in 1997. Till now, several gaseous dielectric fluids have been explored to achieve optimum input and output parameters.

2.2 *Near-Dry-EDM*

It is a machining operation in which a gas and liquid mixture is used as a dielectric medium with the principle of EDM to obtain the desired specification of the work part. Fujiki [4] investigated different machining parameters with the variation of tool lead angle (α), tool tilt angle (β), and dielectric fluid flow rate. In the experiment, a hollow copper electrode with H13 tool steel as the workpiece and a mixture of air and kerosene oil as dielectric fluids were used. It was found that with the decrease in α, MRR decreases, TEWR increases, and fluid flow rate reduces. Also, with the elevation in β, the tool electrode wear ratio increases and MRR decreases. It was investigated that α is expedient for roughing and β is advantageous to avoid scooping in finishing EDM milling.

$$\text{TEWR} = \frac{\Delta m_{\text{electrode}}/\rho_{\text{electrode}}}{\Delta m_{\text{workpiece}}/\rho_{\text{workpiece}}}$$

Masahiro also investigated that the conventional method to control the discharge gap is inefficient because it allows the renunciation of the electrode in the machining path only [5]. Later on, he explored a new gap control strategy that helps to increase the MRR by 30%, while very few effects on TEWR and surface roughness were found. The proposed controller consists of two different controllers. One is the trajectory mode in which the tool electrode advances along the planned path and the other being the ancillary axis mode in which the tool electrode moves in its orientation direction (see Figs. 1 and 2). Taguchi L27 orthogonal array [6] can be used for the optimization of more than one response for the better functioning of the system.

Dhakar and Akshay [7] experimented with copper tube electrode and a mixture of liquid glycerin and air as dielectric fluids. The outcome was quite surprising that the MRR got thrice of the MRR while machining with a combination of water and air as dielectric fluid keeping all the parameters the same. The viscosity of liquid glycerin is more than the viscosity of water, and thermal energy produced at the inter-electrode gap is higher while machining with the glycerin–air mixture as a dielectric fluid.

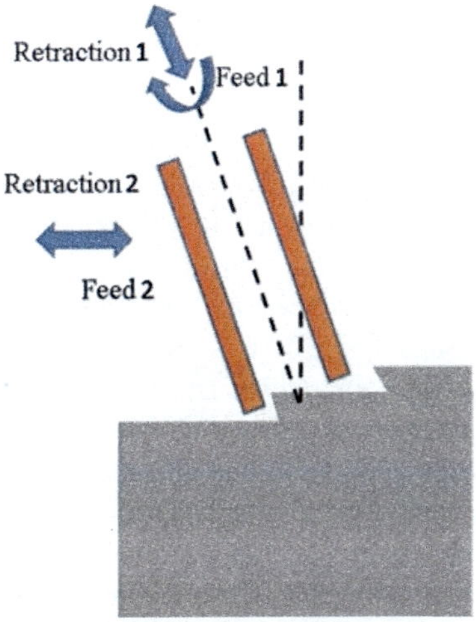

Fig. 1 Retraction of tool in different directions [7]

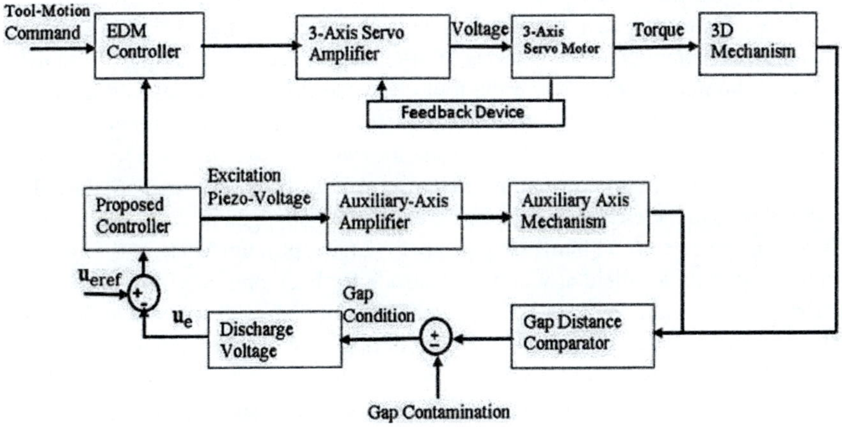

Fig. 2 Block diagram of proposed controller [7]

Discharge energy is directly proportionate to the discharge current and voltage. Hence, MRR increases with a rise in discharge current and discharge voltage [8]. With the surge in air pressure rate, the material removal rate increases as debris removal increases from the vicinity of the work zone. Air pressure also helps in establishing direct contact between current and work part by flushing debris from the inter-electrode gap.

2.3 Dry-EDM

In this machining operation, only gaseous dielectric fluid is used rather than oil as a dielectric fluid in conventional EDM. MRR, TEWR, and surface roughness are affected by the different input machining parameters. One of the important machining parameters is controlling the inter-electrode gap between the electrode tool and the workpiece.

The inter-electrode gap is considerably larger and stable in conventional EDM, but in Dry-EDM this inter-electrode gap is quite less and unstable which affects the MRR and makes Dry-EDM an uneconomic machining process. A piezoelectric actuator was used to control the inter-electrode gap between the tool electrode and the workpiece [9]. In this, the average gap voltage was matched with the *Z*-axis servo reference voltage and the piezoelectric servo reference voltage. A corresponding output is produced by the diode to control the action of different drivers (see Fig. 3).

The influence of tool geometry on MRR and surface roughness was investigated by Saha and Choudhary in 2008 [10]. Two different geometrical factors have been examined named tool external diameter and number of holes available for airflow through the electrode. The conclusion of the experiment was tool having a smaller diameter gives better outcomes in terms of MRR and surface irregularities. With the increase in the number of holes in the tool electrode, MRR increases and later on decreases after a certain number of holes in the tool electrode. The optimal value of holes in the electrode for better MRR and less tool wear rate was realized to be three.

The expansion of plasma was controlled by creating a back pressure with the help of a shield around the electrode [11]. In the experiment, oxygen as dielectric fluid, a hollow copper electrode with reverse polarity to obtain optimum MRR, and a hollow cylindrical aluminum shield around the electrode for back pressure were used. The experiment was conducted on stainless steel 304, and the result was found that the MRR increases linearly with discharge current and speed of the spindle. With the rise in discharge current and voltage, the sparking energy increases, and hence MRR increases. But after a certain value of discharge voltage, MRR falls off because the gap available for sparking increases. The centrifugal force increases with the speed

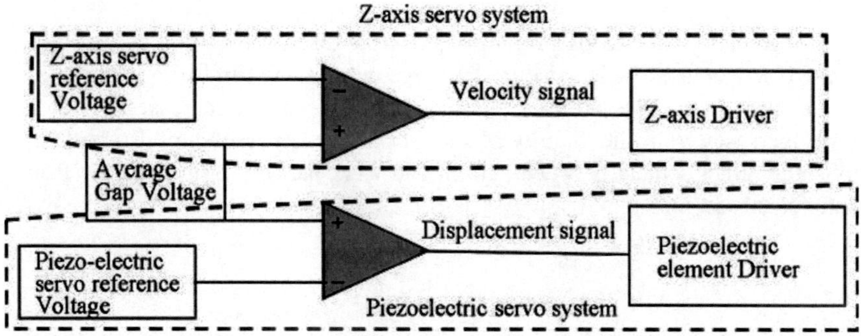

Fig. 3 Block diagram of process control [2]

which helps in easy and efficient flushing of debris from the machined zone. Another method to control the expansion of the plasma region is, by applying a pulsating magnetic field tangential to the plasma generation [12]. The pulsating magnetic field helps in the reduction of magnetic flux losses thereby encouraging an increase in plasma ionization and it helps to control the expansion of plasma. This magnetic field influences the flow of major charge carriers (electrons) because Lorentz force (F) is applied on the charge carriers due to electric field (E) and magnetic field (B)

$$F = q(E + V \times B)$$

where V is the velocity of charge carrier and q is the amount of electric charge. The Lorentz force (F) deflects the electrons toward the sparking region, thereby constraining the motion of electrons which causes a reduction in the mean free path of electrons. In the sparking region, ionization energy increases because of the increase in electron density which leads to an improvement in MRR. It experimented that MRR is high and tool wear rate (TWR) was found to be minimum with magnetic field-aided Dry-EDM as compared to without magnetic field-aided Dry-EDM.

The material removal rate in Dry-EDM can also be enhanced by efficient disposal of debris from the machining area by providing equal-spaced square-shaped (1 mm × 1 mm) slots on the periphery of the hollow copper electrode at the bottom [13]. Several experiments were conducted by varying the number of slots on the bottom edge of the electrode. It experimented that when the number of slots is less than 4, inefficient disposal of debris takes place, and when the number of slots is more than 4, inadequate blowing of plasma takes place by the dielectric medium. Maximum MRR, maximum depth of cut, and least TWR were obtained with the tool having four equal-spaced slots at the bottom edge of electrode.

An experiment was conducted using oxygen gas as a dielectric medium and tipped copper electrode in the existence of the magnetic field to analyze the dimension of the crater produced during the machining operation [14]. The experiment was performed on polished stainless steel 304 workpiece with the least inter-electrode gap. 'Quasi-explosion' condition was maintained by keeping the pulse off time small and constant. With the assistance of the magnetic field, the utmost reduction in crater diameter observed was 81% and the average depression in crater depth observed was 27% in Dry-EDM. It was also observed that the crater surface looks smoother and a more uniform removal mechanism was achieved with the assistance of the magnetic field.

Liqing and Yingjie experimented to advance the functioning of Dry-EDM by using additive mixed dielectric fluid [15]. Oxygen gas was mixed with different gases like argon, nitrogen, and air to produce additive mixed dielectric fluid. Oxygen gas was used as additives because of certain reasons:

(a) Very less tool material deposition takes place on the workpiece in the presence of oxygen.
(b) High electronegativity of oxygen.
(c) Lower ionization energy is required for oxygen.

The experiment was conducted on a quenched 45 carbon steel as a workpiece using a hollow copper electrode. It was examined that MRR increases with the rise in the concentration of oxygen gas with argon, nitrogen, and air. Another experiment was performed on a cryogenically cooled workpiece and found increase in MRR and reduction in surface roughness while machining a cryogenically cooled workpiece.

To increase the MRR, Fattahi and Baseri [16] conducted an experiment using a copper tool having three holes for dielectric flow and used a different combination of dielectric fluid to machine high-speed steel workpiece. It was observed that MRR is more with a combination of argon and air as a dielectric fluid compared to a combination of nitrogen and air as dielectric fluid. Argon has a higher viscosity and density due to which the discharge gap resistance decreases which cause an improvement in MRR. Also, high-viscosity fluid restricts the expansion of the plasma region, thereby focusing discharge to a smaller area.

A hollow graphite electrode and air as dielectric fluid were used to improve the MRR [17]. The thickness of the electrode was kept sufficient to have a high velocity of fluid flow through it, which helps in better disposal of debris. It was examined that MRR is approximately thrice when the workpiece is positive rather than negative. MRR also increases with the rise in discharge current, fluid pressure, tool rotational speed, and pulse duration time. Relative electrode wear rate (REWR) increases during the process with the rise in discharge current and pulse duration time. REWR decreases with an increase in gas pressure and remains approximately constant with electrode rotational speed.

The arc machine and the EDM machine were compounded to make a hybrid machine that was used to enhance the MRR and to reduce surface irregularities [18] with the help of air as a dielectric fluid. The arc machine generates high energy, and the EDM machine generates the high voltage for the discharge gap which causes an increase in the discharge gap helping in easy and efficient disposal of debris material. Hence, dry hybrid machining gives high MRR, low REWR, and good surface finish than Dry-EDM and dry arc machining. With the rise in the airflow rate, the MRR increases and REWR decreases. Surface roughness was degraded with the increase in airflow rate.

A comparative analysis was done by Uhlmann [19] in 2016. A cemented carbide hollow electrode was used to machine composite ceramic which has high hardness, high corrosion resistance, and high-temperature resistance. The time required for the erosion of composite material was more with argon gas, least in oxygen, and intermediate in water as dielectric fluid. The percentage of REWR was least in argon gas, highest with oxygen gas, and intermediate with water provided that all parameters should be the same with a different dielectric medium.

An experiment was conducted by Macedo [20] to explore the dependency of crater formation on the electrical breakdown mechanism. Conical copper electrode and a flat-surfaced copper workpiece were employed with two different dielectric fluids (air and argon) having different breakdown voltages (breakdown voltage of air = 327 V and breakdown voltage of argon = 127 V). The workpiece was made anode, and a single electrical discharge in the air medium was applied with a different pulse duration. Deeper and larger craters were formed by longer discharge in the air compared to craters formed by short pulses. With argon as a dielectric medium, no craters were formed when the tool was made the cathode. Several small craters were formed when the tool was made anode with a single discharge.

Macedo investigated the elementary method of tool electrode wear mechanism with the help of optical emission spectroscopy [21]. Aluminum was used as a work material because of its less ionization energy with the cylindrical pointed copper tool in the air as a dielectric medium. Optical emission spectroscopy gives information about the basic properties of electrical discharge plasma. When the tool was made anode, the percentage of tool material increases with the increase in pulse duration in plasma formation. But with the cathode tool, the constituent of tool material in plasma formation was almost constant.

The dry-EDM process was used for deburring of drilled CFRP composites using copper tool electrode and oxygen and air as dielectric fluid [22]. The workpiece was made cathode throughout the process, and a servo-mechanism was used to control the spark gap between the burrs and electrode. The experiment was conducted in all mediums for a constant amount of time keeping all the parameters the same for all. All of the burrs were removed in oxygen dielectric, only 74% of burrs were removed with air as dielectric fluid, and 42.5% of burrs were removed with conventional EDM.

2.4 Multi-spark Modeling of Dry-EDM

As multi-spark EDM shots the MRR in conventional EDM, it is possible that multi-spark can help to increase the MRR in Dry-EDM. In multi-spark, the tool electrode was divided into several tools to generate multiple discharges by the application of a single pulse to the circuit. To avoid short-circuiting, each tool was insulated from the other. Each tool was coupled with a primary energy source with capacitors in between. The capacitor is used to maintain sufficient discharge current between the electrode tool and the workpiece. With the increase in capacitance of the capacitor, the discharge current increases, and hence the MRR increases [23]. Improvement in MRR was observed with the increase in the number of electrodes [24].

Each electrode in multi-spark Dry-EDM is excited with a separate individual circuit with the capacitor in series. It is well known that with the increase in discharge current, MRR and TWR both increase because heat flux increases with discharge current [25]. TWR can be reduced to a minimum value by controlling the discharge current having the same MRR value in multi-spark Dry-EDM. To maintain a constant MRR in multi-spark Dry-EDM, the gap voltage should be equalized by adaptively changing the polarity of the pulse generator [26]. Transistor-type pulse generator should be employed to obtain high MRR, and RC-type pulse generator is employed to obtain better surface integrity [27].

3 Future Research Scope

As EDM is grounded on the principle of melting and vaporization of material from the work zone, it is obvious that temperature in the vicinity of the arc zone is about the melting point temperature of the workpiece. Temperature developed in the arc zone also depends on the kind of dielectric medium. HAZ of the workpiece will differ in a different dielectric medium. Hence, the mechanical properties, physical properties, electrical properties, etc., will differ in a different dielectric medium which affects the machinability of the workpiece.

4 Conclusion

Different methods were explored to improve the MRR and to reduce the TWR. It has been found that the best material for tool electrode is copper because very less electrode wear takes place with optimum MRR. Oxygen gas is used as a dielectric medium because of low ionization energy and high electronegativity which helps in better material removal rates. MRR is influenced by the process parameters like discharge current, discharge voltage, dielectric fluid, gas pressure, tool material, the polarity of tool electrode, etc. The MRR can also be improved using multi-spark Dry-EDM with very less TWR by dividing the electrode into multiple electrodes. Previous literature has primarily focused on the material removal rate and surface finish of the products. However, the surface integrity of the machined surface which determines the functional life of the component needs to be thoroughly investigated.

References

1. Singh, S., Maheshwari, S., & Pandey, P. C. (2004). Some investigations into the electric discharge machining of hardened tool steel using different electrode materials. *Journal of Materials Processing Technology, 149*, 272–277.
2. Ramani, V., & Cassidenti, M. L. (1985). *Inert-gas electrical discharge machining*. NASA Technical Brief Number npo15660.
3. Kunieda, M., Furuoya, S., & Taniguchi, N. (1991). Improvement of EDM efficiency by supplying oxygen gas into gap. *CIRP Annals Manufacturing Technology, 40*, 215–218.
4. Fujiki, M., Ni, J., & Shih, A. J. (2009). Investigation of the effects of electrode orientation and fluid flow rate in near-dry EDM milling. *International Journal of Machine Tools and Manufacture, 49*, 749–758.
5. Fujiki, M., Kim, G. Y., Ni, J., & Shih, A. J. (2011). Gap control for near-dry EDM milling with lead angle. *International Journal of Machine Tools and Manufacture, 51*, 77–83.
6. Singh Khundrakpam, N., Singh Brar, G., & Deepak, D. (2018). Grey-Taguchi optimization of near dry EDM process parameters on the surface roughness. *Materials Today: Proceedings, 5*, 4445–4451.
7. Dhakar, K., & Dvivedi, A. (2017). Experimental investigation on near-dry EDM using glycerin-air mixture as dielectric medium. *Materials Today: Proceedings, 4*, 5344–5350.

8. Ganachari, V. S., Chate, U. N., Waghmode, L. Y., Mullya, S. A., Shirguppikar, S. S., Salgar, M. M., & Kumbharg, V. T. (2019). A comparative performance study of dry and near dry EDM processes in machining of spring steel material. *Materials Today: Proceedings, 18*, 5247–5257.
9. Kunieda, M., Takaya, T., & Nakano, S. (2004). Improvement of dry EDM characteristics using piezoelectric actuator. *CIRP Annals Manufacturing Technology, 53*, 183–186.
10. Saha, S. K., & Choudhury, S. K. (2009). Experimental investigation and empirical modeling of the dry electric discharge machining process. *International Journal of Machine Tools and Manufacture, 49*, 297–308.
11. Govindan, P., & Joshi, S. S. (2010). Experimental characterization of material removal in dry electrical discharge drilling. *International Journal of Machine Tools and Manufacture, 50*, 431–443.
12. Joshi, S., Govindan, P., Malshe, A., & Rajurkar, K. (2011). Experimental characterization of dry EDM performed in a pulsating magnetic field. *CIRP Annals Manufacturing Technology, 60*, 239–242.
13. Puthumana, G., & Joshi, S. S. (2011). Investigations into performance of dry EDM using slotted electrodes. *International Journal of Precision Engineering and Manufacturing, 12*, 957–963.
14. Govindan, P., Gupta, A., Joshi, S. S., Malshe, A., & Rajurkar, K. P. (2013). Single-spark analysis of removal phenomenon in magnetic field assisted dry EDM. *Journal of Materials Processing Technology, 213*, 1048–1058.
15. Liqing, L., & Yingjie, S. (2013). Study of dry EDM with oxygen-mixed and cryogenic cooling approaches. *Procedia CIRP, 6*, 344–350.
16. Fattahi, S., & Baseri, H. (2017). Analysis of dry electrical discharge machining in different dielectric mediums. In *Proceedings of the Institution of Mechanical Engineers, Part E, 231*, 497–512.
17. Shen, Y., Liu, Y., Zhang, Y., Dong, H., Sun, W., Wang, X., Zheng, C., & Ji, R. (2015). High-speed dry electrical discharge machining. *International Journal of Machine Tools and Manufacture, 93*, 19–25.
18. Shen, Y., Liu, Y., & Sun, W. (2016). High-efficient dry hybrid machining of EDM and arc machining. *Procedia CIRP, 42*, 149–154.
19. Uhlmann, E., Schimmelpfennig, T. M., Perfilov, I., Streckenbach, J., & Schweitzer, L. (2016). Comparative analysis of dry-EDM and conventional EDM for the manufacturing of micro holes in Si3N4-TiN. *Procedia CIRP, 42*, 173–178.
20. Macedo, F. T. B., Wiessner, M., Hollenstein, C., Kuster, F., & Wegener, K. (2016). Dependence of crater formation in dry EDM on electrical breakdown mechanism. *Procedia CIRP, 42*, 161–166.
21. Macedo, F. T. B., Wiessner, M., Hollenstein, C., Kuster, F., & Wegener, K. (2016). Investigation of the fundamentals of tool electrode wear in dry EDM. *Procedia CIRP, 46*, 55–58.
22. Islam, M. M., Li, C. P., & Ko, T. J. (2017). Dry electrical discharge machining for deburring drilled holes in CFRP composite. *International Journal of Precision Engineering and Manufacturing-Green Technology, 4*, 149–154.
23. Shahane, S., & Pande, S. S. (2016). Development of a thermo-physical model for multi-spark wire EDM process. *Procedia Manufacturing, 5*, 205–219.
24. Yang, X., Yang, K., Liu, Y., & Wang, L. (2016). Study on characteristic of multi-spark EDM method by using capacity coupling. *Procedia CIRP, 42*, 40–45.
25. Jadhav, H. P., Mohanty, P. K., & Das, S. (2018). Numerical simulation of multi-spark electric discharge machining analysis for Ti6Al4V alloy drilling. *Materials Today: Proceedings, 5*, 28337–28346.
26. Muto, H. (2000). Technology & Engineering. *Development of Multi-Spark EDM, 49*, 119–122.
27. Huang, R., Yi, Y., Guo, G., & Xiong, X. (2020). Investigation of multielectrode multiloop with series capacitance pulse generator for EDM. *International Journal of Advanced Manufacturing Technology, 109*, 143–154.

An Analysis of Critical Success Factors Using Analytical Hierarchy Process for Implementation of Lean with Industry 4.0 in SMEs

Praveen Saraswat, Rajeev Agrawal, and M. L. Meena

1 Introduction

In today's business competitive environment, small and medium enterprises are willing to transform their operation strategies and practices to achieve long-term goals through technological advancements [1]. Lean production is recognized as one of the most efficient organizational models to enhance performances through the identification of value-added activities and avoiding wastes [2]. But due to the demand for customised products, lean techniques need to be modified to fulfill the present challenges faced by the organizations. Lean production has inadequate applicability in shorter product life cycles and highly customized products [3]. The organizations are dealing with this high demand for customized products by focusing on digital manufacturing or smart manufacturing.

Recently, industry 4.0 technologies become more popular to enhance the performance of the organizations using the combination of humans and information technology. Lean production is transforming to lean automation with these advanced tools and techniques [4]. The process of transforming has been speeded recently due to cheaper hardware and software systems and various developments in the field of automation and artificial intelligence [5]. The integration of lean and industry has emerged as a new area that deals with the potential benefits of both the techniques to identify the overall supportive environment that recognize the successful implementation of both the approach [6, 7]. Critical success factors (CSFs) are a management term that is needed for an organization to attain its goals. CSFs are the focused areas which are important to deal properly for effective implementation of lean with industry 4.0 technologies in organizations. The present study will answer these two research questions:

P. Saraswat (✉) · R. Agrawal · M. L. Meena
Department of Mechanical Engineering, Malaviya National Institute of Technology, Jaipur, India

R. Agrawal et al. (eds.), *Recent Advances in Smart Manufacturing and Materials*, Lecture Notes in Mechanical Engineering,
https://doi.org/10.1007/978-981-16-3033-0_24

1. *What are the most common CSFs for the SMEs to implement lean with industry 4.0 technologies?*
2. *Which are the most significant CSFs to focus more by SMEs in implementation phase?*

Hence, the purpose of this study is to identify the CSFs for lean implementation with industry 4.0 techniques in SMEs specially focusing on process industries so that organization must focus on these areas to implement lean effectively and successfully.

The paper is organized in five sections. Section 1 discusses the general background on lean automation and Industry 4.0 topics and CSFs toward their successful implementation. Section 2 describes literature work based on lean with and industry 4.0 tools and techniques. Section 3 presents the AHP methodology for ranking the CSFs and find the most noticeable one. Section 4 analyze the results and provide the key findings. Section 5 concludes the study and provides the main CSFs which can help other organizations.

2 Lean Manufacturing with Industry 4.0

Lean thinking mainly aims to eliminate all non-value-added activities arises during the manufacturing of the products. So one can eliminate wastes by using lean principles and tools, but in some cases, the wastes look to be unavoidable with the current tools and technologies [8]. There is a need to have more advanced tools and techniques which can integrate with current lean methodology. This gap can be fulfilled with industry 4.0 technologies primarily Cyber-Physical System (CPS), Internet of things based control system, cloud-based manufacturing and radio frequency identification (RFID), etc. The cyber-physical system is the key element which associates with the dynamic requirements of the production system. Cyber-physical systems are also used to improve the efficiency and productivity of the organization [9].

Some research studies have been conducted based on the integration of lean and industry 4.0 Satoglu et al. [10] concluded that Industry 4.0 technologies are not able to handle all problems of an organization, but it works in a very effective way when link with lean activities. Tortorella et al. [11] analyzed the survey data got from Brazilian manufacturing companies to find out the effects of industry 4.0 on lean and operational effectiveness. The results of their hypothesis tests show that for products or services there is a positive moderate relationship between operational effectiveness and lean improvement tools. Khanchanapong et al. [12] investigated the unique effects of the latest technologies and lean tools on the overall performance of the manufacturing plants in Thailand. The results showed that a noteworthy interaction effect occurs between them based on cost, quality of the product, and lead-time. Riezebos et al. [13] studied the influence of various IT technologies on lean system through a managerial approach. Mayr et al. [14] employed a real-time condition monitoring system for a stamping process. The sensor data are stored in the cloud and warning indications will be sent if the value of stamping force and tool wear exceeds the threshold value. They also developed a digital twin- enabled dynamic method using simulation software. A framework is proposed by Dave et al. [15] for the integration of IoT tools into the lean construction management systems to achieve higher real-time monitoring of field tasks. They concluded that the framework is very effective in information flow.

As explained above, the combination of Industry 4.0 technologies and lean principles proved a higher opportunity to implement them in a combined way called lean automation. This integration accelerates the organizations to enhance the performance of the key supplier, customer, process, and control and human factors.

3 Analytical Hierarchy Process Methodology

In the present era, small scale enterprises (SMEs) are looking for strategic decisions to achieve higher customer satisfaction with the proper selection of tools and techniques designed to enhance the enterprise's effectiveness. These decisions are very much important for organization performance. The analytical hierarchy process (AHP) methodology, which was developed by Saaty, is used to solve these types of decision-making problems. AHP divides the complex problem into various subproblems with some hierarchical levels and has its own characteristics. In AHP methodology, various hierarchy levels are linked pairwise and are assigned comparative based rules that show the priority level of each factor over others. The local and global priority weights will help in identifying and give ranking to the alternatives to achieve the final objective of the problem [16, 17]. A case study has been conducted by Belhadi et al. [18] identified 28 success factors critical to lean implementation in SMEs using AHP methodology and found Policy, leadership, and management most noteworthy CSF to implement lean production successfully for SMEs. Thanki et al. [19] application of analytical hierarchy process to analyze the effect of lean and green practices on overall process improvement. Total productive maintenance was found most important lean practice factor, while ISO 14001 was the most significant green practice factor. Singh et al. [20] identified and ranked various CSFs for Industry 4.0 implementation using analytical hierarchy process technique and found cyber-physical system, labor laws, and Internet of Things are the most significant CSFs.

Many case studies are available to show the potential application of the AHP method for prioritizing the factors to lean and industry 4.0 implementation [21–28].

In this paper, the AHP technique is adopted to rank the various CSFs for implementation of lean with industry 4.0 in SMEs. The various steps for implementing the AHP are categorized below.

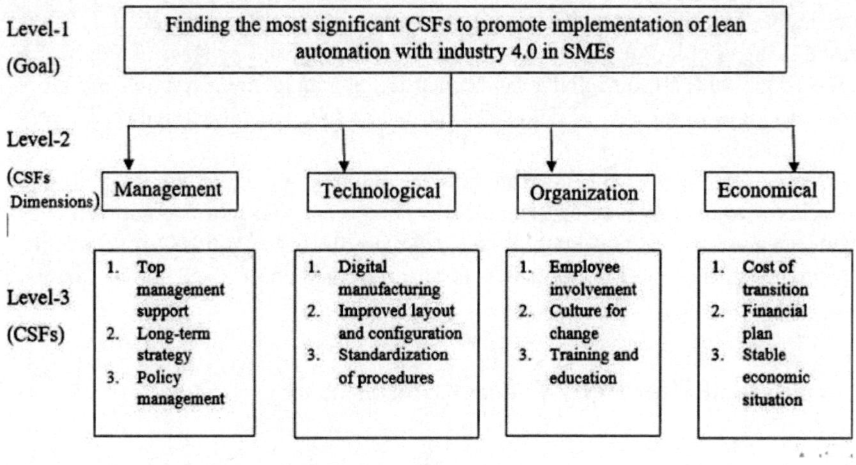

Fig. 1 The proposed hierarchical model of CSFs

3.1 Development of the Hierarchical Structure of CSFs

In this study, the identification of CSFs has been carried out through literature sources. The more focus was on the internal factors within the organization. A total 12 CSFs have been identified which categorized into 4 dimensions. There are three levels in the hierarchical model of CSFs as shown in Fig. 1.

3.2 Pairwise Matrix of Comparison

After the development of the hierarchical structure of CSFs, the next step is the development of pairwise comparison matrices of the main dimensions of CSFs. In this study, a paired comparison matrix showing the relative weight is formed after the discussion of three academics and three industry experts. Table 1 shows the pairwise matrix between the main dimensions of CSFs. Similar comparison matrices are also developed for various CSFs under the main dimensions.

Table 1 Pairwise comparison between main dimensions of CSFs

	Management	Technological	Organization	Economical
Management	1.00	0.50	5.00	2.00
Technological	2.00	1.00	5.00	4.00
Organization	0.20	0.20	1.00	2.00
Economical	0.50	0.25	0.50	1.00

Table 2 Normalized matrix

	Management	Technological	Organization	Economical	Local weight	Rank
Management	0.27	0.26	0.43	0.22	0.30	2
Technological	0.54	0.51	0.43	0.44	0.48	1
Organization	0.05	0.10	0.09	0.22	0.12	3
Economical	0.14	0.13	0.04	0.11	0.10	4

Table 3 Consistency ratios of each binary matrix

S. No.	Comparison matrix	Value of CR
1	Comparison matrix of main criteria or dimensions	0.078
2	Comparison matrix of CSFs under the main dimension of management	0.042
3	Comparison matrix of CSFs under the main dimension of technological	0.035
4	Comparison matrix of CSFs under the main dimension of organizational	0.051
5	Comparison matrix of CSFs under the main dimension of economical	0.075

Table 2 shows normalized matrix with the local and global weight of various dimensions of CSFs, and it is found that technological factor is the most important criteria with 0.48 weight factor and assigned rank. 1. The next dimension which is categorized as management got the second rank, followed by organization.

3.3 Consistency Ratio of Criterion

Consistency ratio is used to know the accuracy of the priority weight evaluated by experts of the comparison matrices. According to the implementation method of AHP, the consistency ratio must be less than 0.1. Table 3 shows the consistency ratios of each binary matrix which indicated the consistency ratios of all comparison matrices are less than 0.1.

3.4 Complete Global Weighting and Ranking of CSFs

From all the above steps, the overall ranking of the main dimensions and CSFs has been done. Table 4 shows ranking of main dimensions and CSFs based on their Global weight for implementing lean with Industry 4.0.

Table 4 Ranking of identified CSFs based on local and global weight

S. No.	Dimensions	Global weight	Rank	CSFs	Local weight	Global weight	Rank
1	Management	0.30	2	Top management support	0.51	0.153	2
				Long-term strategy	0.38	0.114	4
				Policy management	0.10	0.030	8
2	Technological	0.48	1	Digital manufacturing	0.63	0.302	1
				Improved layout and configuration	0.26	0.125	3
				Standardization of procedures	0.11	0.053	7
3	Organization	0.12	3	Employee involvement	0.21	0.025	9
				Culture for change	0.69	0.083	5
				Training and education	0.10	0.012	11
4	Economical	0.10	4	Cost of transition	0.11	0.011	12
				Financial plan	0.70	0.070	6
				Stable economic situation	0.18	0.018	10

4 Result and Discussion

In this study, the various CSFs linked with lean and industry 4.0 for SMEs have been identified and ranked. The 'Technological' dimensions identified as the most important criteria with 0.48 global weight and rank one among all main dimensions, followed by management with 0.30 global weight. In both the approaches, lean and industry 4.0, the most advanced technologies are used. Technology needs to be upgraded concerning time, in smart factories humans need the support of interfacing devices for communication and data flow at the global level. With the help of AHP, organizations can set their priorities and focus on these CSFs for further action plan and implement lean with industry 4.0 in SMEs.

5 Conclusion

SMEs are willing to integrate technological advancements into their processes with well strategy and planning. But there are many challenges for adopting lean with latest IoTs like infrastructure, capital, and technology. This gap can be fulfilled by identify most significance CSFs that so that organizations can move further step by step process and in a systematic manner. The priority wise importance of identified categories of CSFs for lean implementation with industry 4.0 in SMEs stated that the technological dimension holds the highest significance and thus, organizations should look into this dimension to establish their goals. Finally, it can conclude that these types of studies can help organizations to look into the most significant areas and focus on that first for the successful implementation of new techniques. It can be summarized that with the right way implementation of lean with industry 4.0, SMEs can definitely improve their operational effectiveness and sustain in a technological development environment.

References

1. Gunasekaran, A., & Ngai, E. W. T. (2005). Build-to-order supply chain management: A literature review and framework for development. *Journal of Operations Management, 23*(5), 423–451.
2. Belekoukias, I., Garza-Reyes, J. A., & Kumar, V. (2014). The impact of lean methods and tools on the operational performance of manufacturing organisations. *International Journal of Production Research, 52*(18), 5346–5366.
3. Knobloch, J., & Zühlke, D. (2017). Towards a lean automation interface for workstations AU—Kolberg Dennis. *International Journal of Production Research, 55*(10), 2845–2856.
4. Kolberg, D., & Zühlke, D. (2015). Lean automation enabled by industry 4.0 technologies. *IFAC-Papers OnLine, 48*(3),1870–1875.
5. Wollschlaeger, M., Sauter, T., & Jasperneite, J. (2017). The future of industrial communication: Automation networks in the era of the internet of things and industry 4.0. *IEEE Industrial Electronics Magazine, 11*(1), 17–27.
6. Sanders, A., Elangeswaran, C., & Wulfsberg, J. (2016). Finally signs of life for healthcare data bank. *Industry 4.0 Implies Lean Manufacturing: Research Activities in Industry 4.0 Function as Enablers for Lean Manufacturing, 70*(2), 19.
7. Jamwal, A., Agrawal, R., Sharma, M., & Kumar, V. (2021). Review on multi-criteria decision analysis in sustainable manufacturing decision making. *International Journal of Sustainable Engineering*, 1–24.
8. Womack, J. P., & Jones, D. T. (1996). *Lean thinking: Banish waste and create wealth in your corporation.* New York: Simon & Schuster.
9. Lu, Y. (2017). Industry 4.0: A survey on technologies, applications, and open research issues. *Journal of Industrial Information Integration, 6*, 1–10.
10. Satoglu, S., Ustundag, A., Cevikcan, E., & Durmusoglu, M. B. (2018). Lean transformation integrated with Industry 4.0 implementation methodology. In *Industrial Engineering in the Industry 4.0 era* (pp. 97–107). Berlin: Springer.
11. Tortorella, G., Giglio, R., & van Dun, D. H. (2018). Industry 4.0 as a moderator on the relationship between lean and operational performance. In *25th International Annual EurOMA Conference: To Serve, to Produce, and to Servitize in the Era of Networks, Big Data, and Analytics.*

12. Khanchanapong, T., Prajogo, D., Sohal, A. S., Cooper, B. K., Yeung, A. C. L., & Cheng, T. C. E. (2014). The unique and complementary effects of manufacturing technologies and lean practices on manufacturing operational performance. *International Journal of Production Economics, 153*, 191–203.
13. Riezebos, J., Klingenberg, W., & Hicks, C. (2009). Lean production and information technology: Connection or contradiction? *ComputInd, 60*(4), 237–247.
14. Mayr, A., Weigelt, M., Kühl, A., Grimm, S., Erll, A., Potzel, M., & Franke, J. (2018). Lean 4.0-A conceptual conjunction of lean management and Industry 4.0. *Procedia Cirp,72*, 622–628.
15. Dave, B., Kubler, S., Främling, K., & Koskela, L. (2016). Opportunities for enhanced lean construction management using Internet of things standards. *Automation in Construction, 61*, 86–97.
16. Tummala, V. R., & Wan, Y. W. (1994). On the mean random inconsistency index of analytic hierarchy process (AHP). *Computers & Industrial Engineering, 27*(1–4), 401–404.
17. Saaty, R. W. (1987). The analytic hierarchy process—What it is and how it is used. *Mathematical Modelling, 9*(3), 161–176.
18. Belhadi, A., EzahraTouriki, F., & Elfezazi, S. (2019). Evaluation of critical success factors (CSFs) to lean implementation in SMEs using AHP: A case study. *International Journal of Lean Six Sigma, 10*(3), 803–829.
19. Thanki, S., Govindan, K., & Thakkar, J. (2016). An investigation on lean-green implementation practices in Indian SMEs using analytical hierarchy process (AHP) approach. *Journal of Cleaner Production, 135*, 284–298.
20. Singh, J., Garg, D., & Luthra, S. (2018). An analysis of critical success factors for Industry 4.0: An application of analytical hierarchy process. *Industrial Engineering Journal, 11*(9), 1–15.
21. Sevinç, A., Gür, Ş., & Eren, T. (2018). Analysis of the difficulties of SMEs in Industry 4.0 applications by analytical hierarchy process and analytical network process. *Processes, 6*(12), 264.
22. Sony, M., & Naik, S. (2020). Critical factors for the successful implementation of Industry 4.0: A review and future research direction. *Production Planning & Control, 31*(10), 799–815.
23. Rossini, M., Costa, F., Tortorella, G. L., & Portioli-Staudacher, A. (2019). The interrelation between Industry 4.0 and lean production: An empirical study on European manufacturers. *The International Journal of Advanced Manufacturing Technology, 102*(9–12), 3963–3976.
24. Achanga, P., Shehab, E., Roy, R., & Nelder, G. (2006). Critical success factors for lean implementation within SMEs. *Journal of Manufacturing Technology Management, 17*(4), 460–471.
25. Hu, Q., Mason, R., Williams, S., & Found, P. (2015). Lean implementation within SMEs: A literature review. *Journal of Manufacturing Technology Management, 26*(7), 980–1012.
26. Thanki, S., Govindan, K., & Thakkar, J. (2016). An investigation on lean-green implementation practices in indian SMEs using analytical hierarchy process (AHP) approach. *Journal of Cleaner Production, 135*(1), 284–298.
27. Jamwal, A., Agrawal, R., Sharma, M., Kumar, V., & Kumar, S. (2021). Developing a sustainability framework for Industry 4.0. *Procedia CIRP, 98*, 430–435.
28. Fatorachian, H., & Kazemi, H. (2018). A critical investigation of Industry 4.0 in manufacturing: Theoretical Operationalisation Framework. *Production Planning and Control, 29*(8), 633–644.

Sustainability Concerns of Non-conventional Machining Processes—An Exhaustive Review

Kiran, H. Tomar, and N. Gupta

1 Introduction

The machining companies can lower the environmental harms and cut down their production costs, keeping the product the same or less with sustainability principles. The industrial crises are that the production was increased and was not according to the customers' needs. Hence, the produced goods were kept in the warehouses. Conventionally, companies are aimed at short-term financial goals leading to such an issue; a long-term view will help overcome this problem. Sustainable goals can be achieved via long-term vision [1, 2]. The things by which companies can improve various factors such as environmental, economic, and social are:

a. Reducing waste—Increase in wasted re-usage or recycle by less waste generation [3],
b. Efficient usage of resources such as material, water, and others.
c. Improved management of lubricating oils, fluids, hydraulics oils, and swarf—enhanced environment, safety performance, and health,
d. Embracing JIT (Just-In-Time) manufacturing and rest of the sustainable engineering methods [4],
e. Enhance operating statuses,
f. Utilizing the most acceptable method in machining,
g. Sustainable goal-oriented employees [5]

Non-conventional or advanced processes are those techniques by which hard materials can be worked, and complex shapes, different size, high-quality finishing can be achieved quickly. It forms the base of micromachining processes. Electrochemical (ECM), ultrasonic (USM), abrasive jet (AJM), abrasive water jet (AWJM),

Kiran · H. Tomar · N. Gupta (✉)
MED, Delhi Technological University, New Delhi, India
e-mail: Navritigupta@dce.ac.in

R. Agrawal et al. (eds.), *Recent Advances in Smart Manufacturing and Materials*, Lecture Notes in Mechanical Engineering,
https://doi.org/10.1007/978-981-16-3033-0_25

electric discharge (EDM), chemical (CHM), and laser beam (LBM) machining are those non-traditional machining techniques which are in extensive use [6]. Though they are responsible for some significant harms to the environment [7, 8]. The use of chemicals that are harmful to human health, environment-affecting flora and fauna, damaging the soil by increasing, or decreasing the pH level in each of the two later cases, apart from this the disposal of the harmful byproducts in chemical machining is incorrect. Some of disadvantages include generation of explosive hydrogen gas, sludge disposal, and isolation from workers being a hassle, and deadly and eroding property of the substances in an electrolyte such as nitrite and chromate in ECM. It leads to production of hazardous smoke, vapours and aerosols, electromagnetic radiation, difficult disposal of bulky metal waste and noxious liquid, miasma, gases, and substantial rot due to hydrocarbon root. The refined particle sizes of the aerosols, recast layer formation, and the high-intensity laser beam used in LBM also harm both the labour and the environment [9].

The electromagnetic radiation, ultrasonic noises, and the abrasive slurry are the significant impacts of LBM. Recycling parameters of sanders and the extraordinary corruption of abrasives on the use of various abrasive particles as abrasive in AJM is the foremost hurdle mentioned in discussion.

Figure 1a is showing graphical denotation of increase in use of sustainable manufacturing. Figure 1b shows the product life cycle and its impact on environment.

In this paper, major sustainability concerns of unconventional machining are being discussed.

Table 1 highlights the sustainability elements. Sustainable machining is in high demand and is the only key to help in a manufacturing unit's long run. It asks for increased investment in the beginning. There are various advancements introduced in the non-conventional machining trend, which will be discussed in the upcoming sections, which will conclude these developments and discuss the theoretical level solution model.

Green machining should be necessary for the environment, and it will be considered as a need in the upcoming future for manufacturing industries. The dry mechanizing benefits incorporate non-pollution of the outer environment or water, no leftover on the chipping will be displayed in minimized decomposition and sanitizing cost, no threat to well-being, and harmless to the skin and is non-allergenic. It also provides capital reductions in manufacturing. Cutting fluids are decided to be very important in the manufacturing industry as it has in removing the extra heat generated because of friction and also provides the lubrication between the tool and the workpiece. Tool life surface finish also increases, and it stops the formation of multiple edges and facilitates the transportation chips. Noxious vapours, irksome smell, smoke vapours, skin infuriation, bacteria, etc., are caused by cutting fluid that affects the environment and operator. Many environmental problems are arising because of coolant and waste disposal. Perfect time to start dry machining to save the environment as green machining is the machining of the future [11].

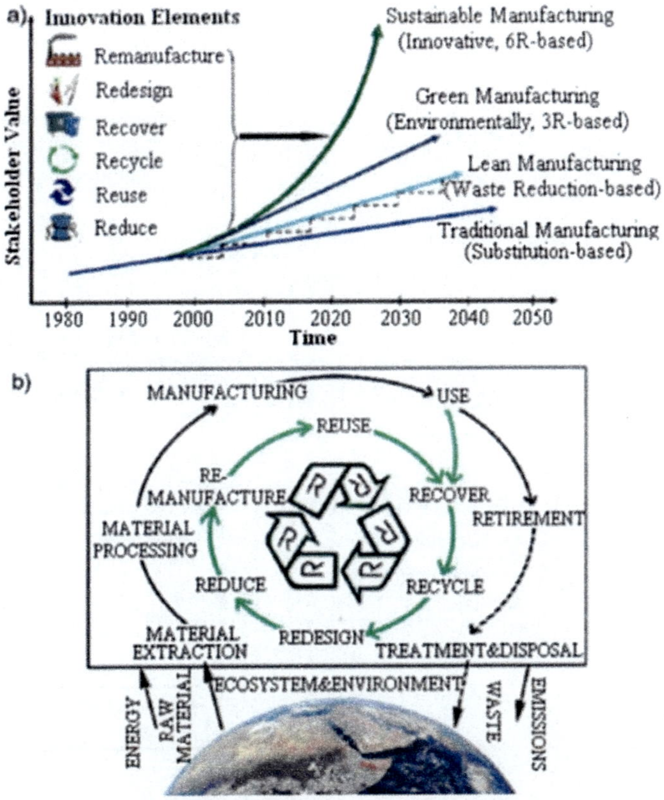

Fig. 1 **a** Graph showing the enhancement in sustainable manufacturing [26], **b** product and life cycle in 6R concept [27]

2 Literature Review

Hike of 60% in material removal rate was noted in almost-dry EDM than conventional EDM in work [12]. Mean surface irregularities enhanced till 40% plainly by the amalgamation of process quantifiers of dry ice machining in contrast with dry machining [13, 14] showed improvements in MRR and surface finish in almost-dry EDM with the assistance of a magnetic field. Figure 2 is showing the advantages of dry and near dry machining.

Taguchi and grey-dependent examination combination technique was efficacious in several methods of parameter hike in green EDM. Use of water faucet as a dielectric medium resulted in higher material removal rates, safer, eco-friendly, and cheaper was shown in separate research work [15]. In a research work, a replacement of hydrocarbon-founded dielectric fluid (kerosene) which was WVO dielectrics was made [16].

Table 1 Important constituents that affect sustainability in non-conventional machining [10]

Key sustainability elements	Determinants key operators	Fundamental operators
Environment	Hazardous scrap	• Utilization of green lubricants, dielectrics, and electrolytes • Process optimization
	Emission of electromagnetic rays	
	Noxious fluid, fumes, vapours	
	Power and substantial inefficiency	
Economy	Energy and resource consumption	• Process enhancement and hybridization • Reducing power and resource usage
	Low fertility	
	Lengthy process method	
	Huge price	
Society	Well-being	• Waste reduction and effective decomposition
	Security	
	Product stock and standards	

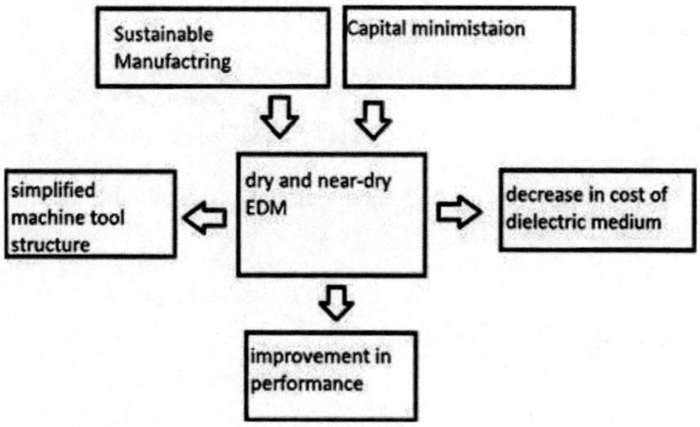

Fig. 2 Advantages of dry and near dry machining [10]

In another research work, a matching result was found in case of WVO dielectrics where using biodiesel as a dielectric liquid in place of traditional fluids which worked more productively and gave better outcomes [17]. Biodiesel, manufactured from utilized vegetable oils, is sustainable and harmless. The most recent of all is done by Mishra and Routara [18]; this year they have formulated a new bio-dielectric fluid, i.e. non edible oil-based Calophyllum Inophyllum (Polanga) which has shown great results. It was capable of increasing smoothness of the surface, increasing the MRR, increasing surface hardness, and decreasing aerosol emission rate by 16.64%, 86%, 6.64%, and 17.33%, respectively, than a conventional hydrocarbon-based dielectric fluid.

In electrochemical machining, there are many advances in the past six years, where regular mineral water in place of conventional electrolytes and it was highly precise, efficient, and inexpensive [19]. In the year 2015, in another research work, the reverse drilling method for the complete elimination of the machining time was proposed [20]. The experiment was held using an electrolyte as citric acid and the whole process and results were examined. In yet another experimental research work, analysing and observing the optimum electrolyte for ECM of stainless steel 304 was done. A complexing agent with neutral salt is the researched electrolyte. The calcium-based electrolyte presents an electrochemical property with high oxidative stability and high ionic conductivity shown in [22]. The AJM process uses abrasives which are not decomposable and are non-separable, so in the year 2011, another research was done in which invention of biodegradable abrasive constitutions consisting of plant-based inborn abrasive means, calcium carbonate, polymeric substance, etc. [23].

3 Discussion

Now in this section, an elaborate discussion over the topics of unconventional machining has been given.

3.1 Non-conventional Machining Process

These advanced processes are responsible for the generation of various hazardous and noxious products in different forms which can affect the environment and the well-being of the operators [24].

3.1.1 Electrochemical Machining (ECM)

ECM is the metal removal process by electrochemical process, and it is used for machining the hard material which are difficult by other machining operations.

Hazards of ECM include [9]:

Hydrogen gas is very explosive, which is released as a product of electrochemical reactions.

$$H_2O + 2e^- \Rightarrow H_2 + 2OH \tag{1}$$

$$Fe + 2H_2O \Rightarrow Fe(OH)_2 + H_2 \tag{2}$$

Equations (1) and (2) are showing the chemical reaction. Electrolyte splash means electrolytes are made up of nitrates and chlorides of sodium, which is non-dissolving in hydroxides and is there as a residue. These are harmful and can cause skin infatuation, so any contact with it must be avoided. A continuous leak proof supply system must be made, or a closed door locking and leak proofing must be done so that any leakage is not possible out of the encapsulated working space.

Chromate: Exposure through the inhalation, injection, or direct contact of the trivalent chromium compounds will affect the skin and organs like kidneys, liver, etc.

Nitrate: Nitrate and nitrite also have several health effects on the human body, such as:

(a). They react with haemoglobin in the blood,
(b). Affect the working of the thyroid gland,,
(c). Efficiency of vitamin A,
(d). There should be a proper exhaust ventilation system to remove all the harmful gases so as not to affect the operator.

Recently, a new ECM technology called recycling ECM has been reported. This technology uses the pulse-ECM concept for recovering copper, alloy steel, and nickel alloy in a stable, metallic form without hydroxides, using intermediate electrolyte processing. Figure 3 gives way to minimize the hazardous effect of ECM process.

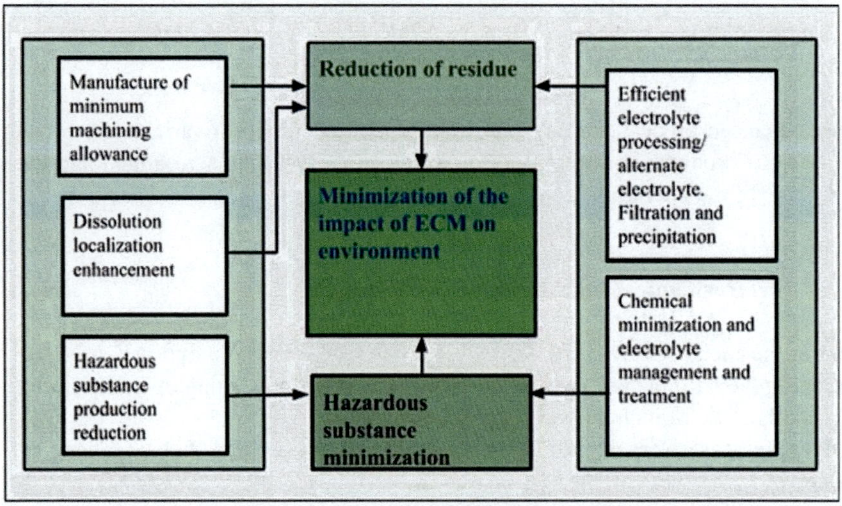

Fig. 3 Different methods to minimize the effect of ECM process [10]

3.1.2 Electrodischarge Machining (EDM)

In EDM, the metal removal process is done by the heat from the spark produced between the two electrodes with dielectric fluid as a medium. EDM process has many hazards produced during machining like [9, 25]:

(a) Hazardous smoke, vapours, and aerosols released in the atmosphere,
(b) Decomposed products and heavy metals traces,
(c) Dielectric fluid containing the hydrocarbons can affect the skin of the operator,,
(d) Possible fire hazard and explosions,
(e) Electromagnetic radiations.

Various factors on which the production of fumes and aerosol depends are material removal process, work material, and dielectric.

(a) Workpieces contain chromium, molybdenum. Nickel is a toxic and health attacking substance.
(b) The composition and the viscosity of dielectric also affect the amount of fumes and vapour. A dielectric having low viscosity produces fewer amounts of fumes.

Using the organic dielectric fluid in EDM generates dangerous vapours such as hydrocarbons in form of polycyclic aromatic compound aerosols and rest of the products produced because of splitting of oil and its additives. There is a demand for finding methods to reduce EDM'S working dielectric exposure. Due to erosion of electrodes and workpieces, various compounds and elements such as titanium carbides, tungsten carbides, nickel, chromium, molybdenum, barium, when released in air, they affect humans who are inhaling them along with air. For reducing the hazards, we have to follow these measures strictly;

(a) Using suitable filters to reduce the number of fumes released into the atmosphere.
(b) Maintain the temperature of the medium around 15 °C below the flashpoint.
(c) Stop the electromagnetic radiation from going out into the atmosphere using proper shielding.
(d) Reduce the chances of a fire in the dielectric fluid.
(e) Use the sensors for monitoring the level of dielectric fluid.
(f) Disposal of waste should be far away from the active area.
(g) Aware the operator about the hazards and risks.

The following are the advancements in the EDM process concluded, a set-up for the process in which the whole EDM process will be carried out in an enclosed environment from which the harmful gases will not be released into the environment directly but will be filtered by a chimney design outlet integrated with another set-up before releasing into the atmosphere. This set-up will be compounded by a supplementary set-up, which will primarily work to separate the metal from the dielectric fluid with the magnetic separator's help and ensure the continuous flow of the fluid.

For the hazardous smoke and vapour, a proper filtration system should be installed to stop them from releasing into the atmosphere. The magnetic separation method to remove the heavy metals from the dielectric can be used. Also, the closed chamber to stop the radiation from going out of the chamber and hence, not affecting the operator can be used. Unconventional processes such as EDM consume more energy than conventional machining, so we should refer to renewable energy sources so that we can save natural resources like coal used in thermal power plants to produce electricity for the industrial area.

3.1.3 Abrasive Jet Machining (AJM)

In the abrasive jet machining process, significant hazards are the dust particle generation, whose source is mainly silica, and the rest are lead, zinc, arsenide, cadmium, and mercury. The operational status in this type of machining process is the use of masks, which are significantly less reliable but still being used, which is highly risky and injurious for health as it is immensely saturated in the atmosphere they work in and are of different sizes from very small to very large. The major one is silica dust elements produced at the time of operation on the workpiece and is very small in size and can cause silicosis (a disease that can cause the lung to get stiff and its disfiguration). The identifying indication is breath shortening, the cough along with pain in the chest. These dust particles are the cause of the deterioration of one's well-being. There are chances that the worker will be prone to having lung cancer. For controlling this health hazard, the main issue of silica dust must be aimed to the filtration of the silica dust is a significantly less effective method, so the use of an abrasive element in place of silica and enclosure of the whole set-up or a perfectly safe working condition set-up can help in overcoming this problem. In the current operating manufacturing units, provision of a safety kit to each worker, encapsulation of the process, and good dust collection followed by recycling and abrasive water jet methods must be implemented strictly. The dust must be wetted before the clean-up, which will reduce the harm. The effect of an abrasive jet involves

Worker—There is a danger to the operator's health and performance. There are chances of stress, weakness, fatigue, dissatisfaction, and defect in working. They must take all the measures strictly.

Product—As the dust particle pollutes the environment, the finished particle surfaces have a high chance of getting contaminated by the settlement of these dust particles on it.

Manufacturing unit—All the mechanical set-up and the machines need proper cleaning and maintenance to work correctly, which will be hindered by the air's dust emission. So, there is a need for a silica dust infiltration set-up and air transfer systems.

Apart from these, the use of biodegradable abrasives will help minimize the problem of the collection, decomposition, and filtration of these abrasives.

The need for high oxidation stability increases as operating temperature increases. Some of the desirable characteristics of cutting fluid are:

(a) Viscosity: The level of dielectric's viscosity is not constant for all applications, but low viscosity helps provide a more comfortable pump and enhanced flushing characteristics.
(b) Acid number: Oil with low levels of acid is preferable for corrosion and cost consideration.
(c) Colour: Dielectric colour has no such significant effect.
(d) Odour: There is no standard measure for the smell, but generally, dielectric oil with minimum odour is preferable to reducing health-related consequences as a factor of sustainability.

4 Future Scope

(a) Use of liquid nitrogen as a dielectric fluid in electric discharge machining, which is being practised yet, and we will continue the experimentation of this hypothesis as our future goal, i.e. to make advancements in liquid nitrogen to be used as a dielectric fluid.
(b) Ca-based fluid can be used as an electrolyte in the ECM process, which has recently been proposed for batteries that can be implemented in ECM.
(c) The experimentation of the solution model will also be carried out.

5 Conclusions

From the discussion above, we can conclude that there are many sustainability issues that affect the environmental conditions and the health of a worker. Need for advancement in various machining operations so that these significant issues can be solved.

For EDM, an enclosed set-up with an integrated solution system which can provide a check at each step of the process to minimize the chances of any harm or hazard to the surroundings. These step-by-step measures are an enclosed chamber with a chimney exit. Similarly, for ECM, the electrolyte must be replaced with a newly harmless alternative along with insulating set-up and real time monitoring systems. And, for AJM, the abrasives need to be replaced with biodegradable and less harmful replacements.

All the processes involve very high power consumption, so we are wasting a lot of non-renewable resources such as coal used in thermal power plants to produce electricity for the industries. So these machining industries should be set-up near the power plants which run on renewable sources like hydropower plants, wind mills, solar energy, etc.

References

1. Pušavec, F., & Kopac, J. (2009). Achieving and implementation of sustainability principles in machining processes. *Advances in Production Engineering & Management, 4*, 151–160.
2. Envirowise, GG 446. (2004) *Sustainable manufacturing: A signposting guide for metal machining companies*. Project documentation, www.envirowise.gov.uk.
3. Balic, J., Cus, F. (2007). Intelligent modelling in manufacturing. *Journal of Achievements in Materials and Manufacturing Engineering, 24*(1), s.340–349, 340–348
4. Zuperl, U., & Cus, F. (2008). Machining process optimization by colony based cooperative search technique. *Strojniski Vestnik, 54*, 751–758.
5. Kopac, J., & Pušavec, F. (2009). Concepts of sustainable machining processes.
6. Davim, J. (2013). *Nontraditional machining processes*. Springer, London. https://link.springer.com/book/10.1007%2F978-1-4471-5179-1
7. El-Hofy, H., & Youssef, H. (2009). Environmental hazards of nontraditional machining. In *Proceedings of the 4th IASME/WSEAS International Conference on Energy & Environment (EE'09)*. World Scientific and Engineering Academy and Society (WSEAS), Stevens Point, Wisconsin, USA, pp. 140–145
8. Gamage, J. R., & DeSilva, A. K. M. (2015). Assessment of research needs for sustainability of unconventional machining processes. *Procedia CIRP, 26*, 385–390. https://doi.org/10.1016/j.procir.2014.07.096
9. El-Hofy, H., & Youssef, H. (2009). Environmental hazards of nontraditional machining, pp. 140–145.
10. Gupta, K., & Gupta, M. K. (2018). Developments in nonconventional machining for sustainable production: A state-of-the-art review. *Proceedings of the Institution of Mechanical Engineers, Part C: Journal of Mechanical Engineering Science, 233*(12), 4213–4232. https://doi.org/10.1177/0954406218811982
11. Jain, A., Kansal, H. (2017). Green machining—machining of the future. In *4th National Conference on Advancements in Simulation & Experimental Techniques in Mechanical Engineering (NCASEme-2017)*
12. Dhakar, K., & Dvivedi, A. (2015). Parametric evaluation on near-dry electric discharge machining. *Materials and Manufacturing Processes, 31*(4), 413–421. https://doi.org/10.1080/10426914.2015.1037905
13. Khanna, N., Agrawal, C. (2019). Titanium machining using indigenously developed sustainable cryogenic machining facility. In *Materials Forming, Machining and Tribology* (pp. 183–205). Springer International Publishing. https://doi.org/10.1007/978-3-030-18854-2_8
14. Gholipoor, A., Baseri, H., Shakeri, M., & Shabgard, M. (2014). Investigation of the effects of magnetic field on near-dry electrical discharge machining performance. *Proceedings of the Institution of Mechanical Engineers, Part B: Journal of Engineering Manufacture, 230*(4), 744–751. https://doi.org/10.1177/0954405414558737
15. Tang, L., & Du, Y. T. (2013). Experimental study on green electrical discharge machining in tap water of Ti–6Al–4V and parameters optimization. *The International Journal of Advanced Manufacturing Technology, 70*(1–4), 469–475. https://doi.org/10.1007/s00170-013-5274-5
16. Valaki, J. B., & Rathod, P. P. (2015). Assessment of operational feasibility of waste vegetable oil based bio-dielectric fluid for sustainable electric discharge machining (EDM). *The International Journal of Advanced Manufacturing Technology, 87*(5–8), 1509–1518. https://doi.org/10.1007/s00170-015-7169-0
17. Ng, P. S., Kong, S. A., & Yeo, S. H. (2016). Investigation of biodiesel dielectric in sustainable electrical discharge machining. *The International Journal of Advanced Manufacturing Technology, 90*(9–12), 2549–2556. https://doi.org/10.1007/s00170-016-9572-6
18. Mishra, B. P., & Routara, B. C. (2020). Evaluation of technical feasibility and environmental impact of Calophyllum Inophyllum (Polanga) oil based bio-dielectric fluid for green EDM. *Measurement, 159*, 107744. https://doi.org/10.1016/j.measurement.2020.107744

19. Yang, Y., Natsu, W., & Zhao, W. (2011). Realization of eco-friendly electrochemical micromachining using mineral water as an electrolyte. *Precision Engineering, 35*(2), 204–213. https://doi.org/10.1016/j.precisioneng.2010.09.009
20. Ryu, S. (2015). Eco-friendly ECM in citric acid electrolyte with microwire and microfoil electrodes. *International Journal of Precision Engineering and Manufacturing, 16*(2), 233–239. https://doi.org/10.1007/s12541-015-0031-3
21. Guodong, L., Yong, L., Quancun, K., & Hao, T. (2016). Selection and optimization of electrolyte for micro electrochemical machining on stainless steel 304. *Procedia CIRP, 42*, 412–417. https://doi.org/10.1016/j.procir.2016.02.223
22. Li, Z., Fuhr, O., Fichtner, M., & Zhao-Karger, Z. (2019). Towards stable and efficient electrolytes for room-temperature rechargeable calcium batteries. *Energy & Environmental Science, 12*(12), 3496–3501. https://doi.org/10.1039/c9ee01699f
23. Imam, S. H., & Canyon, A. (2011). *U.S. Patent No. US 8,008,242 B1.* Washington, DC: U.S. Patent and Trademark Office. https://patents.google.com/patent/US8008242B1/en#patentCitations
24. Hughes, P., & Ferrett, E. (2011). *Introduction to health and safety at work.* Oxford: Butterworth-Heinemann.
25. Rajurkar, K. P., Hadidi, H., Pariti, J., & Reddy, G. C. (2017). Review of sustainability issues in non-traditional machining processes. *Procedia Manufacturing, 7*, 714–720. https://doi.org/10.1016/j.promfg.2016.12.106
26. Jawahir, I. S., & Dillon, Jr., O. W. (2007). Sustainable manufacturing processes: New challenges for developing predictive models and optimization techniques. In *Proceedings of First International Conference on Sustainable Manufacturing*, pp. 1–19
27. Jaafar, I., Venkatachalam, A., Joshi, K., Ungureanu, A., Silva, N., Rouch, K., Dillon, O., & Jawahir, I.S. (2007). Product design for sustainability: A new assessment methodology and case studies. https://doi.org/10.1002/9780470168202.ch2

Progression of EFQM and Deep-Dive into EFQM 2020

M. A. Narasimha Murthy, Kuldip Singh Sangwan, and N. S. Narahari

1 Introduction

Over the years the International community and national industrial bodies are focussed on driving the companies towards Business or Organisation or Service Excellence. The word "Excellence" has posed questions and challenges in the mind of academician, senior executives, practitioners and Industry at large. Due to this, the construct of the model or framework of excellence models is changing and getting evolved. As per the Shewhart Cycle applied for improvements in management practises, it drives home the concept of learning and improvement. Plan-Do-Check-Action-Next Plan is relevant for all aspects of quality including Quality of the organisation through business excellence journey. Even though the EFQM has completed 30 years of journey as a forum and still it is evolving, the look and feel of the model structure and branding were same till 2019, in a newer version of 2020 model announced by EFQM the basic motive itself has changed. During 1992, the EFQM Excellence award was called European Quality Award. In the year 1992, the First Excellence Award of EQA was awarded to m/s Xerox Corporation Limited. The objective of the award was to identify role model organisations, to share their experience and the practises so that others could adopt from such practises. The award process was able to drive the competitiveness of the European region and

M. A. Narasimha Murthy (✉)
Department of Mechanical Engineering, Birla Institute of Technology and Science (BITS), Pilani, Rajasthan 333031, India

K. S. Sangwan
Department of Mechanical Engineering, Birla Institute of Technology and Science, Pilani, Rajasthan 333031, India

N. S. Narahari
Department of Industrial Engineering and Management, RV College of Engineering, Bangalore 560059, India

R. Agrawal et al. (eds.), *Recent Advances in Smart Manufacturing and Materials*, Lecture Notes in Mechanical Engineering,
https://doi.org/10.1007/978-981-16-3033-0_26

helped in economic improvement as well. The good practise has since evolved, tested and started yielding the desired results and has become part of the practise and the norms in many organisations world over. The advances in Organisational excellence achieved by one organisation would inspire other organisations and there was a quality movement globally to record achievements and embrace the excellence Journey. This helped in developing a product, services offerings to the market and business competitive spirit in the whole region.

2 Key Terms and Definitions

It is important to understand the definition of excellence before one can take a deep-dive into the subject of business excellence. Few such examples of these definitions are discussed below;

- The excellence definition provided by Michaela et al. [1], In the dictionary of neologism of the Romanian Academy, the excellence is defined as an "honorary title granted to individuals or a high degree of perfection; the quality of being excellent, very good, exceptionally admirable, wonderful".
- Business excellence models are frameworks that when applied within an organisation can help to focus the thoughts and actions in a systematic and structured way to increased performance.
- Business excellence, as described by Mann et al. [2], refer to "Outstanding practises in managing the organisation and achieving results, all based on a set of eight fundamental concepts".

3 Background of the EFQM Model

Even though officially, EFQM was formed in 1989, Jacques Delors, the president of the European Commission enlightened about the TQM Movements in Japan and introduction of MBNQA in the US and triggered a thought of institutionalising the recognition programme of Quality for Excellence of an organisation. As a part of the EFQM Forum, the leaders in the forum felt the need to building competitiveness in the European Region. With Jacques Delors who led the leaders of 14 European companies, the leaders agreed to establish the European response to the global movement of Quality and Excellence competitiveness in the region. In 1988, the foundation of EFQM signed the agreement and tabled the formation of EFQM.

4 Literature Studies

In the current research, the author's carried out the literature study focussing on a few key papers and reviewed the findings of these papers. These findings are briefed here. Tan and Khoo [3] and Miguel's study [4] deliberated on a worldwide comparison of quality and business excellence programmes in the world. Later, one more key research undertaken on behalf of NIST by Mohammad and Mann [5] highlighted 95 national and quality awards adopted across the globe. Nenadál [6] highlighted the research of new Model EFQM 2020 and the description of certain recommendations by way of guidance points is seemingly superficial and are confusing. From the literature studies, it was evident that researchers have not focussed on the progression of the EFQM model from its stage of infancy to being a more mature model as of today. In this paper, the author focusses on the progression of three cornerstones of the model and other key elements of the EFQM model from its inception to the EFQM model as in 2020.

5 Progression Mapping of EFQM—A Research Study

5.1 Overview of Three Concern Stones of the Model

The philosophy of the framework is based on the three cornerstones of the model: The fundamental concepts, the criteria and the RADAR. This forms the basis for model construct and deployment.

The fundamental concepts define the founding principles which help the organisations to achieve sustainable excellence. The criteria and sub-criteria provide a framework to convert the fundamental concepts and RADAR thinking into practise. RADAR, which stands for result, approach, deployment, assess and refine, is a simple, systematic and structured approach yet powerful tool for driving systematic improvements in all areas of the organisation and it also improves the transparency in the organisation.

5.2 Deliberation on Evolution of EFQM Model Over Three Decades

The research study of the evolution of the EFQM model has been carried based on tracking the changes in the model overtime periods. The authors have phased the model at a broad level into five periods. The initial phase is 1991–1999 and the second phase is the period between 2000 and 2002. The third phase is the period 2003–2010, the fourth phase is defined as the period between 2011 and 2018 and the fifth phase is reckoned from 2019 and period further on.

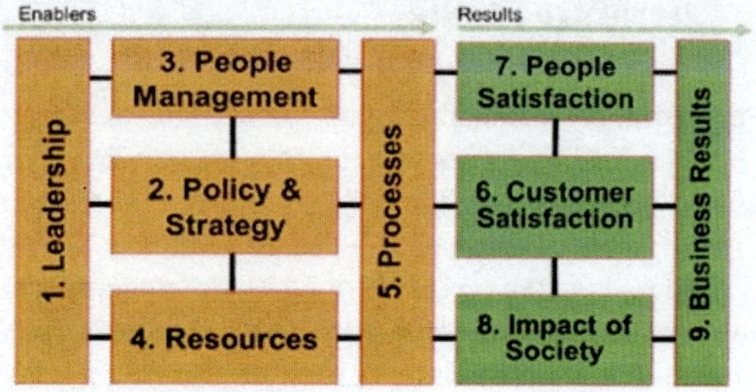

Fig. 1 Initial construct of the model (1991–99). *Source* Nabitz [7]

5.3 Construct of the Model—Criteria

5.3.1 Initial Construct

To facilitate the identification of the Business Excellence organisation in 1991, the forum created a framework consisting of five enablers that represented these under the practises of leadership, policies, processes, resources and people that an organisation uses to develop and deliver their strategies, and the four results criteria track the progress of the organisation against the needs and expectations of the stakeholders. To emphasis organisational role on society, the forum included the "Impact on Society" become a part of the Awards process. This initiative set the tone for expectations from an excellent organisation and formed yardstick for measuring the organisational excellence and the organisational mechanisms that drive excellence and considered society as the key stakeholder. Figure 1 shows the initial construction of the model. In this period, it was realised that the recognition was missing amongst a large part of the public sector organisations.

5.3.2 The Second Phase of the Model

Considering the inputs of award cycle and based on the various academic, research and professionals experiences, the forum started focussing on revamping the structure of the model. During 1997, the model was extended to public sector and in 1999; Formal Release of Fundamental Concepts of Excellence took place and led to the creation of Network of National Partners to associate with a forum for strengthening the adaptability, improvement and enhancement of the EFQM model. In this period, the model got more popular and the forum team size got enhanced. In this period, the resource criteria strengthened and were renamed as "Partnership and Resources". All the result category got relabelled and moved from satisfaction to full-fledged result section. The result categories were termed as customer, people, society and key business results. The changes from the initial version to the second phase are marked in the red-colour font in Fig. 2. To encourage the award process learnings

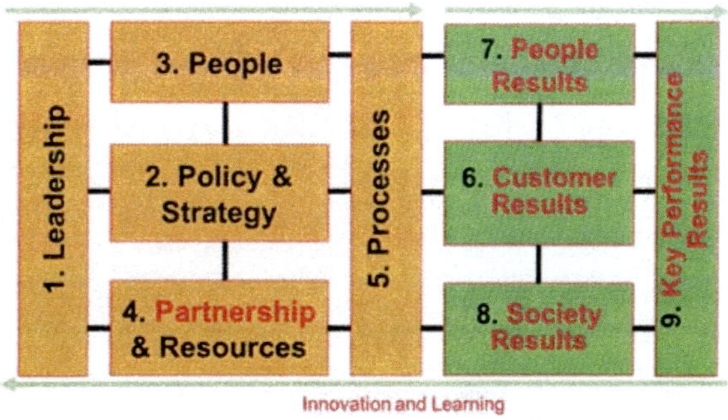

Fig. 2 Second phase construct of the model (2000–02). *Source* Eskildsen [8]

and internal mechanism, innovation and learning got added in the model to focus across the enablers and result category.

5.3.3 The Third Phase of the Model

In this phase, the model started reaching a mature state and the popularity significantly increased, the national network started co-operating well, the model created a great image in the market place. The award process started becoming more competitive and this started putting pressure on the organisation to win the coveted EFQM award. In 2001, EFQM Level Excellence award was introduced a new award category. There are three recognition levels, 3 stars to 5 stars, depending on the maturity observed during the assessment. This step helped thousands of organisations to move toward EFQM Levels of Excellence to support their quality journey. The Criteria-5 got relabelled and enhanced to cover products, processes and services. In the result category, the key business results relabelled to key results. Apart from Learning and Innovation across results and enable category, creativity also added as one of the key ingredients in the model. Figure 3 shows the construct of the model in this phase.

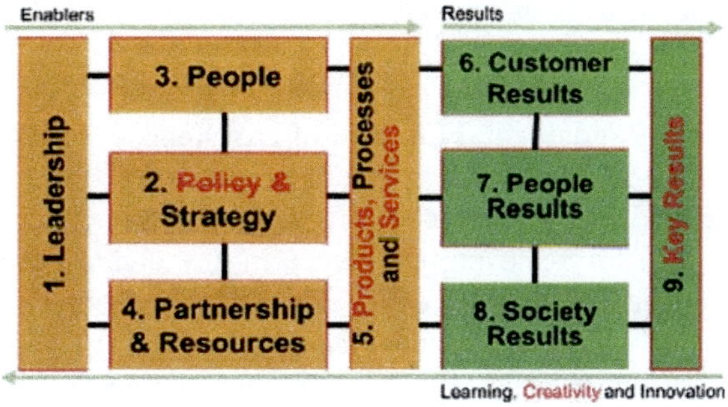

Fig. 3 The phase-3 construct of the model. *Source* EFQM 2010

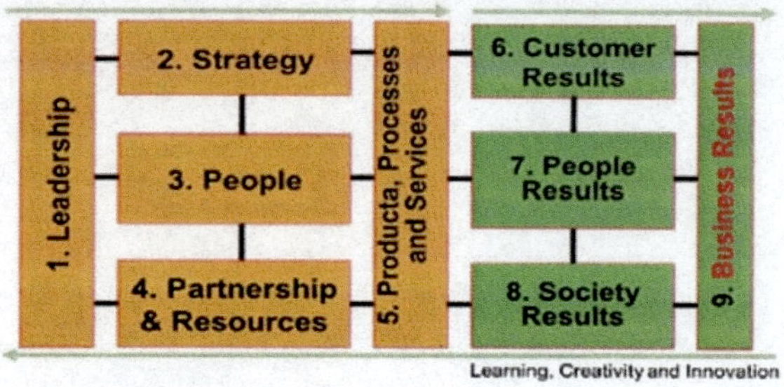

Fig. 4 The phase-4 construct of the model. *Source* EFQM 2012

5.3.4 Fourth Phase of the Construct

In this phase, the model got major penetration and acknowledgement from senior management and executives to adopt this model as a vehicle for an organisational improvement journey. Model and fundamental concepts revised to meet the demands of the changing organisation and business environments. Fine-tuning of the model at sub-criteria and guidance point improvised. However, there are no major drastic changes that happened to the construct of the model. Figure 4 shows the construct of the model in this phase.

5.3.5 Fifth Phase of the Construct, EFQM-2020

The contour of the model got changed significantly; the earlier enabler and result criteria got revamped. The Criteria of five enablers and four results got restructured into five enablers and two result category. The model got sub-divided into three category direction, execution and result. The direction has got two criteria, execution got three criteria and the result got two criteria. The new EFQM Model is a result of a co-creation involving nearly 2000 change experts, 24 organised workshops, face-to-face discussion with leaders in over 60 diverse organisations and created a core team of experts and contributors from across industries and academia. The construct of the model is shown in Fig. 5.

5.3.6 Fundamental Concept of Excellence

The fundamental concept of excellence forms the fulcrum of the model. This shows the path for achieving excellence in any organisation in a sustainable way and describes the attributes of an excellent organisational culture. This also serves as a common language for top management. Taking clues from the European Convention of human rights (1953) and subsequent updates, they become the base for these concepts. During the initial phase, these were "part of the fabric" of the model and there seems to be no written document on this since limited leaders were using the model. The first version of the fundamental concepts, introduced in 1999, consist of

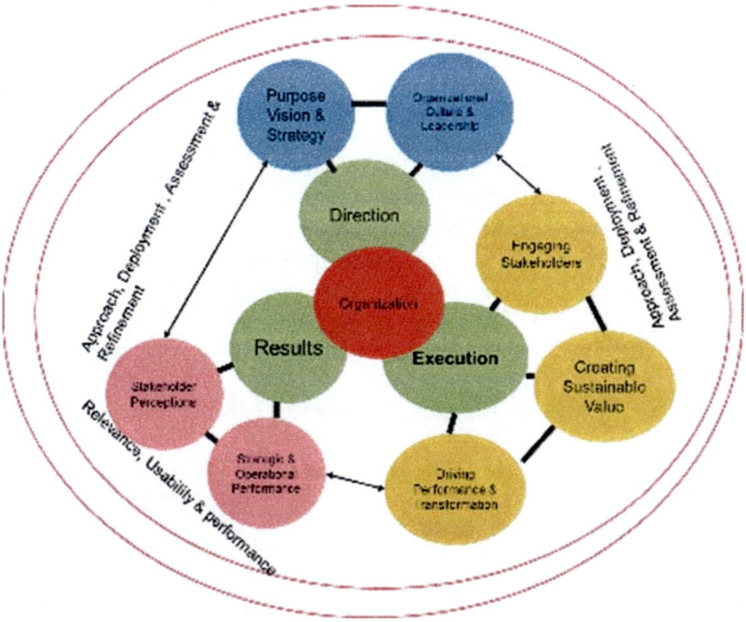

Fig. 5 EFQM construct of 2020. *Source* EFQM 2020

eight titles with high-level descriptions of the essence of the concepts. Subsequently, during the review cycle of the model, to model is oriented towards the emerging trends and challenges faced by community and organisation, the fundamental concept got updated. Table 1 highlights the evolution of the fundamental concepts of excellence.

In the fifth Phase, the concepts moved beyond simple high-level statements and ended up with a duplication of between the criteria and the concepts which started confusing assessor community. Many assessors used of the concepts to structure the executive summary of a assessment feedback report. Since it was confusing to the assessor and applicant organisation. These concepts are not used effectively in the assessment process, hence it is removed in the new version.

5.4 RADAR Logic

The radar logic is built based on the tenets of Deming's PDCA cycle. In simple words, it means that change in an organisation cannot be made simply through noble intentions or exhortations and frequent monitoring. This need to be managed through structured approaches. To assess the excellence, two approaches are generally followed, RADAR for enablers assessment and RADAR for results. In this initial phase, there is no RADAR as a formally defined method, hence it is not presented. In the case of result assessment, the relevance, usability and performance were used for

Table 1 Fundamental concept evolutions

Second phase	Third phase	Fourth phase
1. Results orientation	1. Achieving balanced results	1. Creating a suitable future
2. Customer focused	2. Adding value for customers	2. Adding value for customers
3. Leadership and constancy of purpose	3. Leading with vision, inspiration and integrity	3. Leading with vision, inspiration and integrity
4. Management by processes and facts	4. Managing by processes	4. Managing agility
5. People development and involvement	5. Succeeding through people	5. Succeeding through talents
6. Partnership development	6. Building partnerships	6. Developing organisational capability
7. Corporate social responsibility	7. Taking responsibility for a sustainable future	7. Sustaining outstanding results
8. Continuous learning, innovation and improvement	8. Nurturing creativity and innovation	8. Harnessing creativity and innovation

Source EFQM

assessing. The construct has been fine-tuned for establishing scoring and guidance to assist the assessors, however, there was no change in the framework. The scoring guidelines help the assessor to determine the approach, deployment and assessment refinement of each sub-criterion to arrive at score using a five-level scale of assessment. The parameters used for assessment enablers and results have been highlighted in Tables 2 and 3.

Till 2020, the elements of the RADAR looked at the alignment. No change was noticed in the constituents. The 2020 version of the new RADAR scoring format

Table 2 Comparison of enabler's RADAR evolutions

Elements			Attributes	
	Second phase	Third phase	Fourth phase	Fifth phase—direction and execution
Approach	Sound Integrated	Sound Integrated	Sound Integrated	Sound Aligned
Deployment	Implemented Systematic Measurement	Implemented Systematic Mmeasurement	Implemented Structured Measurement	Implemented Flexible Evaluated and understood
Assessment and refinement	Learning Improvement	Learning Improvement	Learning and creativity Improvement and innovation	Learn and improve

Table 3 Result's criteria RADAR evolution

Elements	Attributes			
	Second phase	Third phase	Fourth phase	Fifth phase—results
Relevance and usability	Scope	Scope	Scope and relevance	Scope and relevance
			Integrity	Usable data
			Segmentation	
	Trends	Trends	Trends	Trends
	Targets	Targets	Targets	Targets
	Comparisons	Comparisons	Comparisons	Comparisons
	Causes	Causes	Confidence	Future focus

has been tried and tested as a part of the recognition process; this would call for fine-tuning over the years.

5.5 *Assessment Weightage of Criteria*

To determine the overall score of the organisation, each sub-criteria assessment was done on the 0–100 scale to arrive at the consensus score as a part of the assessment process. However, in the model weightage was attached to each criterion. To start, with the between enabler criteria and result criteria was equally distributed till 2019 and they were carrying 500 points. In 2020 model, the enabler contribution has gone to 600 points and the result has come down to 400. This is one of the major change in the model. The enablers would play a major role in achieving the result, results are post facto. The organisation understands why it exists and who are the customers it serves. The organisations to focus on creating a culture of success, delivering the best sustainable value and how to make the best use of the results achieved. Table 4 shows the evolution of scoring over phases.

The Criteria-1 is leadership and direction setting these remains to be at 100 as the score. The Criteria-2 which addresses the strategy or policy of the organisation, weightage improved from 80 to 100 point over the years. Criteria-3 moved from 90 to 100 points. Criteria-4 the weighs have moved from 90 to 200 in the recent version. Even though, the model construct is not comparable one to one. The Criteria-5 was focussed on process management, contribution moved from 140 to 100. In this new version, the elements of the process management are considered to fundamental expectation and no specific criteria exist. Criteria-6 weight increased from 140 dropped to 100 and again increased to 200. The EFQM model has been a non-prescriptive, and the newer model provides even greater flexibility. However, EFQM 2020 outlines certain expectations from the management, governance, the link between direction and execution, etc.

Table 4 Evolution of scoring over phases

	Criteria-1	Criteria-2	Criteria-3	Criteria-4	Criteria-5	Criteria-6	Criteria-7	Criteria-8	Criteria-9
Initial phase	100	80	90	90	140	200	90	60	150
Second phase	100	80	90	90	140	200	90	60	150
Third phase	100	80	90	90	140	200	90	60	150
Fourth phase	100	100	100	100	100	150	100	100	150
Fifth phase	100	100	100	200	100	200	200		

Note The Criteria name remains the same till the fourth phase, however, in the new model, it is reduced 7 from 9. Hence, we cannot make a one is to one comparison

Table 5 Sub-criteria level evolution over phases

	Initial phase	Second phase	Third phase	Fourth phase	Fifth phase
Enablers	25	25	24	24	23
Results	4	8	8	8	2
	29	33	32	32	25

In the earlier construct (1–4) phase, the model had four result category namely, customer, people, society and business result. Under each of this result category, the model had two sub-criteria to fulfil. One of the sub-criteria of result across all four categories indicates the performance indicators. And second is the perception data of people, society and customer. In the case of business results, feedback from key stakeholders about their personal experience of dealing with the organisation and their perception. In the new model, this includes customer, people, Business and governing stakeholders, society, partners and supplier perception results. The Criterion 7 focuses on strategic and operational performance to measure the ability of an organisation to deliver the strategy and its fitness for the future.

5.6 *Construct at Sub-criteria Level*

Till now the author discussed the construct at the criteria level and it is important to descend one level down to understand the construct the sub-criteria level. The next stage of criteria is typically addressed as sub-criterion. There are 32 sub-criteria (24 are enablers and eight subcriteria's are from results side). Each sub-criterions are provided with guidance point and are directly linked to the fundamental concepts. In the initial phase, the model had 25 enable criteria and made as 24 enablers in second, third and four. In version 2020, it is revised to 23. In the case of results category, from 8 sub-criteria revised to 5. The sub-criteria 6.b, 7.b, 8.b and 9.b are covered in the new Criteria 7, the new model focussed on futuristic growth and execution of the plan. The results were focussed on strategic goals, performance to satisfy its key stakeholders, future performance sustainability, etc. Table 5 shows a summary of changes over different phases.

5.7 *Drivers for Changes in the Model*

Following are key drivers or influencers for the changes in the model. The authors have kept the previous phase version as a base and identified the changes. Each input from coming like a seed, over ten years (1990–2000), the model started maturing based on the inputs from all concerns. The author identified key drivers of the model change apart from regular feedback of assessment community and core-team.

Phase-1 to 2, changes were driven due to introduction of Fundamental concepts supposed to embedded into the model, Model has moved more perspective for guiding assessment, Construct is modified to suit larger organisation base. In addition to the above, introduction of RADAR and emphasis on innovation and learning, corporate social responsibility concept introduction influenced the modification.

In phase-2 to 3, drivers for these phase changes are due to emerging trends of the market place and also, the model was accepted by local partners and network of the national award, their group feedbacks were taken, different versions for different sector introduced caused some confusion and core team decided to have a single generic model across all sectors.

In phase 3 to 4, constant improvement feedback different channels, focus more agility, scenario planning, organisational development become key process criteria, focus on products, processes and services, agility, flexibility, action orientation, creativity become key ingredients. Key results become business results and RADAR refined.

The 2020 version update was based on Research feedback, Change of organisational operating model and ecosystems, encouraging a co-creation culture, creating sustainable value is the heartbeat of the new model and value proposition for stakeholder focus, focussing on transformational and performance culture in the organisation, provide flexible and adaptive for organisation adoption, hence model was to be redesigned to be more flexible.

6 Highlights of the Research Findings:

The authors have summarised the results and findings on the comparison of the EFQM Framework.

- The construct of the model has changed significantly from 9 to 7 criteria and sub-criteria dropped from 32 to 25. Well settled construct of Five Enablers and four result category dislodged and the names of criteria have changed.
- The official title of the model from "The EFQM Excellence Model" in 2012 becomes the "The EFQM Model". The model becomes more generic and application of could be widespread compare to Excellence. Even though the perspective has changed, it would face challenges in establishing it's identified in the Quality world.
- The model replaces the word products, processes and services with sustainable value. It appears to be generic and scope is broad.
- In the earlier versions of the model, the model appears to be prescriptive due to clear articulation of the guidance point and examples of an excellent organisation; it appears that model moving was towards from its descriptive nature.
- The model has been transformation focussed and futuristic driven. This could help the organisation to focus on Industry 4.0 adoption based on creating sustainable

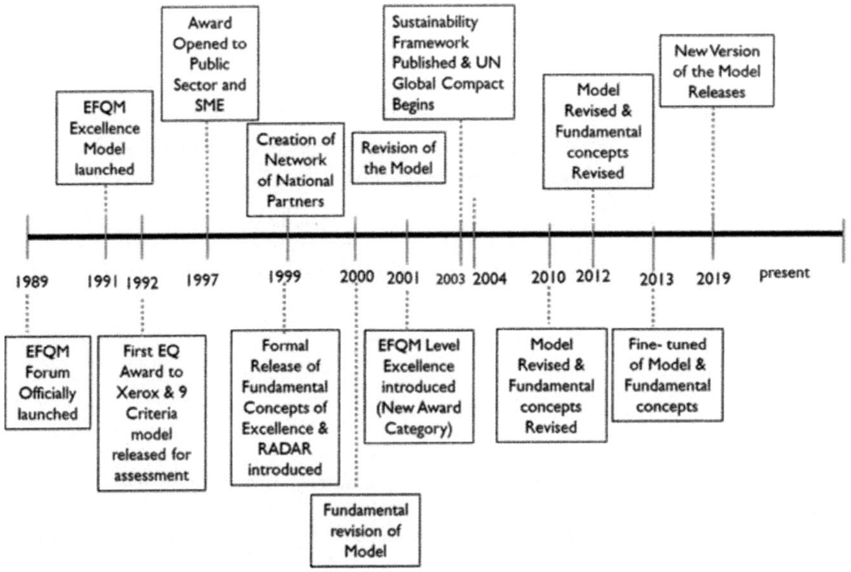

Fig. 6 *Base Source* The EFQM quality model, Kai A. Simon [9] and developed further by the author

value to the stakeholders. Due to this, the model would facilitate in reaching towards digital Quality Era or Quality 4.0.

- In the previous model versions, the technology focus (4.d) and leadership that poses challenges and was addressed in two sub-criteria, in the new model, transformational leadership appears to be more the emphasis. It is evident from the construct, more technological changes are anticipated from the construction point of view.

The authors have traced back the key milestones of the model and depicted in Fig. 6 EFQM Evolution Chart—Bird's Eye View.

7 Conclusions and Limitations

The EFQM model would be moving towards more of the generic model and focussing more on the futuristic model than business excellence are being just a quality award and recognition model. The redesign of the model is based on the research inputs and core decision, hope these specialists have considered the mechanics of the organisation and restructured the model based on quality deliberations. Even though the model is more futuristic or transformative, it is also important to focus on the current state and it is managed properly to secure the future for it to be a sustainable organisational Journey. Outside of the construct, it still appears to be unclear and it is left to the imagination judgement of the assessor and the incumbent organisation. Between descriptive to perspective models, the error of judgement will be high and could prove to be a costly affair. The authors have themselves been involved in the excellence journey of their organisations and understand the drivers of excellence in manufacturing domain specialist and hence can provide expert impressions about the efficacy of the model. Since the model has been newly introduced, it would take some more time to understand the model more closely and apply it in an organisational setup.

References

1. Ghicajanu, M., Irimie, S., Marica, L., & Munteanu, R. (2015). Criteria for Excellence in Business.*Procedia Economics and Finance, 23*, 445–452. ISSN: 2212-5671.
2. Mann, R., Mohammad, M., & Agustin, A. (2004). Understanding Business Excellence, an awareness guidebook for SMEs, 4–8.
3. Tan, K., & Khoo, H. (2003). Worldwide comparison of 27 national quality awards. *Euro Asia Journal of Management, 25*(1), 55–73.
4. Miguel, P. (2005). A comparison of quality and business excellence programs in the world. *Revista De Ciencia & Technologia, 13*, 35–46.
5. NIST research work by Musli Mohammad and Dr Robin Mann. (2010). National Quality/Business Excellence Awards in different countries. Centre for Organizational Excellence Research, www.coer.org.nz. Accessed 21 Sept 2020.
6. Nenadál, J. (2020). The new EFQM model: What is really new and could be considered as a suitable tool with respect to quality 4.0 concept? *Quality Innovation Prosperity, 24*(1), 17.
7. Nabitz, U., Klazinga, N., & Walburg, J. (2000). The EFQM excellence model: European and Dutch experiences with the EFQM approach in health care. European Foundation for Quality Management. *International journal for quality in health care: journal of the International Society for Quality in Health Care, 12*(3), 191–201.
8. Eskildsen, J. K., Kristensen, K. & Juhl, H. J. (2001). The criterion weights of the EFQM excellence model. *InternationalJournal of Quality & Reliability Management, 18*(8/9), 783–795.
9. Simon, K. (1996). The efqm quality model.

Parametric Optimization and Evaluation of Machining Performance for Aluminium-Based Hybrid Composite Using Utility-Taguchi Approach

M. Murali Mohan, E. Venugopal Goud, M. L. S. Deva Kumar, Vivek Kumar, Manish Kumar, and Dinbandhu

1 Introduction

The development of the present era is likely impossible without considering the importance of composite materials. These materials possess many useful qualities, i.e. lightweight, corrosion resistance, and high fatigue strength and because of these qualities, they are widely preferred in several structural applications that incorporate aircraft, automobiles, sports items, electronic packaging, medical equipment, space vehicles, and homebuilding sectors. Composites consist of two and/or several principal ingredients, i.e. matrix phase and reinforcing phase. These phases are amalgamated at the macroscopic level and are mutually insoluble. The matrix phase, which is usually continuous, is high in volume and consists of reinforcing materials, like flakes/fibres, or particles. It holds the entire system of composite material and maintains its integrity. The common composite materials are polymer matrix composites

M. Murali Mohan · E. Venugopal Goud
Department of Mechanical Engineering, G. Pulla Reddy Engineering College (Autonomous), Kurnool, Andhra Pradesh 518007, India

M. L. S. Deva Kumar
Department of Mechanical Engineering, JNTU College of Engineering (Autonomous), Ananthapuramu, Andhra Pradesh 515002, India

V. Kumar
Aryabhatta Center for Nanoscience and Nanotechnology, Aryabhatta Knowledge University, Patna, Bihar 800001, India

M. Kumar
Department of Computer Science & Engineering, Vidya Vihar Institute of Technology, Purnea, Bihar 854301, India

Dinbandhu (✉)
Department of Mechanical & Aero-Space Engineering, Institute of Infrastructure, Technology, Research and Management (IITRAM), Ahmedabad, Gujarat 380026, India

R. Agrawal et al. (eds.), *Recent Advances in Smart Manufacturing and Materials*,
Lecture Notes in Mechanical Engineering,
https://doi.org/10.1007/978-981-16-3033-0_27

(PMC), metal matrix composites (MMC), ceramic matrix composites (CMC), and carbon–carbon composites (CCC)s [1–5].

These days, MMCs are one of the widely used composites due to their several useful characteristics. In these composites, the matrix phase incorporates metal or metal alloys and Al_2O_3, SiC, TiN, TiC, B_4C, AlN, ceramic particulates, etc., as reinforcement materials. The embedment of reinforcing materials in the matrix phase enhances several mechanical characteristics, namely fatigue, wear resistance, specific strength, creep resistance, and stiffness in MMCs. Among the available list of MMCs, aluminium metal matrix composites (AMMC) play a significant role in structural applications. In AMMCs, the matrix phase is made up of aluminium or its alloys, whereas ceramic particulates are embedded as reinforcing materials yielding high-quality structural materials as compared to the conventional engineering materials [6–9].

Laghari et al. [10] discussed several soft computing techniques for machining MMCs. A number of traditional machining operations, namely grinding, milling, turning, and drilling have been explored with respect to soft computing methods. Modelling and optimization of all these processes have been presented using particle swarm optimization, Taguchi methodology, fuzzy logic, genetic algorithm, artificial neural network, and response surface methodology. Pugazhenthi et al. [11] performed a machining operation on the AA7075/TiB_2 composite. PCD inserts were used for cutting tool material. The importance of cutting speed, the effect of TiB_2 on cutting force, and R_a for the prepared AMCs were investigated. The samples were prepared through in situ reaction using KBF_4 and K_2TiF_6 salts in molten aluminium. FESEM was also performed for inspecting the morphology of machined surfaces, cutting tools, and the chips produced during the process. Nicholls et al. [12] studied the machinability aspects of MMCs. The mechanism of tool wear was discussed to evaluate the tool wear rate affecting the overall cutting tool life. The process of optimization was also articulated for obtaining maximum tool life without affecting the quality and quantity of the product. Joel and Xavior [13] presented a review of the machining of MMCs. A brief report has been presented on the traditional machining of composites along with their challenges. The work also details the modern machining techniques adopted for composite machining. Kim et al. [14] performed a hybrid turning process on SiC embedded AMMCs to improve their machinability and surface quality. The obtained results were compared with the traditional turning operation and tangible enhancement was observed in terms of R_a and machining rate. Das et al. [15] evaluated the surface roughness attributes by performing turning on Al 7075/SiCp in dry conditions. Uncoated TiC inserts were used in this operation. Parametric optimization was also done besides performing ANOVA analysis. The result reveals the feed rate as the most influencing parameter succeeded by spindle speed. The study finds the depth of cut as an unimportant variable.

Manufacturing a product includes several engineering processes and one of them is machining. A number of machining parameters such as speed, feed, depth of cut, type of cutting tool, cutting tool diameter, cutting force, affect the machining output characteristics, namely R_a, MRR, and tool wear ratio. These output characteristics influence the end-user performance of the composites. Hence, it is essential to evaluate the machining performance of such materials because of their growing demand in several engineering applications [16–18]. In this context, the present study highlights the influence of metal cutting variables, viz. cutting speed, feed, and depth of cut on MRR and Ra during the turning of the aluminium-based hybrid composites.

2 Materials and Methods

The turning operation was executed on the composite specimens using an all geared semiautomatic lathe (Fig. 1). The trials were performed with different sets of machining variables involving cutting speed (V), feed (f), and depth of cut (d) (Table 1), and their responses have been recorded in terms of MRR and R_a. To do so, the design of the L_9 orthogonal array has been put into operation. The total experimental runs with their respective output responses have been represented in Table 2.

2.1 Utility Concept

The concept of utility functions is well-acknowledged in multi-criteria decision making (MCDM) problems. It is simple and easy to understand by any decision-maker because it does not involve any powerful constraints on the available settings than the aggregation formula.

Fig. 1 Machined composite specimens

Table 1 Ascribing levels to the turning variables

Turning variables	Levels (L)		
	1	2	3
V (in RPM)	700	910	1200
f (in mm/rev.)	0.105	0.132	0.158
d (in mm)	0.2	0.3	0.4

Table 2 Experimental runs and their respective output responses

Trial No.	V	f	d	R_a	MRR
1	700	0.105	0.2	1.83	12
2	700	0.132	0.3	2.32	29
3	700	0.158	0.4	2.78	42
4	910	0.105	0.3	2.31	31
5	910	0.132	0.4	2.59	47
6	910	0.158	0.2	2.22	35
7	1200	0.105	0.4	2.62	48
8	1200	0.132	0.2	2.13	38
9	1200	0.158	0.3	2.54	58

In the utility-based Taguchi process, an MCDM problem is easily transformed into a single response optimization problem using an arbitrary function. This arbitrary function, also known as the response function, acts as an overall utility index and has to be optimized for obtaining the solution [19, 20]. As stated by the utility function approach [21], if A_x is the performance indicator of an output characteristic x and there are k output responses assessing the sample space, then the joint utility function is elucidated as follows:

$$U(A_1, A_2, \ldots A_k) = f\{U_1(A_1), U_2(A_2) \ldots U_k(A_k)\} \quad (1)$$

In Eq. (1), $(U_1(A_1))$ represents the utility of the xth output characteristic. Equation (2) indicates the overall utility function, which will be equal to the summation of individual utilities for individualistic output responses.

$$U(A_1, A_2, \ldots A_k) = \sum_{x=1}^{k} U_x(A_x) \quad (2)$$

Assigning weight to the output responses depends upon their comparative significance and influence on the process. In that case, the overall utility function is elucidated as follows:

$$U(A_1, A_2 \ldots A_k) = \sum_{x=1}^{k} W_x U_x(A_x) \quad (3)$$

In Eq. (3), W_x represents the weightage given to the output characteristic x. The total sum of the weightage given to all the output responses should be 1. The output characteristic is evaluated on the benchmark of lower and higher values. For that, two random arithmetic values 0 and 9 (preference numbers) are allocated. On a logarithmic scale, the evaluation of the preference number N_p can be done using

Eq. (4).

$$N_p = O * \log \frac{A_x}{A'_x} \tag{4}$$

In Eq. (4), A_x represents the value of output characteristic x. A'_x is the lower value of output characteristic x. O is a constant and can be calculated using Eq. (5), only if $A_x = A^*$ (where A^* is the optimal value), then $N_p = 9$. Hence,

$$O = \frac{9}{\log \frac{A_x}{A'_x}} \tag{5}$$

The overall utility is stated as:

$$U = \sum_{x=1}^{k} W_x(N_p). \tag{6}$$

In accordance with the condition:

$$\sum_{x=1}^{k} W_x = 1 \tag{7}$$

Taguchi's *S/N* ratio concept describes several output characteristics, namely higher-is-better (HB), nominal-is-best (NB), and lower-is-better (LB). Out of all these, HB applies to the utility functions. Consequently, when the utility function is maximized, the output attributes taken into consideration for its assessment will be self-optimized (depending upon the case, it will be either minimized or maximized). Figure 2 depicts the intended optimization flow diagram.

3 Result and Discussion

In this section, the impact of the aforementioned variables on their performance characteristics, namely R_a and MRR has been examined. Besides this, the section also includes the intended optimization system for accessing the best ideal solution.

3.1 Influence of Input Variables on R_a and MRR

To examine the consequences of input variables on their output responses, an analysis of variance (ANOVA) was performed. Tables 3 and 4 present the values of ANOVA

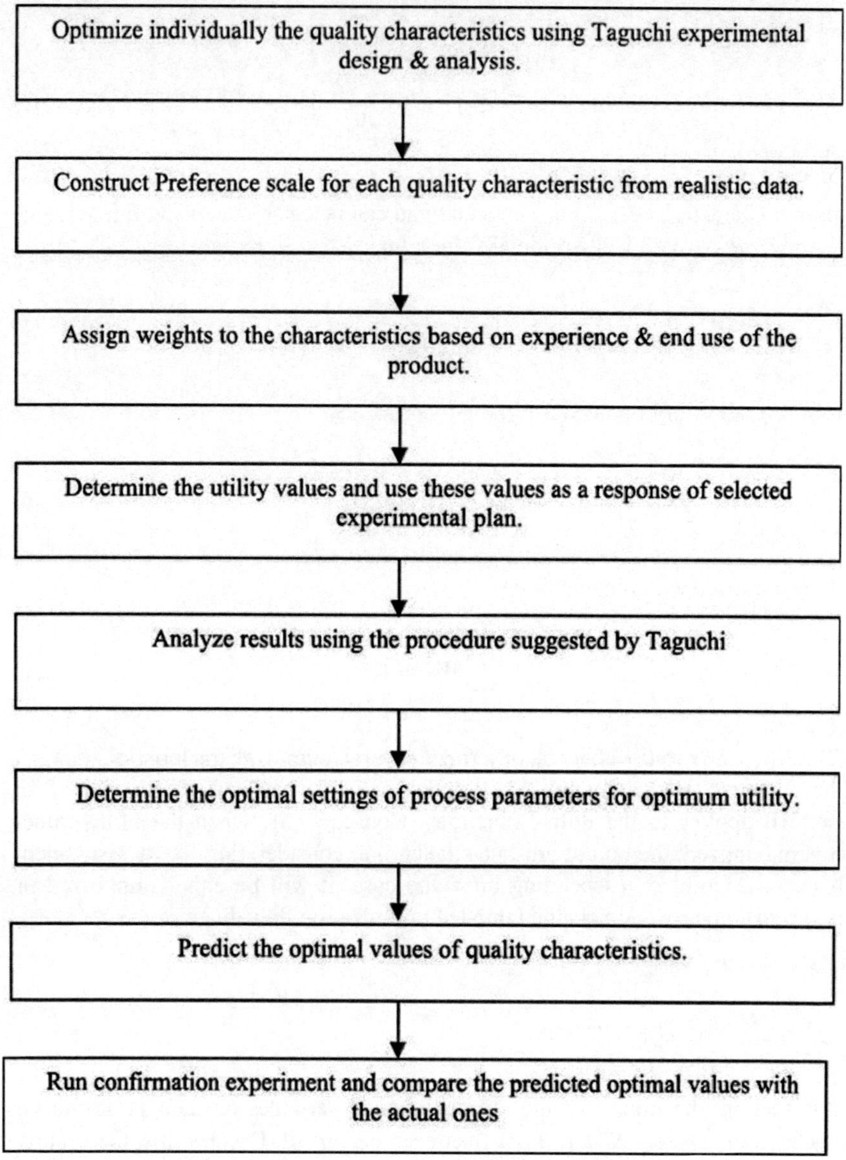

Fig. 2 Flow diagram of the utility function for multi-response optimization

for R_a and MRR, respectively. Figure 3 exhibits the contribution (in percentage) of these variables with respect to their output responses.

Table 3 ANOVA for R_a

Source	DF	Adj SS	Adj MS	F-value	Contribution (%)	P-value
V	2	0.02162	0.010811	1.75	3.15	0.364
f	2	0.10409	0.052044	8.42	15.18	0.106
d	2	0.54762	0.273811	44.32	79.86	0.022
Error	2	0.01236	0.006178		1.80	
Total	8	0.68569				

Table 4 ANOVA for MRR

Source	DF	Adj SS	Adj MS	F-value	Contribution (%)	P-value
V	2	620.222	310.111	90.03	43.94	0.011
f	2	322.889	161.444	46.87	22.87	0.021
d	2	461.556	230.778	67	32.70	0.015
Error	2	6.889	3.444		0.49	
Total	8	1411.556				

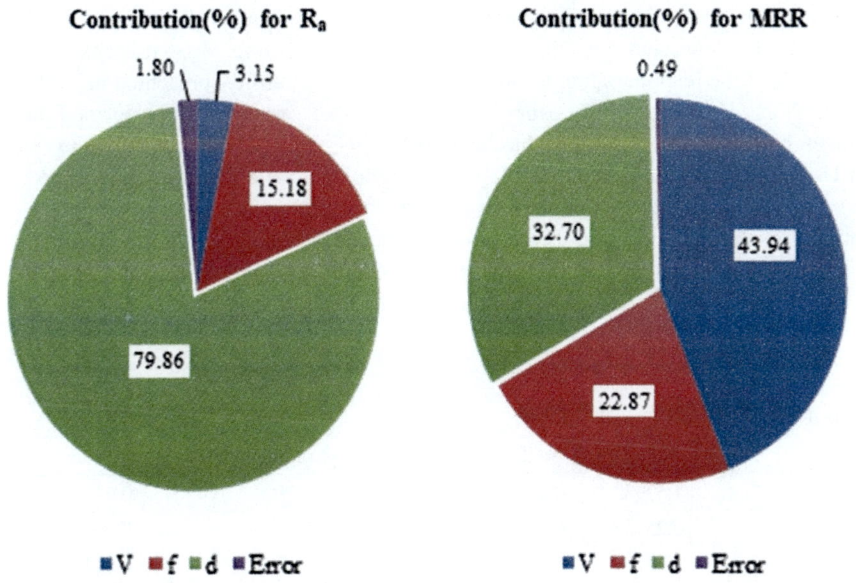

Fig. 3 Contribution (in percentage) of input variables on output performances

3.2 *Intended Optimization Approach*

Experimental values for the performance characteristics, viz. R_a and MRR have been furnished in Table 2. As the performance characteristics are contradictory, so

Table 5 Individual utility and overall utility with the *S*/*N* ratio and predicted *S*/*N* ratio

U-R_a	U-MRR	U-overall	*S*/*N* ratio	*P*-*S*/*N* ratio
9.000000	0	4.500001	13.06425	15.53215
3.89337	4.517704	4.205537	12.47643	
0	6.413964	3.206982	10.12193	
3.986347	4.859154	4.42275	12.91385	
1.523761	6.989835	4.256798	12.58166	
4.841724	5.480504	5.161114	14.25487	
1.275879	7.097625	4.186752	12.43754	
5.732506	5.901551	5.817028	15.29402	
1.943348	8.066516	5.004932	13.98796	

it is essential to convert them into a common scale. Therefore, the utility function approach has been applied in order to aggregate all conflicting criteria into a single index, i.e. overall utility (U-overall). Firstly, an individual utility for aforesaid output responses has been evaluated using Eqs. 4 and 5. For R_a, a lower value is desirable, whereas, for MRR, a higher value is preferable. From Table 5, it is clear that all aforementioned responses are converted into scale of 0–9 where 0 is considered the lowest utility value, and 9 is considered as higher utility value. Now, it is easier to aggregate all the responses into a single index, i.e. overall utility (presented in Table 5) which can be computed by using Eq. 6. Analysis of variance was performed on U-overall (Table 6) and its impact has been shown in Fig. 4. The outcome of ANOVA for U-overall shows the depth of cut (d) as the significant variable followed by speed (V). Thereafter, Taguchi technique was implemented on the overall utility to get the optimal machining condition. From Fig. 5, it has been concluded that the optimal machining condition is $V = 1200$ RPM, $f = 0.132$ mm/rev., and $d = 0.21$ mm. Mean response values have been also computed (Table 7) and the identical outcome, as ANOVA for U-overall, was obtained.

Table 6 ANOVA for U-overall

Source	DF	Adj SS	Adj MS	*F*-Value	Contribution (%)	*P*-value
V	2	1.62983	0.81492	95.26	37.55	0.01
f	2	0.25105	0.12553	14.67	5.78	0.064
d	2	2.44282	1.22141	142.78	56.28	0.007
Error	2	0.01711	0.00855		0.39	
Total	8	4.34081				

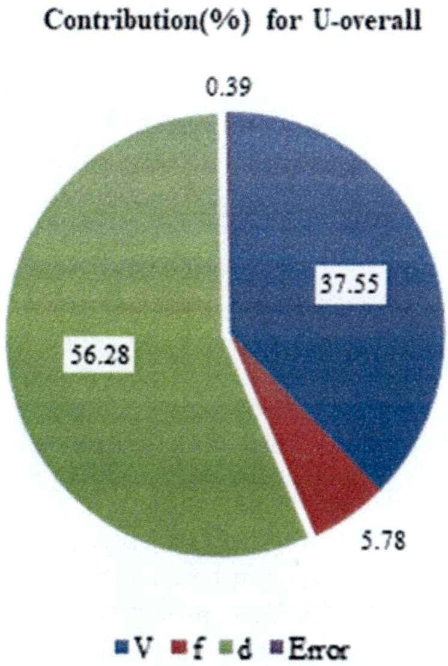

Fig. 4 Contribution (in percentage) of input variables on U-overall

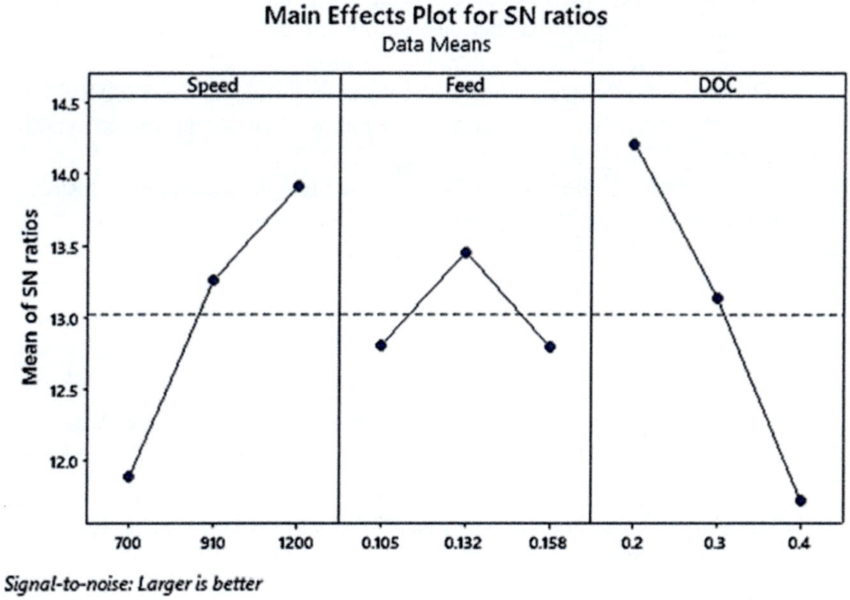

Fig. 5 Main effect plot for U-overall

Table 7 Mean response table for the signal to noise ratios

Level	V	f	d
1	11.89	12.81	14.2
2	13.25	13.45	13.13
3	13.91	12.79	11.71
Delta	2.02	0.66	2.49
Rank	2	3	1

4 Conclusions

The study yields the following conclusions:

i. This study intends a multi-attribute optimization approach utilizing the Taguchi-utility concept to access the optimal parameter settings for machining of aluminium-based hybrid composite.
ii. ANOVA for R_a reveals the depth of cut (d) as a significant variable, whereas, for MRR, the same methodology finds cutting speed (V) as a prominent variable.
iii. ANOVA performed on U-overall depicts the depth of cut (d) as the most influenced variable which is like the outcome of the computed mean response table.
iv. The intended approach yields the ideal parametric settings as $V = 1200$ RPM, $f = 0.132$ mm/rev., and $d = 0.21$ mm for machining of aluminium-based hybrid composite.
v. The proposed methodology may be incorporated into the machining industries to assess favourable machining conditions.

References

1. Thakur, A., Purohit, R., Rana, R. S., & Bandhu, D. (2018). Characterization and evaluation of mechanical behavior of epoxy-CNT-bamboo matrix hybrid composites. *Mater. Today Proc., 5*, 3971–3980. https://doi.org/10.1016/j.matpr.2017.11.655
2. Tripathi, D.R., Vachhani, K.H., Kumari, S., Dinbandhu, & Abhishek, K. (2020). Experimental investigation on material removal rate during abrasive water jet machining of GFRP composites. *Material Today Proceedings, 26*, 1389–1392 (2020). https://doi.org/10.1016/j.matpr.2020.02.280.
3. Bandhu, D., Mohan, M. M., Praveen, G., Yadav, K., Reddy, K. J., & Shavali, G. S. (2016). Mechanical properties of Makhana (Euryale Ferox Salisbury) reinforced composite. *IJSRD—International Journal of Science and Research, 4*, 2321–2613.
4. Shavali, G. S., Venugopal Goud, E., Praveen, G., Yadav, K., & Bandhu, D. (2016). Tensile and flexural characterization of Nomex and E-Glass fibre reinforced epoxy composites. *IJSRD—International Journal of Science and Research, 4*, 2321–2613.
5. Naga Phani Sastry, M., Devaki Devi, K., Bandhu, D. (2016). Characterization of Aegle Marmelos fiber reinforced composite. *International Journal of Engineering Research, 5*, 345–349 (2016). https://doi.org/10.17950/ijer/v5i2/009

6. Venkatesan, K., Ramanujam, R., Shanbhag, V. V., Yalamoori, N. N., & Reddy, D. V. (2014). Preparation, characterization and machinability of Al7075-Al2O3 matrix composite using multi layer coated carbide insert. *Procedia Materials Science, 5*, 1819–1828. https://doi.org/10.1016/j.mspro.2014.07.469
7. Bandhu, D., Thakur, A., Purohit, R., Verma, R. K., Abhishek, K. (2018). Characterization and evaluation of Al7075 MMCs reinforced with ceramic particulates and influence of age hardening on their tensile behavior. *Journal of Mechanical Science and Technology, 32* (2018). https://doi.org/10.1007/s12206-018-0615-9.
8. Surappa, M. K. (2003). Aluminium matrix composites: Challenges and opportunities. *Sadhana—Academy Proceedings in Engineering Sciences, 28*, 319–334. https://doi.org/10.1007/BF02717141
9. Hima Gireesh, C., Durga Prasad, K. G., Ramji, K., & Vinay, P. V. (2018). Mechanical characterization of aluminium metal matrix composite reinforced with *Aloe vera* powder. *Material Today Proceedings, 5*, 3289–3297. https://doi.org/10.1016/j.matpr.2017.11.571
10. Laghari, R. A., Li, J., Laghari, A. A., & Wang, S. (2020). A review on application of soft computing techniques in machining of particle reinforcement metal matrix composites. *Archives of Computational Methods in Engineering, 27*, 1363–1377. https://doi.org/10.1007/s11831-019-09340-0
11. Pugazhenthi, A., Kanagaraj, G., Dinaharan, I., David Raja Selvam, J. (2018). Turning characteristics of in situ formed TiB2 ceramic particulate reinforced AA7075 aluminum matrix composites using polycrystalline diamond cutting tool. *Measurement: Journal of the International Measurement Confederation, 121*, 39–46 (2018). https://doi.org/10.1016/j.measurement.2018.02.039.
12. Nicholls, C. J., Boswell, B., Davies, I. J., & Islam, M. N. (2017). Review of machining metal matrix composites. *International Journal of Advanced Manufacturing Technology, 90*, 2429–2441. https://doi.org/10.1007/s00170-016-9558-4
13. Joel, J., & Anthony Xavior, M. (2018). Aluminium Alloy Composites and its Machinability studies; A Review. *Material Today Proceedings, 5*, 13556–13562. https://doi.org/10.1016/j.matpr.2018.02.351
14. Kim, J., Bai, W., Roy, A., Jones, L.C.R., Ayvar-Soberanis, S., & Silberschmidt, V. V. (2019). Hybrid machining of metal-matrix composite. In *Procedia CIRP* (pp. 184–189). Elsevier B.V. (2019). https://doi.org/10.1016/j.procir.2019.04.162.
15. Das, D., Kumar Thakur, R., Kumar Chaubey, A., & Kumar Sahoo, A. (2018). Optimization of machining parameters and development of surface roughness models during turning Al-based metal matrix composite. *Material Today Proceedings, 5*, 4431–4437. https://doi.org/10.1016/j.matpr.2017.12.011
16. Vincent Sagayaraj, S. A., Vendan, S. T., Professor, A. (2016). Optimization of machining parameters for turning of Aluminium Alloy 7075 using Taguchi Method. *International Research Journal of Engineering and Technology.*
17. Wang, X., & Feng, C. X. (2002). Development of empirical models for surface roughness prediction in finish turning. *International Journal of Advanced Manufacturing Technology, 20*, 348–356. https://doi.org/10.1007/s001700200162
18. Sahoo, A. K., Baral, A. N., Rout, A. K., Routra, B. C. (2012). Multi-objective optimization and predictive modeling of surface roughness and material removal rate in turning using grey relational and regression analysis. In *Procedia engineering*, pp. 1606–1627. Elsevier Ltd (2012). https://doi.org/10.1016/j.proeng.2012.06.197
19. Pavan, M., Todeschini, R. (2009). Multicriteria decision-making methods. In *Comprehensive chemometrics* (pp. 591–629). Elsevier (2009). https://doi.org/10.1016/B978-044452701-1.00038-7.

20. Bagada, C., Damor, H., Prajapati, V., Abhishek, K. (2020). Utility function approach integrated with fuzzy for optimization in milling glass fiber reinforced epoxy composites. In: A. Parwani, P. Ramkumar (Eds.) *Recent advances in mechanical infrastructure. Lecture notes in intelligent transportation and infrastructure* (pp. 85–91). Singapore: Springer. https://doi.org/10.1007/978-981-32-9971-9_10.
21. Deb Barma, J., Roy, J., Saha, S. C., Roy, B. S. (2012). Process parametric optimization of submerged arc welding by using utility based Taguchi concept. *Advanced Material Research, 488–489*, 1194–1198 (2012). https://doi.org/10.4028/www.scientific.net/AMR.488-489.1194

Discussing the Impact of Industry 4.0 in Agriculture Supply Chain

Subhodeep Mukherjee, Manish Mohan Baral, Venkataiah Chittipaka, Sharad Chandra Srivastava, and Surya Kant Pal

1 Introduction

Before Industry 4.0 (I4.0), three industrial revolutions have provoked changes in perspective in the space of assembling. Industry 1.0 began around the 1780s with water and steam power utilization, which helped in mechanical creation. Next, Industry 2.0 is described as when enormous scope assembling and large-scale manufacturing were familiar as the basic and essential methods for creation, all in all. During the twentieth century, Industry 3.0 arose with the digital revolution's appearance, which is like Industry 1.0 and 2.0 the same number of individuals living today think about organizations slanting toward cutting edge advancements in progress. Perhaps, Industry 3.0 was and still is prompt after the impact of the huge improvement in PCs and information and correspondence advancement adventures for certain countries [1]. People are readily adopting latest innovative technologies in the day to day life for better transformation; be that as it may, presently are also compelled to use greetings tech gadgets and its applications, which are speedy transforming into the main factor in their regular working life [2]. I4.0 has been shown as an overall move of digitalization and computerization in all parts of the association, similarly as in the assembling cycle [3]. This fresh start of the Internet time, set apart

S. Mukherjee · V. Chittipaka
GITAM Institute of Management, GITAM (Deemed To Be University), Visakhapatnam, Andhra Pradesh, India

M. M. Baral (✉)
Department of Management, Birla Institute of Technology Mesra, Ranchi, Jharkhand, India

S. C. Srivastava
Department of Industrial and Production Engineering, Guru Ghashidas Vishwavidyalaya (A Central University), Bilaspur, Chhatisgarh, India

S. K. Pal
University School of Business, Chandigarh University, Mohali, Punjab, India

R. Agrawal et al. (eds.), *Recent Advances in Smart Manufacturing and Materials*, Lecture Notes in Mechanical Engineering,
https://doi.org/10.1007/978-981-16-3033-0_28

by an incorporated PC-based mechanization and pervasive figuring frameworks, is also associated with the remote system by the Internet [4]. This paper presents the impact of I4.0 on agriculture supply chain (ASC) and expects to give an idea toward Agriculture 4.0 (A4.0).

The supply chain (SC) can say to be an organization of associations answerable for the creation and dispersion of merchandise and administration from origination to the last shopper while ASC is essentially an improved coordination inside and between different ASC individuals [5]. Then, I4.0 can say to be the data serious change of businesses with the association of information, individuals, measures, administrations, and frameworks through data innovation as methods for getting shrewd industry and biological systems [6]. Execution of I4.0 inside designing SC can prompt a decrease in lead time and cost, arrangement of related dynamic cycle, and improvement in the general presentation of every individual from the SC just as the creation cycle pursuing accomplishing the extraordinary practical objective [7]. Accomplishing maintainability inside any assembling organization begins with the SC; if the SC is all around overseen from the purpose of material sourcing till it gets to the last shopper, at that point, supportability is reachable [8]. Overseeing feasible creation through flexibly change is significant in food fabricating because of the intricacy of food creation that is related with sanitation, food wastage, food security, and so on. Nonetheless, the presentation of specific highlights and parts of I4.0 could guarantee effective reasonable SC inside the food business. The development of I4.0 parts will go far too decidedly affect and improve maintainability in ASC.

Rest of the paper comprises of Sect. 2 which states the I4.0 to A4.0; Sect. 3 gives application of I4.0 to ASC; Sect. 4 states the benefits of I4.0 in agriculture sectors; and Sect. 5 gives the conclusion.

2 Industry 4.0 to Agriculture 4.0

The exchange of the I4.0 to the Agri-food Industry will bring about an exchange of advantages to this industry. The utilization of advanced technologies will permit acting solidly on the elements of creation, catalyzing the development cycle previously presented by the accuracy of the agribusiness worldview [9]. These innovations have then developed to the current day getting progressively refined and dependent on computerized, establishing, about the assembling area, the real I4.0 advances. The full-scale wonder I4.0 has all the earmarks of being upheld by a progression of extremely clear innovative marvels [10]. 63% of the market produced by I4.0, for example, around 1 billion Euros, is legitimately associated with connected factory and IoT. 20% of the market, worth €330 million, is spoken to by industrial analytics. At about 9%, arriving at an estimation of €150 million, we locate a significant marvel, for example, cloud computing. 8%, for 120 million euros, blocks the advanced automation that incorporates self-ruling and communitarian creation and dealing with frameworks, mechanical technology, and collective mechanical technology. About 1%, for an estimation of 15 million Euros, is identified with the advanced human-machine interface which sets up the function of watchers for increased reality, 3D scanners, and wearable and new human-machine interfaces as a contact [11].

The utilization of the standards of I4.0, best spoke to in the smart factory, is seen not just as a genuine component of separation between organizations yet additionally as a catalyst for supportable improvement. Agricultural processing plants and additional benefits associated with the Internet gather and examine information coming from keen items and their savvy applications [12]. Information examination permits organizations to all the more likely characterize market conduct and client needs and by implication recommend giving them new and/or more economical items and administrations. That speaks to a transformative potential for the organization, without question versatile (Deloitte, 2014). The normal advancement is accordingly recognizable in any event in three territories. The smart farm consists of collaboration, coordination, and cooperation for developing efficiency in the over-all operations [13]. The smart SC concerning the administration of streams (material and unimportant) between the different entertainers in the SC. The smart life cycle regarding the universe of item improvement and the board of as long as it can remember cycle [14]. All executed through IoT applications, analytics, cloud stages, and so forth. These advancements will ensure the area higher livelihoods on account of the chance of getting higher creation yields and lower costs or more all lower ecological effect and more prominent sanitation for the buyer [15].

A4.0, in similarity to I4.0, represents the incorporated inner and outer systems administration of cultivating tasks. This demonstrates that advanced structure information is available in agricultural measures like providers' interchanges with end clients performed electronically. A4.0 upheaval measures like information transmission, handling, and examination are computerized. Farmers' fields used with innovation and development bring about reasonable rural creation and efficient development. The investigation presents that farming development, nature of the climate, and other reasonable angles result from A4.0. The effects on work concentrated agribusiness field's cause because of the Fourth Industrial Revolution. This transformation made two advances man-made brainpower which benefits in dynamic capacity and large information which helps in the examination of measurable information gathered by various procedures. These innovations are created by a few hey tech associations under exactness agriculture areas. These areas are utilized in farming exercises like an investigation of soil dampness, the strength of yield, forecast of the careful collect season of harvest, and planning of nuisance control. Framework named the Internet of Things (IoT) making it conceivable to work farmer by means of distantly through cell phones estimates temperature, moistness, and measure of daylight underway homesteads which increment the creation with esteem added. Cultivating these days changing in innovation and creation rehearses business size, asset control, and activity, a plan of action with purchasers and providers [16]. These works give creative cultivating tasks agriculture improvement which brings about new agribusiness in the US. They introduced arrangements for cultivating change to the modern plan of action through expanded quality security and recognizability requests of processors, customers of food items, measure control advancements that encourage the organic assembling of yield and animal items. Industry 4.0 alludes to the keen production line, which comprises of cycles interrelated by the Internet of Things (IoT). Organizations that produce, measure, assembling, sell and serve food, drink, and dietary enhancements are alluded to as food business which contains all phases of the cycle, including plan, development, support, and conveyance of answers for the client [17].

3 Applications of Industry 4.0 in ASC

- Blockchain innovation in farming included smart contracts to put away information from its cause to the last client end guaranteeing the authenticity and inception of every exchange. Blockchain innovation stays away from expected cheats on the nature of items by controlling the treatment surrendered all through the chain to the retail location.
- The expected development of the IoT business is in the agricultural areas. IoT innovation opens worldwide the potential outcomes of conveying IoT gadgets in agriculture with the capacity of detecting information distantly. The cultivating information gathered by gadgets joined to sensors influence the development of yields continuously screens to a measurable study shows further preparing through IoT.
- Agriculture areas are created with A4.0 innovations that had been utilized. Drone examination uses to catch pictures of harvest investigations by enormous information on how the crops are yielding. With the utilization of robots, farmers distinguished which piece of farming required watering or manures, weed expulsion territory without any problem. These days this robot utilized as far off compost vehicles in cultivating so explicit regions or individual plant covers without any problem.
- Different methods were used to gather the information to improve agrarian area advancement. "Big Data" signifies preparing a gigantic measure of information gathered from information communication technologies (ICT) that prompt quick dynamic information for improving profitability. Since all information is accessible from the most recent decade away in regards to trim cycle, yield issues, focuses on every farmer take preventive measures with new devices in A4.0 for various climate conditions.
- Agriculture robots will work in various fields like creation, handling, conveyance, and utilization. These robots recognize the administration air and independently offer shrewd work. The fourth insurgency brings about the usage of robots in many cultivating gear for the determination of suitable items and legitimate appropriation of irritations for bug control. This procedure additionally fixed with the ethereal vehicles used to control the wellbeing with customary checking of organic products, vegetables, and creatures in the farming field.
- The smart cultivating innovation under exactness agriculture is used to quantify crop yield and constitution. Likewise, these upsets in farming areas are utilized

for noticing various sorts of yields, their development during supporting and post-reaping periods. The A4.0 ideas create accuracy agriculture which is utilized for overseeing various exercises in farming.
- The “Global Positioning System” (GPS) is a worldwide satellite framework utilized for situating and route. This framework is initially created for the military, and these days this framework with A4.0 is utilized in different fields in farming. GPS has been utilized in agrarian gear for both fixed units and hand-held gadgets. The errand performed by GPS frameworks like recording positions and developments, self-ruling, or helped driving with applications, for example, monitoring and backing on equal excursions, self-governing driving.

4 Benefits of Industry 4.0 in Agriculture Sectors

I4.0 Technologies and IoT have the potential to transform agriculture in many aspects.

- Data, huge loads of information, gathered by smart farming sensors. Information gathered by the sensor is dissected and states climate conditions, soil quality, yields development progress, or steers wellbeing. This information can be utilized to follow the plants and gear proficiency.
- Better power over the interior cycles and, therefore, lowers creation chances. This new procedure help in getting ready for better item appropriation relies upon yield expectation by information preparing.
- Cost the board and waste decrease on account of the expanded command over creation. With constant checking in I4.0 assists with lessening waste and cost the executives to be utilized for specific farmers. This builds the yield of the harvest.
- Increased business effectiveness through cycle computerization. The usage of shrewd atomized gadgets in most extreme movement underway cycle like water system, treating, and bug control expands the matter of farmers in type of more noteworthy yield.
- Enhanced item quality and volumes. Controlling all the farming cycles and keeping up an elevated requirement of grain quality, profitability is expanded.

5 Conclusions

This research aims to fill the gap in the usage of advancements associated with I4.0 inside the ASC network. It additionally features the uses or applications of I4.0 in the ASC network. A few innovations can bring about both of chances and risks because all the various territories are interconnected. It could have a positive or negative meaning with no specific limits between them contingent upon where it was investigated. This will empower mass customization, permitting organizations to satisfy clients’ needs, making an incentive through continually acquainting new items and administrations with the market. Advances from I4.0 empower the independent mechanizing of whole acquisition forms and the self-governing participation of products and enterprises across hierarchical outskirts. I4.0 to A4.0 can generate lots of technological innovations and advancement in the agro-industries. These technologies will help the farmers a lot as they will be able to get more profits and less wastage.

References

1. Liao, Y., Deschamps, F., Freitas, E. D., & Loures, R. (2017). Past, present and future of Industry 4.0—A systematic literature review and research agenda proposal. *International Journal of Production Research, 55*(12), 3609–3629 (2017).
2. Gorecky, D., Schmitt, M., Loskyll, M., & Zühlke, D. (2014). Human-machine-interaction in the industry 4.0 era. In *12th IEEE International Conference on Industrial Infomatic*, pp. 289–294 (2014).
3. Jamwal, A., Agrawal, R., Sharma, M., & Kumar, V. (2021). Review on multi-criteria decision analysis in sustainable manufacturing decision making. *International Journal of Sustainable Engineering*, 1–24.
4. Federal Ministry of Education and Research (Germany). (2013). *Securing the future of German manufacturing industry: Recommendations for implementing the strategic initiative*
5. Ojo, O. O., Shah, S., Coutroubis, A. (2017). An overview of sustainable practices in food processing supply chain environments. In *IEEE International Conference on Industrial Engineering and Engineering Management, Singapore* (2017).
6. Mangan, J., Lalwani, C., Butcher, T., & Javadpour, R. (2014). *Global logistics and supply chain management* (1st ed.). Chichester: Wiley.
7. Schrauf, S., & Berttram, P. (2016). Industry 4.0: How digitization makes the SC more efficient, agile, and customer-focused. *Strategy & Recuperado de* https://www.strategyand.pwc.com/media/file/Industry4.0.
8. Dani, S. (2015). *Food supply chain management and logistics: from farm to fork* (1st edn.), Konga Page.
9. Balafoutis, T. (2017). Smart farming technologies—Description, taxonomy and economic impact'. In: S. Pedersen, K. Lind (Eds.), *Precision Agriculture: Technology and Economic Perspectives. Progress in Precision Agriculture.* Cham: Springer (2017)
10. Bellini, M. (2017). Effetto Industria 4.0: +25% di crescita a 1,7 MLD Euro', Internet4things.it, 2017. Available: https://www.internet4things.it/industry-4-0/effetto-industria-4-0-25-dicrescita-a-17-mld-euro/. Accessed 13-Dec-2017.
11. Kumar, M., Graham, G., Hennelly, P., & Srai, J. (2016). How will smart city production systems transform SC design: a product-level investigation. *International Journal of Production Research, 54*(23), 7181–7192.
12. Shrouf, F. (2014). Smart factories in Industry 4.0: A review of the concept and of energy management approached in production based on the Internet of Things paradigm. In *2014 IEEE International Conference on Industrial Engineering and Engineering Management*. https://doi.org/10.1109/ieem.2014.7058728
13. Qin, J., Liu, Y., & Grosvenor, R. A. (2016). Categorical framework of manufacturing for Industry 4.0 and beyond. *Procedia CIRP*, 173–178.
14. Okada, T., Namatame, A., & Sato, N. (2015). *An agent-based model of smart SC networks.*
15. Patil, T. G., & Shekhawat, S. P. (2019). Industry 4.0 implications on agriculture sector: An overview. *International Journal of Management, Technology and Engineering*, 9.

16. Vendrell-Herrero, F., Bustinza, O. F., Parry, G., & Georgantzis, N. (2017). *Servitization, digitization, 17*(4), 34–39.
17. Masoni, R., Ferrise, F., Bordegoni, M., Gattullo, M., Uva, E., Fiorentino, M., Carrabba, E., & Donato, M. (2017). Supporting remote maintenance in industry 4.0 through augmented reality. *Procedia Manufacturing, 11*(6), 1296–1302 (2017).

Optimization of Process Parameters in Micro-EDM Through Mixed Flushing and GRA Technique

Saurabh Bhardwaj, Rohit Thakur, C. S. Jawalkar, and Suman Kant

1 Introduction

Among the non-traditional concepts of machining, EDM is widely used to produce dies and many other profiles on any kind of electrically conductive material. The surface workpiece is gradually sparked and eroded by thermal energies through a series of sparks in the gap between electrode and workpiece immersed in a dielectric liquid. Any kind of hard material can be machined with high precision and accuracy. Hence, it is commonly used in finishing parts for aerospace and automotive industry and surgical components. This technique was developed in the 1940s. As soon as the electrode is moved in, the vicinity of workpiece until the gap is too small to ionize the dielectric fluid. There happens continuous initiation of sparks having micro-second duration in the dielectric area between the tool and workpiece material [1]. The material is removed with spark and erosion phenomenon by electrical discharges from tool to workpiece via dielectric medium. In EDM machining, the tool and workpiece are separated by a very small distance known as the inter-electrode space containing the dielectric medium as shown in Fig. 1; thus, chances of mechanical stress and vibrations are reduced to a large extent.

2 Literature Review

Urso et al. [2] have fabricated holes using micro-EDM machining; the investigation focused on the effect of electrodes and workpiece materials on the process performance and was expressed in terms of tool and wear ratio. In particular, the influence of four workpiece materials (titanium, magnesium, SS and brass), three

S. Bhardwaj (✉) · R. Thakur · C. S. Jawalkar · S. Kant
Punjab Engineering College (Deemed To Be University), Chandigarh 160012, India

R. Agrawal et al. (eds.), *Recent Advances in Smart Manufacturing and Materials*, Lecture Notes in Mechanical Engineering,
https://doi.org/10.1007/978-981-16-3033-0_29

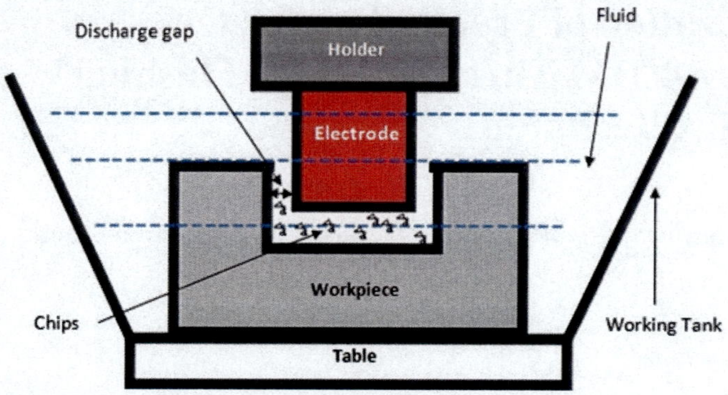

Fig. 1 Schematic of EDM process

electrode materials (tungsten carbide, brass and copper) and two electrode profile were investigated. Munz et al. [3] have found that too much flushing resulted in opposite process behaviour and reduced the speed of drilling; however, on the basis of supplied current to the gap, there was an optimal dielectric flow rate in order to obtain a higher drilling speed. It was also concluded that flow rate of dielectric and material removal was directly proportional in nature. Ayesta et al. [4] used different flushing arrangements inside the slots for this experimental study. A camera having high speed capture rate was used in the tests to observe and analyse the particle movement pattern within the slot. Trials done without flushing, intermittent flushing using a nozzle and a continuous arrangement for flushing were implemented inside the machining zone. In addition to this, electrode jump heights were programmed. The results highlighted the fact that the best debris removal was obtained along with application of continuous dielectric spray in the form of flushing.

Fu et al. [5] had machined the workpiece with different tool materials and further classified them as rough machining, semi-finish machining and finish machining conditions. As the tool feed had a great influence of the tool material, the time for machining was examined to get the detailed tool feed information. Since machining accuracy depends upon machining feed and side gap, their relationship was investigated. Similarly, machining feed and tool wear ration were also studied and investigated along with the investigation of tool material on the wear aspect. Niamat et al. [6] have discussed the effects of factors like pulse-on-time (μs) and current (A) for the response measurement while using fluid materials like water and kerosene as the dielectric medium. A study to compare the responses like MRR and TWR was performed while using water and kerosene separately as the dielectric mediums. Aluminium 6061 T6 alloy has been extensively used in the automobile and aviation industries. Natarajan and Arunachalam [7] used GRA to investigate the optimization of process parameters micro-EDM for SS 316 workpiece and concluded that current came out to be the most influential factor. Pradhan [8] used GRA along with principal component analysis (PCA) for the optimization of EDM machining. Chakravorty et al. [9] used two sets of past experimental data on EDM processes in order to compare four different optimization methods. It was found that utility theory method gave overall better optimization than GRA based and other methods. Dewangan and Biswas [10] used GRA based Taguchi technique to study the optimization process of various responses like MRR and TWR of EDM on a AISI P20 tool steel workpiece material; where a copper tool electrode of cylindrical geometry was used. Priyadarshini and Pal [11] investigated using GRA based PCA optimization resulted better feasible parametric setting which were also in accordance with the validation experiments. This research aims to study the effects of material density of electrodes (copper, aluminium and brass), current (ampere), T_{on} (μs), T_{off} (μs) which were evaluated for the performance measure and analysis by using a combination of through and side flushing. A comparison was performed for different electrode density material for responses like MRR (g/min), TWR (g/min) and micro-images of surface finish.

3 Experimental Set-up and Procedure

Diesinking (SPARKONIX 35A) EDM was used for the experimentation. In micromachining, removal of debris is a major issue with increase in depth of the hole, as soon as the sparking starts and the material starts to erode; there is an urgent requirement to flush-off the machining area, otherwise the eroded material again sticks to the machined surface. The impulse flushing which is the default flushing method fails to remove the debris inside the hole; however, it flushes off the debris in the initial machining time. To rectify this problem, one of the impulse nozzles had been modified to allow the dielectric to pass through the tool electrode, and hence, debris accumulated just beneath the tool electrode gets removed to a certain extent. A mixed flushing arrangement was employed for the micromachining task as shown in Fig. 2 Mild steel plate with thickness 5 mm was used as the workpiece material

Fig. 2 Mixed flushing arrangement

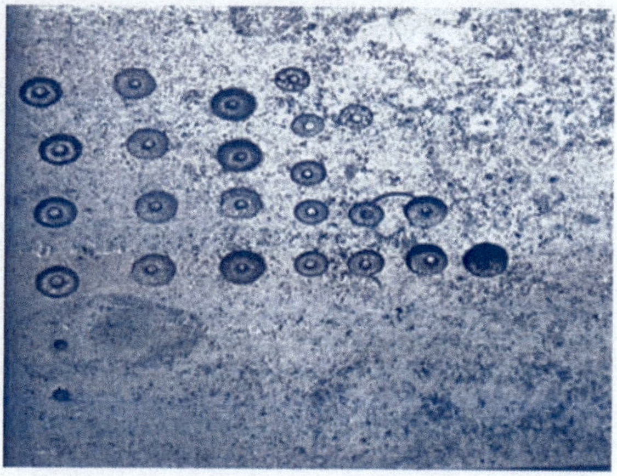

Fig. 3 Mild steel workpiece

Table 1 Electrode materials and density

S. No.	Electrode material	Density (g/cm^3)
1	Copper	8.96
2	Aluminium	2.7
3	Brass	8.73

as shown in Fig. 3 with kerosene oil as dielectric medium. Three different materials (copper, brass and aluminium) were used as the tool electrode, Table 1 shows the density corresponding to each electrode material. The tool electrodes were hollow in geometry with internal diameter 0.8 mm. For each electrode material and workpiece, based on the pilot studies, best machining process parameters were identified and a set of parameters were selected. The effects of material density of electrode (copper, aluminium and brass), current (ampere), pulse-on-time (μs) and pulse-of-time (μs) were evaluated for performance measures using mixed flushing arrangement.

Electrode and workpiece were weighted using digital weight balance prior and post each experimental run. Machining time was obtained by using a stopwatch for a constant 10 min run. The obtained weight differences in electrode and workpiece were divided by time to get MRR and TWR in the unit of mg/min, where "*n*" shows number of experiments and "*y*" shows observed values of MRR and TWR.

All the parameters were set at three levels based on the pilot studies performed in order to get the range as given in Table 2. The experiments were performed by taking in consideration the Taguchi's *L*9 orthogonal array. Further the experimental design is given in Table 3. After getting the output in the form of MRR and TWR, S/N ratio was calculated in order to analyse the overall effects of parameters on responses. *S*/*N* ratio was calculated for MRR using higher the better *S*/*N* ratio as shown in Eq. (1). On the contrast note, lower the better *S*/*N* ratio was employed for TWR as shown in Eq. (2) [11].

Table 2 Factors and levels

Factors	Symbol	Unit	Levels		
			I	II	III
Density	ρ	g/cm^3	8.96	2.7	8.73
Current	I	A	3	6	9
Pulse-on-time	T_{on}	μs	30	50	70
Pulse-off-time	T_{off}	μs	40	60	80

Table 3 Experiment results

Run No.	ρ	I	T_{on}	T_{off}	MRR (g/min)	TWR (g/min)
1	8.96	3	30	40	0.021094	0.000837
2	8.96	6	50	60	0.053069	0.003515
3	8.96	9	70	80	0.061272	0.006361
4	2.70	3	50	80	0.02	0.003333
5	2.70	6	70	40	0.007222	0.010555
6	2.70	9	30	60	0.013889	0.031666
7	8.73	3	70	60	0.001718	0.005326
8	8.73	6	30	80	0.001031	0.013573
9	8.73	9	50	40	0.001546	0.016323

$$\text{Higher the better} \quad \left(\frac{S}{N}\right) = -\log\left(\frac{1}{n}\left(\sum\left(\frac{1}{y^2}\right)\right)\right) \tag{1}$$

$$\text{Lower the better} \quad \left(\frac{S}{N}\right) = -\log\left(\frac{1}{n}\left(\sum(y^2)\right)\right) \tag{2}$$

MRR resulted out to be minimum for brass; however, it increased for aluminium and the highest obtained MRR resulted while using copper electrode. As far as the effect of current is concerned, maximum MRR was seen at the highest level of current while minimum MRR came out to be at the intermediate level. T_{on} and T_{off} had similar effect on MRR, and the intermediate levels of T_{on} and T_{off} resulted in highest MRR. So the optimum set of factors for MRR came out to be $\rho = 8.96$ (Copper), $I = 9$ A, $T_{\text{on}} = 50$ μs and $T_{\text{off}} = 60$ μs. While analysis the effects of factors on the TWR, it was found that lowest TWR was seen in case of copper electrode while highest TWR came out while using the brass electrode. Minimum TWR was found at 9 A while maximum at 6 A current. T_{on} and T_{off} had similar effect on TWR as the intermediate levels yielded lowest TWR (Figs. 4 and 5).

Figures 6, 7 and 8 shows the machined surface with copper, brass and aluminium electrode, respectively. Figure 6 shows the machining through copper electrode; it has yielded better material removal and good surface finish than that of brass and aluminium electrodes. This has been achieved due to proper debris removal

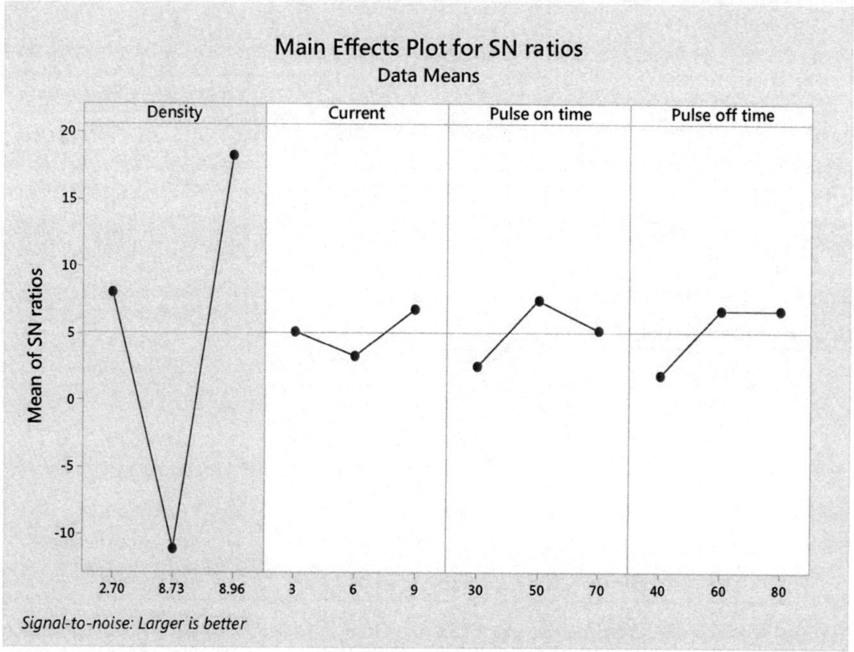

Fig. 4 *S/N* ratio for MRR

which has been achieved due to mixed flushing arrangement and the better electrical conductivity of copper. Figure 7 shows the machining through aluminium electrode; it has yielded significant material removal while surface finish was not as good as that obtained in case of using copper electrode. This could be related to the relatively lower electrical conductivity of aluminium as compared to copper. Figure 8 shows the machining through brass electrode; it has yielded the lowest material removal, and surface finish is not as good as that obtained in case of using copper and aluminium electrodes. This could be related to the low value of electrical conductivity of brass as compared to the other two electrodes. It can be clearly seen that brass debris have sticked in the hole depth.

4 Grey Relational Analysis

There is a limitation in case of Taguchi technique of optimization, and it can either analyse the response with *S/N* ratio as larger is better or smaller is better. In this case, MRR should be analysed with *S/N* ratio as larger is better and TWR with *S/N* ratio smaller is better. To obtain multi-objective optimization, GRA had been used on the response data. GRA technique analysed the optimization by taking into consideration of different response parameters having dissimilar characteristics. Thus, applying

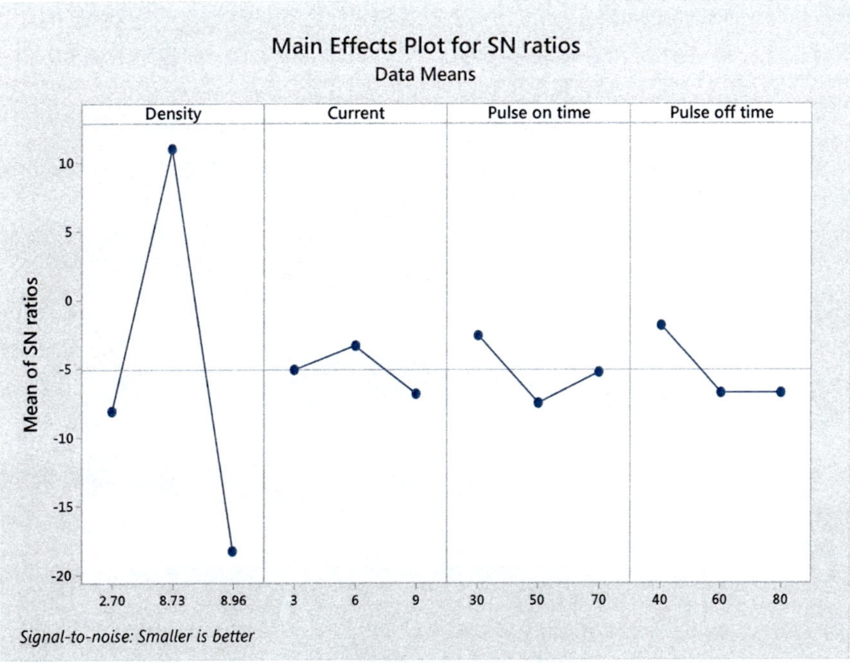

Fig. 5 *S/N* ratio for TWR

Fig. 6 Hole with copper tool

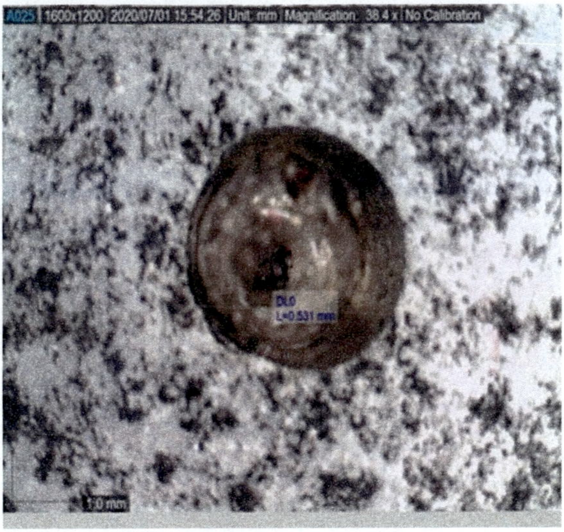

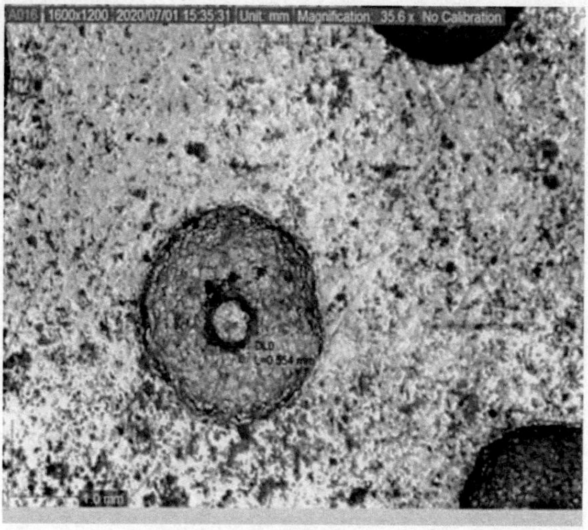

Fig. 7 Hole with aluminium tool

Fig. 8 Hole with brass tool

GRA converts the multi-objective into a single objective as given in Table 4. Steps performed in GRA analysis are listed below [11]:

1. Normalize the response values MRR and TWR in the range 0 to 1 by taking into consideration the nature of response measure (like higher is better or smaller is better). In case of higher the better, the response normalization is done by Eq. (3); while in case of smaller the better, the response normalization is done by Eq. (4) [12].

Table 4 GRA normalizing and deviation sequence

Run No.	Output/response		Normalizing		Deviation sequence	
	MRR	TWR	MRR	TWR	MRR	TWR
1	0.021094	0.000837	0.333041	1	0.66695941	0
2	0.053069	0.003515	0.863829	0.913134	0.13617054	0.086866262
3	0.061272	0.006361	1	0.820818	0	0.179181939
4	0.02	0.003333	0.314884	0.919037	0.68511554	0.08096273
5	0.007222	0.010555	0.102775	0.684777	0.89722522	0.31522268
6	0.013889	0.031666	0.213441	0	0.78655928	1
7	0.001718	0.005326	0.011409	0.85439	0.98859115	0.145609653
8	0.001031	0.013573	0	0.586882	1	0.413117519
9	0.001546	0.016323	0.008557	0.497681	0.99144336	0.502319245

$$X_i^*(k) = \frac{X_i(k) - \min X_i(k)}{\max X_i(k) - \min X_i(k)} \quad (3)$$

$$X_i^*(k) = \frac{\max X_i(k) - X_i(k)}{\max X_i(k) - \min X_i(k)} \quad (4)$$

where $X_i^*(k)$ and $X_i(k)$ are the set of normalized data and the initial observed data for the ith experiment and kth response.

2. Deviation sequences are calculated from the normalized values as obtained in the previous step.
3. Grey relational coefficient (GRC) values $\zeta_i(k)$ are generated which can be used to compare the theoretical and experimental data [12].

$$\zeta_i(k) = \frac{\Delta_{\min} + \zeta\Delta_{\max}}{\Delta_i(k) + \zeta\Delta_{\max}} \quad (5)$$

$\Delta_{\max}$ and $\Delta_{\min}$ refer to the highest and lowest values in the different data series, respectively. The coefficient ζ always ranges in between 0 to 1.

4. Grey relational grade (GRGs) is obtained by taking the mean of the grey relational coefficient values.
5. Based on the GRGs, ranking is done by assigning the highest GRG with rank 1.

Table 4 demonstrates the results of GRA normalizing and deviation sequence calculation, while Table 5 shows the GRCs along with GRGs and the ranks calculated on the basis of these GRGs. Table 6 has been summed up by taking the mean value of GRGs of each factor at level I, II and III, respectively. The difference between maximum and minimum values has been used to predict the rank of the factors in the experiment.

Table 5 GRA rank calculation for experiment run

Run No.	GRCs		GRGs	Rank
	MRR	TWR		
1	0.428463917	1	0.714231958	3
2	0.785952776	0.851982866	0.818967821	2
3	1	0.736179765	0.868089882	1
4	0.421899793	0.860640406	0.6412701	4
5	0.357852116	0.613329354	0.485590735	6
6	0.388633473	0.333333333	0.360983403	9
7	0.335888065	0.774461778	0.555174922	5
8	0.333333333	0.547574643	0.440453988	7
9	0.335245719	0.498843061	0.41704439	8

Table 6 GRA rank for factors

Factors	Level			Range	Rank
	I	II	III		
Density	0.80043	0.495948	0.470891	0.329539	1
Current	0.636892	0.581671	0.548706	0.088186	3
Pulse-on-time	0.505223	0.625761	0.636285	0.010524	4
Pulse-off-time	0.538956	0.578375	0.642135	0.103179	2

5 Conclusion

Based on the above experimental study performed on the mild steel workpiece using different tool electrodes, following observations were noted:

1. Without proper flushing arrangement, the debris could accumulate on the machining area, resulting in poor surface finish inside the hole.
2. The optimum set of factors for MRR came out to be $\rho = 8.96$ g/cm^3 (Copper), $I = 9$ A, $T_{on} = 50$ μs and $T_{off} = 60$ μs.
3. Better hole quality in terms of surface finish and depth was seen in case of machining with the hollow copper tool.
4. Hole quality as obtained by machining with the hollow brass tool resulted in poor surface finish; accumulation of debris damaged the finish of the surface and shortened the hole depth.
5. Tool wear was observed the maximum in case of copper followed by aluminium and brass.
6. Based on GRA, it was observed that experiment number 3 with electrode density 8.96 g/cm^3, current 9 A, T_{on} 70 μs and T_{off} 80 μs was the overall best run in terms of MRR and TWR.

7. GRA study also concluded that density was the most dominant factor followed by T_{off}, current and T_{on}.

References

1. Jaiswal, V. K., Paul, A., Yadav, V., & Singh, V. (2018). Literature review on electrical discharge machining (EDM), *International Journal for Scientific Research and Development, 6*(05).
2. D'Urso, G., & Merla, C. (2014). Work piece and electrode influence on micro-EDM drilling performance. *Precision Engineering, 38*, 903–914.
3. Munz, M., Risto, M., & Haas, R. (2013). Specifics of flushing in electrical discharge drilling. *Procedia CIRP, 6*, 83–88.
4. Ayesta, I., Flaño, O., Izquierdo, B., Sanchez, J. A., & Plaza, S. (2016). Experimental study on debris evacuation during slot EDMing. *Procedia CIRP, 42*, 6–11.
5. Fu, Y., Miyamoto, T., Natsu, W., Zhao, W., & Yu, Z. (2016). Study on influence of electrode material on hole drilling in micro-EDM. *Procedia CIRP, 42*, 516–520.
6. Niamat, M., Sarfraz, S., Aziz, H., Jahanzaib, M., Shehab, E., Ahmad, W., & Hussain, S. (2017). Effect of different dielectrics on material removal rate, electrode wear rate and microstructures in EDM. *Procedia CIRP, 60*, 2–7.
7. Natarajan, N., & Arunachalam, R. M. (2011). Experimental investigations and optimization of process parameters in micro-EDM with multiple performance characteristics. *International Journal Experimental Design and Process Optimization, 2*(4), 336–356.
8. Pradhan, M. K. (2012). Determination of optimal parameters with multi response characteristics of EDM by response surface methodology, grey relational analysis and principal component analysis. *International Journal of Manufacturing Technology and Management, 26*(1/2/3/4), 56–80.
9. Chakravorty, R., Gauri, S. K., & Chakraborty, S. (2013). A study on the multi-response optimization of EDM processes. *International Journal of Machining and Machinability of Materials, 13*(1), 91–109.
10. Dewangan, S., & Biswas, C. K. (2013). Optimization of machining parameters using grey relation analysis for EDM with impulse flushing. *International Journal of Mechatronics and Manufacturing Systems, 6*(2), 144–158.
11. Priyadarshini, M., & Pal, K. (2016). Multi-objective optimization of EDM process using hybrid Taguchi-based methodologies for Ti-6Al-4V alloy. *International Journal of Manufacturing Research, 11*(2), 144–166.
12. Lin, J. L., Wang, K., Yan, B., & Tarng, Y.S. (2000). Optimization of the electrical discharge machining process based on the Taguchi method with fuzzy logics. *Journal of Materials Processing Technology, 102*(1), 48–55.

The Effect of Different Parameters on MRR, Surface Finish While EDM Machining Titanium Alloys: A Review Study

Jatin, Lokesh Kumar, and N. Gupta

1 Introduction

Titanium and its alloys are low machinability metals. The melting point is very high and is very hard and thus difficult to machine [1]. Different researchers have used different machining techniques as hard machining, EDM, USM, etc. to machine them.

Electro discharge spark erosion machining is known as electric discharge machining (EDM). In the EDM process, the metal is removed by the erosion process. Erosion of metal is based on spark discharges [2]. When the two current conduction wires touch each other, an arc is produced during the machining. If we glance closely at the purpose of contact between two wires, a tiny low portion of the metal has been scoured away, effort a tiny low crater [3].

In general, metal removal rate (MRR) has good performance in EDM. It also increased surface roughness. The production of rocket science components like landing gear components, IC engines are made up of good surface finish with EDM. For macro-sinking EDM application, copper and graphite are the best electrode material for use. As of now, graphite becomes the most preferable non-metal, especially during roughing [4]. Because of its properties, graphite is a good conductor of electricity and heat at small amounts of tool wear to obtain the conditions. So, the parameters that obtained profile accuracy is improved [5]. Graphite is up to accomplish anode wear of under 1% corresponding to the profundity of slice while working to more forceful machine boundaries. That is rather than copper, the high electrical marvel and more on-times truly safeguard the carbon conveyor [6]. That is in contrast to copper, the more electrical phenomenon and staying more on intervals really to prevent the carbon conductor. The surface finish of copper is very fine. At the present time, EDM machine technology is civilized, and the surface finish gap

Jatin · L. Kumar · N. Gupta (✉)
Mechanical Engineering Department, DTU, New Delhi, India

R. Agrawal et al. (eds.), *Recent Advances in Smart Manufacturing and Materials*,
Lecture Notes in Mechanical Engineering,
https://doi.org/10.1007/978-981-16-3033-0_30

between graphite and copper is very fine. First of all, we can investigate the specific behaviour in detail and join with many physical properties of graphite material that is used for roughing. After the result, a few supported different conductor materials for a taking machining task is feasible. The titanium alloy is lightweight, corrosion resistant, high toughness, having high weldability and shows high hardness at extreme temperatures [7]. Due to these reasons, it has potential to replace aluminium and its alloys from the aerospace industry. It can be used with composite materials in the aerospace industry. Also, it can be machined accurately and efficiently with the EDM.

The EDM can be used for machining titanium alloys. Mostly, the titanium alloy with 38 grades can be machined through EDM, and for very wide industrial application, titanium alloy of 5 grades is used. Many types of research had been done on machining of titanium alloy with EDM; it is observed that among the different electrode materials used for machining, the graphite electrodes show the highest material removal rate.

The necessary distinction of different Ti6AI4V ELI (grade 23) and Ti6AI4V (grade5) is the decrease in the quantity of gas content to 0.13% (maximum) in grade twenty-three. This refers to increased fracture and ductility properties with some decrease in strength criteria. Ti6AI4V ELI has been mostly preferred in offshore tubular and fracture in the critical airframe structure [7]. Although graphite has some problems using it, graphite has housekeeping dust-related problems and decreases the margin of safety of direct current. Graphite is the most usable material because of its good properties and some other advantages. This research paper builds a comprehensive review on the electric discharge machining of different types of titanium grades and its alloys. In this review, the paper showed the experimental studies and hypothesis and theoretical studies on electric discharge machining. It mainly concentrated to better the process performance, with MRR, the quality of surface, the rate of tool wear. It also tells about the models and processes used to determine the electric discharge machining (EDM) process conditions [8].

2 EDM Process Parameters

There are several parameters on which a process is dependent, termed as input process parameters. Also such parameters are termed as control parameters. Some of the input control parameters are as follows:

2.1 Discharge Current

The amperage of current affects the metal removal rate (MRR) and surface finish. In different stages of power that may be contained by the generator of the electric discharge machining and shown to be the norm of the discharge current intensity.

2.2 Interval of Pulse On

It is allowed the distance of your interval (μs) present to flow per cycle. The unnecessary part of the material removed is directly dependent on the quantity of energy applied throughout the interval of pulse on time. The discharge current is controlled by energy and also the distance of interval pulse on [6].

2.3 Interval Pulse Off

It is the distance of your interval (μs) between the two different sequential flickers (interval pulse off). Now permits the liquefied medium to solidify and to be washed out of the arc gap. This process is to have an effect on the velocity and also the solidity of the cut.

2.4 Duty Cycle

It is a share of the interval pulse on related to the overall cycle interval. Determine this parameter by dividing the interval pulse onto the overall cycle interval (interval pulse on and interval pulse off). The results increased by one hundred for the proportion of potency, known as the duty cycle.

2.5 Dielectric Pressure

This is the flushing pressure of the material jet that removes the chip or rubble made throughout the EDM method removed from the gap zone. This worth of pressure is measured by a gauge existing within the EDM machine.

2.6 Polarity

The machine will run either in traditional polarity or reverse polarity. Thc polarity is always used in straight lines during which negative is the tool and positive is the workpiece, whereas in reverse polarity, the positive is a tool, and the negative is the workpiece [9].

2.7 EDM of Different Titanium Alloys.

In aerospace industries, the cooling hole of turbines and the precise cutting of blade tips are mostly done and manufactured by electro discharge machining (EDM) [10]. The EDM has certain different combinations of workpiece and tools. Many tests had been conducted on different grades of graphite for selecting the tool, and the best grade graphite tool was selected to conduct the experiment on Charmilles RoboForm 41 (manufactured in 1993) for optimizing the material removal rate (MRR). This experiment was done in 2012 by Klocke et al. [11]. This machine tool is applied to different materials workpiece as Inconel 718, Ti-6A-4A, and Ti-X. The Ti-X is also machined with OPS Ingersoll gantry eagle 500 (manufactured in 2010). As per results among the MRR of different titanium alloys as Inconel 718, Ti-6A-4V, Ti-X, highest MRR was observed in Ti-X.

3 Results and Discussion

The interpretations of the experimental study analysed were presented in the following session:

The impact of different EDM parametes on Ti-6Al-4V sing drinking water as dielectric was studied. Many experiments were done on different parameters of EDM. But a recent experiment was done by changing the concentration of graphite powder in drinking water which was used as the dielectric fluid for EDM. This experiment was done in July 2020, by Gugulothu et al. [7]. The composition of the titanium alloy (Ti-6Al-4V) is given as carbon: 0.08 (maximum), aluminium: 5.5–6.5, vanadium: 3.5–4.5, nitrogen: 0.05, oxygen: 0.13, iron: 0.25, hydrogen: 0.01, and the rest is titanium.

In the experiments [7], output parameters taken were material removal rate, surface roughness and average recast layer thickness.

Analysis of MRR and tool wear for roughing in sinking EDM using different graphite grade.

For industrial applications, graphite is one in every of the foremost usable materials. It is used in EDM sinking as electrode/tool, within the optimized parameters results in a really small amount of tool wear rate joined with a large metal removal rate (MRR). After the result, the number of tool electrodes for an exact EDM task will be decreased remarkably resulting in terribly small machining time [12]. A satisfactory discussion for these edges could not be found so far. So, the definite wear behaviour and MRR are analysed thoroughly during this research paper along with graphite material characteristics. There are five different types of graphite chosen to determine the specific electric resistance, thermal conductivity and grain size.

In another experimental study, the MRR was calculated for each graphite material first with the finest grain size [13]. The results proved that electric energy decrease at the tool electrode is smaller if the electric resistance is small.

The impact of different electrodes tools (copper and graphite) on EDM of SS316.

This experiment was done in November 2019 by Selvarajan et al. [14]. In this experiment, they used copper and graphite as their electrode at the positive polarity and SS316 as the workpiece at negative polarity EDM oil as the dielectric and a DC power source for current. They studied the impacts of the electrical parameters like current, the impulse of ON time, the impulse of OFF time and find the results for MRR, electrode roughness and machining time. The composition of the work material is given as carbon 0.08%, manganese 2%, phosphorus 0.045%, sulphur 0.030%, silicon 0.75%, nickel 10–14%, chromium 16–18%, nitrogen 0.10%, molybdenum 2–3% and rest iron. The different parameters were the current levels which were as follows: 10, 12, 14 (A); the impulse on time was as follows: 8, 10, 12 (s); the impulse off time was as follows: 4, 6, 8 (s), and the results shows MRR, machining time and average roughness.

4 Conclusion

Titanium and its alloys are low machinability metals. The melting point is very high and is very hard and thus difficult to machine [1]. Different researchers have used different machining techniques as hard machining, EDM, USM, etc. to machine them. While analysing the experimental research study, some of the conclusions drawn were as follows:

With the same tool and all other parameters to be constant, the MRR of the Inconel 718 was highest.

The electric energy decreases when the tool electrode is smaller as it offers smaller electric resistance.

The MRR is more for graphite electrodes than copper electrodes by keeping all the parameters the same.

The surface roughness of the copper electrode is more than the graphite tool by keeping all the parameters the same.

The machine time of graphite electrodes is less than the copper electrode by keeping all the parameters the same.

References

1. Gupta, N., Walia, R. S., & Agrawal, A. K. (2020). Robust Taguchi based optimization of surface finish during hard turning EN 31 with carbon nanotubes-based nano-coated tip. In *Recent Advances in Mechanical Engineering* (pp. 289–298). Singapore: Springer.
2. Mohan, B., Rajadurai, A., & Satyanarayana, K. G. (2002). Effect of SiC and rotation of electrode on electric discharge machining of Al–SiC composite. *Journal of Materials Processing Technology, 124*(3), 297–304. https://doi.org/10.1016/s0924-0136(02)00202-9

3. Selvarajan, L., Rajavel, R., Gopi, P., GokulKumar, M., & Kasthuri, N. (2018). Investigation on EDM of SS316 alloy material using copper electrode for improving MRR and TWR. *Journal of Manufacturing and Engineering, 13*, 142–147.
4. Selvarajan, L., Rajavel, J., Prabakaran, V., Sivakumar, B., & Jeeva, G. (2018). A review paper on EDM parameter of composite material and industrial demand material machining. *ScienceDirect Materials Today Proceedings, 5*, 5506–5513.
5. Curodeau, A., Marceau, L. F., Richard, M., & Lessard, J. (2005). New EDM polishing and texturing process with conductive polymer electrodes. *Journal of Materials Processing Technology, 159*(1), 17–26. https://doi.org/10.1016/j.jmatprotec.2003.11.004
6. Kumar, S. S., Varol, T., Canakci, A., Kumaran, S. T., & Uthayakumar, M. (2021). A review on the performance of the materials by surface modification through EDM. *International Journal of Lightweight Materials and Manufacture, 4*(1), 127–144. https://doi.org/10.1016/j.ijlmm.2020.08.002
7. Gugulothu, B., Krishna Mohana Rao, G., Hanuantha Rao, D., Kalbessa Kumsa, D., & Bezabih Kassa, M. (2020). Experimental results on EDM of Ti-6Al-4V in drinking water with Graphite powder concentration. *Materials Today: Proceedings*. https://doi.org/10.1016/j.matpr.2020.07.616
8. Yue, X., & Yang, X. (2020). Molecular dynamics simulation of material removal process and mechanism of EDM using a two-temperature model. *Applied Surface Science, 528*, 147009. https://doi.org/10.1016/j.apsusc.2020.147009
9. Nguyen, H.-P., Ngo, N.-V., & Nguyen, Q. (2020). Optimizing process parameters in EDM using low frequency vibration for material removal rate and surface roughness. *Journal of King Saud University—Engineering Sciences*. https://doi.org/10.1016/j.jksues.2020.05.002
10. Klocke, F., Zeis, M., Klink, A., & Veselovac, D. (2012). Technological and economical comparison of roughing strategies via milling, EDM and ECM for Titanium- and nickel-based Blisks. *Procedia CIRP, 2*, 98–101. https://doi.org/10.1016/j.procir.2012.05.048
11. Klocke, F., Schwade, M., Klink, A., & Veselovac, D. (2013). Analysis of material removal rate and electrode wear in sinking EDM roughing strategies using different graphite grades. *Procedia CIRP, 6*, 163–167. https://doi.org/10.1016/j.procir.2013.03.079
12. Mughal, M. P., Farooq, M. U., Mumtaz, J., Mia, M., Shareef, M., Javed, M., Jamil, M., & Pruncu, C. I. (2021). Surface modification for osseointegration of Ti6Al4V ELI using powder mixed sinking EDM. *Journal of the Mechanical Behavior of Biomedical Materials, 113*, 104145. https://doi.org/10.1016/j.jmbbm.2020.104145
13. Liu, J., Li, X., Zhang, Y., Tian, D., Ye, M., & Wang, C. (2020). Predicting the material removal rate (MRR) in surface magnetorheological finishing (MRF) based on the synergistic effect of pressure and shear stress. *Applied Surface Science, 504*, 144492. https://doi.org/10.1016/j.apsusc.2019.144492
14. Selvarajan, L., Rajavel, R., Venkataramanan, K., Elango, T., & Dhinagaran, M. (2020). An experimental investigations and optimization of performance measures in EDM using copper and graphite electrodes. *Materials Today: Proceedings*. https://doi.org/10.1016/j.matpr.2020.02.81617.

Determination of Defect-Free Working Range of Friction Stir Processing for AA6082-T6

Jainesh Sarvaiya and Dinesh Singh

1 Introduction

Friction stir processing (FSP) is a solid-state technique used to modify surface properties, crack repair, remove casting defects, fabricate surface composites, and many more. FSP has environmental benefits as it produces no fumes and consumes lower energy compared to other conventional fusion processes. Mishra et al. [1] first developed the FSP in 1999 to improve microstructural characteristics [2]. The working principle of the FSP is similar to that of friction stir welding (FSW), which was invented by Wayne Thomas and his team to join dissimilar materials at The welding institute (TWI) in 1991 [3]. In FSP, a high-strength non-consumable stirring tool with a pin generates enough heat to plasticize work material, so that it can penetrate. Figure 1 schematically represents the relative movement between the tool and workpiece with its terminology. This localized heat generation results from severe plastic deformation and friction between tool and work material produce ultra-fine grain microstructure [4]. During FSP, the amount of material in front of the tool is forged with the stirring tool pin and shoulder and shifted to the trailing side of the tool as it traverses ahead. However, during the shifting of this plasticized material, some amount of material is lost in the form of flash as a result of insufficient consolidation of material in the stir zone. It is also observed that the proper selection of process parameter levels is crucial for the defect-free surface of FSPed material [5, 6].

There are numerous articles available on the optimum tool pin profile for the effective mixing of material and homogeneous particles distribution [7–10]. Researchers widely utilized square and threaded pin tool profiles as their benefits of pulsating effect and vertical motion of material, respectively [10, 11]. Sharma et al. [12] fabricated Al-SiC surface composites through FSP and optimum level of tool rotational

J. Sarvaiya (✉) · D. Singh
Department of Mechanical Engineering, Sardar Vallabhbhai National Institute of Technology, Surat, Gujarat 395007, India

R. Agrawal et al. (eds.), *Recent Advances in Smart Manufacturing and Materials*, Lecture Notes in Mechanical Engineering,
https://doi.org/10.1007/978-981-16-3033-0_31

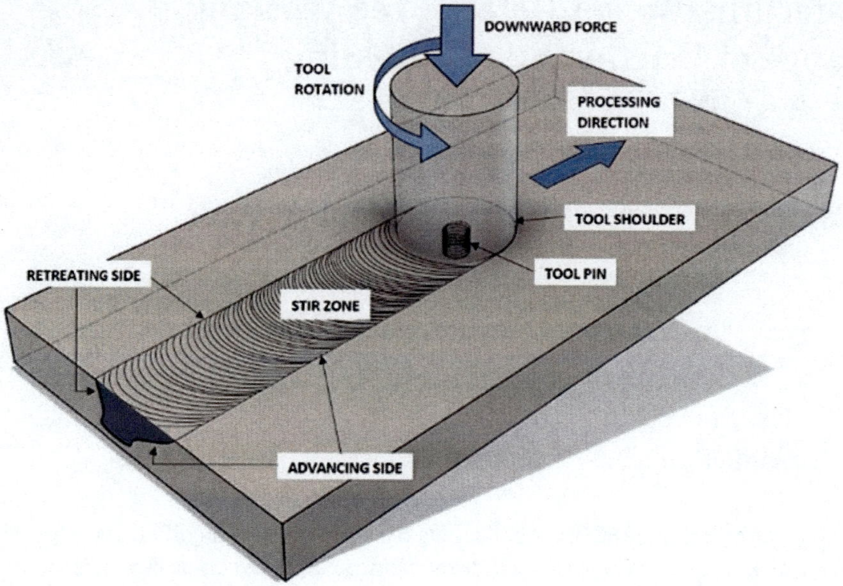

Fig. 1 Schematic diagram of FSP

speed at which excellent mechanical and corrosion properties can be achieved. Beyond or below this rotational speed, the properties of the material deteriorates. Despite achieving homogeneous reinforcing particle distribution at a higher rotational speed, the probability of getting a defect-free stir zone is higher at the lower rotational speed [13]. The insufficient material flow due to selecting inappropriate process parameters or tool pin profiles is responsible for defect formation like a wormhole, tunnel defects, voids, and line defects [14]. The literature reveals that researchers need to perform several trial and error pilot experiments to identify the appropriate process parameters for the defect-free friction stir processed (FSPed) sample [15, 16]. The researcher usually select randomly process parameters level and try to fabricate good-quality FSPed material by altering its levels [17]. The present work has attempted to explore the various process parameters and the proposed standard way for identifying working limits for process parameters in the FSP of aluminum alloys. The experiments were conducted at different levels of process parameters such as tool rotational speed (TRS), processing speed (PS), and tilt angle (TA). This work helps the novice researchers to save their time in the determination of FSP parameters working range.

2 Experimental Procedure

Aluminum alloy AA6082-T6 sheet of thickness 8 mm purchased from local supplier M/s Gautam Metal, Surat, India. Typically, high-stress structural applications like trusses, bridges, cranes, transportation applications, ore skips, milk churn, beer barrels, etc. The purchased sheet is further cut in a rectangular dimension of width 50 mm and length 200 mm with the help of a power hacksaw. The chemical composition of as-received material and its weight percentage is shown in Table 1. Tool material is H13 with a concave shoulder profile of diameter 22.5 mm and a threaded pin profile of length 5 mm and diameter 5 mm (see Fig. 2b). Plunge depth kept constant at 0.12 mm. Surface defects, excessive flash, excessive vibration, and sticking of work material on the tool are the criteria to identify the working range of process parameters like TRS, PS, and TA.

The experiments are conducted on a conventional 3-axis vertical milling machine manufactured by M/s Batliboi Udhna division of model no. BFV-5. The discrete machine control parameter ranges between 45 and 2000 rpm for rotational speed and 5–450 mm/min for transverse speed. The work plate is placed on a machine table with a specially designed and fabricated fixture that avoids movement and vibration of the workpiece during processing (Fig. 2a). The peak temperature is measured using a K-type thermocouple on the advancing side and retreating side at 15 mm away from

Table 1 Chemical composition of AA6082-T6 (in wt. %)

Element	Si	Mg	Mn	Fe	Cr	Zn	Ti	Cu	Al
Standard value	0.7–1.3	0.6–1.2	0.4–1	0–0.5	0–0.25	0–0.20	0–0.1	0–0.1	Bal
Actual value	0.85	0.92	0.26	0.43	0.17	0.18	0.03	0.03	Bal

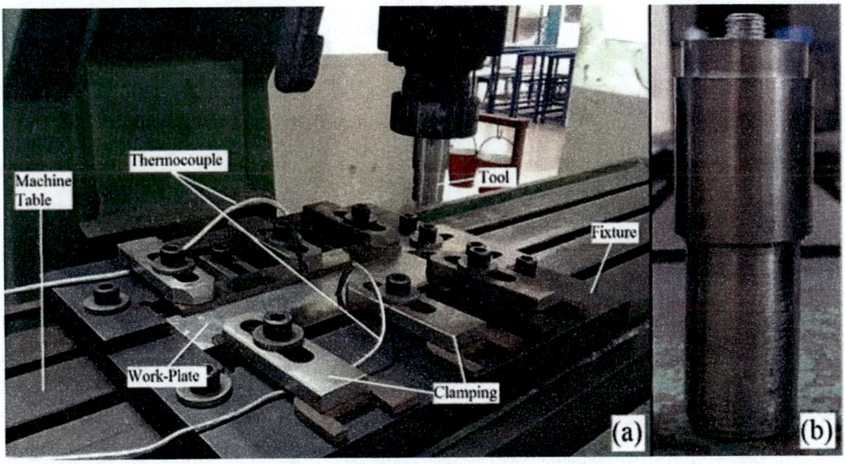

Fig. 2 **a** Vertical milling machine with fixture arrangement and **b** FSP tool

the stir zone center. The thermocouple probe diameter of 1 mm is inserted in 1.5 mm diameter at 4 mm depth of the blind hole.

3 Result and Discussion

3.1 *Identifying the Working Range of Process Parameters*

Tool rotational speed and processing speed. Table 2 shows the experiments performed with different levels of process parameters, peak temperatures, sample images, and observations. The lower limit of TRS is kept at 500 rpm because it is difficult to penetrate the tool pin in work material below that rpm due to insufficient heat generation. The PS upper limits are kept at 112 rpm to prevent excessive vibration and tool failure. The visual and surface inspection found that the higher the value of the TRS/PS ratio more rough and wavy surface. Moreover, the sticking of work material on the tool is witnessed when the peak temperature is higher than 450 °C. Most of the samples processed at 2000 rpm have a rough surface appearance. Additionally, the samples processed with 20 mm/min with TRS other than 500 rpm produced wavy rough surface observed. This poor surface appearance resulted from the high value of the TRS/PS ratio leading to very high heat generation. The TRS of 2000 rpm with any PS causes excessive vibration of the machine table. A similar condition was observed when PS is higher than 112 mm/min. However, this condition can be minimized by proper clamping of the work plate. Figure 3 shows the pictorial overview of the situation of FSPed samples at various process parameters.

As per Frigaard et al. [18] heat model, average heat input during FSP can be predictable by Eq. (1)

$$H = \frac{4}{3}\pi^2\left(\frac{\theta\mu P\omega R^3}{V}\right) \tag{1}$$

where θ is an efficiency of heat input, μ is coefficient of friction, P is pressure in Pa, R is the radius of a tool in meter, ω is TRS, and V is traverse speed. Equation 2 is rewritten as the simplified form of Eq. 1.

$$H = \frac{\omega}{V} \tag{2}$$

Figure 4 shows the comparison graph between peak temperature measured during FSP and TRS/PS ratio. The peak temperature generated during FSP correctly follows the relationship given by Eq. 2.

From Eq. 2, we can say that the two most critical conditions determine the working range of process parameters, and they are as follow.

(a) High TRS and lower PS.

Table 2 FSPed sample of AA6082-T6 at various TRS and PS

Sr. No.	TRS (rpm)	PS (mm/min)	Peak temp. (°C)	TRS/PS ratio	Observation	FSPed surface
1	500	112	329.3	4.46	Smooth surface	
2	500	80	366.4	6.25	Smooth surface	
3	500	40	412.9	12.50	Smooth surface	

(continued)

Table 2 (continued)

Sr. No.	TRS (rpm)	PS (mm/min)	Peak temp. (°C)	TRS/PS ratio	Observation	FSPed surface
4	500	20	436.2	25.00	Smooth surface	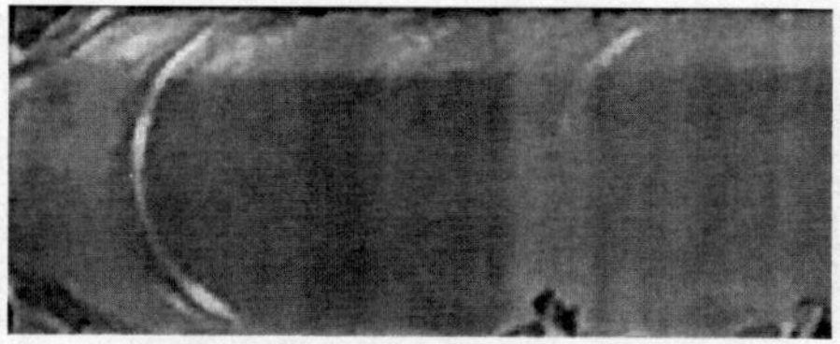
5	1000	112	328.0	8.93	Smooth surface	
6	1000	80	355.2	12.50	Smooth surface	

(continued)

Table 2 (continued)

Sr. No.	TRS (rpm)	PS (mm/min)	Peak temp. (°C)	TRS/PS ratio	Observation	FSPed surface
7	1000	40	389.2	25.00	Smooth surface	
8	1000	20	406.2	50.00	Wavy rough surface	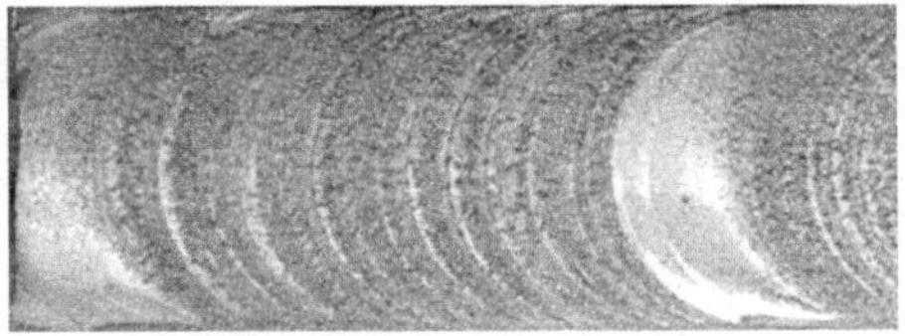
9	1400	112	362.6	12.50	Smooth surface	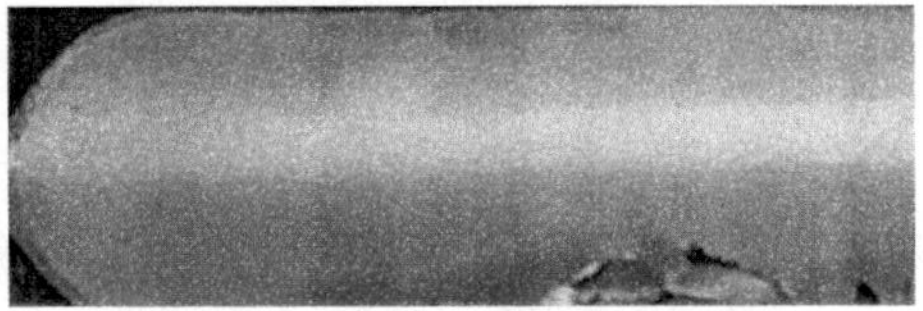
10	1400	80	381.8	17.50	Smooth surface	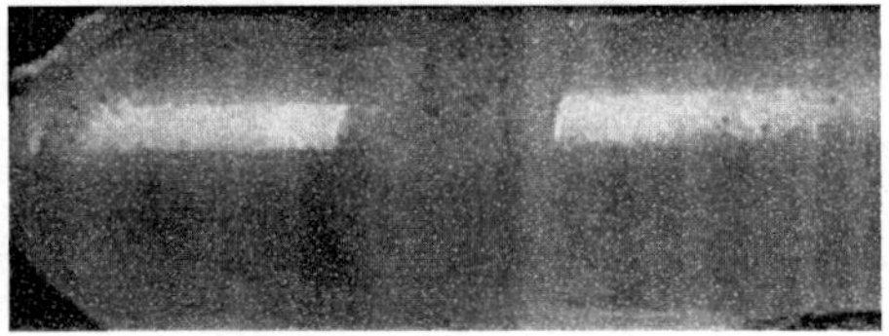

(continued)

Table 2 (continued)

Sr. No.	TRS (rpm)	PS (mm/min)	Peak temp. (°C)	TRS/PS ratio	Observation	FSPed surface
11	1400	40	405.8	35.00	Wavy rough surface	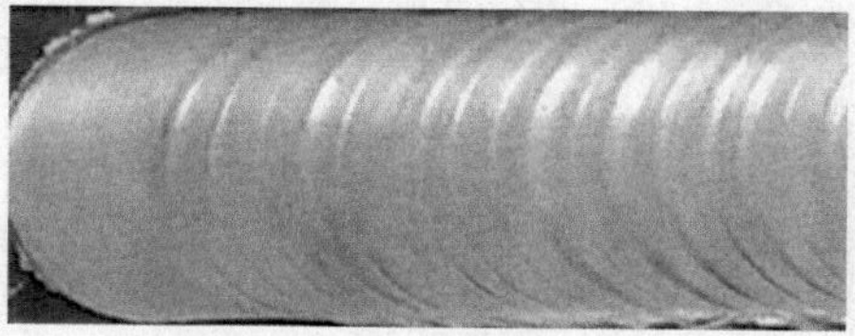
12	1400	20	417.8	70.00	Wavy rough surface	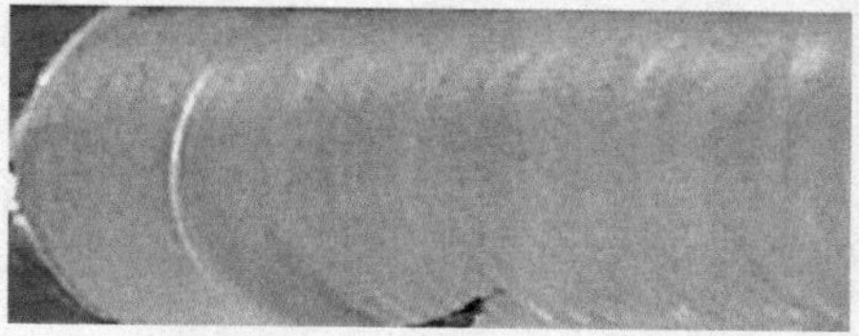
13	2000	112	474.0	17.86	Rough surface	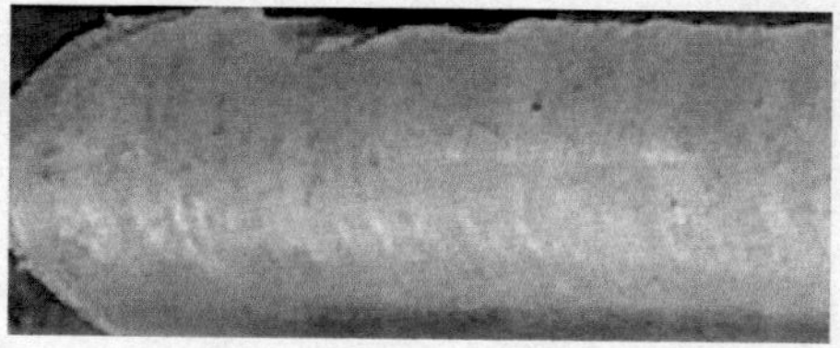

(continued)

Table 2 (continued)

Sr. No.	TRS (rpm)	PS (mm/min)	Peak temp. (°C)	TRS/PS ratio	Observation	FSPed surface
14	2000	80	481.2	25.00	Rough surface	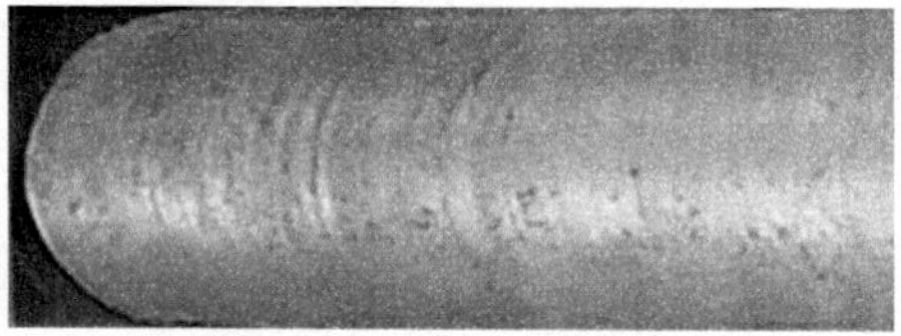
15	2000	40	490.2	50.00	Smooth surface	
16	2000	20	494.6	100.00	Wavy rough surface	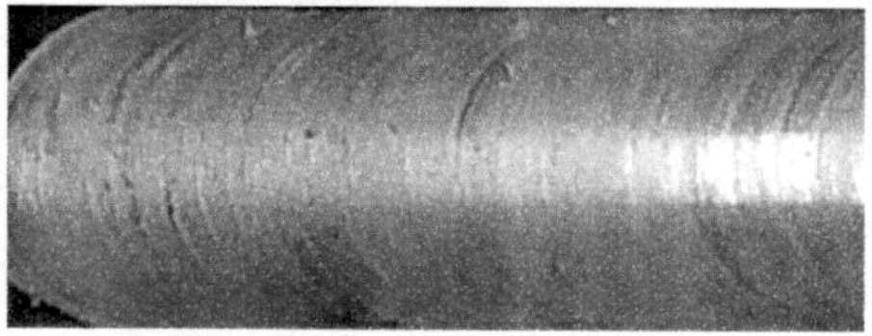

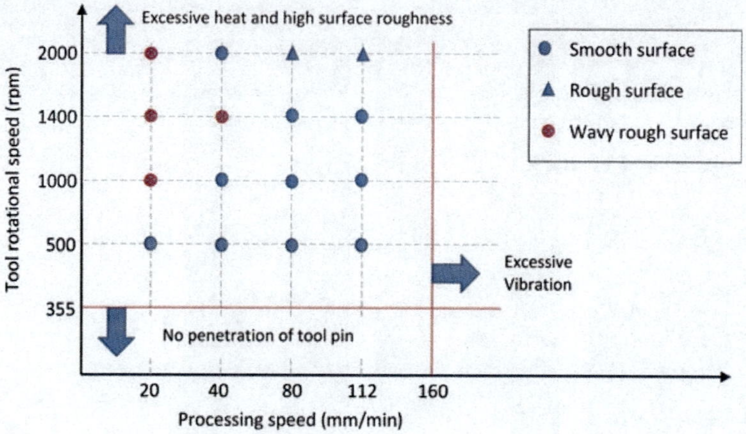

Fig. 3 FSPed AA 6082-T6 sample condition at various levels of process parameters

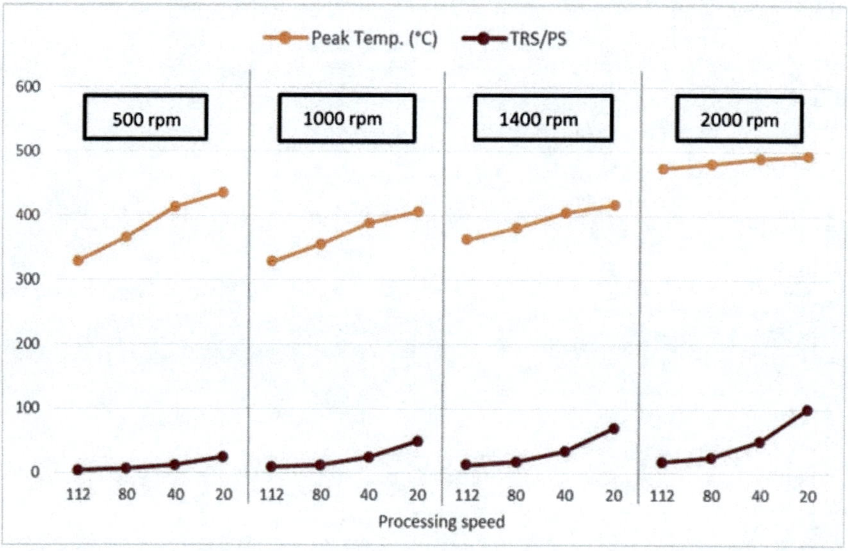

Fig. 4 Average peak temperature and TRS/PS ratio at various processing levels

(b) Lower TRS and high PS.

The first condition produces very high heat input, which increases the flowability of material but waviness and roughness of surface deteriorate its appearance. In the latter case, material flowability drastically reduces as a result of an improper mixture of material, and the tool has to bear high bending stresses. Kim et al. [19] also observed large amounts of flash due to high heat generation, and defects such as cavity or voids are due to insufficient heat or abnormal tool stirring. Thus, the main reason for defect formation is due to excessive heat generation or inadequate heat generation [5, 20]. Therefore, between these two conditions, one can find the working range of process parameters through which defect-free samples can be produced.

Tilt angle. Figure 5 shows that when the TA is 0°, it produces excessive flash only on the retreating side. The full contact of tool shoulder due to 0° TA leads to complete shearing of work material in tool rotational direction. As a result, the material on the leading side slide below the tool shoulder moves from the advancing side to the retreating side. The vector direction of tool rotation and tool traveling is the same on AS, while it is opposite on RS leads to excessive flash and provides no space for the material to further undergo in the stir zone of a rear side of the tool. When the tilt angle is 0°, the shoulder of the tool cannot retain the flow of plasticized material to the tool and squeeze out as flash [21]. However, when the TA is increased by 1°, 2°, and 3°, either very less flash is generated or equally distributed on both sides (i.e., RS and AS). The result is consistent with the finding of past researchers that small tilting of a tool on the trailing side provides better defect-free surface appearance [22] and provides stronger forging action compared to 0° TA [21]. Excessive flash is also responsible for high surface roughness as it fails to consolidate work material appropriately in the stir zone. This condition is usually found in high TRS/PS ratios.

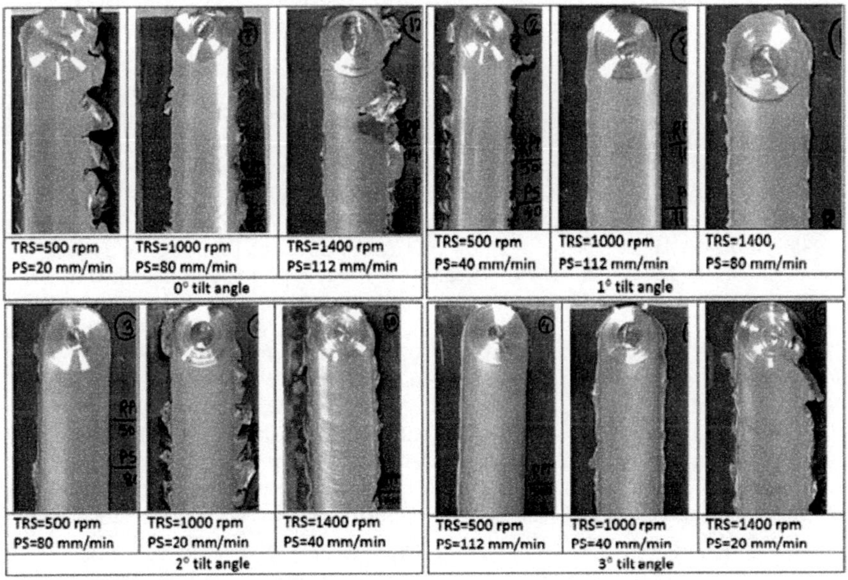

Fig. 5 Formation of flashes in the FSPed sample at different TA

4 Conclusion

In this study, aluminum alloy AA6082-T6 friction stir processed by considering primary process parameters such as TRS, PS, and TA at different values. The influence of these process parameters on surface appearance, surface defects, flash formation, heat generation, and excessive vibration is studied to find the working range of process parameters. The key findings of the study are concluded below.

- At higher TRS (i.e., 2000 rpm), more heat is induced in the stir zone, irrespective of PS, resulting in rough surface appearance and vibration.
- A similar condition appears when PS is lower (i.e., 20 mm/min) irrespective of TRS produce a rough wavy surface. On the other hand, beyond 112 mm/min of PS, excessive machine table vibration is observed.
- The tilted tool provides better forging the material on the trailing side as it advances and produces a good surface appearance compared to the 0° TA.
- Excessive and insufficient heat generation is responsible for the defect formation, which can be evaluated by the TRS/PS ratio. Between these two critical conditions, working range of process parameters can be estimated.

References

1. Mishra, R., Mahoney, M., McFadden, S., Mara, N. A., & Mukherjee, A. K. (1999). High strain rate superplasticity in a friction stir processed 7075 Al alloy. *Scripta Materialia, 42*, 163–168.
2. Mishra, R., Ma, Z., & Charit, I. (2003). Friction stir processing: a novel technique for fabrication of surface composite. *Materials Science and Engineering: A, 341*, 307–310.
3. Thomas, W. M., Nicholas, E. D., Needam, J. C., Murch, M. G., Templesmith, P., Dawes, C. J. (1995). GB Patent Application No. 9125978.8, December 1991 and US Patent No. 5460317.
4. Ebrahimi, M., & Par, M. A. (2019). Twenty-year uninterrupted endeavor of friction stir processing by focusing on copper and its alloys. *Journal of Alloys and Compounds, 781*, 1074–1090.
5. Dialami, N., Cervera, M., & Chiumenti, M. (2020). Defect formation and material flow in Friction Stir Welding. *European Journal of Mechanics, A/Solids, 80*, 103912.
6. Vijayavel, P., Balasubramanian, V., & Sundaram, S. (2014). Effect of shoulder diameter to pin diameter (D/d) ratio on tensile strength and ductility of friction stir processed LM25AA-5% SiCp metal matrix composites. *Materials and Design, 57*, 1–9.
7. Mahesh, V. P., & Arora, A. (2019). Effect of tool shoulder diameter on the surface hardness of aluminum-molybdenum surface composites developed by single and double groove friction stir processing. *Metallurgical and Materials Transactions A: Physical Metallurgy and Materials Science, 50*, 5373–5383.
8. Ugender, S. (2018). Influence of tool pin profile and rotational speed on the formation of friction stir welding zone in AZ31 magnesium alloy. *Journal of Magnesium and Alloys, 6*, 205–213.
9. Jayaseelan, P., Christy, T. V., Vijay, S. J., & Nelson, R. (2019). Effect of tool material, profile and D/d ratio in friction stir welding of aluminium metal matrix composites. *Materials Research Express, 6*(9), 096590.
10. Elangovan, K., Balasubramanian, V., & Valliappan, M. (2008). Influences of tool pin profile and axial force on the formation of friction stir processing zone in AA6061 aluminium alloy. *International Journal of Advanced Manufacturing Technology, 38*, 285–295.

11. Zhang, Y. N., Cao, X., Larose, S., & Wanjara, P. (2012). Review of tools for friction stir welding and processing. *Canadian Metallurgical Quarterly, 51*, 250–261.
12. Sharma, A., Sharma, V. M., Sahoo, B., Joseph, J., & Paul, J. (2019). Effect of exfoliated few-layered graphene on corrosion and mechanical behaviour of the graphitized Al–SiC surface composite fabricated by FSP. *Bulletin of Materials Science, 42*, 1–12.
13. Rathee, S., Maheshwari, S., & Siddiquee, A. N. (2018). Issues and strategies in composite fabrication via friction stir processing: A review. *Materials and Manufacturing Processes, 33*, 239–261.
14. Morisada, Y., Imaizumi, T., & Fujii, H. (2015). Clarification of material flow and defect formation during friction stir welding. *Science and Technology of Welding and Joining, 20*, 130–137.
15. Kadaganchi, R., Gankidi, M. R., & Gokhale, H. (2015). Optimization of process parameters of aluminum alloy AA 2014–T6 friction stir welds by response surface methodology. *Defence Technology, 11*, 209–219.
16. Lakshminarayan, A. K., & Balasubramanian, V. (2009). Comparison of RSM with ANN in predicting tensile strength of friction stir welded AA7039 aluminium alloy joints. *Transactions of Nonferrous Metals Society of China, 19*, 9–18.
17. John, J., Shanmuganatan, S. P., Kiran, M. B., Shanmuganatan, S. P., Kiran, M. B., Senthil Kumar, V. S., & Krishnamurthy, R. (2019). Investigation of friction stir processing effect on AA 2014–T6. *Materials and Manufacturing Processes, 34*, 159–176.
18. Frigaard, Ø., Grong, Ø., & Midling, O. T. (2001). A process model for friction stir welding of age hardening aluminum alloys. *Metallurgical and Materials Transactions A, 32*, 1189–1200.
19. Kim, Y. G., Fujii, H., Tsumura, T., Komazaki, T., & Nakata, K. (2006). Three defect types in friction stir welding of aluminum die casting alloy. *Materials Science and Engineering: A, 415*, 250–254.
20. Bharti, S., Ghetiya, N. D., & Patel, K. M. (2020). A review on manufacturing the surface composites by friction stir processing. *Materials and Manufacturing Processes, 00*, 1–36.
21. Zhai, M., Wu, C., & Su, H. (2020). Influence of tool tilt angle on heat transfer and material flow in friction stir welding. *Journal of Manufacturing Processes, 59*, 98–112.
22. Mishra, R. S., & Ma, Z. Y. (2005). Friction stir welding and processing. *Material Science and Engineering R: Reports, 50*, 1–78.

A Review on the Manufacturing of Aluminum Metal Matrix Composite by Various Stir Casting Method

A. S. Vasava and D. Singh

1 Introduction

The composite materials are made of two phases, i.e., reinforcement and matrix, and both have their separate properties. The reinforcement material is wholly dispersed in the matrix material. Prepared composite performance is directly dependent on the particle distribution in the base metal [1]. The foremost objective of composite material manufacturing is to get enhanced properties to compare to the base metal. During composite manufacturing, the primary challenging work is the homogenous distribution of reinforcing particles in the matrix material to achieve the defect-free composite [2] (Fig. 1).

In the process of stir casting, in which a stirrer is used to mix the reinforcement and matrix material. The following process parameter affects the microstructure and mechanical properties of the base metal.

- Stirring speed
- Size of impeller
- Holding time
- Stirring temperature
- Stirring time
- Holding temperature

2 Literature on Various Stir Casting Methods

Various stir casting methods are used to produce AMMCs. To make the composite, various reinforcement materials are required. For that, in this literature survey, various

A. S. Vasava (✉) · D. Singh
Sardar Vallabhbhai National Institute of Technology, Surat, India

R. Agrawal et al. (eds.), *Recent Advances in Smart Manufacturing and Materials*, Lecture Notes in Mechanical Engineering,
https://doi.org/10.1007/978-981-16-3033-0_32

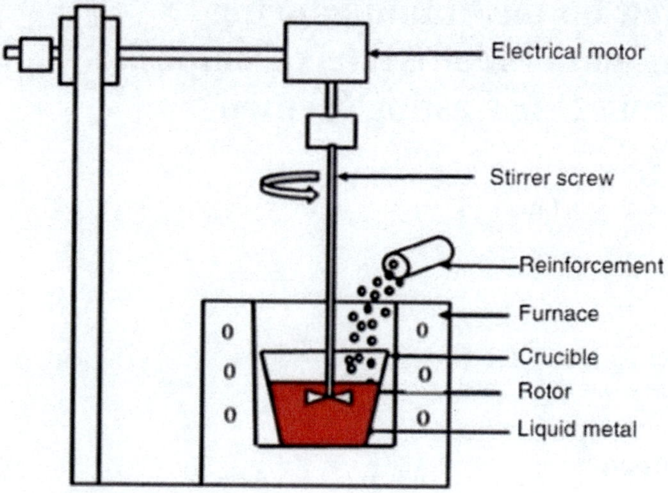

Fig. 1 Conventional stir casting process [3]

reinforcement and their particle size have been discussed. The addition of that different volume fraction used by the previous researchers to enhance the mechanical and tribological properties has been discussed. For the homogenous mixing of both materials, a stirrer is used, so stirring time is also reported (Table 1).

3 Effect of Reinforcement on Mechanical and Microstructure

Mechanical properties have improved due to the homogenous particle distribution. SEM exposed that the crack bridging effect was shown for MWCNT. MWCNT and AIN more with 1 vol% and 0.75 vol%, respectively, showed porosity and cluster formation of the particles [4]. Results displayed significant enhancement in the mechanical properties with 10–25 wt% of reinforcement [5]. OM reveals the homogenous distribution of particles in the composite matrix. Al_2SiO_5 presented superior hardness, coefficient of friction, and superior wear resistance compare to TiO_2 particles. SEM discloses that abrasive and adhesion wear is responsible for composites' wear at different load conditions [6]. Test results exposed that dimples and cracks are revealed. With increasing B_4C—CDA reinforcements showed improved composite properties but with this impact strength reduces [7]. Wear tests reveal that material loss their weight increasing sliding and load conditions. The optimum dry sliding wear was found for load, distance and velocity is 18.1 N, 905.4 m, 4.18 ms^{-1}, respectively. For wear loss calculations, regression model was developed [8].

Table 1 Basic parameters used in various stir casting process

Base metal	Reinforcement, size	Particle size	Volume fraction	Method	Stirring time	Furnace temp	References
A356	Al, Gr, MWCNT, AIN	10 μm 10–20 μm 10–20 mm 2 μm	–	Ultrasonic-assisted stir casting	500 rpm for 3 min	800 °C	[4]
AA6063	Clay particles	250 μm	–	Liquid metallurgy stir casting			[5]
LM27	TiO_2, Al_2SiO_5	1–32 μm	5, 10, and 15 wt%	Stir casting	630 rpm for 5 min	750 °C	[6]
AA7075	B_4C, CDA	50–70 μm 40–60 μm	0–10 wt%	Two stage stir casting	300–400 rpm for 5–8 min	1000 °C	[7]
A336	B_4C, Fly ash	52 μm 36 μm	10 wt% 5 wt%	Stir casting	300 rpm	780 °C	[8]
AA5052	WC, SiC, Gr	60–90 μm	– 5, 10 wt% 4 wt%	Melt-stir casting method	450 rpm for 12 min	650 °C	[9]
AA5083	B_4C	300 mesh	5, 10, 15, 20 wt%	Stir casting	200–300 rpm for 10 min	730 °C	[10]
AA7075	SiC Al_2O_3	30 μm 40 μm	3, 5, 7 and 9 wt% 2 wt%	Stir casting	400–500 rpm for 32 min	780 °C	[11]
LM13	ZrB_2 TiC	25–48 μm 15–35 μm	0, 2.5, 5 and 10 wt.% 5 wt%	Stir casting	650 rpm for 10 min	400 °C	[12]
AA8011	ZrB_2, Si_3N_4	40–50 μm	–	Two stage stir casting	300 rpm for 5 min	700 °C	[13]
AA6063	NFA	76 nm	1, 2, 3 wt%	Ultrasonic-assisted stir casting process	400 rpm for 4–5 min	720 °C	[14]
AA6061	Red mud	–	–	Stir casting	525 °C for 8 h	900 °C	[15]

(continued)

Table 1 (continued)

Base metal	Reinforcement, size	Particle size	Volume fraction	Method	Stirring time	Furnace temp	References
AA6061	SiC, Gr	50, 100, 150 μm	–	Ultrasonic-assisted stir casting technique	80 r/min for 10 min	550 °C	[16]
LM13	TiO_2, MoS_2	–	12 wt% 3 wt%	Stir casting	250 rpm for 10 min	760 °C	[17]
AA6061	FZA	782 μm	0, 5, 10, 15 wt%	Double stir casting	500 rpm for 7 min	900 °C	[18]
AA8011	ZrB_2, Si_3N_4	40–50 μm 50 μm	50: 50 wt%	Two stage stir casting	450 °C for 10 min	750 °C	[19]
AA7075	MoS_2, B_4C	2 μm 10 μm	3 wt% 4, 8, 12 wt%	Stir casting	500 rpm of 15 min	1100 °C	[20]
AA6082	Metakaolin particles	200 μm	0–15 wt%	Ultrasonic cavitation assisted stir casting	700 rpm for 10 min	750 °C	[21]
AA7079	AlN	20 μm	2–12 wt%	Stir casting	400 rpm for 15 min	700–750 °C	[22]
AA6061	SiC, WC	125 μm 3 μm	5, 7.5, 10 wt%	Liquid metallurgy stir casting	400RPM and 5 min	900 °C	[23]
A356	Al_2O_3	50 μm	–	Stir casting	300 rpm for 10 min	700 °C	[24]
AA6351	SiC, RHA	120–150 μm	2, 4, 6, 8 wt.%	Stir casting	400 rpm for 10–12 min	830 ± 30 °C	[25]
AA2024	Red mud	90 μm	5, 10, 15 wt%	Two stage stir casting	700 rpm for 5 min	640 ± 5 °C	[26]
Pure aluminum	Al_2SiO_5	80 μm	5, 10 wt%	Vacuum casting method	500 rpm for 5 min	750 °C	[27]
AA7075	SiC, Al_2O_3 B_4C, TiB_2	–	15 wt%	Stir casting	300 rpm for 12–15 min	800 °C	[28]
AA6063 -T6	Al_2O_3	10 μm	6–14 wt%	Vacuum multi-stir casting process	350 rpm for 10 min	650 °C	[29]
AA6061	ZrO_2	25–30 μm	5–15 wt%	Stir casting	300 rpm	780 °C	[30]

(continued)

Table 1 (continued)

Base metal	Reinforcement, size	Particle size	Volume fraction	Method	Stirring time	Furnace temp	References
AA5052	WC, Gr	75 μm	–	Stir casting	1090 rpm for 10 min	605 °C	[31]
A356	B_4C	100–300 μm	3 wt%	Stir casting	300 rpm for 10, 15, 20 min	750, 850, 950 °C	[32]
AA6061	SiC, Al_2O_3	–	2–3 wt%	Stir casting	600 rpm for 15 min	750 ± 10 or 850 ± 10 C	[33]
AA2024	Eggshell	–	–	Electromagnetic stir casting	200 rpm for 3 min	700 °C	[34]
A356	B_4C	30 μm	5, 10, 15 wt%	Stir casting	400 rpm for 10 min	650 °C	[35]
AA2618	Si_3N_4, AIN, ZrB_2	–	0, 2, 4, 6, 8 wt%	Stir casting	–	850 °C	[36]
AA6082	Gr		3, 6, 9, 12 wt%	Stir casting	500 rpm for 10 min	900 °C	[37]
AA5083	SiC		0, 3, 5, 8,10 wt%	Ultrasonic-assisted stir casting process	5 min	–	[38]
AA6061	B_4C	30 μm	2, 4, 5, 6 and 8 wt%	Ultrasonic-assisted stir casting process	300–350 rpm for 10 min	750 °C	[39]

SiC silicon carbide, *Al* aluminum, *Fe_3O_4* ferrous ferric oxide, *GNPs* graphene nanoplatelets, *HCHCr* high carbon high grapheme steel, *Gr* grapheme, *TiO_2* titanium oxide, *WC* tungsten carbide, *Al_2O_3* aluminum oxide, *B_4C* boron carbide, *RHA* rice husk ash, *PKS* palm kernel shell, *TiB_2* titanium diboride, *CBSA* cow bone fly ash, *ZrB_2* zirconium diboride, *Si_3N_4* silicon nitride, ZrO_2—zirconium dioxide, MoS_2—molybdenum disulfide, *MWCNT* multi-wall carbon nanotubes

The tensile strength was enhanced by 15.12% [9]. Test results exposed that with the increment in the volume fraction of the particles improves the mechanical properties. Volume fraction 20 wt % achieved enhanced mechanical strength compare to other volume fractions [10]. Test results exposed that improvement is shown in tensile and hardness properties [11]. Test results exposed that with increasing in volume of particles and aging time both micro and micro-hardness values improved [12]. Wear test results showed reduction in wear rate by 19%, 28.2%, 22.5% with 50,100 and 150 mm, respectively [13]. Test results exposed that hardness and tensile properties improved by 16.5 and 35%, respectively [14]. Test results exposed that the micro-hardness of the composite with 15% FZA displayed improved hardness 107 VHN and an observed reduction in corrosion [15]. Test results exposed that the wear rate of AA8011/Si_3N_4-ZrB_2 reduces with increases in particle fraction. The COF was observed high in without reinforcement and with reinforcement shows reduced COF [16]. Test results exposed that a higher volume of reinforcement increases the hardness, UTS, and YS of the hybrid composite [17]. The tensile and hardness results increased with increased reinforced particles up to 7.5 wt% and decreased. The T6 heat-treated specimens expressed improvement in mechanical properties compared with cast specimen [18]. Test results exposed that the up to 4% volume porosity content increases. They improved the composites hardness, compressive, and tensile strength improve by 28%, 41%, and 28%, respectively [19]. Test results exposed that the hybrid composite showed enhancement in compressive strength and hardness compared to solid alloy [20]. Test results exposed that SiC expressed better properties compared to RHA [22]. Test results exposed that with increment in particle volume reduces wear rate [23]. Test results exposed that the hardness of the composites is greater than the base alloy. Base alloy with 15% B_4C showed admirable UTS of 261 MPa. With B_4C composite achieved highest impact strength by 11 Nm [25]. Test results exposed that the hardness of Al/Al_2O_3MMC had improved by 1.71 times, and determining tensile strength had improved by 1.27 times [26]. Test results exposed that hardness improved with addition of the reinforcing particles. Wear properties also improves with different loading condition and with reinforcement [27]. Test results exposed that reinforce particle improves the material removal rate of the composite. [28]. Test results exposed that with the fixed stirring time and increase in pouring time, first composite properties are improved and then decreases [29]. Test results exposed that at 75 ° the sample got enhanced tensile properties and with higher content of SiC got enhanced tensile of 181 MPa [30]. Test results exposed that the base metal density is higher compared to eggshell composite [31]. Test results exposed that models developed for the measurement of accuracy of the mechanical properties showed improved results [32]. Test results exposed that reinforce particles enhance the tensile and hardness properties of the composite material [33]. Test results exposed that composites' micro-hardness was reduced from 49.5 to 44 VHN and macro-hardness from 31.6 to 28.3 BHN. [34]. The result shows it upgraded specific UTS and compressive strength at 4 wt% B_4C, 36.32%, and 43.92%, respectively. Composite showed increased specific BHN 50.89% compared to the base metal. SEM and OM reveal that individual distribution of particle and grain improvement achieved at 4 wt% B_4C [35]. The result shows it enhanced specific

UTS and compressive strength improved by adding SiC particles. A rise in hardness reveals that condensed wear and gives extensive life to the component [36].

4 Conclusion

Stir casting proved that it is a growing technique to produce composites and improves material properties. This study reports that researchers used various stir casting methods for the fabrication of the composite material. From the above literature, some critical points are derived and summarized below.

- In stir casting, process parameters affect the composite properties. Numerous studies have been done to examine the composite microstructural, mechanical, and tribological properties. Still, proper selection of the parameters is necessary while performing the experiment work.
- Various studies showed that reinforcing particles improves the composite properties. But, the selection of the volume fraction of the particles is not standardized.
- With the stir casting process, some problems are associated with particle distribution, porosity, and wettability because this process is performed in the open air.
- Many studies are done on the design of a mechanical stirrer, which is used to mix the matrix and reinforce particles. But no standard is available, so more effort has been needed.
- From the last decades, researchers got very interesting and predominant results achieved. All the results are taking into consideration at the laboratory level.

References

1. Prasad, D.S., Shoba, C., Ramanainah, N. (2014). Investigations on mechanical properties of aluminium hybrid composite. Materials Research and Technology, 3, 79–85.
2. Craciun, A. L., Heputţ, T., Pinca-Bretotean, C. (2018). Aspects regarding manufacturing technologies of composite materials for brake pad application. In *IOP Conference Series: Materials Science and Engineering*, (Vol. 294, no. 1, p. 012003). IOP Publishing (2018)
3. Bandil, K., Vashisth, H., Kumar, S., Verma, L., Jamwal, A., Kumar, D., Gupta, P., et al. (2019). Microstructural, mechanical and corrosion behaviour of Al–Si alloy reinforced with SiC metal matrix composite. *Journal of Composite Materials, 53*(28–30), 4215–4223.
4. Logesh, K. P., Hariharasakthisudhan, Arul Marcel Moshi, A., Surya Rajan, B., Sathickbasha, K. (2019) Mechanical properties and microstructure of A356 alloy reinforced AlN/MWCNT/graphite/Al composites fabricated by stir casting. *Materials Research Express, 7*(1), 015004 (2019)
5. Agbeleye, A. A., Esezobor, D. E., Balogun, S. A., Agunsoye, J. O., Solis, J., & Neville, A. (2020). Tribological properties of aluminium-clay composites for brake disc rotor applications. *Journal of King Saud University- Science, 32*(1), 21–28.

6. Jamwal, A., Vates, U. K., Gupta, P., Aggarwal, A., & Sharma, B. P. (2019). Fabrication and characterization of Al_2O_3–TiC-reinforced aluminum matrix composites. In *Advances in industrial and production engineering* (pp. 349–356). Singapore: Springer.
7. Manikandan, R., & Arjunan, T. V. (2020). Studies on micro structural characteristics, mechanical and tribological behaviours of boron carbide and cow dung ash reinforced aluminium (Al 7075) hybrid metal matrix composite. *Composites Part B: Engineering, 183*, 107668.
8. Arunachalam, S., & Chelladurai, S. J. S. (2020). Optimization of dry sliding wear parameters of AA336 aluminum alloy—boron carbide and fly ash reinforced hybrid composites by stir casting process. *Materialwissenschaft und Werkstofftechnik, 51*(2), 189–198.
9. Ebenezer Jacob, D. D. S., Velmurugan, C., Leo Dev Wins, K., & Boopathi Raja, K. P. (2019). Effect of tungsten carbide, silicon carbide and graphite particulates on the mechanical and microstructural characteristics of AA 5052 hybrid composites. *Ceramics International, 45*(1), 614–621 (2019)
10. Das, S., Chandrasekaran, M., Samanta, S., Kayaroganam, P., & Davim, P. (2019). Fabrication and tribological study of AA6061 hybrid metal matrix composites reinforced with SiC/B_4C nanoparticles. *Industrial Lubrication and Tribology*.
11. Singh, R., Shadab, M., Dash, A., & Rai, R. N. (2019). Characterization of dry sliding wear mechanisms of AA5083/B 4 C metal matrix composite. *Journal of the Brazilian Society of Mechanical Sciences and Engineering, 41*(2), 98.
12. Garg, P., Jamwal, A., Kumar, D., Sadasivuni, K. K., Hussain, C. M., & Gupta, P. (2019). Advance research progresses in aluminium matrix composites: manufacturing & applications. *Journal of Materials Research and Technology, 8*(5), 4924–4939.
13. Ramaswamy, A., Perumal, A. V., & Chelladurai, S. J. S. (2019). Investigation on mechanical properties and dry sliding wear characterization of stir cast LM13 aluminium alloy-ZrB2-TiC particulate hybrid composites. *Materials Research Express, 6*(6), 066578 (2019).
14. Raj, N., & Radhika, N. (2019). Tribological characteristics of LM13/Si 3 N 4/gr hybrid composite at elevated temperature. *Silicon, 11*(2), 947–960 (2019)
15. Gosangi, N. R., & Dumpala, L. (2019). Characterization of friction stir welded AL6063/nano fly ash composite. *Materials Research Express, 6*(7), 076564.
16. Panwar, N., & Chauhan, A. (2019). Parametric behaviour optimisation of macro and micro hardness for heat treated Al 6061-red mud composite. *Journal of Materials Research and Technology, 8*(1), 660–669.
17. Prasad Reddy, A., Vamsi Krishna, P., Rao, R. N. (2019). Two-body abrasive wear behaviour of AA6061-2SiC-2Gr hybrid nanocomposite fabricated through ultrasonically assisted stir casting. *Journal of Composite Materials, 53*(15), 2165–2180.
18. Jojith, R., & Radhika, N. (2019). Mechanical and tribological properties of LM13/TiO_2/MoS_2 hybrid metal matrix composite synthesized by stir casting. *Particulate Science and Technology, 37*(5), 570–582.
19. Roseline, S., & Paramasivam, V. (2019). Corrosion behaviour of heat treated Aluminium Metal Matrix composites reinforced with Fused Zirconia Alumina 40. *Journal of Alloys and Compounds, 799*, 205–215.
20. Nayim, S. T. I., Hasan, M. Z., Seth, P. P., Gupta, P., Thakur, S., Kumar, D., & Jamwal, A. (2020). Effect of CNT and TiC hybrid reinforcement on the micro-mechano-tribo behaviour of aluminium matrix composites. *Materials Today: Proceedings, 21*, 1421–1424.
21. Liu, S., Wang, Y., Muthuramalingam, T., Anbuchezhiyan, G. (2019). Effect of B_4C and MOS_2 reinforcement on micro structure and wear properties of aluminum hybrid composite for automotive applications. *Composites Part B: Engineering, 176*, 107329
22. Bright, R. J., Selvakumar, G., Sumathi, M., & Lenin, N. (2019). Development, mechanical characterization and analysis of dry sliding wear behavior of AA6082–Metakaolin metal matrix composites. *Materials Research Express, 6*(12), 126516.
23. Mahesh Kumar, V., & Venkatesh, C. V. (2019). Evaluation of microstructure, physical and mechanical properties of Al 7079–AlN metal matrix composites. *Materials Research Express, 6*(12), 126503.

24. Fenghong, C., Chang, C., Zhenyu, W., Muthuramalingam, T., & Anbuchezhiyan, G. (2019). Effects of silicon carbide and tungsten carbide in aluminium metal matrix composites. *Silicon, 11*(6), 2625–2632.
25. Pourhosseini, S., Beygi, H., & Sajjadi, S. A. (2018). Effect of metal coating of reinforcements on the microstructure and mechanical properties of Al-Al_2O_3 nanocomposites. *Materials Science and Technology, 34*(2), 145–152 (2018).
26. Kumar, A., Arafath, M. Y., Gupta, P., Kumar, D., Hussain, C. M., & Jamwal, A. (2020). Microstructural and mechano-tribological behavior of Al reinforced SiC-TiC hybrid metal matrix composite. *Materials Today: Proceedings, 21*, 1417–1420.
27. Sharma, A., Belokar, R. M., Sanjeev Kumar (2018). Dry sliding wear characterization of red mud reinforced aluminium composite. *Journal of the Brazilian Society of Mechanical Sciences and Engineering, 40*(6), 294.
28. Pethuraj, M., Uthayakumar, M., Rajakarunakaran, S., & Rajesh, S. (2018). Solid particle erosive behaviour of sillimanite reinforced aluminium metal matrix composites. *Materials Research Express, 5*(6), 066514.
29. Bandhu, D., Thakur, A., Purohit, R., Verma, R. K., & Abhishek, K. (2018). Characterization & evaluation of Al7075 MMCs reinforced with ceramic particulates and influence of age hardening on their tensile behavior. *Journal of Mechanical Science and Technology, 32*(7), 3123–3128.
30. Kataria, M., & Mangal, S. K. (2018). Characterization of aluminium metal matrix composite fabricated by gas injection bottom pouring vacuum multi-stir casting process. *Kovove Materialy Metallic Materials, 56*, 231–243.
31. James, J. (2018). Tribological behavior and wear fashion of processed AA6061/ZrO2 composite. *Industrial Lubrication and Tribology.*
32. Dhas, D. S. E. J., Velmurugan, C., & Leo Dev Wins, K. (2018). *Investigations on the effect of tungsten carbide and graphite reinforcements during spark erosion machining*
33. Khademian, M., Alizadeh, A., & Abdollahi, A. (2017). "Fabrication and characterization of hot rolled and hot extruded boron carbide (B 4 C) reinforced A356 aluminum alloy matrix composites produced by stir casting method." *Transactions of the Indian Institute of Metals, 70*(6), 1635–1646.
34. Boopathy, S., & Prasath K. M. (2017). "Structure–Property correlation of Al–SiC–Al 2 O 3 composites: Influence of processing temperatures." *Transactions of the Indian Institute of Metals, 70*(9), 2441–2447.
35. Dwivedi, S. P., Sharma, S., & Mishra R. K. (2017). "Influence of precipitation hardening parameters on the microstructure and mechanical properties of extruded AA2014/eggshells green composites." *Journal of Composite Materials 51,* (30), 4261–4271.
36. Shirvanimoghaddam, K., Khayyam, H., Abdizadeh, H., Akbari, M. K., Pakseresht, A. H., Ghasali, E., & Naebe M. (2016). "Boron carbide reinforced aluminium matrix composite: Physical, mechanical characterization and mathematical modelling." *Materials Science and Engineering: A 658*, 135–149.
37. Kumar, N. M., Kumaran, S. S., & Kumaraswamidhas L. A. (2016). "Aerospace application on Al 2618 with reinforced–Si3N4, AlN and ZrB2 in-situ composites." *Journal of Alloys and Compounds 672*, 238–250.
38. Chaubey, A., Dwivedi, R., Rana, R. S., & Choudhary, K. (2020). "Experimental inspection of aluminium matrix composites reinforced with SiC particles fabricated through ultrasonic assisted stir casting process." *Materials Today: Proceedings.*
39. Gudipudi, S., Nagamuthu, S., Subbian, K. S., & Chilakalapalli S. P. R. (2020). "Enhanced mechanical properties of AA6061-B4C composites developed by a novel ultra-sonic assisted stir casting." *Engineering Science and Technology, an International Journal.*

Comparative Experimental Investigation and Optimization on Rotary Ultrasonic Face Milling and Conventional Grinding on Float Glass

Varun Kumar, Dheeraj Joshi, Praveen Saraswat, and Om Ji Shukla

1 Introduction

Earlier, the conventional machining process was performed on hard and brittle materials for cutting and drilling. Due to some drawbacks in the conventional machining process, non-conventional machining is being preferred to perform on such materials. RUM is most preferable machining process among all non-conventional machining process because of its independent machinability concerning chemical reaction, conductive or non-conductive workpiece, etc. Rotary ultrasonic machining is the feasible and high accurate hybrid process which combines the conventional grinding and ultrasonic machining. Challenge to make cost-effective high-quality products enforcing engineering industry to innovate new hard materials. Transformation of such new hard material into suitable various application creates a new challenge which involves various types of material removal from them. RUM performs an inexpensive and effective way of machining on these materials. In 1927, the first time, the application of ultrasonic was used for machining. The ultrasonic involved first research paper was published after confirming the suitability of the process for drilling in glass slab [1]. Balamuth research that tool type also affects MRR. The research on tool design development results in the improvement of efficiency of machining over brittle and hard materials [2]. Use of abrasives slurry caused limitation in the application of the machine. Tyrell was the first one who revealed the bonding of abrasives material directly on the tool by electroplating. To make a harder and more reliable tool, he used diamond abrasives at the bottom of the tool [3]. After that, Markov extended the research to make that ultrasonic machining to rotary ultrasonic machining (RUM) by using a diamond impregnated tool for slot cut and

V. Kumar (✉) · D. Joshi · P. Saraswat
Department of Mechanical Engineering, SKIT Jaipur, Jaipur, India

O. J. Shukla
Department of Mechanical Engineering, NIT Patna, Patna, India

R. Agrawal et al. (eds.), *Recent Advances in Smart Manufacturing and Materials*, Lecture Notes in Mechanical Engineering,
https://doi.org/10.1007/978-981-16-3033-0_33

deep drilling successfully [4]. Percy Legge develops of rotary ultrasonic machining (RUM) in 1964 [5]. Later, Pei designed a 15° tapered at the bottom face of the milling tool to resolve the longitudinal force problem into lateral force [6]. Pei discussed the MRR of both process (drilling and milling) after which he named to process, rotary ultrasonic face milling (RUFM) [7]. Later, various research found to improve and compare the rotary ultrasonic surface milling to other ultrasonic machining methods.

2 Experimental Setup, Parameters and Design

CNC rotary ultrasonic drilling and slot cut machine is used for the study (Fig. 1). The machine platform consists of various systems like coolant system, ultrasonic system and drive control system. A computer-controlled stepper motor is used to drive ball screws with three DOF in the feed system. Basic G&M codes are used for the experiment.

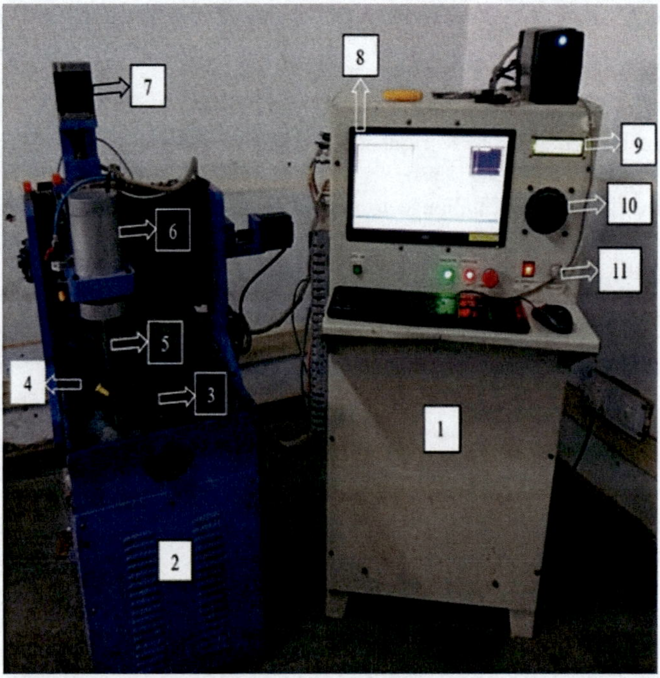

1. Drive control system 2. Coolant tank 3. Workpiece 4. Coolant flow
5. Horn 6. Transducer 7. Stepper motor 8. Display
9. U/S Freq. display 10. U/S Power control 11. Frequency control

Fig. 1 Rotary ultrasonic machining setup

The ultrasonic power and frequency are controlled from the drive control system. The drive system consists of a piezoelectric transducer which converts electric energy into rotation and vibration which is directly transferred to the attached horn. The machine has a maximum value of ultrasonic power is 1000 W and frequency of 25 kHz. The tool is soldered with the horn which also gets the vibration effect of the transducer. A high carbon steel tool which consists of diamond grit implemented at the bottom part is used for the machining. In the coolant system, normal fluid (without abrasives) is used to reduce the temperature and remove waste during the machining.

2.1 Experimental Parameters

The literature review helps to take proper parameters according to the targets of this research. Also, some parameters are taken according to the equipment capability which is used to measure the outputs after the machining like the width of each slot must be more than 3 mm. Also, slot length not less than 25 mm because of limitation of surface roughness measuring equipment. These are the parameters selected for the experiment:

1. General
   i. Slot length and width = 26 and 6 mm
   ii. Gap between slots = 6 mm
2. Inputs
   i. Spindle speed = 1000 rpm (fixed)
   ii. Ultrasonic power = 25 and 75%
   iii. Feed rate = 6, 12 and 18 mm/min
   iv. Depth of cut = 0.5, 1.0 and 1.5 mm
3. Outputs—Surface roughness (SR), material removal rate (MRR), tool wear.

2.2 Experiment Design

Workpieces selection: There is a parameter used to determine the suitability of workpiece on rotary ultrasonic machining that is brittleness. Brittleness (B) is the ratio of shear stress (S_s) and fracture stress (F_s). Brittleness is high for harder material because of its lower fracture stress and vice versa. After keeping all factor like, limited spindle speed, low machinability and material availability, high brittle material, i.e., **float glass** is selected for the experiment.

Design of experiment for rotary ultrasonic face milling (RUFM): MRR varies directly proportion to feed rate and depth of cut of machining. Increment in milling

Table 1 Parametric matrix table (RUFM)

S. No.	Parameters	Level 1	Level 2	Level 3
1.	Feed rate (mm/min)	6	12	18
2.	Depth of cut (mm)	0.5	1.0	1.5
3.	Ultrasonic power (%)	25	75	

Table 2 Parametric matrix table (CG)

S. No.	Parameters	Level 1	Level 2	Level 3
1.	Feed rate (mm/min)	6	12	18
2.	Depth of cut (mm)	0.5	1.0	1.5
3.	Ultrasonic power (%)	0	0	0

resistance cause low surface finish and accelerate tool wear. The designed parameter matrix table with their level is listed in Table 1. The experiment contains milling operation performed of 18 slots of 26 mm length and 6 mm wide each. Also, spindle speed is constant at 1000 rpm and processing frequency of 25 kHz.

Design of experiment for conventional grinding (CG): Theoretically, ultrasonic improved the tool life with milling resistance reduction and surface finish of the workpiece. To analyze the actual impact and utilization of addition of ultrasonic vibration on the grinding process, an experiment for conventional grinding is also performed to compare with rotary ultrasonic face milling results. The parametric matrix table for the conventional grinding (CG) experiment and their level given in Table 2. Value of ultrasonic power is zero during the experiment. Also, spindle speed and processing frequency are the same as RUFM experiment, i.e., 1000 rpm and 25 kHz.

3 Experimentation and Measurement

After clamping workpiece on the worktable with clamps and joining (welding and soldering) tool with horn, milling operation for the experiment is performed, as shown in Fig. 2. The tool and the horn are attached with the transducer which supplies ultrasonic power to it. Tool rotation, ultrasonic vibration and coolant flow are controlled by the computerized drive system. On the same setup, both processes are performed on each float glass slab. During the conventional grinding, ultrasonic power is turned off. A stopwatch was used during the experiment for MRR measurement. After the experiment, a SURFTEST SJ-210 equipment used for surface roughness measurement. Later, the tool was characterized by SEM technology for tool wear analysis.

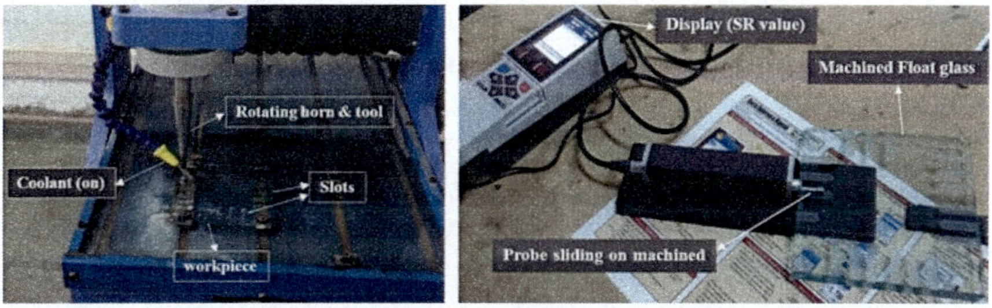

Fig. 2 Milling operation and surface roughness measurement

4 Experimental Results and Discussion

4.1 Experiment Results for RUFM on Float Glass

Standard Taguchi orthogonal array does not contain the full combination of each level on parameters for the experimentation. Hence, a simple full combination of each level of parameters is used for the total experiment with 18 slots during RUFM (Fig. 3). All 18 slots result in orthogonal array format represented in Table 3.

Effect of process parameters on MRR and SR during RUFM: The mean effect graph (Fig. 4) shows no effect of ultrasonic power on MRR. Where the depth of cut and feed rate proportionally affects MRR during RUFM process. The maximum value for MRR found at the highest level of all these parameters because of the large

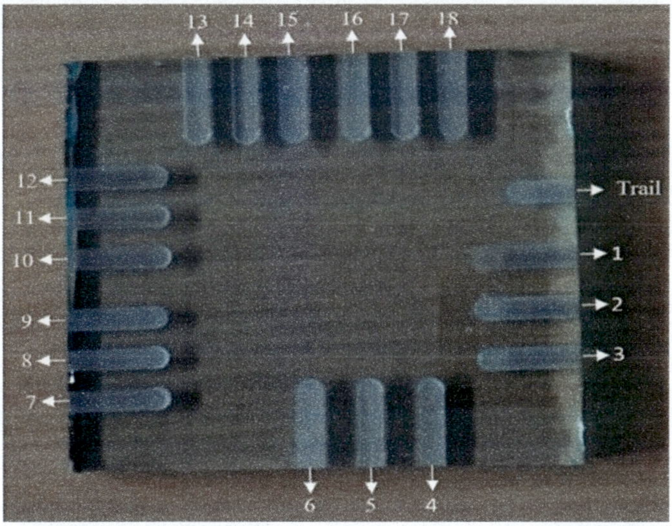

Fig. 3 Machined float glass during RUFM

Table 3 Experimental result for RUFM

Exp. No.	UP (%)	FR (mm/min)	DOC (mm)	MRR (mm^3/min)	SR (μm)
1.	1	1	1	17.82	0.964
2.	1	1	2	35.65	1.302
3.	1	1	3	53.49	1.074
4.	1	2	1	35.66	1.542
5.	1	2	2	71.32	1.358
6.	1	2	3	106.99	1.231
7.	1	3	1	53.72	1.219
8.	1	3	2	107.44	1.172
9.	1	3	3	161.16	2.493
10.	2	1	1	17.84	1.130
11.	2	1	2	35.68	1.425
12.	2	1	3	53.52	1.383
13.	2	2	1	35.64	1.617
14.	2	2	2	71.30	1.356
15.	2	2	3	106.97	1.394
16.	2	3	1	53.71	1.703
17.	2	3	2	107.45	1.403
18.	2	3	3	161.18	1.903

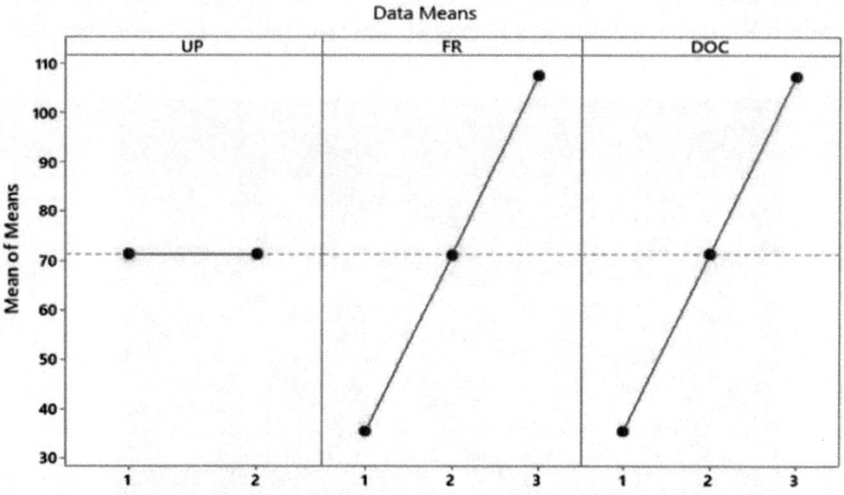

Fig. 4 Mean effects graph for MRR during RUFM

area of contact of the tool to workpiece. But for SR, ultrasonic power shows some linear effect, unlike MRR, as shown in Fig. 5. Also, the other two parameters show the different effect on SR.

ANOVA Results during RUFM on float glass: ANOVA table for MRR and SR during RUFM (Table 4) shows that the negligible contribution of ultrasonic power where feed rate and depth of cut both are significant factors in the change of MRR during RUFM. Also, all factors are not significant in the change in SR. Contribution within variance is more than 50% compared to between variance. Feed rate has a maximum contribution on SR during RUFM.

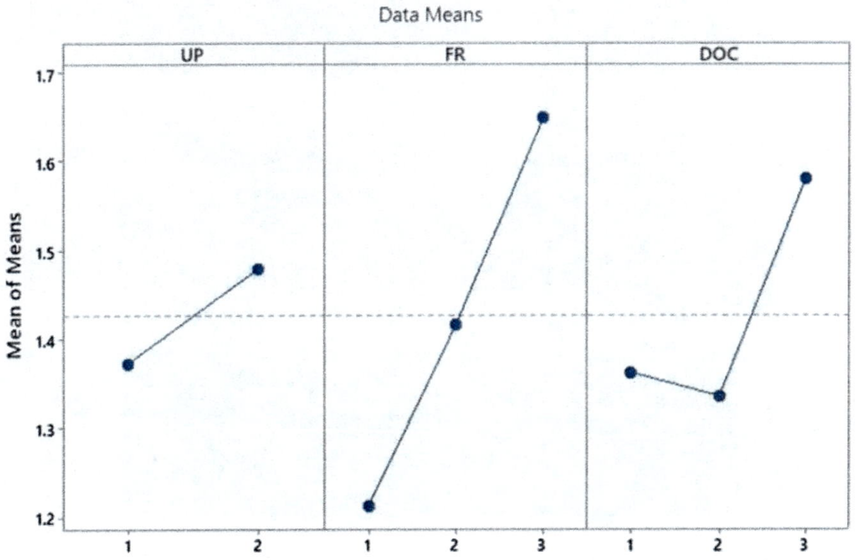

Fig. 5 Mean effects graph for SR during RUFM

Table 4 ANOVA results for MRR and SR during RUFM

		MRR			Contribution (%)	SR			Contribution (%)
Source	DOF	SS	*F*	*P*		SS	*F*	*P*	
UP	1	0.0	0.0	0.99	0.0	0.0510	0.49	0.49	2.44
FR	2	15,455	36.0	0.00	46.32	0.5707	2.73	0.10	27.28
DOC	2	15,331	35.7	0.00	45.95	0.2144	1.03	0.38	10.25
Error	12	2576				1.2550			
Total	17	33,363				2.0913			

4.2 Experiment Results for CG on Float Glass

Similarly, a total of 9 number of slot cuts were performed during CG (Fig. 6). Table 5 represents the results of all 9 slot cuts during conventional grinding on float glass.

Effect of process parameters on MRR and SR during CG: The mean effect graph (Figs. 7 and 8) shows the maximum value of MRR found at the highest level of parameters. For SR, feed rate shows a uniform rising slope throughout the process

Fig. 6 Machined float glass during CG

Table 5 Experimental result for CG

Exp. No.	UP (%)	FR (mm/min)	DOC (mm)	MRR (mm^3/min)	SR (μm)
1.	0	1	1	17.86	1.050
2.	0	1	2	35.68	0.946
3.	0	1	3	53.47	1.103
4.	0	2	1	35.66	1.168
5.	0	2	2	71.33	1.464
6.	0	2	3	106.98	1.675
7.	0	3	1	53.72	1.900
8.	0	3	2	107.43	1.816
9.	0	3	3	161.15	1.846

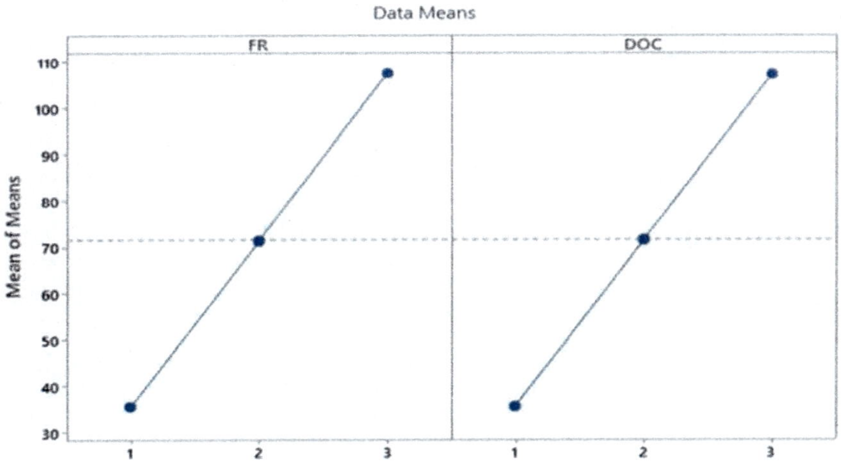

Fig. 7 Mean effects graph for MRR during CG

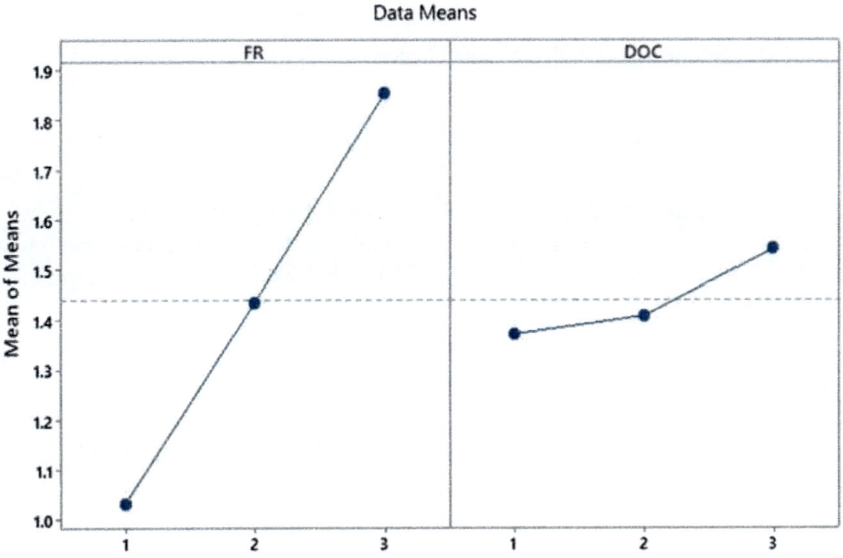

Fig. 8 Mean effects graph for SR during CG

while the depth of cut shows very less range of variation in all levels. But SR increases in both factors. Minimum SR found at 1st level of both parameters.

ANOVA results for CG on float glass: ANOVA table for MRR and SR during CG (Table 6) shows both process parameters are significant factors in the change in MRR. Also, the feed rate is an only significant factor in the change in SR. Contribution of feed rate on SR is 87.37% which shows a major effect during the process.

Table 6 ANOVA results for MRR and SR during CG

		MRR			Contribution (%)	SR			Contribution (%)
Source	DOF	SS	*F*	*P*		SS	*F*	*P*	
FR	2	7725	12.0	0.02	46.33	1.0111	20.5	0.01	87.37
DOC	2	7658	11.9	0.02	45.93	0.0473	0.96	0.45	4.1
Error	4	1290				0.0987			
TOTAL	8	16,673				1.1572			

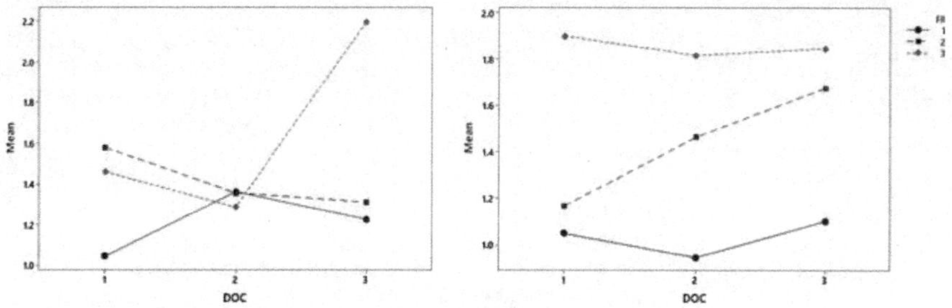

Fig. 9 Means interaction plot for SR during RUFM and CG

4.3 Comparison Between CG and RUFM

Mean effects graph and interaction graph for MMR during CG and RUFM show similar results because of the same machining condition and value of process parameters. Mean effects graph for SR during RUFM and CG shows a different level of process parameters for optimum point.

Interaction plot for SR of RUFM and CG: Figure 9 shows means interaction plot for SR on float glass during RUFM and CG during RUFM, means of SR at 2nd level of depth of cut is lower than other levels. During CG, the lowest mean value for SR in the graph is at the 1st level of both parameters. Error% for MRR during RUFM and CG is found same, i.e., 11.16%. For SR, Error% during RUFM and CG are 17.7%, 8.12%.

4.4 Tool Wear Analysis

Tool wear increases with increment in feed rate, depth of cut and milling resistance. RUFM includes vibration which reduces the resistance and enhances the tool life, but with the same machining condition, CG causes more tool wear. A tool after completion of both processes was characterized and highly magnified on SEM which indicated the tear-out of diamond grits form tool bottom surface, as shown in Fig. 10.

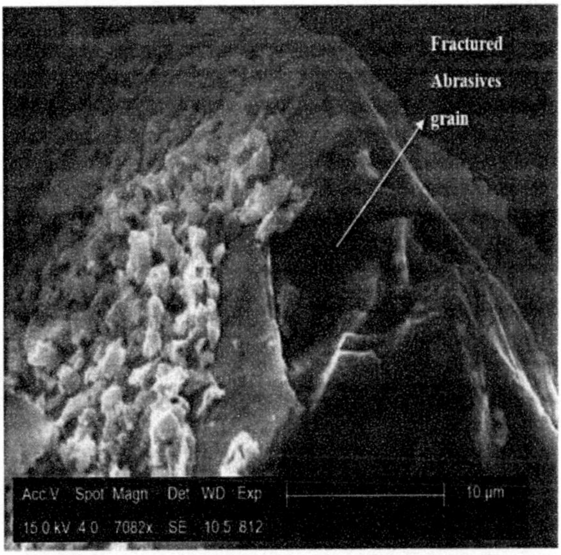

Fig. 10 Tear-out of diamond grit from the tool

5 Conclusions

1. MRR during RUFM process largely depends on the feed rate and the depth of cut than ultrasonic power in these selected parameters. Also, process parameters show the equal contribution of MRR, during CG.
2. Effect of parameters, ANOVA results and means predicted error% for both processes are found same on MRR during milling because of the same machining conditions with same values of each parameter.
3. During RUFM, ultrasonic power shows a minor effect on SR. SR increases with increment in ultrasonic power and feed rate. SR rise after downfall with increment in depth of cut.
4. Feed rate and depth of cut shows same effect on SR during CG on float glass, the only difference is that SR increases in very less range with depth of cut.
5. Within variance causes major contribution than between variance on SR during RUFM. But during of CG, feed rate contributes major (87%) in the ANOVA table.
6. Comparison of both processes shows the same optimum settings for MRR but different for SR. Also, error% by means predicted optimum value gives equal value for both processes on MRR. But for SR, CG gives the minimum error%.
7. Ultrasonic vibration reduces the milling resistance cause improvement in tool life but also somewhere it reduces the surface finish as compared to conventional milling.

References

1. Baek, D. K., Ko, T. J., & Yang, S. H. (2013). Enhancement of surface quality in ultrasonic machining of glass using a sacrificing coating. *Journal of Materials Processing Technology, 213*, 553–559.
2. Wang, J., & Zhang, J. (2017). Damage formation and suppression in rotary ultrasonic machining of hard and brittle materials: A critical review. *Ceramics International*.
3. Tyrrell, W. R. (1970). *Rotary ultrasonic machining*. SME Technical Paper.
4. Markov, A. I. (1977). Ultrasonic drilling and milling of hard non-metallic materials with diamond tools. *Machine and Tooling, 48*, 45–47.
5. Singh, B., Kumar, J., & Kumar, S. (2015). Influences of process parameters on MRR improvement in simple and powder-mixed EDM of SiC composite. *Materials and Manufacturing Processes*, 303−312.
6. Pei, Z. J., & Ferreira, P. M. (1991). An experimental investigation of rotary ultrasonic face milling. *International Journal of Machine Tools and Manufacture, 39*(8), 1327–1344.
7. Pei, Z. J., Prabhakar, D., & Ferreira, P. M. (1995). Rotary ultrasonic drilling and milling of ceramic. *Ceramic Transactions, 49*, 185–196.

Blockchain Technology Adoption in Healthcare Sector for Challenges Posed by COVID-19

Prasun Sarote and Om Ji Shukla

1 Introduction

Industry operations in health care have majorly developed in the past decade, with policies gradually accepting technological changes brought by digitalization. However, our healthcare operational structure is not designed to deal with an unpredictable crisis and uncertainties such as the COVID-19 pandemic. Therefore, these operations need a digital tool to streamline industry services/processes to render optimal results. Blockchain is one such effective tool. Although blockchain is at a very nascent stage, it is slowly making its place in various manufacturing industries, MNCs and tech-startups.

A block can be compared to a repository chest that is designed to store data, i.e., information in them. These blocks are secured by a cryptographic hash, referenced to one another. A hash is a cryptographic code that identifies a certain block [1]. It is often compared with a fingerprint as it is unique and essential for identification of a block. The hash to the previous block is linked with the parent block, creating a chain of information that is methodically associated with each other. Therefore, a continuous systematic system of referenced blocks forms a typical 'blockchain.' Blockchain technology is a distributed ledger technology which is based on a peer-to-peer (P2P) network [2]. In simple words, the server is decentralized and interlinked with end users/nodes on a blockchain network. Thus, distributed copies of data (information) can be accessed by each participating node without the fear of theft. A tamper/change in one block changes the code to it which means disturbing the whole chain, over millions of computers on the blockchain network, making it practically impossible. It is called immutability of blockchain. Consensus mechanism is used to determine the truth of the record stored in the blockchain [3]. Therefore, a decision in this non-authoritative distributed network is, achieved through consensus, that can

P. Sarote (✉) · O. J. Shukla
Department of Mechanical Engineering, National Institute of Technology Patna, Patna, India

R. Agrawal et al. (eds.), *Recent Advances in Smart Manufacturing and Materials*, Lecture Notes in Mechanical Engineering,
https://doi.org/10.1007/978-981-16-3033-0_34

Table 1 Key features and advantages of BCT

Key features	Advantages
Decentralization	Fraud prevention, faster transaction time, effective store of value
Consensus mechanism	Fault tolerant, proof-of-stake, multi-agent
Distributed ledgers	Elimination of intermediaries, safer communication
Immutability	Tamper resistant, efficient auditing, data integrity

be provided and validated through proof-of-work, proof-of stake, proof-of-authority [4] by validators (miners) holding the same data secured by cryptographic algorithms over the network. Thus, this process ensures the records (information on the chain) are safe and true. Blockchain is divided into three types—public, private, and consortium [5]. A public blockchain is open to everyone with their transactions which are permanently recorded. A private blockchain allows only a group of verified entities to have insights to the network (partial centralization). Protocols modified to make two or more enterprises responsible for a blockchain make up a consortium network [6]. Blockchain technology (BCT) extends key features that offer several advantages and benefits. Some of the most essential features of BCT are described in Table 1.

This paper aims to put forward a framework for blockchain implementation in healthcare industry. It can be argued that advantage of blockchain technology implemented in each industry through 4As—Aggregation–Analysis–Arrangement–Action framework can mitigate risk and streamline the healthcare industry to achieve optimization in performance. The 4As framework make use of crucial blockchain implemented lineaments—smart contracts, certification, transparency, tokenization to modernize industry operations of healthcare sector.

2 Application Tools Associated with Blockchain Technology

2.1 Smart Contracts

These are digital contracts encrypted with lines of codes that are executed automatically if predetermined conditions for the contract are completed between the parties [7]. Agreement over a set of data uploaded on the peer-to-peer network ensures security over transfers, as the code will only be executed automatically if it is validated/accepted by every entitled node.

2.2 *Certification*

In the healthcare industry, certification through BCT can be used for authentication of medical professionals to be hired, research advancements (patents), partners in the pharma chain, quality of tools and equipment, drugs, etc.

2.3 *Transparency*

As blockchain offers a unified source of data, it can facilitate the tracking of the location of medical supplies across partners in the pharma supply chain [8]. Integration of BCT with IoT can create a highly efficient traceable structure, combined with real-time monitoring of factors such as temperature and humidity for vaccines (especially, in cold chains).

2.4 *Tokenization*

Tokenization is creating value for physical assets through digital representation store of value. The digital representative of the asset on the blockchain is called a 'token' [6]. In order to improve payment procedure, tokenization can remove sensitive cardholder information. This can be used to acquire working capital from non-liquid assets.

3 Blockchain-Enabled Framework for Handling Challenges Posed by COVID-19

Healthcare is concerned with a structure that incorporates health systems and functions which aim to provide, maintain, and improve medical aid to the people. However, India's operational structure is not designed to deal with an unpredictable crisis and uncertainties such as the COVID-19 pandemic.

Escalating cases in COVID-19 pandemic could not be sustained by India's healthcare facilities, compromising public welfare and safety. In such cases, forecasting cannot be reliable as the historical data itself is circumstantial and not accurate although peaks in cases caused by an epidemic can be integrated with rise in total number of testing to analyze a substantial amount of planning to manage staff and resources accordingly. Resource management problems are both internal and external. Internally, it consists of ward allocation, bed management, operational theaters, and heavy machinery, etc. Thus, to source and allocate without a structured system becomes disoriented, even with precisely calculated demand. Necessary

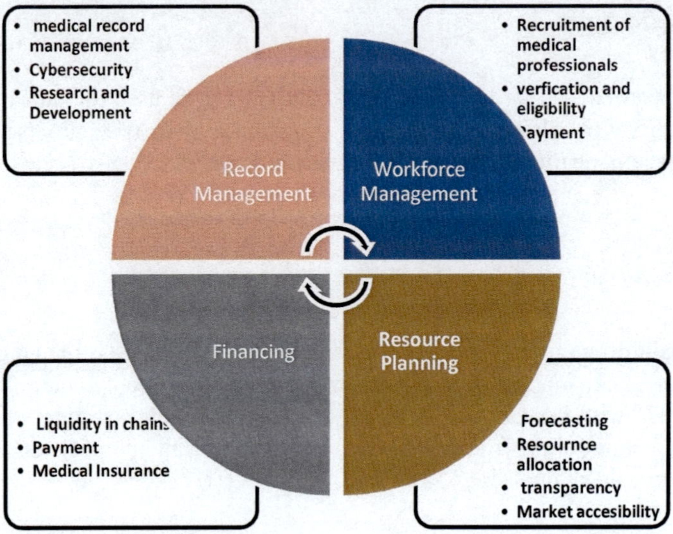

Fig. 1 Blockchain-enabled framework

resources are to be transported from bigger cities to small towns, alarming domestic SCs.

These factors of uncertainties can be categorized in four components within a healthcare facility, i.e., records, workforce, resource, financing. Therefore, these operations need a digital tool to streamline industry services to render optimal results. Blockchain is one such effective tool.

A blockchain-enabled cyclic framework of 4As—Aggregation–Analysis–Arrangement–Action incorporated with BCT application tools described in Sect. 2, can deliver an optimized performance of the industry as a whole. The blockchain-enabled framework is shown in Fig. 1, can largely justify productivity, quality of service, and use of physical resources to optimize industry performance. Leveraging BCT in this methodology will restructure health care digitally.

3.1 Aggregation of Records

A falsely interpreted data between any two departments in a healthcare facility will lead to failure of the other. Therefore, every patient record entered in a system must be categorized. Uncertainties such as an epidemic make departmental facilities more complicated as a clear screening, monitoring, and testing procedure has to be installed for each patient entry in either of these departments. Therefore, as blockchain can facilitate a single distributed source of data (medical records, patient entries, etc.) it can provide improved quality of database of patient health records as well as transaction records. Blockchain-enabled electronic health records (EHR) [9] authorize patients to share their health records with hospital networks. It is nearly impossible to change and steal data (information, in this case medical records) as it is immutable. Thus, it ensures security of stored health records and clinical reports while providing access for patients and medical professionals in a Health Insurance Portability and Accountability Act (HIPAA) [10] compliant manner. To record, process and report patient entries into a blockchain supported system will establish an appropriate record management system governed by a robust clinician credentialing management structure. While the development of new drugs incurs substantial costs related to trials to evaluate the safety and efficacy of the drug, the use of smart contracts allows to facilitate the procedure of the informed consent as well as improve identity management and data quality [10]. Hence, BCT can also be used as a tool to oppose the sale of counterfeit drugs.

Data aggregation for research and innovations in the field of health care is crucial as it delivers a base for further developments (modern equipment, vaccines, research study, etc.). BCT can base a strong infrastructure for a research network that allows organizations and researchers to share and proactively develop solutions to major health challenges and diseases. For example, pharmaceutical and research companies can access the database of patients who have agreed to be contacted, and information provided to these research companies can be used to develop vaccines, that are streamlined accordingly as per the needs. A patient's health record can be easily extracted from the private BC to facilitate, verified prescriptions to be uploaded under the patient' name registered on the network. Moreover, BCT can also support fast acknowledgment and registration of medical professionals and verified skilled volunteers to ensure constructive attention to a patient's needs.

3.2 Analysis of Workforce

The goal is to attend to more patients at a time, coordinated staff management and certified health workers must be ensured. Workforce must be allocated and appointed in a structured manner to treat and care for a continuously growing patient entry. Blockchain is an effective system to ease the recruitment process for medical professionals. BCT through decentralization of authority can provide real-time service monitoring including verifiable digital identity records as well certifications to avoid fraud. Candidates and volunteers can then easily upload their verified degree and important documents to share it with the hospitals. As blockchain is immutable, it is impossible to backdate or alter credentials. Therefore, it maneuvers the issue of legitimacy of an employee's identity and credentials by offering 'security to identity' [11] and verification of credibility through certification on the distributed network.

3.3 Arrangement and Allocation of Resources

Pharma supply chains deeply impacted whenever encountered with a crisis due to loss of information along the value chain. Therefore, functional misconfiguration of the value chain can endanger, performance of the industry. As logistic scale is scattered geographically, it involves the need for authentication, coordination, traceability (provenance). BCT provides four key features that enhance integration and coordination among the members of the chain [6]: (1) transparency, (2) validation, (3) automation, and (4) tokenization [12]. Facilitation of movement of goods (PPEs, test-kits, tools, equipment, medicines, etc.) through BCT implementation is capable of establishing a trusted network making each participant in the chain authoritative and responsible, and it can provide a network for healthcare, notable suppliers, and manufacturers, encompassed through blockchain's transparency factor. Combining transparency and validation with automation via smart contracts, one could envision a supply chain that operates on an automated basis on pre-specified rules [12]. It can surface allocation of real-time information on goods along the chain, through permissioned network. In case of an uncertainty, BCT can provide uncomplicated commerce of information and establishes a platform that connects medical suppliers/distributers and hospitals to avail necessary PPEs and medical supplies promptly to meet the demand. As it promotes just-in-time solutions, financial loss is avoided.

3.4 Action to Finance

COVID-19 Pandemic imposed lockdowns everywhere, thus freezing the supply chains. Therefore, a lot of money is frozen in these pharma supply chain. This is observed even in normal working scenarios. As valuable assets become inaccessible, working capital is restricted. Whereas, blockchain can facilitate a huge amount of capital back into the chains by cutting off interest on credits, encompassing payment orders as collateral [8]. Thus, blockchain in a way can increase productivity by ensuring elasticity in the supply chain through collateralized assets [8] through tokenization. The utilization of smart contracts as policy agreements can eliminate inherent processing delays of traditional paper-based policies [13]. Blockchain eliminates third-party intermediaries [13]. Hence, a blockchain-enabled system can be set up at a low cost to grant insurance records, accessed within no time for medical insurance.

4 Conclusion

Understanding and fundamental implementation of blockchain technology in healthcare industry can facilitate a resiliently structured paradigm of industry operations in order to optimize performance as a whole, in response to an uncertainty in managerial architecture of the structure. Classification of these stated cyclic uncertainties, i.e., record management, workforce management, resource allocation, and financing; in the healthcare industry, jointly blow down performance rate hinged on factors, i.e., productivity, quality of service, and use of resources. Blockchain-enabled framework presented in this paper, homogeneously integrate application tools of BCT lineaments, i.e., smart contracts, certification, transparency, tokenization to restructure healthcare operations digitally, and optimize performance by providing a distributed skeletal structure to overcome some of the crucial challenges posed by COVID-19.

References

1. Seebacher, S., & Schüritz, R. (2017). Blockchain technology as an enabler of service systems: A structured literature review. In *International Conference on Exploring Services Science* (pp. 12–23). Springer.
2. Hastig, G. M., & Sodhi, M. S. (2020). Blockchain for supply chain traceability: Business requirements and critical success factors. *Production and Operations Management, 29*(4), 935–954.
3. Kirkman, S., & Newman, R. (2017). Using smart contracts and blockchains to support consumer. In *International Conference on Grid, Cloud, & Cluster Computing* (Vol. 17). GCC.
4. Perera, S., Nanayakkara, S., Rodrigo, M. N. N., Senaratne, S., & Weinand, R. (2020). Blockchain technology: Is it hype or real in the construction industry. *Journal of Industrial Information Integration, 17*, 100125.
5. Nasir, Q., Qasse, I. A., Abu Talib, M., & Nassif, A. B. (2018). Performance analysis of hyperledger fabric platforms. *Security and Communication Networks, 14.*
6. Babich, V., & Hilary, G. (2018). Distributed ledgers and operations: What operations management researchers should know about blockchain technology. *SSRN Electronic Journal.*
7. Pilkington, M. (2016). Blockchain technology: Principles and applications. In *Research handbook on digital transformations*. Edward Elgar Publishing.
8. Agrawal, S., Jamwal, A., & Gupta, S. (2020). Effect of COVID-19 on the Indian economy and supply chain, 2020050148. Preprints.
9. Zhang, P., Schmidt, D. C., White, J., & Lenz, G. (2018). Blockchain technology use cases in healthcare. In *Varian medical systems* (pp. 1–41). Elsevier.
10. Prokofieva, M., & Miah, S. J. (2019). Blockchain in healthcare. *Australasian Journal of Information Systems, 23.*
11. Hossein, K. M., Esmaeili, M. E., & Dargahi, T. (2019). Blockchain-based privacy-preserving healthcare architecture. In *IEEE Canadian Conference of Electrical and Computer Engineering (CCECE)* (pp. 1–4). IEEE.
12. Blossey, G., Eisenhardt, J., & Hahn, G. (2019). Blockchain technology in supply chain management: An application perspective. In *52nd Hawaii International Conference on System Sciences* (pp. 6885–6893). HICS.
13. Kalla, A., Hewa, T., Mishra, R. A., Ylianttila, M., & Liyanage, M. (2020). The role of blockchain to fight against COVID-19. *IEEE Engineering Management Review, 48*(3), 85–96.

Effects of Al Content on Corrosion Behavior of High Entropy Alloys—A Review

Parth Vaidya

1 Introduction

Most of the conventional alloys are based on primary elements, and additional elements are added as per desired properties to improve the mechanical properties. With increasing alloying elements and minor elements, conventional alloys have the disadvantage of forming unstable and fragile intermetallic phases [1, 2]. Ye et al. [3] proposed a concept of a new class of material called that is called "high entropy alloys" (HEAs) about 16 years ago. High entropy alloys consist of 5 or more elements with a concentration of around 5–35% in the equimolar or nearly equimolar ratio [3, 4]. The effects such as high entropy, sluggish diffusion, lattice distortion, and cocktail are generally observed in HEAs. Out of these effects, the most significant effect is high entropy effect that is more in case of HEAs compared to conventional alloys. This high entropy will not make the formation of the intermetallic compound and favor the formation of disordered solution states [4]. HEAs possess some superior properties compare to conventional alloys such as high hardness, good elastic modulus, elastic strain, good wear resistance, fracture strength, thermoelectric property, catalytic property, and corrosion resistivity [1, 5–8]. Although the cost of HEA may be higher than typical conventional alloys because it consists of multiple elements, it is still cheaper than many superalloys and Ti alloys [4].

Generally, HEAs are solid solutions that consist of a simple face-centered and body-centered cubic structure or a mix of both [1, 4]. HEAs are applicable for special applications such as engine materials, nuclear materials, tool materials, hard-facing materials, marine structures, chemical plants, and functional coatings [9, 10]. HEAs are manufactured by many techniques such as laser cladding [2, 5], arc melting [11, 12], and powder metallurgy [13]. For more improvement in surface performance,

P. Vaidya (✉)
Mechanical Engineering Department, S V National Institute of Technology Surat, Surat, Gujarat 395007, India

R. Agrawal et al. (eds.), *Recent Advances in Smart Manufacturing and Materials*, Lecture Notes in Mechanical Engineering,
https://doi.org/10.1007/978-981-16-3033-0_35

cladding is important, which provides another layer on substrate materials without reforming the composition of bulk material. Laser cladding is widely used for HEAs coating, having advantages over a large heat-affected zone, poor process stability, and low deposition rate [1].

Corrosion is the gradual deterioration of material mainly because of chemical or electrochemical reaction between the material and the surrounding environment. More than three percent of GDP is affected by corrosion [13]. Corrosion is responsible for 25% of induced failure of total equipment failures in the oil and gas industry [14]. HEAs and coatings are alternative to conventional alloys and show good corrosion resistance and mechanical properties. The effect of corrosion on HEAs mainly depends on composition and its molar proportion. So, it is necessary to investigate the effect of a particular composition and molar fraction. Previous researchers already investigated the effect of various compositions and their molar fractions, such as Al [5, 6, 15–17], Ti [17], Mo [18], and Cu [19]. Al is a more preferable element in the design of HEAs as it has a low melting point temperature and also has a large atomic radius and can be easily dissolved in Fe, Co, Ni, and other elements [20]. In this paper, the corrosion behavior of HEAs coatings by varying Al content in the past decade has been summarized. The present study will help the researchers to work in this domain.

2 Corrosion Measurement Technique

Corrosion measurement is the practice of studying and acquiring the information of corrosion damage to the material. There are numerous techniques to measure corrosion behavior such as weight-loss technique, surface analysis techniques, polarization technique (Tafel polarization technique), electrical resistance technique, and non-destructive technology such as visual inspection, liquid penetrate inspection, magnetic particle testing, and radiographic inspection.

Potentiodynamic polarization is one of the standard techniques to study corrosion behavior [13] because it is faster than a weight-loss method, and it separates cathodic reactions from anodic reactions; it is also a qualitative method. The important parameters such as corrosion current density (I_{corr}), corrosion potential (E_{corr}), pitting potential (E_{pit}), and passive current density (I_{pass}) are shown in potentiodynamic curve in Fig. 1. I_{corr} is a more detrimental parameter than $E_{corr,}$ and it is used to calculate the average corrosion rate. Corrosion potential is a thermodynamic value, and it will change quickly with different corrosive environments compare to corrosion current density [5]. Corrosion potential gives only the corrosion tendency of the solution, while the corrosion rate is directly proportional to corrosion current density [13]. The intersection of the anodic Tafel line and cathodic Tafel line gives a point and projection of this point on the X-axis and the Y-axis gives I_{corr} and $E_{corr,}$ respectively.

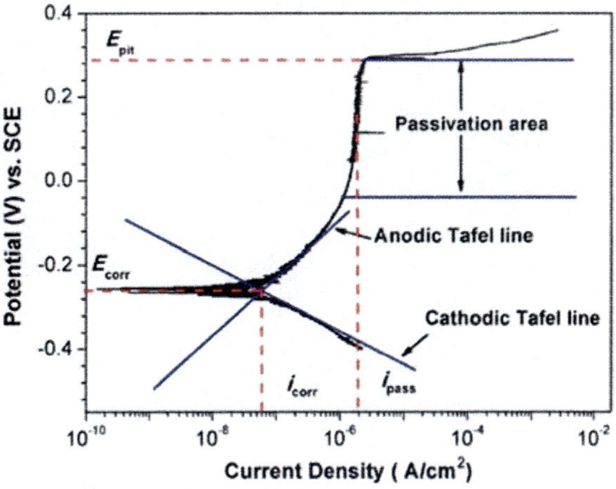

Fig. 1 Potentiodynamic polarization curve in 3.5% NaCl showing parameters [13]

3 Literature Review of HEAs Showing Effect of Al on Corrosion Behavior

Corrosion is the result of interactions between metals and the environment which results in degradation and wastage of material. Metals are extracted from ores, and pure metallic form is achieved. Metals are unstable in this form and tend to go back to ore form. Hence, corrosion is considered as the reverse of metal extraction.

Li and Shi [5] carried out the research on the corrosion behavior of $Al_xCrFeCoNiCu$ HEA coatings on AA5083. Laser cladding was used to prepare $Al_xCrFeCoNiCu$ HEA coating with different molar ratios of Al (x from 0 to 1.8) on the AA5083 surface, and the polarization test was carried out in a 3.5% NaCl medium to study corrosion behavior. It has been observed that with increasing Al content, initially, free corrosion current density (I_{corr}) decreases and then increases. I_{corr} is the deciding factor that gives an idea about corrosion rate, and its lower value indicates that material is having high corrosion resistance. Figure 2 shows the polarization curve of $Al_xCrFeCoNiCu$ HEA coatings by varying Al content, and it has been observed that up to $x = 0.8$, I_{corr} decreases, and then it shows increasing trends. The value of I_{corr} for Al content up to 0.8 is on the order of 10^{-7}, and then it increases to 10^{-6} while substrate material has I_{corr} on the order of 10^{-5}, larger compared to coating samples. Passive films fail mainly by Cl^- ions, the active Cl^- ions first attack at the interface and then penetrate through the oxide films [5, 13]. In HEAs, Cu segregation to the interdendritic region causes intergranular corrosion. Increasing Al content up to 0.8 promotes uniform distribution of Cu, which improves corrosion resistance. Further increasing Al content increases AlNi phase (B2), and its corrosion resistance is poor compare to BCC1 phase, so additional increasing Al content results in decreasing corrosion resistance [5].

Lee et al. [21] synthesized work on corrosion behavior of $Al_xCrFe_{1.5}MnNi_{0.5}$ HEA in aqueous environments (NaCl and H_2SO_4) for different Al content. Increasing Al

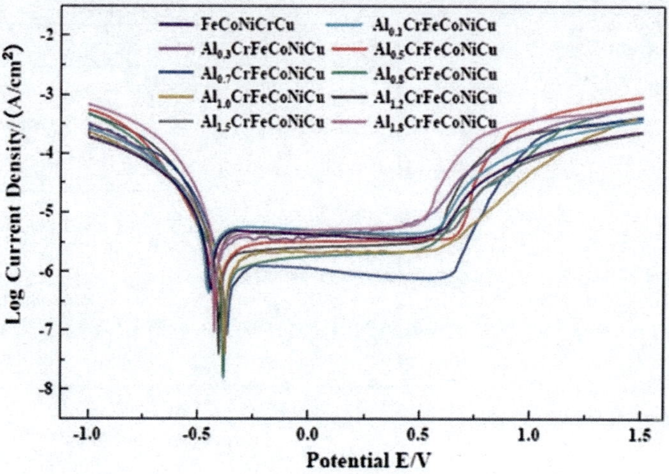

Fig. 2 Polarization curve for $Al_xCrFeCoNiCu$ HEA coatings [5]

content in acidic and chloride environment drops the corrosion resistance compared to Al-free alloys. It was observed that there is no pitting corrosion in Al-free HEA ($x = 0$) while in $Al_{0.3}CrFe_{1.5}MnNi_{0.5}$ HEA, there is a small surface attack of pits. An increasing trend of pitting corrosion was observed with increasing Al content, which indicates deep localized corrosion ($x = 0.5$). Al tends to form porous oxide films and increases localized corrosion. Corrosion current density and passive current density of $Al_xCrFe_{1.5}MnNi_{0.5}$ were measured lower in aqueous environments [21]. Sathyanarayana Raju et al. [12] successfully fabricated $Al_xCoCuFeNiTi$ by varying Al content from $x = 0$ to 1. It was observed that increasing Al content up to 0.25%, I_{corr} decreases, and then it increases. The corrosion rate will be lower in $Al_{0.25}CoCuFeNiTi$ compare to other HEA samples. Also, it was observed that as Al content increases in HEAs, the melting point of the alloy also increases and the density of the alloy decreases [12]. Ye et al. [2] carried out the research on corrosion behavior of $Al_xFeCoNiCuCr$ HEAs coatings on AISI1045 substrate material prepared by laser cladding. The corrosion test was performed in 0.05 M HCL electrolyte at room temperature. The current density for $x = 1, 1.3, 1.5, 1.8, 2$ was 29.2, 47.9, 31.4, 7.6, and 31.7, respectively. $Al_{0.3}FeCoNiCuCr$ shows the highest corrosion current density of 47.9 $\mu A\ cm^{-2}$, which indicates a high corrosion rate where no passivation region, while $x = 1.8$ shows the lowest corrosion rate. It has been concluded that with increasing Al content, corrosion behavior exhibits little regularity in 0.05 mol/HCL electrolyte [2]. Muangtong et al. [22] studied the corrosion behavior of CoCrFeNi-X (X = Cu, Al, Sn) in chloride medium by adding Cu, Al, and Sn alternatively and compared with SS304 and SS316. It has been observed that CoCrFeNiSn possesses good corrosion resistance properties. Sn addition shows a wide passivation region and highest E_{corr}, which indicates good corrosion resistance. Al addition in HEAs shows weak corrosion property because it forms oxide of Al_2O_3, weak in chloride solutions [22]. Zang et al. [23] performed a corrosion test of $Al_xCoCrFeNiTi_{0.5}$ HEAs by varying Al content in aqueous solution. The aqueous solution was of 0.5 M H_2SO_4 at room temperature. As Al content increases, E_{corr} shows an increasing trend, and I_{corr} shows a decreasing trend, which concludes that pitting corrosion resistance increases significantly. $Al_{1.0}CoCrFeNiTi_{0.5}$ exhibits better corrosion resistance compared to other HEA samples. This trend was observed because BCC phase proportion increases as Al content increases and BCC plays better corrosion resistance [23].

Liu et al. [15] studied about corrosion of $Al_xCoCrCuFeNi$ in acidic, alkaline, and saline solutions, and the corrosive medium were HCL, NaOH, and NaCl, respectively. Figure 3 is showing polarization curve using different mediums for every attempt to study corrosion behavior. It was observed that free corrosion potential and free corrosion current values −0.378 V and 6.231×10^{-5} A cm^{-2}, respectively. From Fig. 3a, it is concluded that $Al_2CoCrCuFeNi$ corrodes faster in an acidic medium than in the alkali environment [15]. From Fig. 3b, it is observed that as Al content increases from 0.5 to 2.0, corrosion potential decreases from −0.230 to −0.378 V while corrosion current density increases from 5.237×10^{-7} A cm^{-2} to 6.231×10^{-5}. The corrosion test shows almost the same trend for rising and falling in E_{corr} and I_{corr} density in an alkali environment, as shown in Fig. 3c. As shown in Fig. 3d, both E_{corr} and I_{corr} decrease, and passivation film growth was observed with increasing

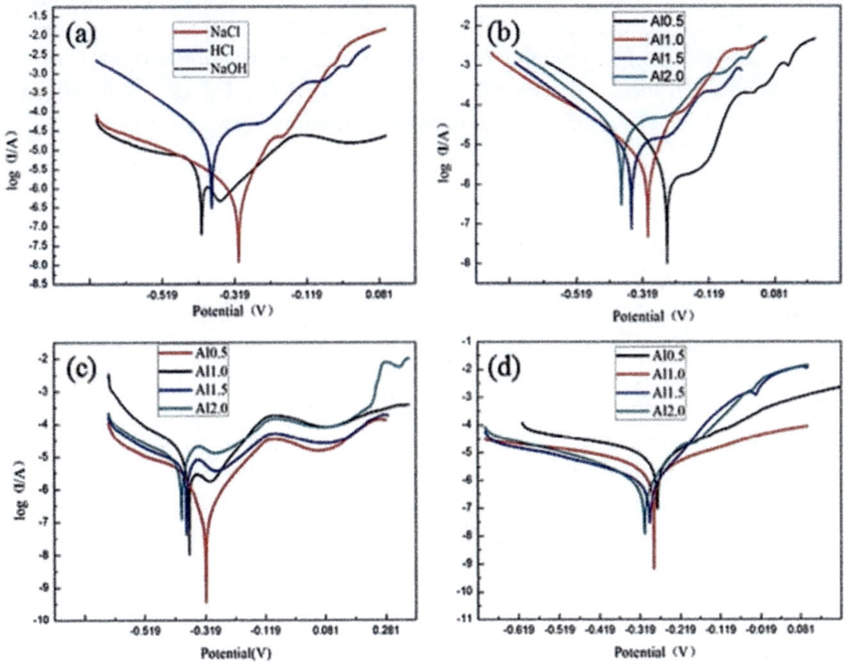

Fig. 3 **a** $Al_2CoCrCuFeNi$ HEA in various solutions; **b** $Al_xCoCrCuFeNi$ HEAs in (HCL + NaCl) solution; **c** $Al_xCoCrCuFeNi$ HEAs in (NaOH + NaCl) solution; **d** $Al_xCoCrCuFeNi$ in NaCl solution of 3.5 wt% [15]

Al content. Refinement of grain, dislocation, and increase grain boundary helps for the growth of the initial nucleation position in aspect to passivation film [15].

Bachani et al. [6] successfully deposited coatings of VNbMoTaWAl with different Al molar fractions on 304 SS and studied corrosion behavior in 0.5 M H_2SO_4. As increasing Al content, corrosion resistance of coatings can be improved significantly. The sample made of 2.37% Al contained VNbMoTaWAl coating was observed polarization resistance of 768,754 Ω cm^{-2} that is highest, and a good passivation range was also observed [6]. Qiu et al. [17] studied corrosion property of Al_xCoCrFeNi in 0.6 M NaCl at 25 °C by potentiodynamic polarization test (x = 0.3, 0.6, 0.9). Passive window ($E_{corr} - E_{pit}$) was seen decreasing with increasing Al content because of the lower concentration of Cr in HEAs as Al increases. However unexpectedly, $Al_{0.6}$CoCrFeNi and $Al_{0.9}$CoCrFeNi show more E_{corr} and low I_{corr} which means better corrosion resistance than $Al_{0.3}$CoCrFeNi and Al-free alloys [17]. Raza et al. [24] successfully fabricated Al_xCrFeMoV HEAs by powder metallurgy, and samples were formed by varying Al content (x = 0, 0.2, 0.6 and 1.0). It has been observed that with increasing Al content by 0, 0.2, 0.6, and 1. E_{corr} values were observed – 397, – 410, – 460, and – 307 mV$_{SCE,}$ respectively. Interestingly, $Al_{1.0}$CrFeMoV was observed with the highest E_{corr} and lowest I_{corr} values. There is inconsistency in an observed trend due to volume fraction variation of the BCC2 phase [24]. Shi et al. [16] studied the corrosion behavior of Al_xCoCrFeNi in two conditions, namely forged and equilibrated, and observed that as fraction of Al content increases, E_{corr} decreases and I_{corr} increases. Further, increasing Al content also decreases critical pitting potential (E_p) and increases passivation current density (I_{pass}) indicates that localized corrosion rate is less in Al-free HEAs. For localized corrosion, E_p is a determining factor to measure corrosion behavior [16].

4 Conclusions

This paper reviews the effect of varying Al content in corrosion behavior of different types of HEAs and coatings of HEAs during the past decade. As volume fraction of the Al-rich phase increases in HEAs which results in decreasing volume fraction of other contents such as Cr and Cu. BCC phase is prone to attack by Cl^- ions in Cr depleted phase and results in decreasing corrosion resistance, while in HEAs containing Cu, corrosion resistance increases up to some level with increasing Al content because of uniform distribution of Cu in dendrite and interdendritic region but more addition of Al causes a change in microstructure and corrosion resistance decrease. From past research analysis, there is no clear trend that can be predicted for corrosion behavior as it depends on many factors such as elements of HEAs, electrolyte medium, and volume fraction of content although HEAs show better corrosion resistance and mechanical properties compared to conventional alloys. The coating of HEAs over conventional alloys or utilization of pure HEAs is advisable for better corrosion resistance property applications.

References

1. Menghani, J., Vyas, A., Patel, P., Natu, H., & More, S. (2020). Wear, erosion, and corrosion behavior of laser cladded high entropy alloy coatings—A review. In *Materials Today: Proceedings*. Elsevier Ltd.
2. Ye, X., Ma, M., Cao, Y., Liu, W., Ye, X., & Gu, Y. (2012). The property research on high-entropy alloy $Al_xFeCoNiCuCr$ coating by laser cladding. In *Physics Procedia* (pp. 303–312). Elsevier Ltd.
3. Ye, Q., Feng, K., Li, Z., Lu, F., Li, R., Huang, J., & Wu, Y. (2017). Microstructure and corrosion properties of CrMnFeCoNi high entropy alloy coating. *Applied Surface Science, 396*, 1420–1426.
4. Chuang, M. H., Tsai, M. H., Wang, W. R., Lin, S. J., & Yeh, J. W. (2011). Microstructure and wear behavior of $Al_xCo_{1.5}CrFeNi_{1.}5Tiy$ high-entropy alloys. *Acta Materialia, 59*(16), 6308–6317.
5. Li, Y., & Shi, Y. (2020). Microhardness, wear resistance, and corrosion resistance of $Al_xCrFeCoNiCu$ high-entropy alloy coatings on aluminum by laser cladding. *Optics and Laser Technology, 134*(7089), 106632.
6. Bachani, S. K., Wang, C. J., Lou, B. S., Chang, L. C., & Lee, J. W. (2020). Microstructural characterization, mechanical property and corrosion behavior of VNbMoTaWAl refractory high entropy alloy coatings: Effect of Al content. *Surface and Coatings Technology, 403*, 126351.
7. Sun, X., Zhu, H., Li, J., Huang, J., & Xie, Z. (2019). Influence of aluminum content on the microstructure and properties of the in-situ TiC reinforced $Al_xFeCoNiCu$ high entropy alloy matrix composites. *Materials Science and Engineering, 743*, 540–545.
8. Yan, X., & Zhang, Y. (2020). Functional properties and promising applications of high entropy alloys. *Scripta Materialia, 187*, 188–193.
9. Murty, B. S., Yeh, J. W., Ranganathan, S., & Bhattacharjee, P. (2019). *High entropy alloys* (2nd Ed.). Elsevier.
10. Murty, B. S., Yeh, J. W., Ranganathan, S., & Bhattacharjee, P. (2014). *High entropy alloys* (1st Ed.). Butterworth-Heinemann.
11. Shi, Y., Yang, B., Xie, X., Brechtl, J., Dahmen, K. A., & Liaw, P. K. (2017). Corrosion of Al_x CoCrFeNi high-entropy alloys: Al-content and potential scan-rate dependent pitting behavior. *Corrosion Science, 119*, 33–45.
12. Sathyanarayana Raju, C. H. V., Venugopal, D., Srikanth, P. R., Lokeshwaran, K., Srinivas, M., Chary, C. J., & Ashok Kumar, A. (2018). Effect of aluminum addition on the properties of CoCuFeNiTi high entropy alloys. In: *Materials today: Proceedings* (pp. 26823–26828). Elsevier Ltd.
13. Shi, Y., Yang, B., & Liaw, P. K. (2017). Corrosion-resistant high-entropy alloys: A review. *Metals, 7*(2), 1–18.
14. Wang, Q., Ai, M., Shi, W., Lyu, Y., & Yu, W. (2020). Study on corrosion mechanism and its influencing factors of a short distance intermittent crude oil transmission and distribution pipeline. *Engineering Failure Analysis, 118*, 104892.
15. Liu, Y. Y., Chen, Z., Shi, J. C., Wang, Z. Y., & Zhang, J. Y. (2019). The effect of Al content on microstructures and comprehensive properties in AlxCoCrCuFeNi high entropy alloys. *Vacuum, 161*, 143–149.
16. Shi, Y., Collins, L., Feng, R., Zhang, C., Balke, N., Liaw, P. K., & Yang, B. (2018). Homogenization of $Al_xCoCrFeNi$ high-entropy alloys with improved corrosion resistance. *Corrosion Science, 133*, 120–131.
17. Qiu, Y., Thomas, S., Fabijanic, D., Barlow, A. J., Fraser, H. L., & Birbilis, N. (2019). Microstructural evolution, electrochemical and corrosion properties of $Al_xCoCrFeNiTi_y$ high entropy alloys. *Materials & Design, 170*, 107698.
18. Dai, C., Zhao, T., Du, C., Liu, Z., & Zhang, D. (2020). Effect of molybdenum content on the microstructure and corrosion behavior of $FeCoCrNiMo_x$ high-entropy alloys. *Journal of Materials Science and Technology, 46*, 64–73.

19. Thorhallsson, A. I., Csáki, I., Geambazu, L. E., Magnus, F., & Karlsdottir, S. N. (2020). Effect of alloying ratios and Cu-addition on corrosion behaviour of CoCrFeNiMo high-entropy alloys in superheated steam containing CO_2, H_2S and HCl. *Corrosion Science, 178*, 109083.
20. Kumar, A., Arafath, M. Y., Gupta, P., Kumar, D., Hussain, C. M., & Jamwal, A. (2020). Microstructural and mechano-tribological behavior of Al reinforced SiC-TiC hybrid metal matrix composite. *Materials Today: Proceedings, 21*, 1417–1420.
21. Lee, C. P., Chang, C. C., Chen, Y. Y., Yeh, J. W., & Shih, H. C. (2008). Effect of the aluminium content of $Al_xCrFe_{1.5}MnNi_{0.5}$ high-entropy alloys on the corrosion behaviour in aqueous environments. *Corrosion Science, 50*(7), 2053–2060.
22. Muangtong, P., Rodchanarowan, A., Chaysuwan, D., Chanlek, N., & Goodall, R. (2020). The corrosion behaviour of CoCrFeNi-x (x = Cu, Al, Sn) high entropy alloy systems in chloride solution. *Corrosion Science, 172*, 108740.
23. Zhang, J. J., Yin, X. L., Dong, Y., Lu, Y. P., Jiang, L., Wang, T. M., & Li, T. J. (2014). Corrosion properties of $Al_xCoCrFeNiTi$ 0.5 high entropy alloys in 0.5M H_2 SO_4 aqueous solution. *Material Research Innovations, 18*(4), 756–760.
24. Raza, A., Abdulahad, S., Kang, B., & Ryu, H. J. (2019). Applied surface science corrosion resistance of weight reduced AlxCrFeMoV high entropy alloys. *Applied Surface Science, 485*, 368–374.

Effect of Microstructure on the Tribo-mechanical and Wear Behaviour of Thin PVD Films: A Review

Alok Vats, Amar Patnaik, M. L. Meena, and Dinesh Shringi

1 Introduction

Physical vapour deposition (PVD) coating techniques are usually deployed by manufacturers and researchers to increase the component life. PVD is a one of the branch vacuum coating technologies. Other processes like magnetron sputtering process also belong to the same class. These processes are used to deposit the hard materials over the substrate. These hard substances include both the oxide (TiO_x) coatings, the nitride (TiN, TiAlN) coatings and carbide (TiC) ones. They are grouped into three classes which include the cathode arc evaporation (CAE), magnetron sputtering and combined magnetron and arc sputtering. The difference between these coatings is the established on how the metallic or non-metallic target materials are converted from solid to vapour form. In the magnetron sputtering process, the Ar atoms are ionized using a strong magnetic field. They are bombarded on to the target. It causes the target atoms to be ejected towards the substrate material for deposition. This method is very energy efficient as compared to arc methods [1, 2]. While in the case of arc methods, the cathode area, which is very small, is provided with a high energy arc. The energy from this arc converts the solid target into vapour form. The schematic of PVD equipment has been shown in Fig. 1 while PVD deposition process is shown in Fig. 2. Enhanced mechanical properties make PVD a better choice for industrial purposes. These properties include superior hardness, thermal stability and better hardness-to-elastic modulus ratio (*H*/*E* ratio). Due to the enhancement in the mechanical and wear behaviour, the PVD coatings are used in marine, high-temperature oxidation and excessive wear environments. As a result, the application areas of PVD coatings are many and quite diversified. They include aircraft industry, power plants,

A. Vats (✉) · D. Shringi
Mechanical Engineering Department, MBM Engineering College, Jodhpur, India

A. Patnaik · M. L. Meena
Mechanical Engineering Department, MNIT Jaipur, Jaipur, India

R. Agrawal et al. (eds.), *Recent Advances in Smart Manufacturing and Materials*, Lecture Notes in Mechanical Engineering,
https://doi.org/10.1007/978-981-16-3033-0_36

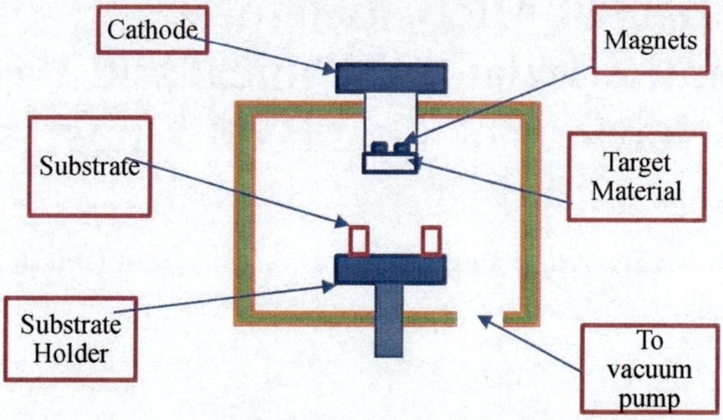

Fig. 1 Schematic PVD coating process setup

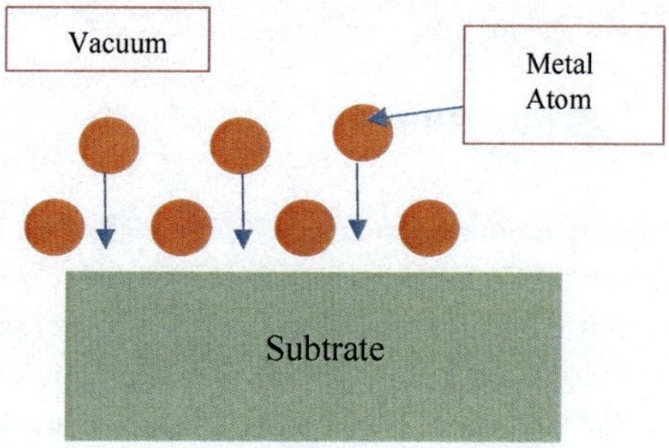

Fig. 2 PVD deposition process

automobile and transportation sector, hydraulic machinery, petroleum, printing and paper industry, manufacturing, metal processing and textile machinery. The reason for enhanced features of PVD coatings is their compact and hard microstructure with relatively less porosity. It is this microstructure which provides the PVD coatings its well-known coating characteristics. The microstructure of the PVD coatings depends on the deposition and geometrical parameters like pressure in vacuum chamber, arc current, arc temperature, nitrogen flow rate. The microstructure of PVD coatings may also contain the precipitates, hard phases and salts like nitrides, oxides and carbides. The conformance of these phases is done by XRD, EDAX, while the surface characterization is done by SEM analysis. In this paper, a review on the influence of microstructure on the tribo-mechanical and wear behaviour of PVD-coated layers is done.

2 Phase Composition and Microstructure

In a study, involving WC-Co-Cr powder, WC was the main phase and Co_3W_3C was the secondary phase. There was the formation of the solid solution of Co and Cr, as confirmed by the XRD peaks. Co_3W_3C was the product of decarburization of WC in the presence of Co and Cr [3]. Similar results were obtained in a study involving the cemented carbides in which the WC phase appeared bright and the Co binder phase appeared dark. The softness and ductility of Co binder, as well as the hardness and wear resistance of WC particles, provided a good combination to enhance overall performance. It improved the ability of the tools to sustain high-temperature effectively [4].

The spherical or flake α phase microstructure of Ti-6242S resulted in better fatigue strength, good thermal stability and fracture toughness. The duplex microstructure of Ti-555 with few α phase b/w β phases provided it strength, plasticity and toughness. There was micro-chipping for the first 5 min on (Ti-Al)N + Ti-N tools [5]. A columnar microstructure with high density and clear grain boundaries was observed while studying the behaviour of TiN coatings [6].

In a similar study, the CrN (thickness 1.80 μm) displayed a monolayer cubic microstructure. There was a primary deformation zone in near proximity of the interface b/w the chip and tool. Grain refinement was observed in this zone [7]. The uniform and dense microstructure of PVD coatings was obtained in Al-TiN PVD 1 (2.5 μm monolayer) and PVD 2 (5 μm monolayer) [4]. In a study involving Ti-Nb alloy coating, the Nb was deposited by triode plasma nitriding process. The coating had a dense columnar (cubic) structure with 33 atomic% Nb (Ti:Nb 2:1). The degree of coating crystallinity on Ti-AVM substrate was higher [8].

In a polymeric study, PTFE was found to consist of a long chain of C-atoms. The F-atoms surrounded the C-atoms (ratio 1:2), and there existed a great bonding b/w them. PTFE thin film was deposited as the top layer. PTFE was present above the Al_2O_3 thin film [9]. The α-Zr phase (HCP crystal structure) was present in the Zr-4 alloy [10].

In another study, the attached droplets in TiAlN coatings were loosely bonded to the coatings' surface. The subsequent adhesion of WS_2 soft coatings suffered. The presence of micro-cracks at the WS_2/Ti-Al-N interface supported this phenomenon. These micro-cracks acted as nuclei that resulted in fractures and cracks along the interface of WS_2/TiAlN coating system in early stages [11]. PVD technique could also be used to deposit materials like carbon (Avg. 26 μm thick) on the Ti-6Al-4V alloy system. In the XRD analysis of D-1250-1.6-70 sample, the peaks of TiN, TiO_2 and Ti_2AlC phase were found [12].

The XRD patterns of the ($Cr_{47}Al_{53}$) N-coated samples had a cubic CrN and a cubic AlN phase. By adding Mo and Cu, there was a shift in the Cr-N and Al-N peaks at 2θ of 37.5° and 37.8°. The shift with Cu was less. And the formation of cubic Mo phases was favoured [13]. In another study, the bond and top coats were deposited during the coating process. They had different percentages of elements in them. The top coat had fairly large % of Zr, Y and O (small). While the bond coat had large % of Ni, Cr, Al, O and C & Y (small). The presence of melted particles in the stationary substrate was more compared to the substrate rotating at 20 rpm [14]. The nitride coating like TiN grew on orientation plane (220) [15]. Crystalline phases like TiN (111) and TiO_2 (112) were found in the coatings deposited on Si (100) substrate at various angles [16].

The grain growth in the columnar microstructure decreased the hardness and mechanical strength of coatings. The multilayer state of the coatings prevented this

[17]. Certain PVD coatings had the presence of nitride salt like TiN. In some of the PVD coatings, a layered architecture was observed. It was seen that the modulation period decreased from 60 to 30 nm with the introduction of Ni [18].

The SEM was also used to study the interfacial area. This particular area exhibited either sharp (50 nm thick) or blurred (125 nm) images depending on coating thickness [19]. There was fine carbide particle formation, in the Co matrix in Al-TiN (2.7 μm) and Al-CrN (3.1 μm) coatings. The micro-defects were in the form of micro-holes, micro-droplets and macro-defects. The oxide phase of CoO was evident as peaks at 2θ angles of 36.49°, 42.38°, 61.49° and 73.65° respectively [20]. The features of PVD coatings like porosity and defects were exhibited in Fe_3Al coatings (120 μm thick). The presence of melted and un-melted particles was also observed. The CrN and DLC films (100 μm thick) were deposited on substrate [21].

3 Mechanical Aspects

3.1 Scratch Test

Scratch test is usually conducted by means of a diamond stylus. In the case of softer coatings like WS_2, there was a low value of COF stabilization. After that, excessive plastic deformation increased the value of COF. Due to continued rubbing, the WS_2 layer was scraped off and TiAlN coating was exposed [11]. In the adhesion test of TiN coating (SS substrate), the first crack was at critical load of L_{c1} 13.6 N. The coating was up rooted at L_{c2} 17.2 N [6].

The maximum adhesive strength was observed at $L_c = 14.1$ N for coating no. 6. It was followed by detachment of the coating from substrate. A cohesive failure was observed in coating no. 8 at a critical load of 2.5 N. In this case, there was no detachment [22].

In the scratch test of the CrC hard coating, on Cu alloy, a gradual wear process was indicated. The fine scratch pattern with no sharp abrasive scratches of the CrC-coated surface was visible [23].

3.2 Coefficient of Friction

In a study, there was a drop in the COF of the T-T-T samples. The middle layer was exposed. Diffusion took place up to atomic level (2500 cycles). Plasma nitriding greatly improvement the wear resistance of nano-structured TiN coating [24]. In the case of [9], the wear mechanisms of cold-sprayed (higher friction) and the sputtered (PVD) samples surfaces were different. This variation in COF was because of the production of wear debris. It increased the wear rate. The increase in the applied load, and consequently, the loss of energy owing to excessive deformation (plastic) caused an increase in COF [9].

The evolution of the COF with sliding time for the steel–steel and the steel–TiN system was studied. Lubrication of glycerol (4 N load; 1.16 GPa contact stress) was used during the test. For uncoated steel, COF was 0.19 for the S-S (steel) system and 0.04 for steel + TiN system. The wear rate of uncoated steel disc was 8.3×10^{-5} mm^3 N^{-1} m^{-1}. In a steel+ TiN system, COF had a high value only during initial sliding. After 1000 s, it was gradually reduced to 0.018 [13]. For the C-C samples, after the run-in stage the average value of COF was in the range of 0.70–0.85. However, the COF increased to > 1.0 (4000 s), because of de-lamination of Ti-Al-N samples [11].

The COF was described as the function of sliding distance as in the case of untreated Ti-6Al-4V and D-1250-1.6-70 samples. A normal load of 10 N was applied. Average COF of the D-1250-1.6-70 sample was reduced up to 66%, as compared to the Ti-6Al-4V sample [12].

The COF for coatings $(Cr_{37}Al_{50}Mo_{13})N$, $(Cr_5Al_7Mo_{88})N$ and $(Cr_{23}Al_{31}Mo_{40}Cu_6)N$ was in the range 0.08–0.1. For the $(Cr_{34}Al_{47}Cu_{19})N$ coating, COF increased from 0.10 to 0.11, within the first 70 m. It was followed by a decline in the COF. After sliding distance of 300 m, the COF was around 0.10, indicating strong fluctuation in COF. Surface roughness also influences the COF. The variation of mean surface roughness of the nitride coatings containing Mo and Cu is shown in Fig. 3 [13].

In the study of ion-bombarded TiN samples, the frictional force was quite unstable because of the wave-like pattern. In the scratch test, the TiN samples (low bombarding time) had higher adhesive properties, when compared to the high bombarding timed ones [15].

During the initiation of sliding, COF values were minimum but, they increased suddenly in the stage-II. The COF of Ti-6Al-4V substrate was 0.6 (unstable). The distance of sliding was 100 m, and it caused excessive wear. The COF for C_1 specimen was between $0.2 < \text{Friction}_{\text{co-efficient}} < 0.3$ before 50 m and became $0.3 < \text{Friction}_{\text{co-efficient}} < 0.4$ later on. Its average value was 0.38. The steady state was reached much earlier C_2 and C_3 [16]. In Al-Cr-N coatings, initially there was a peak in $\text{Friction}_{\text{co-efficient}}$. After that, the COF value was stabilized to 0.4 (after 0.05 m of

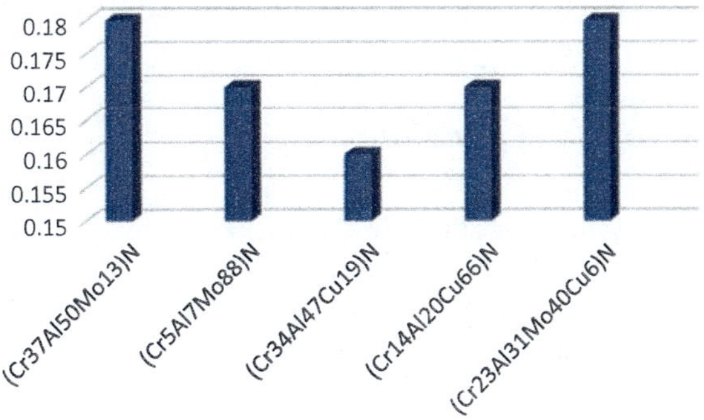

Fig. 3 Surface roughness of the Mo and Cu containing nitride coating

sliding). However, for Cr-W-N samples, the friction underwent peak initially and was stabilized to 0.45 [25]. There was a temperature function, involved in the COF of $Al_{0.66}Ti_{0.33}N$ coating by arc PVD method. The graph between $\text{Friction}_{\text{co-efficient}}$ and coating distance was drawn in Ar jet environment. A COF of 0.2 was obtained at room temperature (10,000 cycles) [26].

There was a higher coefficient of friction in un-reinforced PA (no G.F.). It was because of higher adhesion of PA to steel substrate. There was a reduced adhesion because of lesser polymeric contacts [27].

3.3 Composite and Multilayer PVD Coatings

The TiAlN layers had FCC structure (NaCl type) with (220) plane orientation. So, Al substituted for Ti atoms in the TiN lattice and the solid solution of c-Ti(Al)N was formed. The peaks for Ti phase were detected in ML1 and ML2 coatings. The mass gain of SL1 coating increased steadily (80 cycles). After that, it began to increase rapidly. The mass gain of the other three coatings were consistently lower than that of SL1 coating during the cyclic oxidation process. The N_2 inward diffusion caused the formation of a compound layer composed of TiN and Ti_2N on the surface of the substrate, followed by an Al-rich layer and a N_2-stabilized Ti layer [28].

Excess of O_2 existed at the top surface of TiO_xN_y-TiN composite layered film due to the parent oxides. As the time for etching was increased, the N (atom level) conc. also increased, while O_2 concentration decreased. In the XRD patterns of CrN and TiO_xN_y-TiN sealed CrN (FCC structure) coatings, the CrN phase was present in planes of (111), (200) and (220). There was a red shift in the diffracted peaks for CrN, after the samples were sealed with ALD-TiN (360 °C) [29].

There was an interface formation b/w the inner layer of SiC and outer layer of ZrB_2-$MoSi_2$-SiC. Cracking did not take place. SiC nano-whiskers were present in the coating structure. The length of these nano-whiskers was less than 5 μm, and their diameter was about 50–100 nm. ZrB_2 and $MoSi_2$ particles were also present. In the C/SiC/ZrB_2-$MoSi_2$-SiC samples, the two coats were there, namely the top coat (ZrB_2-$MoSi_2$-SiC and 70 μm thick) and a bond coat (SiC overlay and 200 μm thick). White (secondary phase) particles were also present in the top coat of coating [30].

3.4 Surface Roughness

In a study, involving the Ti or Ti-Nb coatings, the nitriding treatment led to an increase in the R_a values. Consequently, the PVD coatings deposited before the nitriding treatment had low R_a values. During the micro-abrasion testing, a greater R_a led to higher pressure of contact at the surface asperities. It fastened the detachment process of the (nitride) asperities. A reduced R_a value proved better for the Ti and Ti-Nb coating system in terms of the wear performance [8].

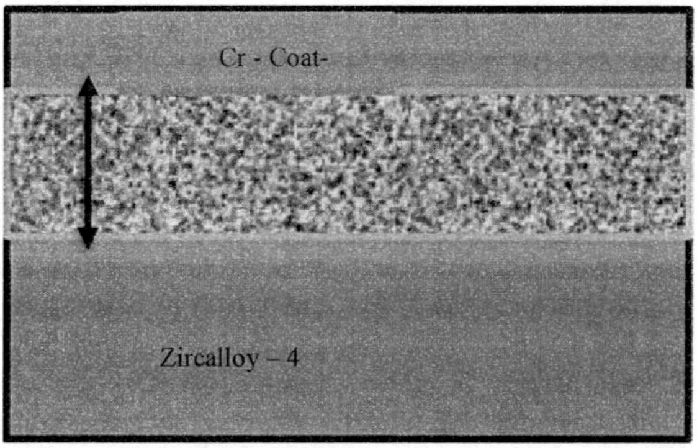

Fig. 4 Diagram depicting the structure and thickness of Cr coatings

The substrate and the coating (29.0 ± 2.0 μm) had a strong interface. Mechanical inter-locking enhanced the bonding between the two. The Cr layer for AR-Zr_4-Cr-PVD samples was 6.48 ± 1.41 μm thick. The R_a value was 0.434 μm for the substrate AR-Zr_4 and 1.222 μm for the AR-Zr_4-Cr-CS samples. The observed R_a of the coated samples was greater, as compared to the substrates. The structure and thickness of Cr coatings are shown in Fig. 4 [10].

There was a reduction in the peaks of blasted samples as compared to those of un-blasted ones. It was because of the abrasive action of media (MB) which levelled the droplets. There was an improvement in surface finish after micro-blasting process. Droplets' density at the top surface was also decreased. An optimal combination of micro-blasting and ion bombardment was necessary for enhanced wear resistance. It was observed that excessive micro-blasting alone was in-sufficient for better wear resistance [15].

The deposited films had very low R_a value. The R_a value of C_1 was higher in comparison with C_2 and C_3. There was an increase in the columnar size of TiO_2 layer (upper C_2 or C_3), as compared to TiN layer (upper C_1) [16].

Very low roughness values were found for LST + Al-TiN and LST + Al-CrN samples. The uniform growth of coating and a reduction in coating macro-particles' size were the reasons for low R_a of LST surfaces. As a result of it, the available surface area for coating was increased. The LSP samples had high R_a values for both coatings [20].

The indentation resistance was for CrN (0.048 ± 0.006 μm) coatings in CrN/HSS samples. Initially, there was residual impression (10 mN). The residual impressions dominated the scene, on increasing the force to 30 mN. Cracks in form of ring were formed at the force of 50 mN. There occurred cracking and removal of material b/w 30 and 50 mN. Such cracks depended on the energy of impact [31].

3.5 Hardness

The movement of the indenter with respect to coating thickness and the penetration depth influence the hardness level in case of Cr-coated samples. The AR-Zr_4-Cr-PVD and AR-Zr_4-Cr-CS layers were deposited on AR-Zr_4 substrate. 5 and 10 mN loads were employed for hardness measurement. The Cr coatings were harder, as compared to that of the substrate [10]. The micro-hardness of the samples treated with laser was 2.5–4.5 times that of the substrate. The highest micro-hardness was in Ti-1.6-1250-D samples. The generation of hard and brittle compounds like TiC and TiN was the reason for this behaviour [12].

The influence of substrate movement (rotation) on hardness was also observed in certain coating types. The hardness of stationary substrate was 878.94 HV, while that of the rotating substrate was 834.66 HV [14]. In the multi-layer PVD coatings, it is the thickness of layers and their number which affects the micro-hardness. There is a direct relationship between the total coating thickness and hardness to be achieved [22]. However, there was a little influence of increase in coating thickness on mechanical properties like hardness, elastic modulus and mechanical strength [17]. The thicker alumina layer in TiAlN was α-Al_2O_3 modification [32]. Various treatment of the coating surfaces can also influence hardness. There was an increase in the hardness of the WC surface after LSP treatment. However, with the LST treatment, no change in hardness was observed for both un-coated and coated WC surfaces [20]. TiN (hardest coating) was studied in pre- (207 VHN) and post-heat treatment (613 VHN) modes [33].

3.6 Coating Thickness/Porosity/Stress

There was a little variation in thickness in the stationary substrate (207.59 mm thick) coatings, as compared to the rotating ones (200.40 mm thick). The speed of rotation was 20 rpm. Lesser number of pores were present in the surface of stationary substrate [14].

A residual state of stress existed for all coatings. However, the residual stress was low for T_1–T_3 coatings. The deposition parameters also influenced the state of stress. The SFC technology and a low bias voltage were the reasons for low residual stress. The increased thickness effected the tool geometry and mechanical properties of the coated inserts. Edge radius was increased as the coating became thicker [17].

4 Wear Performance

4.1 Tool Wear

The coated tools show an enhanced wear resistance. The various factors that influence the wear were the feed, spindle speed, friction property, cutting forces, cutting temperature and time for the test. Flank wear resistance was improved (43% reduction) by the TiAlN-coated micro-drills [34].

Improper friction and large forces of cutting in uncoated tool led to micro-chipping. During milling of Ti-6242S (uncoated tool), the wear mechanisms were chipping, diffusion and adhesion while in case of Ti (C, N) + Al_2O_3 + TiN-coated tool micro-chipping and adhesion were main wear mechanisms. In (Ti, Al)N + TiN-coated tool micro-chipping, diffusion and adhesion were (rake and flank face) predominant [5].

There was a reduction in the resultant forces in TiAlN-1 and TiAlN-2 coatings. The value of decrease was 27.46 and 23.51% for 90 m/min and 34.85 and 21.52% for 120 m/min, respectively. The cutting temperatures of TiAlN-1 and TiAlN-2 were also decreased, as compared with uncoated tool. During turning operation, the suitability of PVD $Ti_{0.55}Al_{0.45}N$-coated tool was established for Inconel 718 at high cutting speeds of 90–120 m/min [35]. No detachment was seen in the CrN monolayer PVD coating on $TiAl_6V_4$ alloy. The CrN coating had the ability to form thermal barrier films of Cr_2O_3. This influenced the intensity of crater wear strongly, leading to an increase in tool life [7]. Better plasticity and load carrying capacity rendered CVD 1 as the best performing coating. The tribological properties like CCR, metal flow and shear band were also enhanced. A combination of Co binder with TiC in CVD 1 reduced hardness and improved fracture toughness. There was not much effect of the thickness on the wear behaviour of PVD coatings [4].

The erosion wear amongst CrN, TiAlN and Ti-Al-N/CrN ($Ti_{45}Nb$ substrate) coatings was maximum for CrN coating (8×10^{-3} mm^3/N m). Also, the erosion wear rate amongst Ti-Al-N (1), Al-Cr-N (2) and Ti-Al-Si-N (3) was maximum (1.5×10^{-7} mm^3/N m) for (1) and (2). The greatest hardness was for TiN-coated AISI 304 because of the absence of Al. As the Si content was increased, the hardness also got improved (maximum value of 18.7 GPa at 5.09% Si) [24].

There was a change in the wear mechanism of Ti-6Al and Ti-10Al from adhesion to abrasion. The self-lubricating film (thin) was removed from the wear track completely. The primary wear mechanism for PVD sputtered Ti-6Al and Ti-10Al-coated samples was abrasion. The inter-particle cracks were present on the worn surface. For cold-sprayed sub-surfaces, de-lamination and cohesion mechanisms dominated under high loads. The friction and wear were increased due to this [9].

The de-lamination dominated in PVD samples. No de-lamination was there in CS-coated Zr-4 alloy. The width and depth of the scratch were increased until the load reached critical length, L_c at 312 μm [10].

The ternary coatings like Ti-Al-N covered by WS_2 film (CC + W 6100 s; CM + W 7000 and 2850 s; CP + W 7800 and 4300 s; CF + W 7900 and 4400 s; CPF

+ W 8800 and 5200 s) exhibited a better wear resistance, as compared to the bare Ti-Al-N samples (CC 4000 s) [11].

On the samples ($(Cr_{37}Al_{50}Mo_{13})N$-LVO-$100Cr_6$ (sample I), $(Cr_5Al_7Mo_{88})N$-FVA_3 + S-$100Cr_6$, $(Cr_{37}Al_{50}Mo_{13})N$-FVA_3 + S-Si_3N_4 and $(Cr_{23}Al_{31}Mo_{40}Cu_6)N$-$FVA_3$ + S-Si_3N_4 slight wear tracks were detected. There was a strong co-relation b/w the increase in wear track's width (up to 6.60×10^{-2} μ), with the evolution of COF on sample I [13].

There was a direct relationship b/w the coefficient of wear and media pressure. The higher media pressure (upto 5 bar) and exposure time (upto 30 s) deteriorated the wear resistance property. Hence, optimal media pressure and time of exposure are desirable. A graph between the coefficient of wear in TiN coating with time of exposure is plotted in Fig. 5 [15].

As the erodent particles were impinged on the coating, there started a reduction in the thickness of coating. It occurred during the time of re-strained wear. Material loss was less in coating no. 8 as compared to the wear of coatings no. 1 and 2. The greatest stress (shear) τ_{max} existed at the edges of cleavages and micro-cracks. The lower the stress (shear) level, the better was the resistance of erosion. A graph of the erosion wear has been plotted in Fig. 6, depicting the above-said phenomena [22].

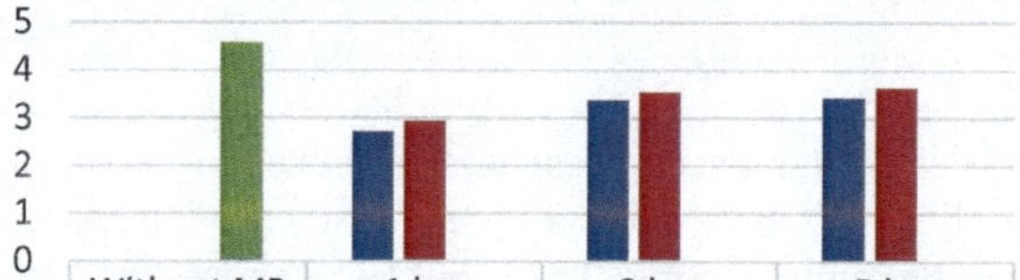

	Without MB	1 bar	3 bar	5 bar
10 Sec	0	2.72	3.38	3.43
30 Sec	0	2.93	3.54	3.64
Without MB	4.58	0	0	0

■ 10 Sec ■ 30 Sec ■ Without MB

Fig. 5 Graph of the coefficient of wear of TiN coating with time of exposure

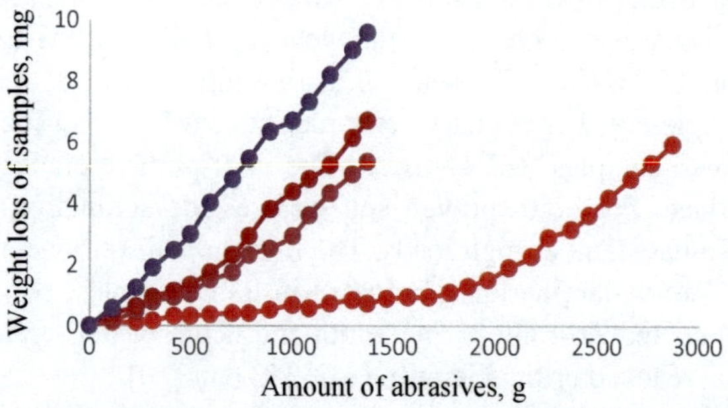

Fig. 6 Erosion wear of coatings

The stick and slip phenomena took place during the cutting process. The thicker, the coating, the better is the ability to withstand wear. It was also observed that T_2 coating underwent minimum wear [17].

In a particular study involving Cr/CrN layers, the cracks were formed. The generation of chips (both internal and external) was also observed at an increasing rate. Despite all these phenomena, the anti-wear properties of coatings were intact [36].

4.2 Abrasion Wear

The coating properties like hardness, fracture toughness and plasticity influenced the wear mechanism strongly. In a study, the Ti-Al-Mo-N coating had high fatigue resistance, good hardness and enough plasticity. As a result, the coatings suffered less wear and good resistance to the initiation and propagation of cracks [18].

There was a better exhibition of wear performance by TiAlTaYN samples, as compared to Ti-Al-Ta-N, Ti-Al-Y-N and Ti-Al-N counterparts. There were grooves located within the wear tracks, in all these coated samples. The wear track was wider for coating containing Y. The debris were mostly oxide particles [32].

The ternary coatings like Cr-W-N exhibited a higher wear resistance than the Al-Cr-N coating. Pressure of contact was 5 MPa, sliding velocity 0.1 ms^{-1} and sliding distance 0.075 m. Surface pores (less than micron size) generated lots of small wear debris. The 22Mn-B5 alloy steel deposited with Al-Si had brittle mechanical behaviour due to the formation of the complex entities like Fe-Al-Si. There was a distinct interface between the Fe-Al layer of inter-metallic and PVD layers. This was mechanically attached and the homogenous in appearance suggesting that it was well-bonded to the surface. There was large Al content in the inter-metallic complexes [25].

In a certain study, the coats were studied after a duration of 600 (stage-1) and 1200 (stage-2) cycles. In stage-1, Al-Cr-O-coated tools had the lowest wear (0.4 mm^3). It was followed by the DLC (0.8 mm^3). The other two coatings Ti0.57 Al0.43 N/Mo-Si-B and TiN/Mo-Si-B had greater material loss. In stage-2, the build-up volume increased for the DLC but reduced for the Al-Cr-O coating. The $Ti_{0.57}Al_{0.43}N$/Mo-Si-B multilayer coating also experienced a reduction in wear [19].

There were parallel grooves inside the wear track of the Fe_3Al/SS304 coating, as compared to the CrN/Fe_3Al/SS304 and the DLC/Fe_3Al/SS304 duplex coatings. Large pits were observed at this stage. A schematic showing wear tracks is shown in Fig. 7 [21].

There was more wear in the areas where the carbide was exposed. The location of wear was in between WC grains. It covered partly the Co and consisted of both Cu and Zn. The transferred material on CrN coating was rich in the partly oxidized Zn [23].

The salt nitrided samples had a wear track of 4 μm while its value was less than 1 μm in the ionic nitrided samples. Nitriding experienced most of the wear in

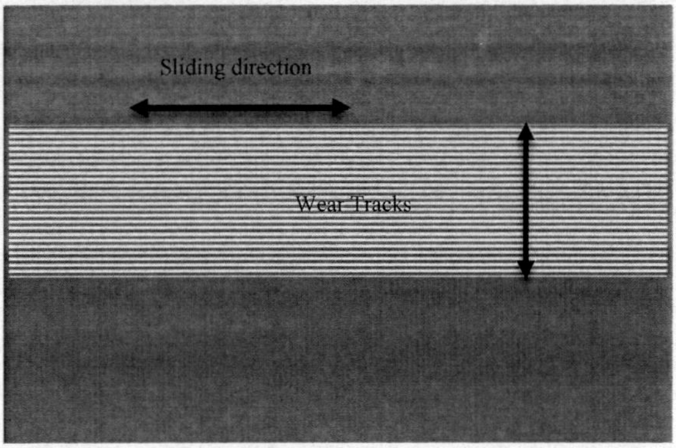

Fig. 7 Schematic of the surface of the wear track

the initial discharge. It was followed by a less loss of material. The formation of compound and diffusion layers was responsible for such behaviour [27].

References

1. Mubarak, A., Hamzah, E., & Toff, M. R. M. (2005). Review of physical vapour deposition (PVD) techniques hard coating.
2. Physical vapor deposition (PVD) technology. Areco Engineering.
3. Song, B., Murray, J. W., Wellman, R. G., Pala, Z., & Hussain, T. (2020). Dry sliding wear behaviour of HVOF thermal sprayed WC-Co-Cr and WC-Cr_xC_y-Ni coatings. *Wear, 442–443*, 1–10. https://doi.org/10.1016/j.wear.2019.203114
4. He, Q., Paiva, J. M., Kohlscheen, J., Beake, B. D., & Veldhuis, S. C. (2020). An integrative approach to coating/carbide substrate design of CVD and PVD coated cutting tools during the machining of austenitic stainless steel. *Ceramics International, 46*(4), 5149–5158. https://doi.org/10.1016/j.ceramint.2019.10.259.
5. Qinglong An, M. C., Chen, J., Tao, Z., & Ming, W. (2019). Experimental investigation on tool wear characteristics of PVD and CVD coatings during face milling of Ti-6242S and Ti-555 titanium alloys. *International Journal of Refractory Metals and Hard Materials*, 108653. https://doi.org/10.1016/j.ijrmhm.2019.105091.
6. Fu, X., Cao, L., Qi, C., Wan, Y., & Xu, C. (2020). Ultralow friction of PVD TiN coating in the presence of glycerol as a green lubricant. *Ceramics International.*
7. Chowdhury, M. S. I., Bose, B., Yamamoto, K., Shuster, L. S., Paiva, J., Fox-Rabinovich, G. S., & Veldhuis, S. C. (2020). Wear performance investigation of PVD coated and uncoated carbide tools during high-speed machining of TiAl6V4 aerospace alloy. *Wear, 446–447*(August 2019). https://doi.org/10.1016/j.wear.2019.203168.
8. Yumusak, G., Leyland, A., & Matthews, A. (2020). The effect of pre-deposited titanium-based PVD metallic thin films on the nitrogen diffusion efficiency and wear behaviour of nitrided Ti alloys. *Surface & Coatings Technology, 394*(October 2019). https://doi.org/10.1016/j.surfcoat.2020.125545.
9. Ng, C., Rao, J., & Nicholls, J. (2019). The role of PVD sputtered PTFE and Al_2O_3 thin films in the development of damage tolerant coating systems. *Journal of Materials Research and Technology*, 1–12.
10. Umretiya, R. V., Elward, B., Lee, D., Anderson, M., Rebak, R. B., & Rojas, J. V. (2020). Mechanical and chemical properties of PVD and cold spray Cr-coatings on Zircaloy-4. *Journal of Nuclear Materials.* https://doi.org/10.1016/j.jnucmat.2020.152420.

11. Zhang, K., Guo, X. Wang, C., Liu, F., & Sun, L. (2020). Effect of plasma-assisted laser pretreatment of hard coatings surface on the physical and chemical bonding between PVD soft and hard coatings and its resulting properties. *Applied Surface Science, 509*(8).
12. Bahiraei, M., Mazaheri, Y., Sheikhi, M., & Heidarpour, A. (2020). A new approach to synthesis Ti_2AlC MAX phase using PVD coating and postlaser treatment. *Surface & Coatings Technology, 385*(December 2019).
13. Bobzin, K., Kalscheuer, C., Technology, C., & Kalscheuer, C. (2019). Arc PVD (Cr,Al,Mo)N and (Cr,Al,Cu)N coatings for mobility applications.
14. Kadam, N. R., Karthikeyan, G., & Kulkarni, D. M. (2020). Effect of substrate rotation on the microstructure of 8YSZ thermal barrier coatings by EB-PVD. *Materials Today: Proceedings*, 8–13.
15. Mundotia, R., Kothari, D.C., Kale, A., Mhatre, U., Date, K., Thorat, N., & Ghorude, T. (2020). Effect of ion bombardment and micro-blasting on the wear resistance properties of hard TiN coatings. *Materials Today: Proceedings.*
16. Bahi, R., Nouveau, C., Beliardouh, N. E., Ramoul, C. E., Meddah, S., & Ghelloudj, O. (2020). Surface performances of Ti-6Al-4V substrates coated PVD multilayered films in biological environments. *Surface & Coatings Technology, 385*(January).
17. Abdoos, M., Yamamoto, K., Bose, B., Fox-Rabinovich, G., & Veldhuis, S. (2019). Effect of coating thickness on the tool wear performance of low stress TiAlN PVD coating during turning of compacted graphite iron (CGI). *Wear, 422–423*(January), 128–136. https://doi.org/10.1016/j.wear.2019.01.062.
18. Sergevnin, V. S., Blinkov, I. V., Volkhonskii, A. O., Belov, D. S., & Chernogor, A. V. (2019). Structure formation of adaptive arc-PVD Ti-Al-Mo-N and Ti-Al-Mo-Ni-N coatings and their wear-resistance under various friction conditions. *Surface & Coatings Technology, 376*(September), 38–43. https://doi.org/10.1016/j.surfcoat.2018.09.068.
19. Aschauer, E., Riedl, H., Koller, C. M., Bolvardi, H., Arndt, M., Polcik, P., Mayrhofer, & P.H. (2019). Adhesive wear formation on PVD coated tools applied in hot forming of Al-Si coated steel sheets. *Wear, 430–431*(March), 309–316. https://doi.org/10.1016/j.wear.2019.05.019.
20. Mishra, S. K., Ghosh, S., & Aravindan, S. (2019). Physical characterization and wear behavior of laser processed and PVD coated WC/Co in dry sliding and dry turning processes. *Wear, 428–429*(March), 93–110. https://doi.org/10.1016/j.wear.2019.03.008.
21. Pougoum, F., Qian, J., Martinu, L., Klemberg-Sapieha, J., Zhou, Z., Li, K. Y., Savoie, S., Lacasse, R., Potvin, E., & Schulz, R. (2019). Study of corrosion and tribocorrosion of Fe_3Al-based duplex PVD/HVOF coatings against alumina in NaCl solution. *Surface & Coatings Technology, 357*, 774–783. https://doi.org/10.1016/j.surfcoat.2018.10.060.
22. Erosive wear behavior of Ti_Ti(V,Zr)N multilayered PVD coatings for Ti-6Al-4V alloy.
23. Heinrichs, J., Mikado, H., Kawakami, A., Wiklund, U., Kawamura, S., & Jacobson, S. (2019). Wear mechanisms of WC-Co cemented carbide tools and PVD coated tools used for shearing Cu-alloy wire in zipper production. *Wear, 420–421*(October 2018), 96–107. https://doi.org/10.1016/j.wear.2018.12.075.
24. Gupta, G., Tyagi, R. K., Rajput, S. K., Saxena, P., Vashisth, A., & Mehndiratta, S. PVD based thin film deposition methods and characterization/property of different compositional coatings—A critical analysis. *Materials Today: Proceedings.*
25. Mozgovoy, S., Alik, L., Hardell, J., & Prakash, B. (2019). Material transfer during high temperature sliding of Al-Si coated 22MnB5 steel against PVD coatings with and without aluminium. *Wear, 426–427*(September 2018), 401–411. https://doi.org/10.1016/j.wear.2018.12.042.
26. Mondragón-Rodríguez, G. C., Chipatecua-Godoy, Y., Camacho, N., & Espinosa-Arbeláez, D. G. (2019). Effect of thermal treatments in high purity Ar on the oxidation and tribological behavior of arc-PVD c-$Al_{0.66}Ti_{0.33}N$ coatings. *Surface & Coatings Technology, 362*(January), 44–56 [Online]. Available: https://doi.org/10.1016/j.surfcoat.2019.01.070.
27. Zabala, B., Fernandez, X., Rodriguez, J. C., López-Ortega, A., Fuentes, E., Bayón, R., Igartua, A., & Girot, F. (2019). Mechanism-based wear models for plastic injection moulds. *Wear, 440–441*(October 2018). https://doi.org/10.1016/j.wear.2019.203105.

28. Cheng, Y. (2020). Cyclic oxidation behaviour of Ti/TiAlN composite multilayer coatings deposited on titanium alloy (January).
29. Yato, K., Doi, S., Ishihara, A., Mitsushima, S., Kamiya, N., & Ota, K. (2006). Improved corrosion protection of CrN hard coating on steel sealed with TiO_xN_y-TiN composite layers. *Journal of Hydrogen Energy Systems Society of Japan, 31*(1), 58–65.
30. Abdollahi, A., Valefi, Z., & Ehsani, N. (2019). Erosion mechanism of ternary-phase SiC/ZrB_2-$MoSi_2$-SiC ultra-high temperature multilayer coating under supersonic flame at 90° angle with speed of 1400 m/s (Mach 4) (December).
31. Shi, X., Li, H., Beake, B. D., Bao, M., Liskiewicz, T. W., Sun, Z., & Chen, J. (2020). Dynamic fracture of CrN coating by highly-resolved nano-impact. *Surface & Coatings Technology, 383*(December 2019). https://doi.org/10.1016/j.surfcoat.2019.125288.
32. Aninat, R., Valle, N., Chemin, J.-B., Duday, D., Michotte, C., Bourgeois, L., & Choquet, P. (2019). Addition of Ta and Y in a hard Ti-Al-N PVD coating: Individual and conjugated effect on the oxidation and wear properties. *Corrosion Science, 156*(December 2018), 171–180. https://doi.org/10.1016/j.corsci.2019.04.042.
33. Motru, S., Hussain, N., Ali Khan, Z., & Avinash (2019). Tribological studies of high surface finish ceramic coatings for low friction and adhesive wear resistant applications. *Materials Today: Proceedings, 27*, 2208–2212. https://doi.org/10.1016/j.matpr.2019.09.098.
34. Azim, S., Gangopadhyay, S., Mahapatra, S. S., Mittal, R. K., & Singh, R. K. (2020). Role of PVD coating on wear and surface integrity during environment-friendly micro-drilling of Ni-based superalloy.
35. Zhao, J., & Liu, Z. (2020). Influences of coating thickness on cutting temperature for dry hard turning Inconel 718 with PVD TiAlN coated carbide tools in initial tool wear stage. *Journal of Manufacturing Processes, 56*(June 2019), 1155–1165. https://doi.org/10.1016/j.jmapro.2020.06.010.
36. Gronostajski, Z., Kaszuba, M., Widomski, P., Smolik, J., Ziemba, J., & Hawryluk, M. (2019). Analysis of wear mechanisms of hot forging tools protected with hybrid layers performed by nitriding and PVD coatings deposition. *Wear, 420–421*(January), 269–280. https://doi.org/10.1016/j.wear.2019.01.003.

Fatigue Crack Initiation and Life Prediction of Rail Weldment Under the Effect of Vertical and Lateral Load Conditions in Indian Railways

Prakash Kumar Sen, Mahesh Bhiwapurkar, and S. P. Harsha

1 Introduction

Nowadays, evolution of railways industries in the world is experiencing the need of much efficient working and along with development of new and more flawless design approach to minimize the costs and maximize the reliability and safety. Between all the sub-structures and segments which are part of a rail–track, the rail–wheel contact interface and joint connections are the most sensitive for performance analysis of the train and security perspectives as well.

The above consideration justifies the monitoring of these components and the need for frequent non-destructive control operations to monitor the condition of the rail and the welding surface on rail [1–4]. These checks, however, are a bit costly and if possible, may be reduced. This is only feasible if some new design methods could be introduced which can predict precisely the growth of failure processes and thereafter show the requirement of regular non-destructive inspection.

Fatigue is most frequent and dangerous among slow wear deterioration process. A sudden fracture is developed by fatigue in wheel, rail, rail weldment, and loss of surface material. Such deficiencies can lead to damage to weldments on rail specially, which ultimately tends to damage serious derailment of train having a big loss in infrastructure and lives. In order to correctly define the condition of stresses at rail–wheel contact, Hertz theory of contact is very simple to utilize if the problem satisfies the Hertz contact assumptions [5]. Numerical approaches for contact analysis, such as FEM and BEM, are often generally used to cope with the complexity resulting in the analytical approach from some decades. Myers et al. [6] attempted a 2D FEM model and a fatigue model based on multiaxial study created by Chau-Cho et al.

P. K. Sen (✉) · M. Bhiwapurkar
Department of Mechanical Engineering, O.P. Jindal University Raigarh, Raigarh, India

S. P. Harsha
Department of Mechanical and Industrial Engineering, IIT Roorkee, Roorkee, India

R. Agrawal et al. (eds.), *Recent Advances in Smart Manufacturing and Materials*,
Lecture Notes in Mechanical Engineering,
https://doi.org/10.1007/978-981-16-3033-0_37

[7] for RCF investigation in bearing which can be also utilized for finding contact stresses on wheel–rail/weld [8, 9].

Rail Defect Manual [10] and Yu and Keer [11] made a study with Hertz contact hypothesis to compute the stress outcomes and fatigue results, considering multiaxial analysis as a uniaxial analysis problem; thus, only one directional component of principle stress and strain was used for fatigue calculation. UIC, Leaflet 518 [12] proposed a semi-analytic approach to calculate the stresses, by using 3D FEM, but it was under the contact stresses as per theory of Hertz [13].

In this work, a 3D FE model with fine mesh method is implemented at contact zone. To find the life under the fatigue stress of the weld, this paper has used the numerical simulation and stress responses under contact of weldment and wheel specifically on weld under rolling.

The criteria Von Mises, which is most often used for multiple axis fatigue of materials with ductile properties, is assumed here. The modified Goodman theory is then opted to better understand the impact of moderate stresses developed. The outcomes with multiple parametric values are calculated on the basis of data collected, and several results are presented graphically.

In general, primary and secondary loading elements are under service in rail and its welding. Under rolling contact, there is bending stresses due to wheel–axle vertical loading, axial stresses due to creep and also stresses considering Hertzian contact theory.

The bending stresses are developed by the axle vertical load, which is typically between 8 and 22.5 tons (it is proposed to be up to 25 tons for heavy transport in India) along with moving vehicle dynamic magnification factor. A little bit of bending stresses also comes in to account due to self-load of the rail. Defects on the rail surface, like joints, twist, and dips and abnormalities in the wheel like flats and non-roundness play a significant act in fatigue. When train starts and stops, due to irregularities in track structure and dynamic movement of the train, there is a growth of axial stresses. In case of early crack formation phase, rolling contact load plays an important part. We must note here that for tracks having curves, switches, and crossings, there is an extra load in the lateral direction. With an increase in rpm, these forces are also dynamically intensified. In vehicles having tilting technology, the irregularities in local track, especially in small radius of curvatures, impose an important role to increase the lateral loads on track and on wagon too [14], noting that lateral bending is the main problem for rails in switches. Boudnitski and Edel [15] indicated about the occurrence of majority of rail failures in Europe at or a little above 0 °C in temperatures. Large tensile thermal stresses are associated with comparatively low rail material toughness properties at this temperature to induce catastrophic collapse.

In rails, the residual stresses initially induced by the manufacturing, mainly from the heat treatment and straightening with rolling operations [16, 17], and the maintenance activity of two standard rails like welding, etc. A maximum axial residual stress of tensile in nature about 200 MPa has been measured on rail weldment beneath the running surface center line in vertical direction and in the other rail section's region with rail foot over where those are specified by residual stresses in compressive in nature. Natural differences have been observed in distribution for residual stresses in the HH rail sections due to method used in welding for joining two rails [18]. It is found that the condition of residual stresses at the rail sections ends at the position of joint is different from its central section because of the variations in procedures adopted for straighten the rail beam other than rolling straightening method and fracture repair [19].

2 Crack Types in Rail and Rail Weldment

Cracks may be induced at or below to rail surfaces as well as on the rail weldment. Cracks on surfaces are getting started because of high tractive forces at rapid rail movement and continues to flow influenced by lubrication at inclination angle between direction of motion of vehicle and in the direction of the rail. Transverse spreading of its branches may further approach to the ultimate weld fracture. There is an intention of beginning of under surface crack below the weld top surface. It is presumed that crack starting position is 12–14 mm from side of the rail weld and about 8–10 mm down from rail head surface [20, 21]. After penetration due to contact, these crack linings are supposed to flow toward the surface of weld section and pretends as cracks on surface plane [22].

3 Methodology and Simulation Discussion

This work has used 3D FEM models of rail having a weld on it with grinded head surface and wheel made on ANSYS 19 Student Edition's software under SpaceClaim, which is its model drafting module. The description for procedure is stated below.

The UIC60 section profile rail and coach wheel profile are used for contact analysis [23, 24]. The length of a rail is appropriate for the distance of two sleepers. Analysis setting has been made as fixed supports on two cross-sectional areas at the ends and its bottom in all directions of the rail. After modeling the rail and wheel, a weld is prepared having a real look of extra extruded part of AT welding in side of rail of 25 mm of length with bonded contact as in Fig. 1. Figure 2 shows the final assembly of this all.

Since the analysis is nonlinear in nature, it will be better to apply fine mesh of about 2 mm average in size at the close contact region. Concider a node as a pilot node at the centre of the wheel which is an important location and here this pilot point is connected to the wheel using a rigid link element. At this pilot point, all the boundary conditions and external loading of the wheel/axle are to be applied. It is possible to obtain these loading and boundary conditions by field calculation or through computational simulation of motion analysis of the track structure.

Fig. 1 An alumino-thermit weld on rail

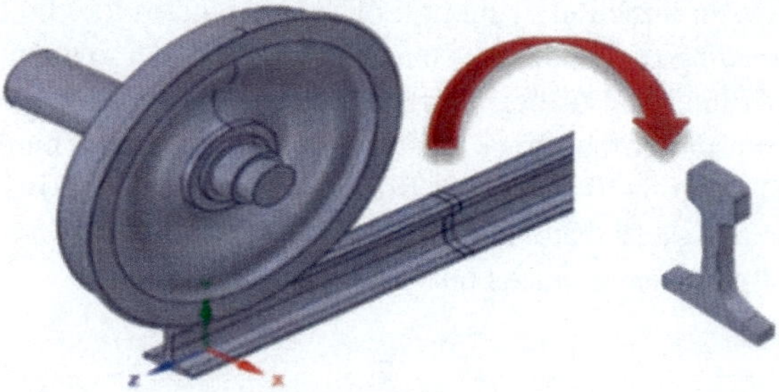

Fig. 2 Rail–wheel weld in ANSYS

Contact elements as target (Green) and contact (Red) are used for the meshing of wheel and rail geometry at the possible contacting faces of the weld head and wheel flat. Pre-existing augmented Lagrangian algorithm method (as in ANSYS) is used for contact analysis. The Coulomb frictional model is used to provide the frictional effect on the material property of the contact zone. Elastic properties of the weldment material is given in Table 1, in which the frictional coefficient is taken as 0.3, assembly setup is chosen as kinematic harden, and then, analysis based on quasi-static assumptions is performed with recording the outcomes of each step at

Table 1 Parameter of elastic plastic weld model [26]

Mechanical property	Value
Ultimate tensile strength	996.7 MPa
Young's modulus	207 GPa
Poisons ratio	0.3
Yield strength	675.7 MPa
Percentage elongation	3.09
Percentage reduction of area	4.22

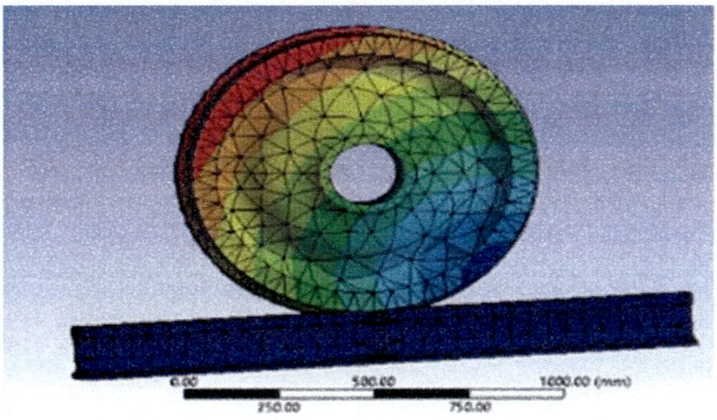

Fig. 3 FE modeling of wheel and rail contact with weldment simulation observation

contact surface area, and in some depth, fine type mesh is applied for obtaining results accurately. The accuracy of the findings can be compared with physical examination.

Assembly contains wheel diameter about 0.90 m having flange contour as per Indian Railways criteria. Overall design load is 146.2 kN in this simulation work considering it as vertical load. Assuming rail and wheel properties equal to 205 MPa for Young modulus, 500 MPa as yielding strength, coefficient of friction as 0.3, 4000 MPa tangent module, and hardening model considering linear kinematics.

The rail length is chosen to be 800 mm in this work [25] and welds as alumino thermites with a weld length of 25 mm. Initially, it is believed that contact point exists on the mid of rail weld's upper surface and the rolling wheel surface. Under static load analysis, the assembly simulation (Fig. 3) gives the result as shown in Figs. 4 and 5 with contact patch. The Von Mises stresses on the weld from various sectional views are there in Fig. 4. Figure 5 indicates shear stress on the weld. Figure 4 represents the occurrence of maximum Von Mises stresses just under the weld surface, as depth becomes higher, the stress value descends quickly. High stresses are developed on the rail weld and the wheel under a limited contact area. Figure 4 also shows that the maximum stresses is experienced by rail weld below its surface at a distance about 6–8 mm. The presence of underlying crack nucleation on the weld may be

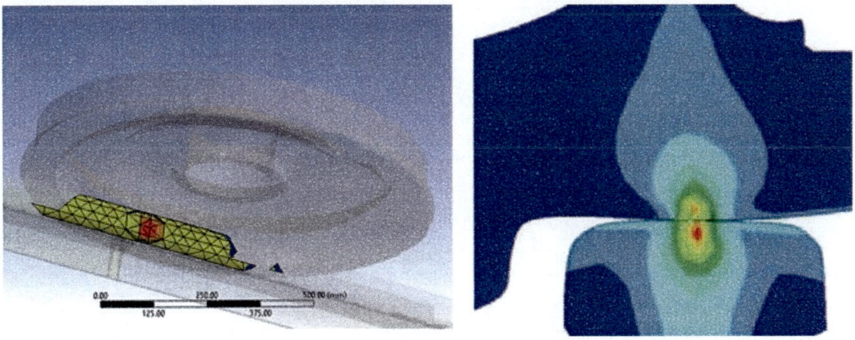

Fig. 4 Weldment Von Mises stress distribution at contact area

Fig. 5 Weldment shear stress distribution at contact

the controlling factor for this. The numerical outcomes precisely established the corresponding laboratory outcomes of the crack sight. Though in other parts of rail have no stresses observed, it is better to simulate only a small portion of assembly which helps to reduce computing time for unnecessary zero contact stresses calculation elsewhere. Shear stress τ_{xy} is found in Fig. 5 which is like a butterfly pattern. The highest value of τ_{xy} often exists at down around 6–8 mm under surface of the rail weld from the contact location of rail weld–wheel assembly.

As indicated in Figs. 4 and 5, the stress pattern represents several contact points between wheel–weld contact interfaces. The Von Mises principle is used after obtaining the stress record of the contact on weld zone for the multiaxial fatigue measurement and evaluating the initiation of fatigue life. Mean stress effects on model can be explained by modified Goodman method. Rail weld is under constant vertical load for each step and ranges between the highest value to zero. To study the results for loads applied under varying steps, the magnitude ranges between 55 and 155 kN, which can be seen in Fig. 7a.

For fatigue crack propagation, the critical location on the weld is designated as the position of the highest Von Mises stress. Figure 6 shows plot of S–N curve for fatigue under torsional and uniaxial loading. Plot on fatigue damage distribution,

Fig. 6 Stress-life plot for torsional and uniaxial load

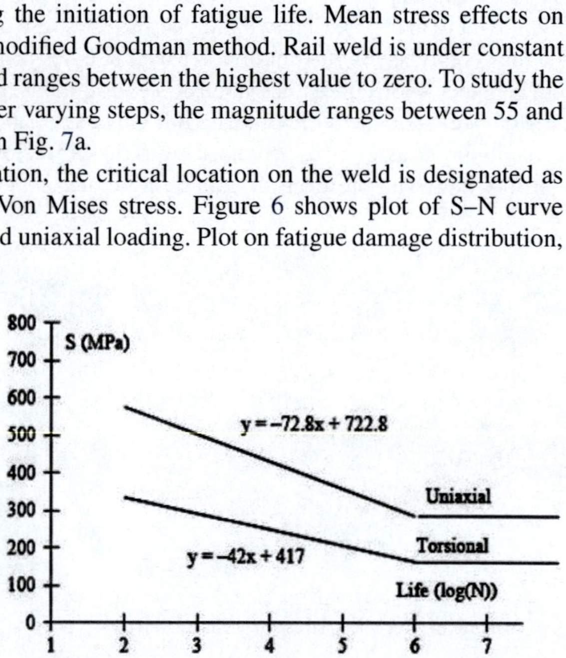

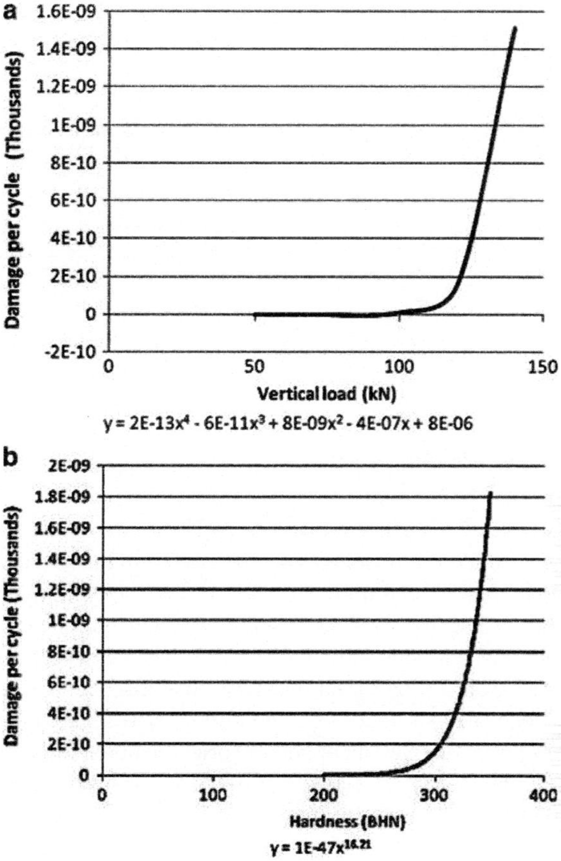

Fig. 7 **a** Vertical loading versus damage per cycle. **b** Hardness versus damage per cycle

hardness, and vertical load is shown in Fig. 7a, b. By applying different vertical loads on the wheel–rail–weld component, their respective values are measured and plotted in Fig. 7a on the basis of the cumulative damage rate. The damage compilation rate rises as per prediction, while increasing the vertical loads. In this case, in load below 110 kN in vertical direction, the endurance limit is higher than the equivalent stress amplitude. Figure 7b shows that as the hardness increases the damage compilation rate also increases. This may conclude as a result that reduction in contact area between the rail–wheel model creates maximum stresses on area of contact.

Figure 8 demonstrates the influence of lateral movement on the wheel when the maximum stress observed in rail weld. Stresses will significantly boost in this region; it is observed that region of contact is shifted toward wheel flange to left. Furthermore, it is seen that there is a rise in stresses where the weld and wheel are in closely in touch at wheel's varying gradient. As a result, it can be stated that the weakest position for wheel and weld contact seems to be very close to the rail flange, hence transmitting the inclination changing zone inside wheel surface which will be quite good.

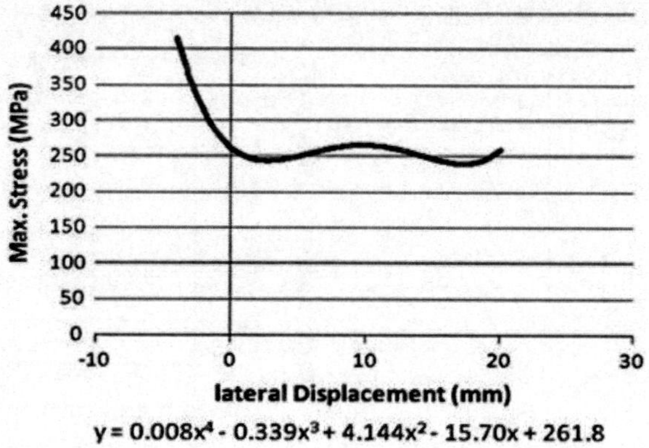

Fig. 8 Lateral displacement versus max. stress on weldment

4 Conclusion

Study has been carried out to find distribution of mechanical stresses and fracture damage risk of rail weldment under different loading conditions to perform just like actual service condition. Not like most of the earlier studies, the present work used a 3D CAD model and FEM analysis under nonlinearity to compute stress values. Finite element simulations of models were successfully simulated and performed stress visualization for hardness, vertical loads, and various damage levels.

Load analysis under static condition was executed for predicting the damage risk of railways rail weldment and historical data for stress and then used to evaluate the life under fatigue. The study of effects of various parameters such as fatigue strength, rail–wheel contact on weldment, vertical loads, material stiffness and damage due to fatigue in the rail component has been conducted using the Von Mises model. Figures plotted show the maximum stress distribution for Von Misses theory under rolling contact with vertical and lateral load due to lateral movement conditions of the wheel, respectively.

In the current study, the outcomes based on various parameters have been studied individually. Because the wheel–rail/weld contact problems are of highly nonlinear in nature, so the forthcoming research must need the interactive effects of above-stated parameters.

Other issues such as rail and its welding residual stress due to manufacturing and servicing, braking loads, thermal loads, effect due to impact and dynamic loadings, flows in material, etc., may be adopted under above-discussed technique for detailed simulation and recognition of the principal source of welding failure and its service life.

References

1. Hosseini Tehrani, P., & Saket, M. (2009). Fatigue crack initiation life prediction of railroad. *Journal of Physics: Conference Series, 181*, 012038. https://doi.org/10.1088/1742-6596/181/1/012038.
2. Kim, J.-K., & Kim, C.-S. (2002). Fatigue crack growth behavior of rail steel under mode I and mixed mode loadings. *Materials Science and Engineering: A, 338*(1–2), 191–201. https://doi.org/10.1016/S0921-5093(02)00052-7.
3. Low, A. (2015). The design of railway viaducts without rail joints. *Structural Engineering International, 25*(2), 218–223. https://doi.org/10.2749/101686614x14043795570651
4. Mutton, P. (2000). Material aspects of weld behaviour in wheel-rail contact. In *5th International Conference on Contact Mechanics and Wear of Rail/Wheel Systems*, Tokyo, Japan.
5. Franklin, F. J., Widiyarta, I., & Kapoor, A. (2001). Computer simulation of wear and rolling contact fatigue. *Wear, 251*(1–12), 949–955.
6. Myers, J., Geiger, G. H., & Poirier, D. R. (1982). Structure and properties of thermite welds. *Welding Journal Research Supplement, 61*, 258–268.
7. Chau-Cho, Y., Keer, L. M., & Steele, R. K. (1997). Transactions of the ASME. *Journal of Tribology, 119*, 660–666.
8. Journet, B. G., & Pelloux, R. M. (1987). *Theoretical and Applied Fracture Mechanics, 7*, 19–22.
9. Korea Railroad Research Institute. (1998). A Study on the Properties of the Welded Part in Rail Steel, No. 98-35.
10. Rail Defect Manual. (1968). Compiled by Sperry Rail Service, A Division of Automation Industries, Inc.
11. Yu, C.-C., & Keer, L. M. (1996). Edge effect on the rolling/sliding contact. *Computational Mechanics, 18*, 259–268.
12. UIC, Leaflet 518. (2009). Testing and approval of railway vehicles from the point of view of their dynamic behaviour—Safety—Track fatigue—Ride quality.
13. Mirković, N., Popović, Z., Pustovgar, A., Lazarević, L., & Zhuravlev, A. (2018). Management of stresses in the rails on railway bridges. *FME Transactions, 46*(4), 636–643. https://doi.org/10.5937/fmet1804636m.
14. Arslan, M. A., & Kayabaşi, O. (2012). 3-D rail-wheel contact analysis using FEA. *Advances in Engineering Software, 45*, 325–331.
15. Boudnitski, G., & Edel, K.O. (2000) Spannungsintensitcatsfactoren in Schienen. In: *Proceedings of Internationales Symposium Schienenfehler*, Brandenburg, Germany [chapter 11].
16. Skyttebol, A., Josefson, B. L., & Ringsberg, J. W. (2005). Fatigue crack growth in a welded rail under the influence of residual stresses. *Engineering Fracture Mechanics, 72*(2), 271–285.
17. Pyrgidis, C. (2009). *Railway transport systems* (Zitis Ed.) (available in Greek language).
18. Iwnicki, S. (2006). *Handbook of railway vehicle dynamics*. Taylor & Francis Group, LLC.
19. Szablewski, D., LoPresti, J., & Sammon, D. (2013). *Premium rail testing at FAST*. Technology Digest TD-13-016.
20. Zhu, M., Xu, G., Zhou, J. H., Wang, R. M., & Gan, X. L. (2017). Study on service performance of 880 MPa-grade and 980 MPa-grade rail steels. *IOP Conference Series Materials Science and Engineering, 244*, 012025.
21. Krauss, G. (2005). *Steels—Processing, structure and performance*. ASM.
22. Zerbst, U., Schödel, M., & Heyder, R. (2009). Damage tolerance investigations on rails. *Engineering Fracture Mechanics, 76*, 2637–2653.
23. Zhang, F., Lv, B., Hu, B., & Li, Y. (2007). Flash butt welding steel crossing and carbon steel rail. *Materials Science & Engineering A, 454–455*, 288–292.

24. Zerbst, U., & Beretta, S. (2011). Failure and damage tolerance aspects of railway components. *Engineering Failures Analysis, 18*, 534–542.
25. Rails, Korean Standards Association. (1994). KS B 8106-94.
26. Wang, Y., Zhou, H., Shi, Y., & Feng, B. (2012). Mechanical properties and fracture toughness of rail steels and thermite welds at low temperature. *International Journal of Minerals, Metallurgy and Materials, 19*(5), 409.

Effect of Process Parameters on Weld Bead Geometry, Microstructure, and Mechanical Properties in Submerged Arc Welding

Vinod Aswal, Jinesh Kumar Jain, Tejendra Singh Singhal, Rajeev Agrawal, and Sundeep Kumar

1 Introduction

The welding process is a joining process of two comparable or non-comparable materials. Welding has affected various enterprises by increasing productivity and the life of the product. Welding is a process to produce a good quality weld by varying the process parameters [1]. SAW is a fusion welding process which used over other processes because of its high deposition rate, smooth finishing, deeper penetration, and high reliability [2]. Due to these properties, the SAW process is utilized in pressure vessels, offshore structures, LPG cylinders, and pipelines [3]. SAW process comprises a formation of an arc between the electrode tip and workpiece at the joint area. This joint area completely covered by a flux contains various chemical compounds such as calcium silicate and manganese used to remove phosphorous from weld pool, the addition of SiO_2 and TiO_2 reduces the mechanical properties and increases the heat input, ammonium chloride, and borax flux which is used to reduce the melting point of the unwanted iron oxide. Thus, the use of these chemical compounds depends on the workpiece material to produce an efficient weld that is free from atmospherical contaminations [4]. The weld produced in the SAW process depends on different parameters such as voltage, current, welding speed, wire feed rate, and electrode polarity. Gunaraj and Murugun [5, 6] studied that by increasing arc length, the gap between the workpiece and electrode tip increases which gives more melting of the workpiece surface and shallow penetration, and by increasing

V. Aswal · J. K. Jain (✉) · T. S. Singhal · R. Agrawal
Department of Mechanical Engineering, Malaviya National Institute of Technology Jaipur, Jaipur 302017, India
e-mail: jineshjain.mech@mnit.ac.in

S. Kumar
Department of Technical Education, Centre for Electronic Governance, Government of Rajasthan, Jaipur 302004, India

R. Agrawal et al. (eds.), *Recent Advances in Smart Manufacturing and Materials*,
Lecture Notes in Mechanical Engineering,
https://doi.org/10.1007/978-981-16-3033-0_38

welding current, the heat input increases, which results in deeper penetration [7]. Similarly, the effect of various welding parameters is studied in this paper, and their effects are observed. Yang et al. [8] explained the impact of parameters on the weld pool of the SAW process, and for a particular value of heat input, welds made using direct current electrode negative at high welding speed and low voltage produce large weld-deposited area [9]. Ghosh et al. [10] studied that the thermal diffusivity of workpiece makes a significant impact on the heat-affected zone.

Higher diffusivity provides higher cooling rate which gives smaller heat-affected zone and when diffusivity is low, lowers the metal cooling rate, which results in larger HAZ. Some properties of the weld pool region are significantly affected by varying percentage dilution such as ductility, strength, resistance to weld cracking, and corrosion resistance. In welding applications, for the joining of similar types of material, maximum percentage dilution is required, and for joining of dis-similar types of materials (like cladding and hard facing), required minimum dilution.

2 SAW Process Parameters

The selection of appropriate parameters is necessary for producing a good quality welded joint. Welding process parameters may be variable or held constant. Parameters of SAW are discussed below.

2.1 Welding Voltage

In the SAW process, the voltage measured between the electrode and the workpiece at a joint area at the time of initiation of an arc is known as the welding arc voltage. Systems for automatic welding can maintain consistent control of arc voltage, but in manual arc welding processes, control and monitor of welding voltage are difficult. Welding voltage is an important parameter for producing a good quality weld joint. It controls the consumption of flux, arc length, and weldment properties. Welding arc voltage having a positive effect on arc length and by increasing voltage arc length increases and gives shallow penetration. The arc length makes a significant impact on weld bead width because as the arc length increases weld bead width increases. Junior et al. [11] studied that with an increase in welding voltage, penetration decreases and weld bead width increases, but loss in penetration is lesser than the weld bead width, which results in increases in the weld pool size. Hence, the percentage of dilution increases. Gunaraj and Murugun [5, 12] studied that the welding voltage decreases the penetration and reinforcement of the weld pool and increases the weld bead width and percentage dilution.

2.2 Welding Current

In a circuit, the term electric current is used as a transfer of energy from one terminal to other. The type of electrode polarity, the intensity of current, and composition of the wire covering significantly affect the melting rate [13]. Variation in current significantly affects the weld pool. As welding current increases, melting rate increases and results in deeper penetration, and also increase in welding current increases the consumption of flux. Too much reduction in welding current produces an unstable arc. Cho et al. [14] studied that with an increase in welding current, the heat content of the droplets increases, and molten droplet quickly striking the weld pool causes more heating of the base plate, results in deeper penetration. Balasubramanian et al. [15] studied that with an increase in welding feed rate, arc current also increases, which results in higher heat input and metal deposition rate. Hence, the reinforcement of the weld bead increases. Mahapatra et al. [16] observed that welding current and a voltage having more impact on reinforcement and with 625 A and 35 V better reinforcement is achieved.

2.3 Wire Feed Rate (WFR)

In manual arc welding processes, the amount of filler material deposited on the weld pool depends on the wire feed rate which is manually operated. To prevent the problem of incomplete fusion also done by adjusting the WFR. By increasing WFR, the welding current and the metal deposition rate increase which results in deeper penetration and large bead width. Cho et al. [14, 17] studied that increasing the wire feed rate, welding current increases, and improved heat input results in increasing weld bead width. By increasing WFR, penetration, weld bead width, and reinforcement increased, but these factors decrease with an increase in welding speed [5, 12]. Om and Pandey [18] studied that using DCSP and increasing welding speed with lower WFR gives a smaller heat-affected zone in the SAW process.

2.4 Welding Speed

The electrode wire travels along the weld pool region is termed as welding speed. The precise welding speed produces a good weld pool appearance. By increasing the welding speed, heat input decreases, and less filler material is applied to the weld pool resulting in shallow penetration, smaller bead width, and lesser reinforcement. Penetration depth is more affected by welding speed other than welding current. Too large welding speed may cause undercutting. Gunaraj and Murugun [6, 19] studied that by increasing the welding speed, the torch travels at a higher speed, which results in a reduction in metal deposition rate and heat input. Hence, the percentage dilution and weld bead width decrease. Om et al. [20] studied that percentage dilution is generally affected by the weld travel speed and electrode polarity. Khallaf et al. [21] studied that the crack length decreases by increasing welding speed results in low heat input. Hence, penetration decreases.

2.5 Contact Tip to Work Distance (CTWD)

The distance between the contact tip to the workpiece surface is known as a contact tip to work distance. CTWD significantly influenced the weld bead properties. Gunaraj and Murugun [5, 6] studied that by increasing the CTWD, it results in expansion of the arc cone. Similarly, because of the Joule's heating effect, at higher values of *N*, the metal fusion rate increases. Hence, bead width increases. By increasing the WFR, welding current increases, which results in wire melting, and deposition rate increases. Therefore, the bead width increases. Shen et al. [23] studied that by increasing CTWD, heat input decreases and results in reinforcement of weld bead decreases. The stand-off distance (SOD) increases arc length, results in the heat supplied over a wide range of the joint area of the workpiece surface. Hence, the percentage of dilution decreases [19].

Although, these parameters are making a significant impact on weld bead. Some more parameters like wire diameter, flux composition, electrode polarity, and preheat current also make a huge impact on weld bead properties. The electrode diameter is an important parameter to perform any welding operation. The selection of the diameter of the electrode depends on the thickness of the workpiece material. For achieving a high deposition rate, thicker electrode is selected. The weld pool region is completely covered by a flux that contains various chemical compounds to protect it from environmental contaminants. Fluxes deoxidize and clean the weld pool and improve the mechanical properties. Submerged arc fluxes are available in fused agglomerated or bonded types for welding stainless alloys. Some agglomerated or bonded fluxes contain alloys such as chromium, nickel, molybdenum, or niobium to compensate for element loss across the arc. Thus, the use of these chemical compounds depends on the workpiece material to produce an efficient weld that is free from atmospherical contaminations [24]. Preheat is used to bringing the workpiece material to a specified minimum temperature before welding begins, is used to achieve a variety of operating factors, including reducing the hardness of weld metal and HAZ, reduce residual stresses, and achieve the desired microstructure or avoid an undesired microstructure. Cho et al. [17] studied that DCSP without changing OCV and WFR, a quick reduction in dilution takes place, and this is because of a reduction in current. Yang et al. [25] studied that constant current power source with acidic flux has no impact on bead width but when the basic flux is used, bead width increases. Om et al. [20] studied that percentage dilution is generally affected by the weld travel speed and electrode polarity. Changing polarity from electrode positive to electrode negative results in a decreasing percentage dilution at the similar OCV, welding speed, and welding feed rate. For electrode positive, the dilution level increases with the WFR and for electrode negative, around 20% reduction in the level of percentage dilution observed. Balasubramanian et al. [15] studied that an increase in torch oscillation frequency, the torch speed is also increased and covers a wide area of the workpiece material, results in decreasing percentage dilution. At a lower feed rate of powder, less heat is used for melting the powder and more heat is utilized for melting the workpiece surface which results in higher percentage dilution.

3 Effect of Welding Parameters on Weld Bead Characteristics

The weld bead width is the maximum width of the weld pool on the workpiece surface. Cho et al. [14, 17] studied that by increasing the wire feed rate, welding current increases, result in improved heat input, and by increasing current, the droplets containing heat increases which rapidly striking the weld joint causes more heating of the base plate, which results in increasing weld bead width. Gunaraj and Murugun [5, 6] studied that by increasing the CTWD, resulting in expansion of the arc cone. Similarly, because of the Joule's heating effect, at higher values of N, the metal fusion rate increases. Hence, bead width increases. By increasing the WFR, welding current increases, which results in wire melting, and deposition rate increases. Therefore, the bead width increases. Yang et al. [25] studied that constant current power source with acidic flux having no impact on bead width, but when the basic flux is used, bead width increases.

Penetration is the maximum distance from the workpiece to fusion depth. Cho et al. [14, 17] studied that by increasing the wire feed rate, welding current increases, which result in improved heat input, and by increasing current, the droplets containing heat increases which rapidly striking the weld joint causes more heating of the base plate, results in deeper penetration. In single DC welding, a different type of weld bead geometries measured by varying the torch angle results in deeper penetration achieved in negative torch angle and vice-versa. Gunaraj and Murugun [5, 6] studied that increasing the arc length results in more melting at the workpiece surface gives shallow penetration, and by increasing speed, heat input decreases and the welding torch travels at a higher speed over the workpiece surface results in a lower rate of metal deposition on the weld joint. So, because of a lower metal deposition rate and less heat input, penetration decreases. And with increasing wire feed rate, welding arc current increases and the rate of metal deposition also increases. Hence, penetration increases. Balasubramanian et al. [15] studied that lower torch oscillation frequency covers a narrow region of the workpiece surface, results in deeper penetration. Khallaf et al. [21] studied that the crack length decreases by increasing welding speed results in low heat input. Hence, penetration decreases.

The ratio of the area of melted base metal to the total area of weld bead measured across its cross section, known as percentage dilution. Some properties of the weld pool region are significantly affected by varying percentage dilution such as ductility, strength, resistance to weld cracking, and corrosion resistance. In welding applications, for the joining of similar types of material, maximum percentage dilution is required, and for joining of dissimilar types of materials (like cladding and hard facing), minimum percentage dilution is required. Om et al. [20] studied that percentage dilution is generally affected by the weld travel speed and electrode polarity. Changing polarity from electrode positive to electrode negative results in a decreasing percentage dilution at the similar OCV, welding speed, and welding feed rate. For electrode positive, the dilution level increases with the WFR and for electrode negative, around 20% reduction in the level of percentage dilution observed. All the welding parameters such as an increase in wire feed rate, welding speed, and OCV favor higher percentage dilution. Gunaraj and Murugun [6, 19] studied that by increasing the welding speed, the torch travels at a higher speed, results in a reduction in metal deposition rate and heat input. Hence, the percentage dilution decreases. The stand-off distance (SOD) increases arc length, results in the heat supplied over a wide range of the joint area of the workpiece surface. Hence, the percentage dilution decreases. Balasubramanian et al. [15] studied that to an increase in torch oscillation frequency, the torch speed is also increased and covers a wide area of the workpiece material which results in decreasing percentage dilution. At a lower feed rate of powder, less heat is used for melting the powder and more heat is utilized for melting the workpiece surface results in higher percentage dilution. Khallaf et al. [21] studied that the travel speed increases crack length, which results in heat input decreases. Hence, the percentage dilution decreases.

Reinforcement is a distance between the workpiece surfaces to weld pool height. Gunaraj and Murugun [5, 12] studied that the welding voltage decreases the reinforcement of the weld pool. Balasubramanian et al. [15] studied that with increasing WFR, arc current increases results in a higher rate of metal deposition and heat input. Hence, reinforcement increases. Shen et al. [23] studied that with increasing CTWD, the heat input decreases results in reinforcement decreases and vice-versa.

The heat-affected zone (HAZ) contained a uniform and fine-grain microstructure of pearlite and ferrite with a large amount of carbide distributed evenly throughout the zone, which could guarantee a high hardness at this zone. Ghosh et al. [10] studied that the thermal diffusivity of the workpiece makes a significant impact on the heat-affected zone, higher diffusivity provides a higher cooling rate gives a smaller heat-affected zone and when diffusivity is low, lower the metal cooling rate, results in larger HAZ. Om and Pandey [18] studied that using DCSP and increasing welding speed with lower WFR gives a smaller heat-affected zone in the SAW process.

4 Effect of Welding Parameters on Microstructure and Mechanical Properties

Some properties of the weld pool region are significantly affected by varying the percentage dilution such as ductility, strength, resistance to weld cracking, and corrosion resistance. Qiu et al. [26] studied that the fine-grained heat-affected zone has fine productive grains and a high density of mis-orientation grain boundaries, which makes a significant impact on crack propagation and increasing heat-affected zone toughness. Kolhe and Datta [27] while performing a detailed study on the microstructure and mechanical properties of SAW multi-pass joint found that welding heat input can control the percentage of phase formation in the welded joint which influences the microstructure. Microstructure and wear property study of alloy cladding by SAW revealed that the presence of increased content of retained austenite in the microstructure results in lower hardness [28]. Crack susceptibility of carbon steel plates in SAW decreases as current decreases and welding speed or WFR increases [21]. Tarang et al. [29] during the determination of SAW process parameters in hard facing observed that GRA can be used to convert optimization of multiple characteristics into a single characteristic called gray relational grade. Jesus et al. [30] while studying the influence of the SAW process in the mechanical behavior of steel derived energy life relation by using the strain life data. Funderburk [31], these relations indicated that fatigue resistance of weld metal is defined in terms of strain energy of the cycle and is lower than as observed for the base metal. Heat input affects the weld bead, cooling rate, and mechanical properties of the weld pool region [32]. Kanjilal et al. [33] when focused on the combined effect of parameters and flux on mechanical properties of SAW and observed that electrode polarity has a greater effect on the chemical composition of the weld joint. Further welding parameters mainly determined the hardness and yield strength, whereas flux mixtures variable determined the impact toughness. Suitable regression models are found to be useful in expecting the transfer of elements such as oxygen, manganese, silicon, and sulfur across the weld joint and chemical reaction related to SAW fluxes which affected the metal transfer [34]. Kaçar and Baylan [35] studied microstructural properties in dissimilar welding between austenitic SS and martensitic and observed that the tensile strength of the weld joint for electrode E2209-17 was lesser than that of electrode E308L-16. The impact toughness for both the electrode was accepted at low temperature and exhibits ductile fracture. Singhal and Jain [36] investigated various process parameter to control corrosion in welded and cladded parts. Study shows diffrent processes, microstructure obtained and their effect on corrosion resistance.

5 Conclusions

Various parameters and their effects on the weld bead, mechanical properties, and microstructural properties were studied from the study, we have found that the penetration kept increasing with increasing current and wire feed rate but keeps on decreasing with welding speed and nozzle to plate distance. Similarly, width and reinforcement increase with the wire feed rate and decreases with dropping voltage, whereas the width also increases with high voltage.

We have also found that with more heat input the dilution kept on increasing, and high dilution is good for weldments but a limited limit. Apart from dilution, HAZ has also kept increasing with high heat input, but this is generally not acceptable for the application of the weldments. As HAZ is the weakest part of the weldment, welds are prone to break from this portion. Also, most of the weldments studied lied in the austenite and martensite forms, which lower the tensile properties of the weld.

References

1. Nadkarni, S. V. (1988). *Modern arc welding technology.* IBH Publication.
2. Houldcraft, P., & John, R. (1989). *Welding and cutting—A guide to fusion welding and associated cutting processes.* Industrial Press Inc.
3. Viano, D. M., Ahmed, N. U., & Schumann, G. O. (2000). Influence of heat input and travel speed on microstructure and mechanical properties of double tandem submerged arc high strength low alloy steel weldments. *Science and Technology of Welding and Joining, 5*(1), 26–34.
4. Sharma, L., & Chhibber, R. (2020). Study of weld bead chemical, microhardness & microstructural analysis using submerged arc welding fluxes for line pipe steel applications. *Ceramics International, 46*(15), 24615–24623. ISSN 0272-8842. https://doi.org/10.1016/j.ceramint.2020.06.250.
5. Gunaraj, V., & Murugan, N. (1999). Application of response surface methodology for predicting weld bead quality in submerged arc welding of pipes. *Journal of Materials Processing Technology, 88*(1–3), 266–275.
6. Gunaraj, V., & Murugan, N. (2000). Prediction and optimization of weld bead volume for the submerged arc process—Part 1. *Welding Journal, 79*, 286–294.
7. Dallam, C. B., Liu, S., & Olson, D. L. (1985). Flux consumption dependence of microstructure and toughness of submerged arc HSLA weldments. *Welding Journal*, 140s–151s.
8. Yang, L. J., Chandel, R. S., & Bibby, M. J. (1993). The effects of process variables on the weld deposit area in submerged arc welds.
9. Mortazavian, E., Wang, Z., & Teng, H. (2020). Repair of light rail track through restoration of the worn part of the railhead using submerged arc welding process. *The International Journal of Advanced Manufacturing Technology, 107*, 3315–3332.
10. Ghosh, A., Chattopadhyaya, S., Das, R. K., & Sarkar, P. K. (2011). Assessment of heat affected zone of submerged arc welding process through digital image processing. *Procedia Engineering, 10*, 2782–2785.
11. Junior, R. C., Esteves, L., Santos, N. F., Oliveira, I. R., Mendes, D. S., Lins, V. F. C., & Modenesi, P. J. (2019). Influence of heat input and cold wire feeding rate on pitting corrosion resistance of submerged arc welding duplex stainless steel welds. *Journal of Materials Engineering and Performance, 28*, 1969–1976.

12. Murugan, N., & Gunaraj, V. (2005). Prediction and control of weld bead geometry and shape relationships in submerged arc welding of pipes. *Journal of Materials Processing Technology, 168*(3), 478–487.
13. Sonia, P. (2019). Review on heat and mass transfer in submerged arc welding (SAW) and gas metal arc welding (GMAW). *International Journal of Engineering and Advanced Technology, 8*, 503–512.
14. Cho, D.-W., Song, W.-H., Cho, M.-H., & Na, S.-J. (2013). Analysis of submerged arc welding process by three-dimensional computational fluid dynamics simulations. *Journal of Materials Processing Technology, 213*(12), 2278–2291.
15. Balasubramanian, V., Lakshminarayanan, A. K., Varahamoorthy, R., & Babu, S. (2009). Application of response surface methodology to prediction of dilution in plasma transferred arc hard facing of stainless steel on carbon steel. *Journal of Iron and Steel Research, International, 16*(1), 44–53.
16. Mahapatra, M. M., Datta, G. L., Pradhan, B., & Mandal, N. R. (2006). Three-dimensional finite element analysis to predict the effects of SAW process parameters on temperature distribution and angular distortions in single-pass butt joints with top and bottom reinforcements. *International Journal of Pressure Vessels and Piping, 83*(10), 721–729.
17. Cho, D.-W., Kiran, D. V., Song, W.-H., & Na, S.-J. (2014). Molten pool behaviour in the tandem submerged arc welding process. *Journal of Materials Processing Technology, 214*(11), 2233–2247.
18. Om, H., & Pandey, S. (2013). Effect of heat input on dilution and heat affected zone in submerged arc welding process. *Sadhana, 38*, 1369–1391.
19. Murugan, N., Parmar, R. S., & Sud, S. K. (1993). Effect of submerged arc process variables on dilution and bead geometry in single wire surfacing. *Journal of Materials Processing Technology, 37*(1–4), 767–780.
20. Pandey, N. D., Bharti, A., & Gupta, S. R. (1994). Effect of submerged arc welding parameters and fluxes on element transfer behaviour and weld-metal chemistry. *Journal of Materials Processing Technology, 40*(1–2), 195–211.
21. Khallaf, M. E., Ibrahim, M. A., El-Mahallawy, N. A., & Taha, M. A. (1997). On crack susceptibility in the submerged arc welding of medium-carbon steel plates. *Journal of Materials Processing Technology, 68*(1), 43–49.
22. Karaoglu, S., & Seçgin, A. (2008). Sensitivity analysis of submerged arc welding process parameters. *Journal of Materials Processing Technology, 202*(1–3), 500–507.
23. Shen, S., Oguocha, I. N. A., & Yannacopoulos, S. (2012). Effect of heat input on weld bead geometry of submerged arc welded ASTM A709 Grade 50 steel joints. *Journal of Materials Processing Technology, 212*(1), 286–294.
24. Jain, A., Sonia, P., Kumari, S., & Pushp, P. (2020). Study of intermetallic compound (IMC) formed in welding of steel with magnesium: A review. *Materials Today: Proceedings, 26*, Part 2, 1159–1166. ISSN 2214-7853. https://doi.org/10.1016/j.matpr.2020.02.232.
25. Yang, L. J., Chandel, R. S., & Bibby, M. J. (1992). The effects of process variables on the bead width of submerged-arc weld deposits. *Journal of Materials Processing Technology, 29*(1–3), 133–144.
26. Qiu, C., Lan, L., Zhao, D., Gao, X., & Du, L. (2013). Microstructural evolution and toughness in the HAZ of submerged arc welded low welding crack susceptibility steel. *Acta Metallurgica Sinica, 26*, 49–55.
27. Kolhe, K. P., & Datta, C. K. (2008). Predication of microstructure and mechanical properties of multi pass SAW. *Journal of Material Processing Technology, 197*, 241–249.
28. Lu, S. P., Kwon, O. Y., Kim, T. B., & Kim, K. H. (2004). Microstructure and wear property of Fe-Mn-Cr-Mo-V alloy cladding by submerged arc welding. *Journal of Material Processing Technology, 147*, 191–196.
29. Tarang, Y. S., Juang, S. C., & Chang, C. H. (2002). The use of grey based Taguchi methods to determine submerged arc welding process parameters in hardfacing. *Journal of Material Processing Technology, 128*, 1–6.

30. Jesus, D., Ribeiro, S. A., & Fernandez, A. (2007). Influence of the submerged arc welding in the mechanical behaviour of the P355NL1 steel. Part-II: Analysis of the low /high cycle fatigue behaviour. *Journal of Material Science, 42*, 5973–5981.
31. Funderburk, S. R. (1999). Key concepts in welding engineering. *Welding Innovation, XVI*(1).
32. Sonia, P., Jain, J. K., & Saxena, K. K. (2020). Influence of severe metal forming processes on microstructure and mechanical properties of Mg alloys. *Advances in Materials and Processing Technologies*. https://doi.org/10.1080/2374068X.2020.1802554
33. Kanjilal, P., Pal, T. K., & Majumdar, S. K. (2006). Combined effect of flux and welding parameters on chemical composition and mechanical properties of submerged arc weld metal. *Journal of Material Processing Technology, 171*, 223–231.
34. Kanjilal, P., Pal, T. K., & Majumdar, S. K. (2007). Prediction of element transfer in submerged arc welding. *Welding Journal, 69*, 135–146.
35. Kaçar, R., & Baylan, O. (2004). An investigation of microstructure/property relationships in dissimilar welds between martensitic and austenitic stainless steels. *Materials & Design, 25*(4), 317–329. ISSN 0261-3069. https://doi.org/10.1016/j.matdes.2003.10.010.
36. Singhal, T., S., Jain, J., K., (2020). GMAW cladding on metals to impart anti-corrosiveness: Machine, processes and materials. *Materials Today: Proceedings 26*, 2432–2441.

Condition Monitoring of Surface Grinding Process Using Low-Cost Vibration Sensor Module

Shivam Seth, Avinash Bhashkar, Prameet Vats, Tushar Singh, Farooqui Rizwan Ahmed, and Anupam Keshari

1 Introduction

The grinding process is considered as a critical surface finishing process which accounts for about 70% within the spectrum of precision machining [1]. Grinding adds the surface quality value to the parts which are already of high value, thus required a continuous and expertise condition monitoring over the processes, either by manual efforts, mathematical and process physics understanding or through sensor-based real-time monitoring and control. Moreover, it is highly required to balance two phenomena: quality/tolerances and productivity/cost.

There are various types of devices/sensor systems (dynamometer, power sensor, Industrial accelerometer, acoustic emission sensor setup, etc.) which are researched for the monitoring of different type of grinding processes [2–6], but the cost of such devices/sensor systems is considerably high and mostly does not fit with industries economic concerns. Thus, there is a strong need to develop low-cost devices/sensor system which can enable the monitor the different type of grinding processes, or can be utilized with a high-cost industrial sensor as supporting device/system. Here, a low-cost acceleration and vibration sensor-based setup is developed, which is utilized for real-time condition monitoring of the surface grinding process. The developed setup is equipped with low-cost ADXL-335 three-axis accelerometer, microcontroller, data storage card and Node MCU module [7]. It is portable and facilitates the data storage to the cloud as well.

Arun et al. [2] determined the condition of grinding by utilizing acoustic emission sensor and a machine learning approach. Wherein the elastic waves are produced during the process. Acoustic signals are sensed by the piezoelectric sensor after which piezoelectric crystal converts the mechanical signal into an electrical signal

S. Seth (✉) · A. Bhashkar · P. Vats · T. Singh · F. R. Ahmed · A. Keshari (✉)
Centre for Advanced Studies, Dr. A.P.J. Abdul Kalam Technical University, Lucknow, Uttar Pradesh 226031, India

R. Agrawal et al. (eds.), *Recent Advances in Smart Manufacturing and Materials*, Lecture Notes in Mechanical Engineering,
https://doi.org/10.1007/978-981-16-3033-0_39

and predicts the surface roughness of the grinding surface. Lin et al. [3] demonstrated the characterization of grinding wheels (soft wheel, normal wheel and hard wheel) using the AE signals. By use of AE signal characteristics (root mean square (RMS) and ratio of power (ROP) statistics using frequency band) and frequency bands, grinding wheel was characterized. D'Addona et al. [4] demonstrated a method to distinguish the different dressing wear condition by the use of vibration signal analysis and artificial neural networks (ANN). The input data such as epochs, number of nodes, training function and network function this features are extracted from vibration signal. The output of which is coded into three dressing conditions, i.e. Class 1—New Dresser Condition, Class 2—Half Life Dresser Condition and Class 3—Worn Dresser Condition. Cheng et al. [5] utilized a deep convolutional neural network (DCNN) to monitor the wear of the abrasive belt using vibration signal. The experimental setup uses a ceramic abrasive belt. The wear is classified as fracture, adhesion, fall off and abrasion. Baban et al. [8] build a fuzzy logic-based process planning and wheel maintenance scheme for automated grinding lines. The three-axis vibration monitoring scheme is developed to monitor the grinding conditions. Devendiran and Manivannan [6] developed a method to detect worn out wheel which works on acoustic emission signals which extracted the features such as wavelet transform and statically which can be used with RMS and standard deviation for wavelet decomposition, and then data are classified using decision tree C4.5 data mining technique. Oliviera et al. [1] demonstrated many challenges regarding the grinding processes and their industrial implementations. Besides the grinding precision, there are many other challenges such as inventory reduction, need for automation and batch production problems on which researchers should be focused for improving overall production efficiency. For some grinding situations, utilization of advanced grinding wheels is promoted, i.e. use of electroplated wheels in more precise grinding operations, implementation of CBN wheels for grinding operations, self-cooling grinding wheel, etc. Moia et al. [9] proposed a method to monitor the dressing of grinding wheel by using the AE sensor after which by the use of statistics neural network is used to classify the grinding wheel as sharp or dull.

2 Experimental Setup

All the experiments are performed on a horizontal surface grinding machine, where the axis of the rotating wheel is parallel to the reciprocating electromagnetic table. The maximum speed of the rotating wheel is 2840 RPM. The workpiece is placed to the electromagnetic table. It is moved beneath the grinding abrasive wheel by a power feed (Fig. 1).

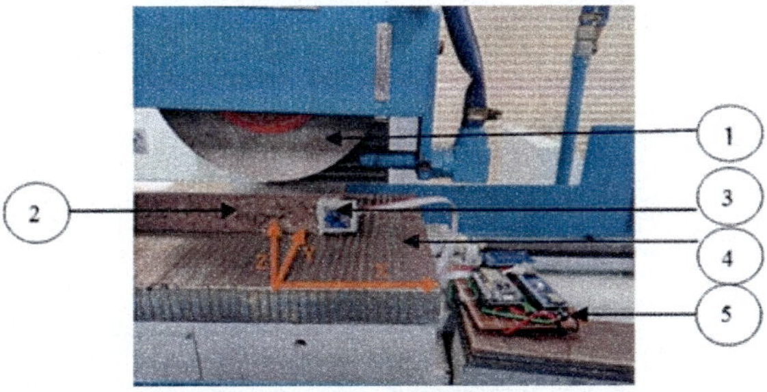

Fig. 1 Condition monitoring of surface grinding process using low-cost acceleration/vibration sensor setup, (1) grinding wheel (carborundum); (2) mild steel workpiece; (3) reciprocating table with electromagnetic platform; (4) ADXL-335 sensor; (5) data acquisition and logging system

3 Low-Cost Vibration and Acceleration Sensor Module

The low-cost vibration and acceleration sensor module comprises of an ADXL-335 acceleration sensor, Node MCU (ESP8266), Arduino Nano, data loggers and data storage cards (SD card), as shown in Fig. 2. All the components are installed in the zero PCB board along with 7.4 battery which is behind the PCB board, and the sensor is fixed over a 3D printing tiny box-shaped structure. The SD card is inserted inside the data logger for the collection of data. It is low-cost and a portable setup, which also gives possibilities to upload the sensor data to the cloud.

Arduino Nano is a low-cost small microcontroller board based on ATmega328p (Arduino Nano V3.x) which is compatible and flexible in working with different sensors [1, 10]. It has the same functionality as Arduino Uno; the only difference is small in size. There is no prior requirement of any device to use Arduino Nano, and all you need is Arduino IDE software along with mini USB cable for transfer of program between devices which runs both online and offline. Node MCU is an ESP module along with ESP8266 chip which contains 32-bit LX 106 microprocessor. The operating frequency of the system is 80–160 MHz of which microprocessor supports real-time operating system (RTOS). Node MCU contains 128 KB RAM and 4 MB of flash memory. It is suitable to be used for IoT applications. A block diagram of

Fig. 2 Low-cost acceleration and vibration sensor setup

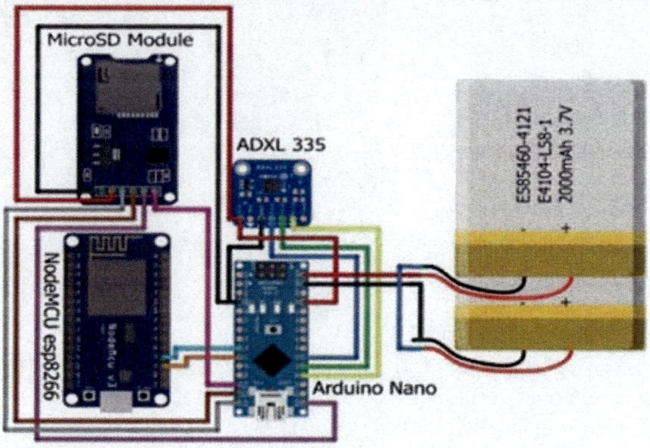

Fig. 3 Connection between different sub-modules used in the low-cost sensor setup

the connection between different elements is shown in Fig. 3. There is a switch that is attached on the side of the data logger for the power connection.

3.1 Calibration and Data Transformations

ADXL-335 acceleration sensor is an analog sensor, but the values are changed to digital form with the help of Arduino Nano. The raw data collected from *X*-, *Y*- and *Z*-directions are in voltage form which can be converted to other meaningful forms [10]. ADXL-335 data sheet shows all the important technical details about the accelerometer. The acquired signals are digital values (V_r) in mV. We are using Arduino Nano in our experiment with 10-bit ADC ($2^{10} = 1023$).

Acceleration in $g = \{((V_r \times V_s)/1023) - (V_s/2)\}/\text{sensitivity}$ [10].

Sensor parameters	Specification
ADXL-335 tri-axial	Analog sensor
Sensitivity (mV/g)	330
Sampling rate (Hz)	10
Voltage supply (V_s)	3.3 V
Measurement range	± 3 g (gravitational constant)

4 Case Studies

Surface grinding processes are carried out for two materials, i.e. mild steel and aluminium. Two grinding parameters such as depth of cut and table speed are varied during the experimentation. The grinding wheel speed (2840 rpm) is constant during the whole experiment. Some experiments are conducted when the condition wheel is good. During the later experiments, the grinding wheel was in worn condition (rough or irregular). ADXL sensor is mounted on the workpiece, and the vibration data are stored in the data logger after the signal processing through the Arduino Nano. There are possibilities to store the data on the cloud, but for this work, data are stored into SD card only (Fig. 4).

Experimental details of the grinding experiments: Mild steel case

Table Speed: 4.81, 5.15 and 5.53 m/min,
Grinding Wheel: Carborundum grinding wheel, AA60 K5 V8 white wheel, Size 200 mm × 20 mm × 31.75 mm,
Depth of Cut: 30, 40, 50 and 60 (μm) (Table 1).

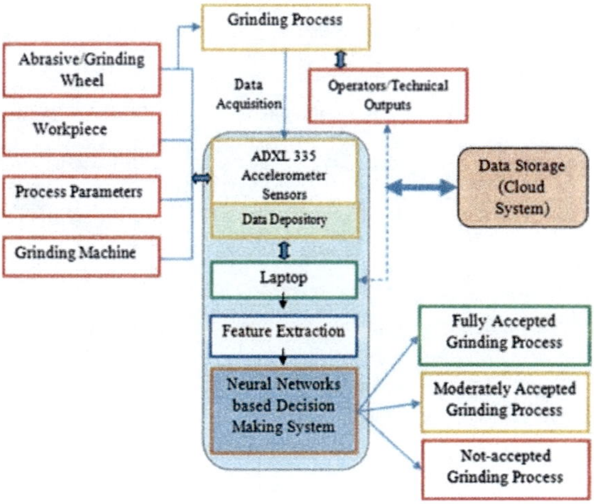

Fig. 4 Block diagram representing the grinding process condition monitoring system

Table 1 Examples showing three different process conditions along with the obtained surface finish

Depth of cut (μm)	Table speed (m/min)	Workpiece material	Wheel speed (rpm)	Process condition (surface finish shown through camera picture)
30	4.81	Mild steel	2840	
40	5.15	Mild steel	2840	
50	5.53	Mild steel	2840	

Experimental details of the grinding experiments: Aluminium 6061 case

Table Speed: 4.81, 5.15 and 5.53 m/min,
Grinding Wheel: Carborundum grinding wheel, AA60 K5 V8 white wheel, size 200 mm × 20 mm × 31.75 mm,
Depth of Cut: 10, 20, 30 and 40 (μm).

A total of 48 experiments have been performed for each case study. Seventy percent data were used to train the NNs, and rest 30% data were used for validation and testing. Pattern recognition NN model is developed for the predictions; the NN configuration is shown in Fig. 5. In initial grinding experiments, the grinding wheel is fresh. But in later experiments (grinding passes), the wheel is in worn condition. For example, three data samples obtained in different experiments are shown in Table 2. Acceptance (output) of the process conditions is decided on the basis of obtained surface finish, operator's recommendation regarding the process stability and sound generated during the grinding.

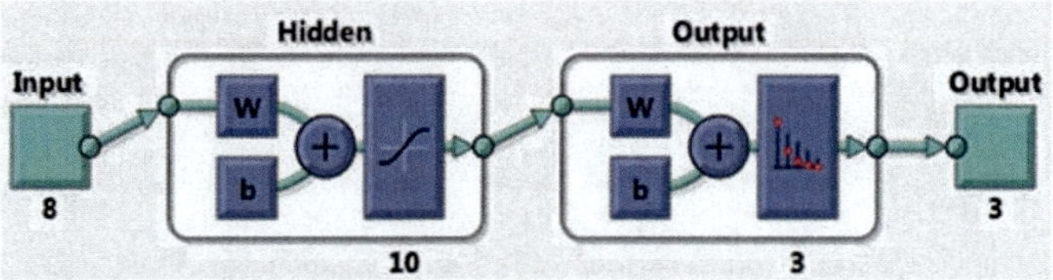

Fig. 5 Pattern recognition neural network configuration

Table 2 Example of some data samples utilized to train the NNs

Data sample (inputs)								Data sample (outputs)
X-axis vibration signal (μ_x)	*X*-axis vibration signal (σ_x)	*Y*-axis vibration signal (μ_y)	*Y*-axis vibration signal (σ_y)	*Z*-axis vibration signal (μ_z)	*Z*-axis vibration signal (σ_z)	TS (m/min)	DoC (μm)	Output label
331.45	0.66	334.95	0.724	400.551	1.355	4.81	30	Accepted (2)
331.24	1.80	334.92	1.69	399.61	3.36	4.81	50	Moderately accepted (1)
329.52	13.89	342.94	3.88	416.58	18.344	5.15	60	Not-accepted (0)

Note Where μ_x, μ_y and μ_z represent the mean of the sensorial signals in *x*-, *y*- and *z*-directions, respectively. σ_x, σ_y and σ_z represent the standard deviation of the sensorial signals in *x*-, *y*- and *z*-directions, respectively
TS Table speed, *DOC* depth of cut
Accepted, moderately accepted and not-accepted outputs are coded as 2, 1 and 0.

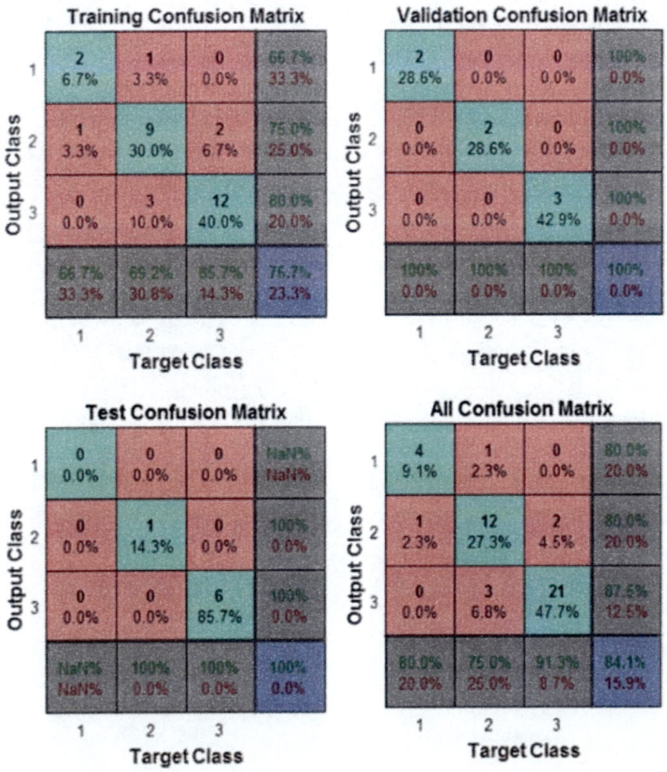

Fig. 6 Confusion plot obtained through the NNs during the grinding of aluminium workpiece

5 Results and Discussions

After taking the data from the data logger, pattern recognition techniques are applied for classification of data into three categories—accepted, moderately Accepted and not-accepted cases. Figures 6 and 7 show the results in the form of confusion matrix for aluminium and mild steel material case studies, respectively. Wherein the achieved success rates are about 84.1 and 86.6%.

6 Conclusions and Remarks

Day by day, low-cost sensor modules are gradually being advanced towards its applicability, efficiency, availability; even in some studies, these low-cost sensors have shown its remarkable usability in industrial applications as well. This study demonstrates the utility of low-cost acceleration sensor for grinding process condition monitoring. The accelerometer provides the *x*-, *y*- and *z*-axis vibration signals

Fig. 7 Confusion plot obtained through the NNs during the grinding of mild steel workpiece

generated during the surface grinding process. And using the real-time acceleration data, condition monitoring is performed by a pattern recognition-based NN model. A low-cost acceleration sensor setup includes an ADXL-335 acceleration sensor, Node MCU (ESP8266), Arduino Nano, data loggers and data storage cards (SD card). Its portability and connection reliability are ensured through a compatible construction, and a suitable frame is designed so that the sensor can easily/reliably be placed at vibration sensitive areas. In presented case studies, the success rate of the developed system is quite good. The research explores that even if the sensitivity/sampling rates of these sensors are inferior to the industrial sensors, but with the help of cognitive/computational support. These sensors can be capable of providing compromised but acceptable solutions for many industrial applications.

References

1. Oliveira, J. F. G. D., Silva, E. J. D., Guo, C., & Hashimoto, F. (2009). Industrial challenges in grinding. *CIRP Annals, 58*(2), 663–680.

2. Arun, A., Ramesh Kumar, K., Unnikrishnan, D., & Sumesh, A. (2018). Tool condition monitoring of cylindrical grinding process using acoustic emission sensor. *Materials Today: Proceedings, 5*(5), 11888–11899.
3. Lin, Y. K., Wu, B. F., & Chen, C. M. (2018). Characterization of grinding wheel condition by acoustic emission signals. In *2018 International Conference on System Science and Engineering (ICSSE)* (pp. 1–6). IEEE.
4. D'Addona, D. M., Matarazzo, D., de Aguiar, P. R., Bianchi, E. C., & Martins, C. H. R. (2016). Neural networks tool condition monitoring in single-point dressing operations. *Procedia CIRP, 41*, 431–436.
5. Cheng, C., Li, J., Liu, Y., Nie, M., & Wang, W. (2019). Deep convolutional neural network-based in-process tool condition monitoring in abrasive belt grinding. *Computers in Industry, 106*, 1–13.
6. Devendiran, S., & Manivannan, K. (2013). Condition monitoring on grinding wheel wear using wavelet analysis and decision tree C4.5 algorithm. *International Journal of Engineering and Technology, 5*(5), 4010–4024.
7. Hsu, C. N., Lin, Y. C., Yang, C. C., Tsai, H. Y., Huang, K. C., Tseng, S. F., & Hsiao, W. T. (2019). Low-cost vibration and acceleration sensors module for the drilling processes monitoring. In *2019 IEEE Sensors Applications Symposium (SAS)* (pp. 1–5).
8. Baban, M., Baban, C. F., & Moisi, B. (2018). A fuzzy logic-based approach for predictive maintenance of grinding wheels of automated grinding lines. In *2018 23rd International Conference on Methods & Models in Automation & Robotics (MMAR)* (pp. 483–486).
9. Moia, D. F. G., Thomazella, I. H., Aguiar, P. R., Bianchi, E. C., Martins, C. H. R., & Marchi, M. (2015). Tool condition monitoring of aluminum oxide grinding wheel in dressing operation using acoustic emission and neural networks. *Journal of the Brazilian Society of Mechanical Sciences and Engineering, 37*(2), 627–640.
10. Alhassan, A. (2019). Re: Conversion of ADXL 335 raw data into g? Retrieved from: https://www.researchgate.net/post/conversion_of_ADXL_335_raw_data_into_g/5cc1de964f3a3e0d8b1823f3/citation/download.

Biocompatibility Enhancement of Magnesium Alloys via Surface Modification Method: A Review

Mahesh Choudhary, Jinesh Kumar Jain, Toshit Jain, Rajeev Agrawal, and Sundeep Kumar

1 Introduction

1.1 A Subsection Sample

Bone defect repair implants that would resorb after completion of the healing process is still a challenging task in the orthopaedic clinic [1, 2]. A perfect resorbable implants would minimize the chances of secondary operation required to remove the defected implant from the body and hence prevent from the cost of multiple operations [3–5]. For an ideal bone regeneration environment suitable bone substitutes and fixation devices are required [6–8]. A very popular standard in routine clinics is autogenous bone graft in which, the bone is obtained from the same patient, but donor site complication donor shortage and secondary operation are some of the shortcomings associated with autogenous bone graft [9, 10]. Another popular approach for bone defect repair is an allogeneic bone graft in which the bone tissues are obtained from a deceased donor. Synthetic or natural biomaterials such as metals, ceramics, polymers are widely used in bone defect repair and bone fracture fixation and bone defect repair in both orthopaedic and dentistry [6, 10, 11]. A perfect resorbable material implant should resorb after complete healing of bone, i.e. not earlier than 12 weeks. The resorbable material implant should not create any adverse effect on the biological function after degradation and should not be toxic, i.e. resorbable material should be biocompatible with the biological fluid and helps in gone regeneration and cell

M. Choudhary · J. K. Jain (✉) · T. Jain · R. Agrawal
Department of Mechanical Engineering, Malaviya National Institute of Technology, Jaipur 302017, India
e-mail: Jineshjain.mech@mnit.ac.in

S. Kumar
Department of Technical Education, Center for Electronic Governance,, Government of Rajasthan, Jaipur 302004, India

R. Agrawal et al. (eds.), *Recent Advances in Smart Manufacturing and Materials*,
Lecture Notes in Mechanical Engineering,
https://doi.org/10.1007/978-981-16-3033-0_40

growth. For a resorbable material as a load-bearing implant, it should provide sufficient strength to healing bone [12]. Magnesium is a suitable resorbable material implant because of its biocompatibility and non-toxicity and sufficient mechanical strength (40–55 GPa) adjacent to natural bone (10–20 GPa). Magnesium is the naturally occurring element in the human body, our body contains 25 gm magnesium, and about 50% to 60% is found in human bone. The stress shielding due to implants of higher mechanical strength (Young's modulus) in comparison with the bone, can be avoided in magnesium resorbable implant materials [13], however, the problem with magnesium is its fast degradation rate which will release hydrogen gas on contact with fluids [8, 14]. In water, magnesium hydroxide accumulates on the surface of the magnesium implant to form a corrosion layer, also known as a degradation layer. While this film slows corrosion under aqueous conditions, it reacts with chlorine ions present in the blood to produce highly soluble $MgCl_2$ and hydrogen gas. Hydrogen gas affects primary implant stability in bone by interferes with normal tissue healing [14]. Due to the corrosion of magnesium, the release of magnesium ions will increase the pH level, for the proper function of biological function the pH level should be appropriate [15, 16].

2 Biodegradable Magnesium Implant

The metallic biomaterials are used to provide structural support and replace the broken or damaged human bone, so the biomaterials should possess sufficient mechanical strength and Young's modulus adjacent to the human bone and it should not release any toxic elements by reaction with living cells or body fluid. Magnesium and its alloys are the advancements in biomaterials because magnesium is the part of the bone composition and help in metabolism. Magnesium is the second most abundant element next to Na in the hydrosphere and the fourth most abundant cation in the living organism (human body: Ca > K > Na > Mg) [17]. Also, the physical and mechanical properties of magnesium are similar to human bone. Other metals such as titanium and its alloy (Ti_6Al_4V), stainless steel (316L) [18], Chromium and its alloy and other metals and their alloy are widely used as biomaterials and commercially available in the market, but the problem with this alloy is their high Young's modulus (> 100 GPa) leads to stress shielding, stress shielding is a phenomenon in which the bone-implant mechanical strength is more than bone's strength and the more load is bear by the implant and the bone in contact with implant losses its density and leads to wear due to less load is exposed to the adjacent bone. Whereas the mechanical properties of magnesium are similar to the bone, i.e. the Young's modulus of bone is 10–27 GPa, whereas Young's modulus for magnesium is approximately 42 GPa which is adjacent to bones and the problem of stress shielding can be avoided. Another reason which makes magnesium as new generation biomaterial is its biodegradability, i.e. the magnesium will degrade itself with time and there is no need to secondary operation to remove the implant from the body, whereas in case of other metals the degradation rate is very low and due to release of some toxic

Table 1 Comparison of the mechanical and physical properties of the natural bone and various biomaterials

Material	Density (g/cm^3)	Elastic modulus (GPa)	Yield strength (MPa)	Fracture toughness ($MPa\ m^{1/2}$)	References
Human cortical bone	1.80–2.10	3–20	130–193	3–6	[14]
Mg	1.74–2.00	41–45	65–100	15–40	[21]
AZ91	1.81	45	160	N/A	[22]
WE43	1.84	44	170	N/A	[23]
Mg–6Zn	N/A	42.3	169.5	N/A	[24, 25]
Mg–1Ca–Zn	N/A	45.3	67	N/A	[7, 26]
Ti_6Al_4V	4.40	115	900	N/A	[17]
Ti alloy	4.40–4.50	110–117	758–117	55–115	[18, 27]
XLPE	0.47–1.26	0.005–0.69	20	N/A	[28]
Co–Cr alloy	8.3–9.20	230	450–1000	N/A	[29, 30]
Synthetic-HA	3.10	73–117	600	0.7	[31]

elements after some time approximately twelve months to some year due to reaction with a living organism. But the problem with magnesium is its very fast degradation rate (4–6 week). Magnesium is biocompatible and osteoclast in nature which enhance the new bone generation and cell growth on the artificial implant, the new bone generation rate on the implant is less than the degradable rate of magnesium which is a serious problem related to magnesium and its alloy as biomaterials and a lot of research is going on this topic.

Magnesium and its alloys are widely used as biodegradable orthopaedic implant due to its mechanical property closer those of natural bone. Table 1 summarises the physical and mechanical properties of different biomaterials and natural bones. A comparison is trying to show in this table to better understand the various physical and mechanical properties, similarities and difference between biomaterials and natural bone.

2.1 The Reactions Involved in the Corrosion of Pure Magnesium

Corrosion of magnesium and its alloy are mainly influenced by the characteristics by the surface film. The reaction involved in the corrosion of pure magnesium in contact with water is showing through the equation. [19, 20]

$$Mg \rightarrow Mg^{2+} + 2e^{-}(\text{anodic reaction}) \tag{1}$$

$$2H_2O + 2e^- \rightarrow H_2(\text{anodic reaction}) \quad (2)$$

$$Mg^{2+} + 2(OH)^- \rightarrow Mg(OH)_2(\text{product formation}) \quad (3)$$

$$Mg^{2+} + 2Cl^- \rightarrow MgCl_2 \quad (4)$$

The main oxide layer which is not compact and present on the surface of Mg is $Mg(OH)_2$. Hence pitting corrosion on the alloy surface due to solution infiltrate the film. The thermodynamic equilibrium of Eq. (3) can be break by chloride ions, which is the dynamic equilibrium through the reaction (4). The increase in chloride ion concentration results in the formation of $MgCl_2$. Therefore, the corrosion of Mg alloys exits the formation and dissolution of oxide layer containing $Mg(OH)_2$. The in vitro corrosion properties are tested by immersion and electrochemical tests, whereas animal models are used to assessed vivo corrosion properties.

2.2 *Problem Due to Mismatch Mechanical Properties of Adjacent Bone and Implant Materials*

For a load-bearing implant an appropriate strength and stability are required, but when there is a mismatch between the elastic moduli of adjacent bone and the scaffold materials Young's moduli than stress shielding will occur and make the operation unsuccessful or failure of the operation. Stress shielding is phenomena which occur when there is a mismatch between the adjacent bone and scaffold materials elastic moduli, due to high-mechanical strength (elastic moduli) of implant material more load with being transferred to the implant and less to the adjacent bone due to this the adjacent bone will become less dense and bone bearing capacity of the bone will decrease. When fatigue load is subjected to the adjacent bone the bone may break and result in the failure of the operation, this phenomenon is called stress shielding. To prevent stress shielding the mechanical strength of adjacent bone and the implant materials should be closer to each other. Amongst all possible biomaterials, magnesium is a better biomaterial and consideration of attention due to its mechanical strength (40–45 GPa) closer to adjacent bone(3–20 GPa). Whereas the mechanical strength of titanium and steel is greater than 100 GPa which will create stress shielding. During the last decade's researchers, interest is increasing towards magnesium due to its potential biodegradation, which avoids secondary operation and appropriate mechanical strength which prevent stress shielding.

To reduce the mismatch of mechanical strength of materials with adjacent bone researchers have developed another technique to reduce the difference of mechanical strength mismatch. Researchers have made porous scaffold to overcome the stiffness mismatch. The porous implant enhances the biocompatibility by increasing bone regeneration on the scaffold surface, cell growth and fixation of bone in their place, but due to increase in porosity the surface area of the scaffold will increase and the density of scaffold will decrease which results in the decrease in fatigue strength of the scaffold. The mechanical properties were effective by the change in pore size pore volume and number of pores. By increasing porosity compression strength, fatigue strength, Young's modulus and yield strength of the scaffold will decrease.

3 Biocompatibility Enhancement Technique of Magnesium

Magnesium is the biodegradable biomaterials and the essential elements of the human body. Magnesium is used as an orthopaedic implant due to its mechanical properties is close to the mechanical properties of natural bone. Due to corrosion, the mechanical strength and fatigue strength of the implant will reduce and make the operation fail. The problem associated with the use of pure magnesium as an implant is due to its fast degradation rate and excess release of magnesium ions which will increase the pH level of the surrounding environment and may damage the host tissues. To improve the corrosion resistance of the magnesium, and enhance the biocompatibility various magnesium alloys are used as bio-implants which does not create any adverse event on the biological fluid and help in cell proliferation and bone regeneration. Another technique to improve the biocompatibility or to improve corrosion resistance is to use of surface modification method, i.e. coating of different materials on magnesium scaffold because corrosion is governed by the surface characteristics.

The two different surface modification processing techniques of the deposited coating or conversion coating are used to increase the corrosion resistance of the implant material, and the surface modification is done before the implant is inserted into the body. The materials used for coating must be corrosion protection nontoxic biocompatible and Osseointegration and control degradability. The processing techniques in which the electrochemical interaction or in situ interaction of magnesium with environment and results in the formation of inorganic ceramic coating on the magnesium scaffold surface is called conversion coatings. Different methods such as spraying, painting, dip coating, spin coating or immersion materials coated on the magnesium scaffold are mostly organic coating called deposited coating.

4 Coating of Different Bioactive Materials and Their Results

Hydroxyapatite similar chemical and structural nature as bone and due to its exceptional bioactivity, biocompatibility property it is widely used as biomaterials implants. The corrosion resistance of magnesium scaffold is enhanced by a coating of hydroxyapatite. Different coating techniques such as electrophoretic deposition, biomimetic deposits, sol–gel process and the sputtering process can be used to coat

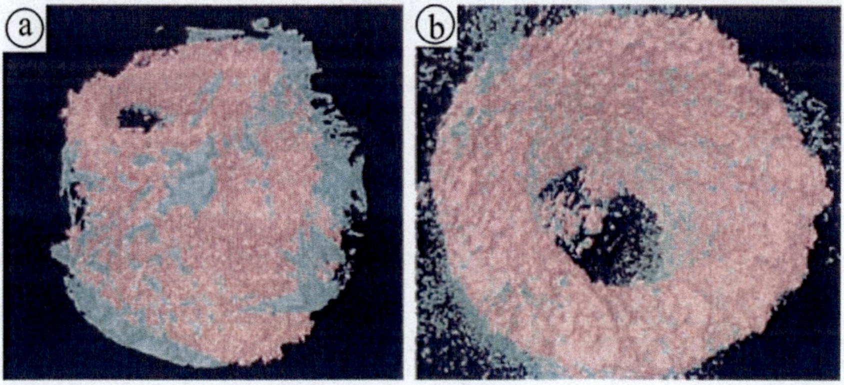

Fig. 1 Micro-CT image of **a** magnesium scaffold and **b** hydroxyapatite scaffold as a control group. The new bone tissue in-growth is shown in green colour [32]

hydroxyapatite materials and to enhance cell proliferation cell growth and bone regeneration on the magnesium hydroxyapatite implant. Since a high temperature is required to densify the coating on the magnesium substrate and magnesium is poor heat resistant material, none of these technologies will give an effective result. Therefore, electrochemical deposition techniques are used to surface modification of magnesium scaffold in electrochemical deposition low temperature is required to control and uniform thickness and chemical composition coating of hydroxyapatite. Hydroxyapatite brittleness and low strength limit are its application in the load-bearing application. Loose coating of hydroxyapatite helps in cell adhesion and accelerate the healing process by new bone formation on the implant. In an, in vivo assessment, the biocompatible behaviour of pure magnesium scaffold and hydroxyapatite are studied, and it was observed that the new bone formation was observed on both the scaffold which is shown in Fig. 1. The new bone formation is shown in the green colour.

Fluorine is also used as corrosion-resistant materials to enhance bioactivity and biocompatibility of magnesium scaffold. Fluorine is an essential element of teeth and the human skeleton. Li et al. reported review where enhancement to the corrosion resistance and biocompatibility of magnesium alloy by using a vacuum evaporation deposition technique to coat Mg–1Ca alloy with MgF_2 is been done [19]. It was observed that the corrosion resistance of Mg–1Ca was increased and cell adheres to the implant surface hence enhance biocompatibility. Thus, the experiment demonstrates that the corrosion resistance of Mg–1Ca was increased by surface modification coating of MgF_2 and enhanced biocompatibility was also observed.

Calcium phosphates (CaP) are also used as surface modification materials to make a ceramic coating on the magnesium substrate to increase corrosion resistance and new bone formation cell proliferation and cell growth on the coated magnesium implant in contact with the host tissue. One of the different phases of calcium phosphates is hydroxyapatite, both are the essential part of human bone. The effect of

Table 2 Surface modification by organic and polymer coating on magnesium scaffold and their results

Methodology	Coating material	Substrate material	Result	References
Hydrothermal treatment	$CH_3(CH_2)16COOH$	Mg	Corrosion resistance enhance four order magnitude, bone growth	[31]
Dip-coating technique	nHAp/PCL	Magnesium alloy (AM50)	Good corrosion resistance in SBF medium, enhancement of cell adhesion, proliferation and rapid CaP layer formation	[33]
Dip-coating technique	Polyvinyl acetate (PVAc)	AM50	The corrosion rate of the coated alloy is stabilised	[34]
Micro-arc technique	MAO/PLLA	WE42	Coating layer prevents WE42 from corrosion and decrease the corrosion rate	[29]
Direct deposition	PCL and dichloromethane membrane	AZ91	Produce several pores to reduce the corrosion rate of magnesium alloy implant	[35]

surface modification by organic and polymer coating on the magnesium and its alloy is shown in Table 2. The coating of biocompatible organic and polymer coating results in increase corrosion resistance and new bone formation on the substrate was observed hence result in enhancement of biocompatibility of magnesium scaffold.

5 Conclusion

Bone fracture caused by accident or diseases which are too complex to treat by external medical treatment and surgical operations are required to replace, repair or provide structural support to damaged bones. The traditional method used inert permanent materials as medical implants of titanium, steel or other metals or ceramic or polymer, but these materials required a secondary operation to remove these implant after some years of operation. Hence to avoid the secondary operation or to prevent the cost of multiple operation degradable implants are the new scope of research. Magnesium is used as biodegradable biomaterials, but the problem with its fast degradation rate which increases the pH level and releases excess hydrogen gas under the skin which may damage the host tissue, therefore to increase the corrosion resistance and biocompatibility of magnesium and its alloy surface modification techniques are used to deposit a coating of different bioactive and biocomposite materials to enhance cell adhesion and new bone formation on coated implant surface in contact with host tissue. The coating materials should not create any adverse effect on the body fluid and host tissues.

References

1. Yang, H., et al. (2020). Alloying design of biodegradable zinc as promising bone implants for load-bearing applications. *Nature Communications, 11*(1), 401.
2. Traverson, M., et al. (2018). In vivo evaluation of biodegradability and biocompatibility of $Fe_{30}Mn$ alloy, *31*(01), 010–016.
3. Kraus, T., Fischerauer, S. F., Hänzi, A. C., Uggowitzer, P. J., Löffler, J. F. Weinberg, A. M. (2012). Magnesium alloys for temporary implants in osteosynthesis: In vivo studies of their degradation and interaction with bone, *8*(3), 1230–1238.
4. Denkena, B., Lucas, A., Thorey, F., Waizy, H., Angrisani, N., & Meyer-Lindenberg, A. (2011). Biocompatible magnesium alloys as degradable implant materials-Machining induced surface and subsurface properties and implant performance. In *Special issues on magnesium alloys*. IntechOpen.
5. Charyeva, O., et al. (2016). Biocompatibility of magnesium implants in primary human reaming debris-derived cells stem cells in vitro, *17*(1), 63–73.
6. Sheikh, Z., Najeeb, S., Khurshid, Z., Verma, V., Rashid, H., & Glogauer, M. (2015). Biodegradable materials for bone repair and tissue engineering applications. *Materials (Basel), 8*(9), 5744–5794.
7. Zhang, Y., et al. (2019). Investigation on the microstructure, mechanical properties, in vitro degradation behavior and biocompatibility of newly developed Zn-0.8% Li-(Mg, Ag) alloys for guided bone regeneration, *99*, 1021–1034.
8. Hasan, M. L., Kim, B., Padalhin, A. R., Faruq, O., Sultana, T., & Lee, B. T. (2019). In vitro and in vivo evaluation of bioglass microspheres incorporated brushite cement for bone regeneration. *Materials Science & Engineering, C: Materials for Biological Applications, 103*, 109775.
9. Zhao, D. et al. (2016). Vascularized bone grafting fixed by biodegradable magnesium screw for treating osteonecrosis of the femoral head, *81*, 84–92.
10. Araújo, P. P., Oliveira, K. P., Montenegro, S. C., Carreiro, A. F., Silva, J. S., & Germano, A. R. (2013). Block allograft for reconstruction of alveolar bone ridge in implantology: A systematic review, *22*(3), 304–308.
11. Sterio, T. W., Katancik, J. A., Blanchard, S. B., Xenoudi, P., Mealey, B. L., & Dentistry, R. (2013). A prospective, multicenter study of bovine pericardium membrane with cancellous particulate allograft for localized alveolar ridge augmentation, *33*(4).
12. Zhang, R., et al. (2015). Silver nanoparticles promote osteogenesis of mesenchymal stem cells and improve bone fracture healing in osteogenesis mechanism mouse model, *11*(8), 1949–1959.
13. Emily Walker, M. H. (2015). Magnesium, iron and zinc alloys, the trifecta of bioresorbable orthopaedic and vascular implantation—A review. *Journal of Biotechnology & Biomaterials, 05*(02).
14. Kruppke, B., et al. (2016). Gelatine modified monetite as a bone substitute material: An in vitro assessment of bone biocompatibility. *Acta Biomaterialia, 32*, 275–285.

15. Dorst, K., Rammelkamp, D., Hadjiargyrou, M., & Meng, Y. (2014). The effect of exogenous zinc concentration on the responsiveness of MC3T3-E1 Pre-Osteoblasts to Surface Microtopography: Part II (Differentiation). *Materials (Basel), 7*(2), 1097–1112.
16. Gambling, L., & McArdle, H. J. (2004). Iron, copper and fetal development. *Proceedings of Nutrition Society, 63*(4), 553–62.
17. Putra, N. E., Mirzaali, M. J., Apachitei, I., Zhou, J., & Zadpoor, A. A. (2020). Multimaterial additive manufacturing technologies for Ti-, Mg-, and Fe-based biomaterials for bone substitution. *Acta Biomaterialia, 109*, 1–20.
18. Khosravi, F., et al. (2020). Development of a highly proliferated bilayer coating on 316L stainless steel implants. *Polymers (Basel), 12*(5).
19. Li, L., Zhang, M., Li, Y., Zhao, J., Qin, L., & Lai, Y. (2017). Corrosion and biocompatibility improvement of magnesium-based alloys as bone implant materials: A review, *4*(2), 129–137.
20. Virtanen, S., & Fabry, B. (2011). Corrosion, surface modification, and biocompatibility of Mg and Mg alloys. In *Magnesium Technology 2011* (pp. 409–412). Springer.
21. Lee, J.-W., et al. (2016). Long-term clinical study and multiscale analysis of in vivo biodegradation mechanism of Mg alloy, *113*(3), 716–721.
22. Prasad, K., et al. (2017). Metallic biomaterials: Current challenges and opportunities, *10*(8), 884.
23. Schinhammer, M., Hanzi, A. C., Loffler, J. F., & Uggowitzer, P. J. (2010). Design strategy for biodegradable Fe-based alloys for medical applications. *Acta Biomaterialia, 6*(5), 1705–1713.
24. Xiao, C., et al. (2018). Indirectly extruded biodegradable Zn-0.05 wt% Mg alloy with improved strength and ductility: In vitro and in vivo studies, *34*(9), 1618–1627.
25. Murni, N., Dambatta, M., Yeap, S., Froemming, G., & Hermawan, H. (2015). Cytotoxicity evaluation of biodegradable Zn–3Mg alloy toward normal human osteoblast cells, *49*, 560–566.
26. Pan, S., An, L., Meng, X., Li, L., Ren, F., & Guan, Y. (2017). $MgCl_2$ and $ZnCl_2$ promote human umbilical vein endothelial cell migration and invasion and stimulate epithelial-mesenchymal transition via the Wnt/beta-catenin pathway. *Experimental and Therapeutic Medicine, 14*(5), 4663–4670.
27. Datta, S., Raza, M. S., Saha, P., & Pratihar, D. K. (2019). Effects of process parameters on the quality aspects of weld-bead in laser welding of NiTinol sheets. *Materials and Manufacturing Processes, 34*(6), 648–659.
28. Shuai, C., Li, S., Peng, S., Feng, P., Lai, Y., & Gao, C. (2019). Biodegradable metallic bone implants. *Materials Chemistry Frontiers, 3*(4), 544–562.
29. Wang, Z. Q., et al. (2018). Effect of Ni content on the microstructure and mechanical properties of weld metal with both-side submerged arc welding technique. *Materials Characterization, 138*, 67–77.
30. Yang, H., et al. (2018). Enhanced osseointegration of Zn–Mg composites by tuning the release of Zn ions with sacrificial Mg-rich anode design, *5*(2), 453–467.
31. Bohner, M., Santoni, B. L. G., & Dobelin, N. (2020). beta-tricalcium phosphate for bone substitution: Synthesis and properties. *Acta Biomaterialia, 113*, 23–41.
32. Liu, Y., Yang, Z., Tan, L., Li, H., & Zhang, Y. Z. (2014). An animal experimental study of porous magnesium scaffold degradation and osteogenesis, *47*(8), 715–720.
33. Wu, L., et al. (2017). Dip-coating process engineering and performance optimization for three-state electrochromic devices, *12*(1), 1–15.
34. Thongsuriwong, K., Amornpitoksuk, P., & Suwanboon, S. (2013). Structure, morphology, photocatalytic and antibacterial activities of ZnO thin films prepared by sol–gel dip-coating method, *24*(1), 275–280.
35. Orinakova, R., et al. (2015). Study of electrochemical deposition and degradation of hydroxyapatite coated iron biomaterials, *10*, 659–670.

Sustainable Natural Bio-composites and its Applications

Ankit and Moti Lal Rinawa

1 Introduction

In recent years, a strong awareness of emerging goods from green energy has been raised by customers. Green ads, new recycling directives, social effects, and cognitive transition have driven customers to eco-friendly goods. Both Synthetic fibers and Natural fabrics have been appealing to researchers and have been tremendously developed/Studied [1–3]. Recently the focus of study on natural fibers and their applications has generated tremendous interest due to environmental considerations and rising synthetic material expense. In particular, because of their similar properties in certain instances in glass, polymer, carbon fibers, etc. the study of lingo cellulosic fibers attracted researchers [4]. Today the production of renewables, biodegradables, and environmental goods which are better affordable from natural sources is becoming increasingly environmental conscious. It was discovered that natural fibers, also possible to substitute synthetic fibers, can be used in many engineering applications. Natural fibers are commonly referred to as vegetable fibers in the composites industry, though. Natural fibers like sisal, hemp, and Jute were used as reinforcing materials, but for most high-end applications, in particular in the tireless and high-temperature settings, they lack insufficient strength [5]. Composites have been created and revamped, in particular, to strengthen, evolve and sustainably, and safely incorporate traditional goods. Other qualities, including high water absorption and low polymer matrix compatibility, made it impractical for all technological applications. Furthermore, synthetic resin systems have strengthened lignocellulosic fibers, which impair the biodegradability of the material. Bio-based materials remain a significant focus of current research in the composite industry, considering all the challenges that remain, in large part due to the rise in synthetic raw materials and the

Ankit (✉) · M. L. Rinawa
Department of Mechanical Engineering, Government Engineering College Jhalawar, Jhalawar, Rajasthan, India

R. Agrawal et al. (eds.), *Recent Advances in Smart Manufacturing and Materials*,
Lecture Notes in Mechanical Engineering,
https://doi.org/10.1007/978-981-16-3033-0_41

decline in vehicle, aerospace (indoor), civil, and textile resources costs [6]. Study into natural fibers-enhanced composites has gained more attention in recent years. The main objective of the research was to incorporate plant fibers as a reinforcing agent in composites to an environmentally sustainable environment. Compounds reinforced with natural plant fibers can be used as parts for reinforced glass-fibers composites in structural applications, and the car industry and construction industries have taken a keen interest [7]. Biocomposite materials, which are made up of natural fibers reinforced by polymer matrix, are considered composite materials. In the automotive field with different fillers and/or reinforcements, polymer composites are widely used. Due to the benefits of minimizing the total lower weight of a vehicle and improving the sustainability of production processes, several such compounds have been developed for internal and outside use in recent years. The wide-ranging manufacturing of composites of natural fibers poses the problem of poor adhesion between fibers and matrices and the presence of fibers. The key features of CO_2 include avoiding, reusing, limiting, and minimizing. In last decades, textiles were manufactured from natural fibers from daily textiles to patriotic and ceremonial [8]. In recent years, the awareness of the need for sustainable production has increased, which has raised concern to use natural fibers as a reinforcement for synthetic fibers in polymer composites. Weight saving benefits may also be reached by the substitution of existing composites with fabrics with natural composites. A viable means for mitigating environmental effects and for promoting the sustainable growth of the transport sector can be used with less denser recycled natural product by using thermoplastic and thermoset composite materials [9]. The use of natural composites in the shopper's goods has grown greatly in recent years in the growth of industries. The distributed knowledge and variations in mechanical properties are one of the problems of natural fibers. The lack of standards in respect of the methods of gather, packaging, and post-process natural fibers, for both manufacturers and consumers of these product contributes to the difficulty of choosing. In fact, these concerns are crucial deterrents to the widespread use of natural fibers. Brief Study of various mechanical characteristics of natural fibers and their applications is discussed in this paper to fill this void.

This paper includes the following Sections. In Sect. 2, A brief introduction to Natural fibers and bio-composites like bamboo, Sisal, Jute, and many more have been discussed. In Sect. 3, Applications of Natural fibers Bio-Composites in Sustainable Development. Conclusion was eventually addressed.

2 Natural Fibers Bio-composites

Simple-defined natural fibers are fibers that are not manmade or synthetic and are graded according to origin from sources of animals, minerals or plants. Certain natural fibers, including herbs, cotton, and many others, are ready-made. Thus, fiber is collected from various portions of plants that belong to diverse places. The cellulose fibers can be categorized by plants in fiber bases, such as Flax, hemp, jute, leaf fibers, seed fibers (coir or cottons), and all other kinds, as well as fibers making plants with cellulose fibers. Many uses in the raw materials field for example sisal, flax, jute, hemp, and bamboos. Cellulose, lignin, hemicelluloses, pectin, and wax are the major chemical constituents of natural plant fibers [10]. Organic composite materials, where at least one of the elements is derived from natural products, is rejected as composites. In terms of comparatively sustainable materials, the benefits of natural fibers over synthetic fibers consist of quantity, less risk to manufacturing facilities, compact weight, low expense, strong relative mechanical properties such as tensile modulus, bending modulus, and better surface finish for composite components. The bio-composite fabrics from natural fibers blends, such as reinforced polymers and bio-polymers, are non-biologically degradable and are not entirely ecological synthetic fiber-reinforced polymers such as gas and glass. Bio-polymers are naturally-generated polymers composed of monomeric units covalently bound to larger structures. The melting low indices, impact properties, hardness, vapor transfer characteristics, frictional coefficient, and decomposition differ between bio-polymers. Thermoset, thermoplastic, and elastomer can be bio-polymers [11]. Bio-polymers can and will be applied in a wide variety of areas such as automobile sector, medical/pharmacy products, hygiene, and many other sectors [12]. The industry includes many different kinds of polymers. The properties and efficiency of products constructed from composites of natural fibers depend on the methods of manufacturing and the characteristics of their different components and their compatibility and interfacial bonding. Many inconveniences of composites such as higher water absorption, lower resistance to air, and less mechanical character of natural fibers will limit their application as against synthetic fibers. The method will further increase the physical and mechanical properties and the air resistance of the natural fibers, while the moisture absorption by changing the surface and the binding agents can be minimized. A crucial consideration for superior fibers reinforcement material properties is the attachment strength between the composite fibers and the polymer matrix. Coupling agents serve as a barrier to and strengthen communication between polymers and natural fibers. Polyester or polypropylene are the most commonly used polymers, while lax, hemp, and sisal are the natural fibers [13]. Many cars and building industries have taken a great deal of exposure to natural fibers composites. However, the use of natural composites in this industry is motivated by costs, weight reduction, and the marketing rather than technological criteria as depicted in Table 1 [14].

3 Applications in Sustainable Development

There are many sectors like automobile industry, building, electricity, and aerospace, which for example, are forced to make environmentally sustainable goods and to reduce their reliance on fossil fuels by society and governments as shown in Table 2 [15]. In applications which also include automobile, aerospace, maritime, sport,

Table 1 Various mechanical properties of fibres

Fiber	Density (g/cm^3)	Diameter (μm)	Length (μm)	Tensile Strength (N/m^2)
Jute	1.23	15	3.4	480
Sisal	1.2	27	4.4	681
Bamboo	0.85	56.5	2.75	566
Coir	1.2	18.5	1.65	175
Hemp	1.47	30.5	30	845
Flax	1.38	21.5	37.5	689

Table 2 Applications of fibers in sustainable development

Fiber	Applications
Jute	Covered door panels, and seat backs
Sisal	In the interior door linings, panels, and door panels
Bamboo	Modular house construction
Coir	Car seat covers, mattresses, doormats, and rugs
Hemp	Carrier for covered door panels, racing bicycle, and cases for musical instruments
Flax	In the interior door linings, panels, door panels, package trays, door panel inserts, racing bicycle, cases for musical instruments, and green wall panel

and electronics industry, natural fibers reinforced composites evolve very quickly as potential replacements for metal or ceramic-based materials.

3.1 Application in Automobile Sector

Currently, above 50% of the car's interior contains numerous polymeric materials [16]. Fuel performance development and lower emissions are under immense demand from automakers and associations. The best approach is to reduce the total vehicle weight that can be replaced by lightweight composite materials by metal [17]. The application of natural fibers emerges in automobile, in 1940s when H. Ford developed the 1st compound component that used the hemp fibers. The authors took steps to create and use natural sustainable products as part of composites. In interior components such as door panels, dashboard parts, packing racks, seat coiling, and backrests, cable lining are commonly used to manufacture naturally fiber-reinforced compositional composites. External uses are constrained because of the high mechanical strength requirement.

3.2 *Wind Turbine Blade*

Different sources of biomass, petroleum, coal, and today's products are used for electricity production, including coke, natural gas, nuclear material, etc. [18]. When population, civilization, and industrialization grew greatly, energy consumption increased several times. In the scenario of today, this lack of balance in the climate has led to an increasing interest in green and alternative energy sources through increased environmental consciousness and strict national laws and policies. The continuous search for renewable sustainability has contributed to wind energy development with a minimal emission and increased efficiencies, leading to energy reduction. It is a leading source of clean energy for humanity that can contribute to addressing the issues of global energy [19]. The industry of wind energy is increasing in popularity and highly efficient solutions are available that transform kinetic wind energy into mechanical or electrical energy.

3.3 *Structural Applications*

A variety of fibers that have proven their value in the thermal and thermoplastic matrix, such as oil sisal, jute, wheat, flax straw, cotton, silk, and bamboo, have been found to be a strong and efficient enhancement [20]. Natural fiber composites have been used for a number of structural uses because they have a high specific strength and anti-metal modulus. These technologies vary from household applications to more critical and advanced areas such as satellites and aircraft. The use of natural fibers in composite materials is expected to be a growing market [21]. The key arguments for the use of natural fibers are sustainable pricing, combined with growing understanding of environmental problems such as 'renewable energy,' 'recycling' and 'decreasing carbon dioxide emissions,' etc. The challenges for both of these regions can easily be met by natural fibers [22].

4 Conclusions and Future Scope

Natural fibers bio-composites may be used to replace synthetic fibers with environmental and cost advantages. Compound construction composites based on organic fibers (cement, lime, hydraulic lime, gypsum, etc.) are often exposed to wet conditions during their lifespan, such as bamboo, flax, jute, sisal fibers. Technical hemp is becoming a significant subject of interest because of the health benefits of green housing. Bio-composites have various pros on synthetic fiber such as lower coefficient of friction, environmentally friendly, and lower density. This makes these bio-composites more popular and has already created a tremendous amount of scientific awareness. This paper addresses the various uses of natural fibers and its composites. This study concludes that natural fibers composites are one of the new fields for use in the materials sciences. Practical explanations for environmental efficiency considering bio-materials must therefore be established. This helps artists to make educated opinions and lose time and money without performing exhaustive experimental work. Such approaches which involve designing the more desired nature eristics of components of bio-composites that designers may consider for sustainable designs to increase potential performance.

References

1. Nabi Saheb, D., & Jog, J. P. (1999). Natural fiber polymer composites: A review. *Advanced Polymer Technology*. https://doi.org/10.1002/(SICI)1098-2329(199924)18:4<351::AID-ADV6>3.0.CO;2-X.
2. Faruk, O., Bledzki, A. K., Fink, H. P., & Sain, M. (2012). Biocomposites reinforced with natural fibers: 2000–2010. *Progress in Polymer Science*. https://doi.org/10.1016/j.progpolymsci.2012.04.003
3. Nyström, B. (2007). Natural fiber composites: A review. *Engineering*.
4. Bledzki, A. K., & Gassan, J. (1999). Composites reinforced with cellulose based fibres. *Progress in Polymer Science (Oxford)*. https://doi.org/10.1016/S0079-6700(98)00018-5
5. Fuqua, M. A., Huo, S., & Ulven, C. A. (2012). Natural fiber reinforced composites. *Polymer Reviews*. https://doi.org/10.1080/15583724.2012.705409
6. Sathishkumar, T. P., Navaneethakrishnan, P., Shankar, S., Rajasekar, R., & Rajini, N. (2013). Characterization of natural fiber and composites—A review. *Journal of Reinforced Plastics and Composites*. https://doi.org/10.1177/0731684413495322
7. Sood, M., & Dwivedi, G. (2018). Effect of fiber treatment on flexural properties of natural fiber reinforced composites: A review. *Egyptian Journal of Petroleum*. https://doi.org/10.1016/j.ejpe.2017.11.005
8. Bogoeva-Gaceva, G., et al. (2007). Natural fiber eco-composites. *Polymer Composites*. https://doi.org/10.1002/pc.20270
9. George, J., Sreekala, M. S., & Thomas, S. (2001). A review on interface modification and characterization of natural fiber reinforced plastic composites. *Polymer Engineering & Science*. https://doi.org/10.1002/pen.10846
10. Cellulose fibers: bio- and nano-polymer composites (2011).
11. Herrera-Franco, P. J., & Valadez-González, A. A study of the mechanical properties of short natural-fiber reinforced
12. Babu, R. P., O'Connor, K., & Seeram, R. (2013). Current progress on bio-based polymers and their future trends. *Progress in Biomaterials*. https://doi.org/10.1186/2194-0517-2-8
13. Hossain, S., Rahman, M. M., Jamwal, A., Gupta, P., Thakur, S., & Gupta, S. (2019, September). Processing and characterization of pine epoxy based composites. In *AIP Conference Proceedings, 2148*(1), 030017.
14. Elanchezhian, C., Ramnath, B. V., Ramakrishnan, G., Rajendrakumar, M., Naveenkumar, V., & Saravanakumar, M. K. (2018). Review on mechanical properties of natural fiber composites. https://doi.org/10.1016/j.matpr.2017.11.276.
15. Sanjay, M. R., Arpitha, G. R., Naik, L. L., Gopalakrishna, K., & Yogesha, B. (2016). Applications of natural fibers and its composites: An overview. *Natural Resources*. https://doi.org/10.4236/nr.2016.73011
16. Holbery, J., & Houston, D. (2006). Natural-fiber-reinforced polymer composites in automotive applications. *JOM Journal of the Minerals Metals and Materials Society*. https://doi.org/10.1007/s11837-006-0234-2

17. Friedrich, K., & Almajid, A. A. (2013). Manufacturing aspects of advanced polymer composites for automotive applications. *Applied Composite Materials*. https://doi.org/10.1007/s10443-012-9258-7
18. Brøndsted, P., Lilholt, H., & Lystrup, A. (2005). Composite materials for wind power turbine blades. *Annual Review of Materials Research*. https://doi.org/10.1146/annurev.matsci.35.100303.110641
19. Katnam, K. B., Comer, A. J., Roy, D., Da Silva, L. F. M., & Young, T. M. (2015). Composite repair in wind turbine blades: An overview. *Journal of Adhesion*. https://doi.org/10.1080/00218464.2014.900449
20. González, C., Vilatela, J. J., Molina-Aldareguía, J. M., Lopes, C. S., & LLorca, J. (2017). Structural composites for multifunctional applications: Current challenges and future trends. *Progress in Materials Science*. https://doi.org/10.1016/j.pmatsci.2017.04.005.
21. Asmatulu, E., Twomey, J., & Overcash, M. (2014). Recycling of fiber-reinforced composites and direct structural composite recycling concept. *Journal of Composite Materials*. https://doi.org/10.1177/0021998313476325
22. Dweib, M. A., Hu, B., O'Donnell, A., Shenton, H. W., & Wool, R. P. (2004). All natural composite sandwich beams for structural applications. *Composite Structures*. https://doi.org/10.1016/S0263-8223(03)00143-0

A New Constrained-Based Multiobjective Optimization Method for Electric Discharge Machining

Neeraj Agarwal, Nitin Shrivastava, and M. K. Pradhan

1 Introduction

EDM is an unconventional machining procedure. An electrically conductive workpiece is submerged in the dielectric fluid. Two electrodes are used to regulate the machining operation (workpiece and tool). There is a usage of straight polarity, i.e., the workpiece is made positive electrode, while the tool is a negative electrode. Both electrodes are held at a certain distance and are subject to a potential differentiation. Sparking between electrodes is generated by this potential gap. Between the workpiece and the electrode, there is a sequence of sparks [1]. The temperature of the workpiece increases past the melting point of the sparking spot owing to the sparking of electrical energy transferred into thermal energy. At the point of flickering, there is a method for flushing the dielectric. Any material from the workpiece is melted and separated by evaporation and, due to successful flushing, any molten material is moved away [2]. The strength-to-weight ratio of titanium alloy is strong, so it is commonly used in the aerospace industry [3]. Machining of the Titanium alloy is difficult, hence EDM is chosen for machining [4]. EDM has many input parameters, such as discharge voltage, peak current (I_p), pulse on time (Ton), voltage (V), duty factor (t) the polarity of the electrode, a distance of the electrode, dielectric flushing [5]. For the machining of titanium alloy 685, four significant input parameters 'I_p', 'Ton', 'V', and 't' is chosen as the control parameter. An electrode is chosen from a 10-mm diameter copper rod.

Response surface methodology (RSM) is used in the model generation of material removal rate (MRR) and surface roughness (Ra) [6]. MRR to be maximized to increase productivity. Better surface finishing increases the product's longevity,

N. Agarwal (✉) · N. Shrivastava
UIT, Rajiv Gandhi Proudyogiki Vishwavidyalaya, Bhopal, India

M. K. Pradhan
Maulana Azad National Institute of Technology, Bhopal, India

R. Agrawal et al. (eds.), *Recent Advances in Smart Manufacturing and Materials*, Lecture Notes in Mechanical Engineering,
https://doi.org/10.1007/978-981-16-3033-0_42

thereby minimizing Ra is also required [7]. There are also sophisticated strategies of computation, such as the Bat algorithm, GA, PSO, Firefly algorithm, bacteria foraging, Sine–cosine algorithm, adaptive dynamic computing, optimization of teaching–learning-based optimization, and many more [8, 9]. A new heuristic optimization technique 'Jaya Algorithm' is gaining popularity among researchers [10]. The Jaya algorithm is simple to implement because it is an algorithm without parameters [11]. Multiobjective optimization is the strategy to optimize more than two quality measures [12]. Multiobjective optimization allows MRR and Ra to be configured simultaneously [13]. The aim is to optimize MRR and to concurrently reduce Ra. Both goals are conflicting, i.e., Ra also rises with the rise of MRR, and MRR declines with a reduction of Ra. In general, multiobjective optimization requires a single objective function; all answers are allocated separate weights in this method [14]. This gives way to a single blend of the two objective functions (quality measures) [15].

The multiple options of multiobjective optimization are presented in this article. This provides many variations using multiobjective optimization with restriction Ra for MRR and Ra. Thus, the customer has a vast range of combinations to select from. This is a special component of this paper and so far no study has been published. This is going to be helpful for the prosecutor. In this paper, Ra is restricted and attempts to optimize MRR (while Ra is fixed). Several consequences are produced in such a way. The customer has finished with the best approach with a wide range of options (a mixture of Ra and MRR).

2 Experimental Details

A plate of Titanium alloy 685 with size (40 × 40 × 5 mm) is chosen as a workpiece and a 10-mm diameter copper rod is picked as a tool electrode. In a dielectric solution, the workpiece is completely submerged. As an input control parameter, four significant input parameters 'I_p', 'Ton', 'V', and 't' are selected. Total 30 numbers of machining operations were performed with the RSM design of experiments. The duration of each is for 30 min. As process performance, MRR and Ra are chosen. Surface roughness is measured with Tesa Rugosurf 10G Profilometer. The weight of the workpiece is measured using a digital weighing machine with 0.1 mg resolution. Table 1 represents a selected range of input parameters. The duty factor is the ratio of sparking time to the total cycle time. 'Ton' is the duration of each spark in microseconds. 'I_p'

Table 1 Input parameter range

Parameter	I_p (A)	V (V)	Ton (μs)	t (% duty factor)
Level 1	4	40	50	8 (25.0%)
Level 2	6	70	100	12 (37.5%)
Level 3	8	100	150	16 (50.0%)

is the maximum current for each spark. 'V' is the potential difference between the workpiece and the electrode. Table 2 represents the experimental values of MRR and Ra. The MRR is calculated as follows

$$\text{MRR} = \frac{\text{Mf} - \text{Mi}}{\text{Tp}} \tag{1}$$

Table2 Experimental observations of MRR and Ra

S. No.	I_p (A)	V (V)	Ton (μs)	t (%)	MRR (mg/min)	Ra (μm)
1	8	40	150	16	4.183	5.351
2	8	40	50	8	3.735	3.416
3	4	40	50	8	2.442	3.188
4	8	40	50	16	3.187	4.926
5	6	70	100	12	3.230	4.065
6	6	70	100	12	3.433	4.004
7	6	70	100	12	3.435	3.168
8	4	40	50	16	2.828	3.206
9	4	40	150	16	2.810	4.383
10	4	100	150	8	1.577	4.222
11	4	40	150	8	2.502	3.521
12	4	100	50	16	1.957	3.763
13	4	100	50	8	2.005	3.253
14	8	100	50	16	2.570	9.480
15	6	70	100	12	3.163	4.033
16	8	100	150	16	3.462	4.756
17	8	100	150	8	3.167	4.909
18	8	40	150	8	4.460	4.271
19	4	100	150	16	2.165	3.982
20	8	100	50	8	2.930	3.816
21	6	70	50	12	3.555	3.363
22	6	40	100	12	3.998	4.068
23	6	100	100	12	2.565	3.489
24	6	70	100	16	2.853	4.149
25	6	70	100	8	2.968	3.478
26	6	70	100	12	3.237	4.401
27	6	70	150	12	3.359	3.429
28	8	70	100	12	3.990	3.519
29	6	70	100	12	3.282	3.870
30	4	70	100	12	2.598	4.038

Where Mf = Final weight of workpiece after maching, Mi = Initial weight of workpiece, Tp = Time period of machining.

Software package Minitab 18 is used to develop a response surface equation of MRR and Ra, after elimination of not significant terms final response surface model of MRR (in mg/min) given as Eq. (2) and response surface model of Ra is given as Eq. (3).

$$\begin{aligned}\mathrm{MRR} = {} & 1.276 + 0.3099 * I_\mathrm{p} - 0.01435 * V \\ & - 0.01417 * \mathrm{Ton} + 0.721 * t - 0.02767t * t \\ & + 0.001893 * I_\mathrm{p} * \mathrm{Ton} - 0.01660 * Ip * t \\ & + 0.000464 * \mathrm{Ton} * t \end{aligned} \tag{2}$$

$$\begin{aligned}\mathrm{Ra} = {} & 4.03 - 1.27 * I_\mathrm{p} - 0.0426 * V + 0.0577 * \mathrm{Ton} \\ & + 0.004 * t + 0.078 * I_\mathrm{p} * I_\mathrm{p} + 0.000349 * V * V \\ & + 0.00424 * I_\mathrm{p} * V - 0.00316 * \mathrm{Ip} * \mathrm{Ton} \\ & + 0.0543 * I_\mathrm{p} * t - 0.000218 * V * \mathrm{Ton} \\ & - 0.00192 * \mathrm{Ton} * t \end{aligned} \tag{3}$$

3 Jaya Algorithm

This is an optimization technique. Let 'i' as generation number 'j' as a number of control variables, population size as 'k', $X_{j,\mathrm{best},i,}$ and $X_{j,\mathrm{worst},i}$ are marked best and worst solution. If $X_{j,k,i}$ is the current value of jth design variable, kth population, for ith generation; and the modified solution is represented by $X'_{j,k,i}$ as per Eq. (4). Two random numbers are considered as $r_{1j,i}$ and $r_{2j,i}$.

$$\begin{aligned} X'_{j,k,i} = {} & X_{j,k,i} + r_{1,j,i}\left(X_{j,\mathrm{best},i} - \left|X_{j,k,i}\right|\right) \\ & - r_{2,j,i}\left(X_{j,\mathrm{worst},i} - \left|X_{j,k,i}\right|\right) \end{aligned} \tag{4}$$

4 Multiobjective Constraint Optimization (User Selection)

There is a limitation of a classical multiobjective optimization that it generates only one solution. If one wants to increase one objective value by compromising to other objective function values, this would not possible. This limitation is overcome by constraint optimization. In this case, we constraint Ra to a certain limit and attempt to optimize MRR. A large number of combinations is created for MRR and Ra in this section. The flow diagram of multiobjective optimization with constraint Ra is shown in Fig. 1.

The target is to concurrently optimize MRR and minimize Ra. The measured spectrum of Ra is (2.40, 4.3) in the experiment.

The restriction for the process

$$\text{Ra} \leq 2.40 \tag{5}$$

$$h(x) = \text{Ra} - 2.40 \leq 0 \tag{6}$$

where Ra from Eq. (3).

Let's say that I_p, V, Ton, and t are design variables, and population size 5 is condsidered. The goal is to maximize MRR according to Eq. (2), while Ra should be satisfied in conjunction with Eq. (5). Here, penalty is assigned if Eq. (6) breaks. The penalty is added to the objective function (2). If $h(x) \leq 0$ then the assigned penalty value would be 0 and if $h(x) \geq 0$ then $1000 * h(x)^2$ would be assigned as the penalty. In comparison, the penalty could be $25 * h(x)^2$, $75 * h(x)^2$, $200 * (x)^2$, $250 * h(x)^2$. To address the application, the higher value of the penalty is required. The penalty is $1000 * h(x)^2$ in this situation. Now assigned value of the penalty is substracted from the objective function (2). The updated objective function as Eq. (7) is prepared. A pseudo-objective function is called the function $f'(x)$. The new objective function is as below

$$f'(x) = f(\text{MRR}) - \text{penalty} \tag{7}$$

where f(MRR) from Eq. (2)

$$\begin{aligned} &\text{and penalty} = 1000 * h(x)^2 \\ &\text{if } h(x) \leq 0;\ \text{Penalty} = 0 \\ &\text{else penalty} = 1000 * h(x)^2 \end{aligned} \tag{8}$$

Step 1: Let us consider population size 5 into account. As shown in Table 1, within the given range, the initial population is generated randomly and the corresponding objective function value (MRR) is calculated. As per Eq. (3), Ra is determined as well. As per Eq. (3), for candidate 1, Ra = 3.2352. Value measured as 3.2352 – 2.40 = 0.8352 that breaks the (6) function, i.e., $h(x) \geq 0$, so the penalty can be given and value is taken as $1000 * h(x)^{2,}$ i.e., $1000 * (0.8352)^2 = 697.5818$. The updated objective function is determined as 3.0574 – 697.5818 = – 694.5244 (Penalty assigns 0 if $h(x) \leq 0$) as per Eq. (7). Penalty = $1000 * (3.4008 - 2.4)^2 = 1001.6379$ for the second nominee and changed target feature as 3.6364 – 1001.6379 = – 998.0015. Penalty = $1000 * (3.1642 - 2.4)^2 = 583.9571$ for the third nominee and changed objective function as 3.0360 – 583.9571 = – 580.9211. Penalty for fourth applicant

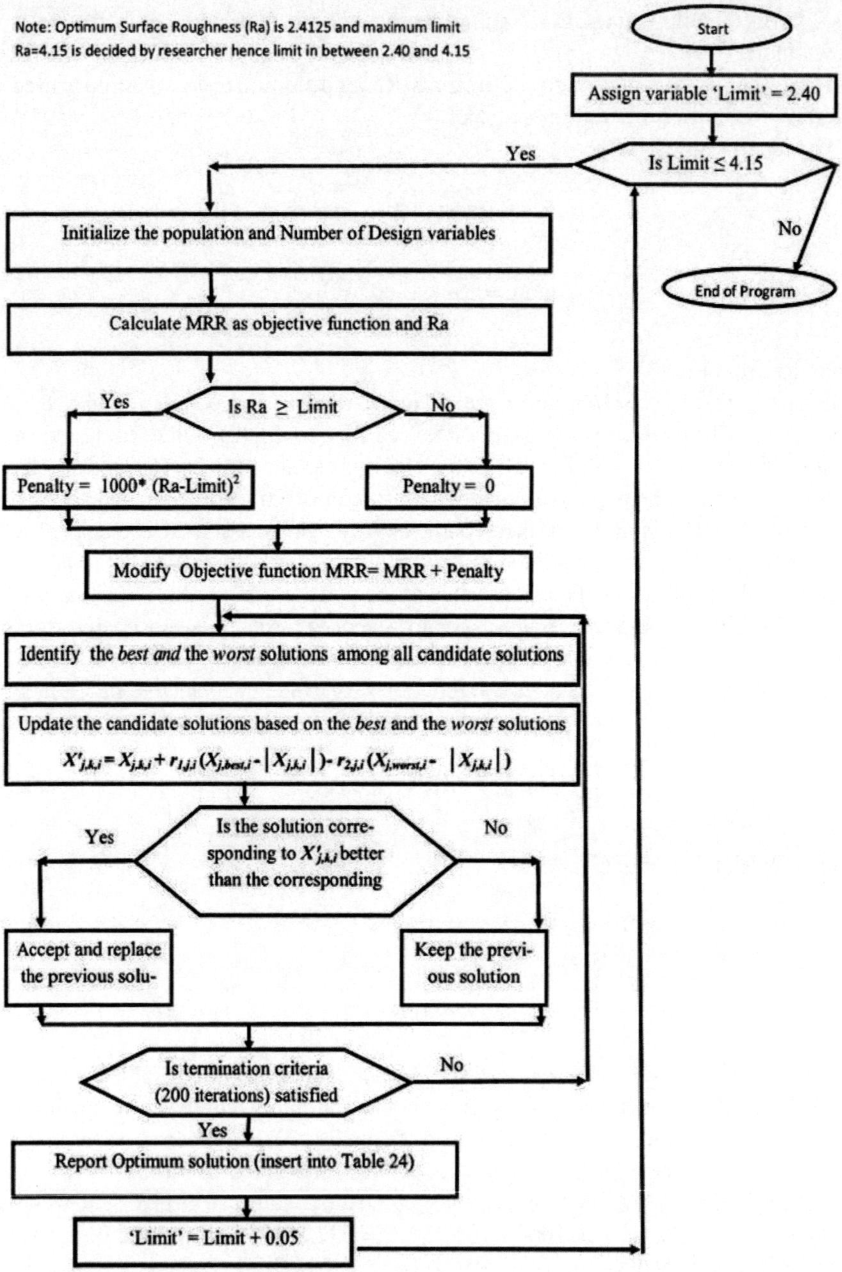

Fig. 1 Flow diagram of multiobjective optimization with constraint Ra

$= 1000 * (3.8628 - 2.4)^2 = 2139.6441$ and target feature updated as $4.0623 - 2139.6441 = -2135.5818$. Penalty $= 1000 * (3.7259 - 2.4)^2 = 1757.9990$ for the fifth nominee and changed target feature as $2.3938 - 1757.9990 = -1755.6051$.

Step 2 Because maximization is the case, the highest $f'(x)$ for the best value is selected and the lowest $f'(x)$ is selected as the worst. Candidate 3 is the strongest candidate, and candidate 4 is the weakest candidate.

Best $I_p = 5.1102$, Best Ton $= 60.3670$, Best $t = 10.8977$, Best $V = 66.3564$, Worst $I_p = 7.2751$, Worst Ton $= 145.1366$, Worst $t = 10.3032$, worst $V = 55.6996$ random variables are now generated as $r1 = 0.9619$, $r2 = 0.7556$ for I_p; $r1 = 0.8778$, $r2 = 0.5111$ for Ton; $r1 = 0.6935$, $r2 = 0.3221$ for t; $r1 = 0.5592$, $r2 = 0.2894$ for V. New value for candidate 1 from Eq. (4) is calcutead as follow

$$
\begin{aligned}
I_p &= 4.9188 + 0.9619 * (5.1102 - |4.9188|) \\
&\quad - 0.7556 * (7.2751 - |4.9188|) \\
&= 3.3224 (= 4 \text{ because the lower bound is } 4) \\
\text{Ton} &= 85.4961 + 0.8778 * (60.3670 - |85.4961|) \\
&\quad - 0.5111 * (145.1366 - |85.4961|) \\
&= 32.9555 \ (= 50 \text{ because lower bound is } 50) \\
t &= 10.54131 + 0.6935 * (10.8977 - |10.54131|) \\
&\quad - 0.3221 * (10.3032 - |10.54131|) = 10.8663 \\
V &= 59.8420 + 0.5592 * (66.3564 - |59.8420|) \\
&\quad - 0.2894 * (55.6996 - |59.8420|) = 64.6835
\end{aligned}
$$

Calculate the new value of MRR and Ra from the Eqs. (2) and (3). Since Ra-2.4 $= (2.9081 - 2.40) = 0.5081$, it is higher than 0, hence penalty would assigned as $1000 * (2.9081 - 2.40)^2 = 258.1916$ and $f'(x) = 2.8035 - 258.1916 = -255.3882$.

Similar calculations for all candidates have been made. The changed values are now inserted into Table 4.

Step 3 Since Table 4 candidate 1 is higher than Table 3 candidate 1, hence Table 4 candidate 1 is then placed into Table 5 as candidate 1. Similarly, in Table 5, candidates (2, 3, 4, 5) from Table 4 are added. The first version has just finished.

Step 4 Repeat a similar procedure for 100 iterations (step 1–step 3) and insert optimum value in Table 6.

Step 5 Now start a new equation with an increment of 0.05 (maybe 0.1, 0.2) for Ra. Modify Eqs. (6)–(9), and repeat the same procedure from step 1–4. Repeat 100 iterations and insert results (Maximum MRR for Ra ≤ 2.45) into Table 6.

$$
\text{Ra} \leq 2.45 \tag{9}
$$

Table 3 Initial population for Jaya algorithm

Candidate	I_p (A)	Ton (μs)	T (%)	V (V)	MRR (mg/min)	Ra (μm)	Penalty	$f'(x)$
1	4.9188	85.4961	10.5431	59.8420	3.0574	3.2352	697.5818	– 694.5244
2	6.9529	103.7071	8.9708	54.0208	3.6364	3.4008	1001.6379	– 998.0015
3	5.1102	60.3670	10.8977	66.3564	3.0360	3.1642	583.9571	– 580.9211
4	7.2751	145.1366	10.3032	55.6996	4.0623	3.8628	2139.6441	– 2135.5818
5	5.5376	146.5337	8.0434	92.1662	2.3938	3.7259	1757.9990	– 1755.6051

Step 6 Now, change the equation to Ra $\leq$ 2.50 (in the series of 0.05 or other) and repeat the same procedure. After 100 iterations insert results in Table 6. Repeat the similar procedure till Ra $\leq$ 4.20 (as per user requirements). Table 6 depicts combination of MRR and Ra.

5 Results and Discussion

Table 6 shows a combination of MRR and Ra with the control variable's value. If the researcher agrees to consider the Ra value as 2.4134 μm, then the maximum MRR value 2.9898 mg/min will be attained. Corresponding control parameter values are given in the first row of Table 6. Corresponding to the second row, if a researcher agrees to increase the acceptable limit of Ra to 2.4514 μm, the MRR will be increased to 3.2287 mg/min (second row). Table 6 shows a combination of MRR and Ra with a corresponding control parameter's value. Figure 2 represents a combination of MRR and Ra.

6 Conclusion

A new constrained-based multiobjective optimization using the Jaya algorithm is successfully applied for multiobjective optimization. The user has a large number of choices, while another multiobjective optimization technique produces a single optimum value. A computer program in MATLAB is used to handle the mathematical calculations involved in this algorithm.

Table 4 Modified values of all variable with a fitness function

Cand	I_p (A)	Ton (μs)	t (%)	V (V)	MRR (mg/min)	Ra (μs)	Penalty	$f'(x)$
1	4.000 (3.322)*	50.0000 (32.955)*	10.8663	64.6835	2.8035	2.9081	− 258.1916	− 255.388
2	4.9370	50.0000	9.8780	60.4325	2.9873	2.8119	− 169.6706	− 166.683
3	4.0000	50.0000	11.0892	69.4407	2.7509	2.9849	− 342.0941	− 339.343
4	5.1927	70.7274	10.7155	61.6585	3.1182	3.1750	− 600.6685	− 597.550
5	4.0000	71.6123	9.2951	88.2883	2.2269	3.2813	− 776.6150	− 774.388

*means adjusted to upper bound or lower bound

Table 5 Updated value for all variable after fitness comparison

Cand	I_p (A)	Ton (μs)	T (%)	V (V)	MRR (mg/min)	Ra (μm)	Penalty	$f'(x)$	Remark
1	4.0000	50.0000	10.8663	64.6835	2.8035	2.9081	– 258.1916	– 255.3882	
2	4.9370	50.0000	9.8780	60.4325	2.9873	2.8119	– 169.6706	– 166.6833	Best
3	4.0000	50.0000	11.0892	69.4407	2.7509	2.9849	– 342.0941	– 339.3432	
4	5.1927	70.7274	10.7155	61.6585	3.1182	3.1750	– 600.6685	– 597.5503	
5	4.0000	71.6123	9.2951	88.2883	2.2269	3.2813	– 776.6150	– 774.3881	Worst

Table 6 Combination of MRR and Ra with design variable

S. No.	Limiting Ra (μm)	MRR (mg/min)	I_p (A)	Ton (μs)	t (%)	V (V)
1	2.4134	2.9888	5.2364	50	8	44.0712
2	2.4514	3.2287	5.9043	50	8	40
3	2.5009	3.3343	6.2927	50	8	40
4	2.5507	3.4097	6.5705	50	8	40
5	2.6006	3.4717	6.7985	50	8	40
6	2.6505	3.5256	6.9967	50	8	40
7	2.7005	3.5738	7.1742	50	8	40
8	2.7504	3.6179	7.3365	50	8	40
9	2.8004	3.6588	7.4869	50	8	40
10	2.8504	3.6971	7.6277	50	8	40
11	2.9004	3.7332	7.7606	50	8	40
12	2.9503	3.7674	7.8866	50	8	40
13	3.0003	3.7998	8	50.3454	8	40
14	3.0503	3.8279	8	56.3406	8	40
15	3.1003	3.856	8	62.3358	8	40
16	3.1503	3.8842	8	67.2935	8.028	40
17	3.2003	3.9128	8	71.352	8.0823	40
18	3.2503	3.9419	8	75.4248	8.1391	40
19	3.3003	3.9717	8	79.5495	8.1976	40
20	3.3503	4.002	8	83.7317	8.2578	40
21	3.4003	4.0329	8	87.9716	8.32	40
22	3.4503	4.0646	8	92.3026	8.3833	40

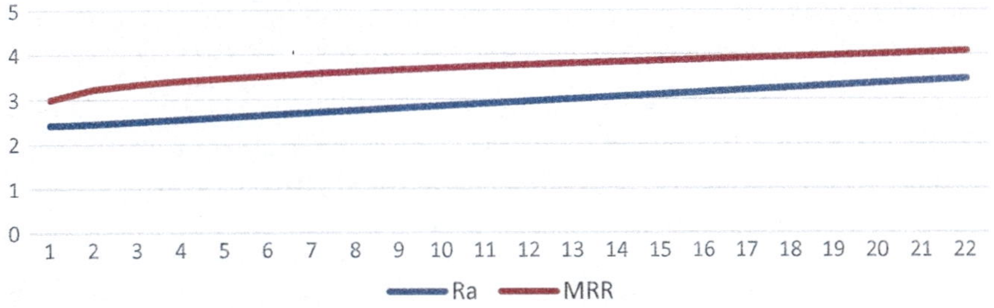

Fig. 2 MRR and Ra combination

References

1. Mohd Abbas, N., Solomon, D. G., & Fuad Bahari, M. (2007). A review on current research trends in electrical discharge machining (EDM). *International Journal of Machine Tools and Manufacture, 47*, 1214–1228. https://doi.org/10.1016/j.ijmachtools.2006.08.026
2. Ho, K. H., & Newman, S. T. (2003). State of the art electrical discharge machining (EDM). *International Journal of Machine Tools and Manufacture, 43*, 1287–1300. https://doi.org/10.1016/S0890-6955(03)00162-7
3. Warlimont, H. (2018). Titanium and titanium alloys.https://doi.org/10.1007/978-3-319-69743-7_7.
4. Abu, Q. J. E., Mourad Abdel-Hamed I., Ziout Aiman, Haider, A. M., & Ahmed, E. (2018). Electric discharge machining of titanium and its alloys: A review. https://doi.org/10.1504/IJMMM.2012.044922.
5. Gostimirovic, M., Kovac, P., Skoric, B., & Sekulic, M. (2011). Effect of electrical pulse parameters on the machining performance in EDM. *Indian Journal of Engineering and Materials Science, 18*, 411–415.
6. Kung, K. Y., & Chiang, K. T. (2008). Modeling and analysis of machinability evaluation in the wire electrical discharge machining (WEDM) process of aluminum oxide-based ceramic. *Materials and Manufacturing Processes, 23*, 241–250. https://doi.org/10.1080/10426910701860616
7. Shrivastava, P. K., & Dubey, A. K. (2013). Experimental modeling and optimization of electric discharge diamond face grinding of metal matrix composite. *International Journal of Advanced Manufacturing Technology, 69*, 2471–2480. https://doi.org/10.1007/s00170-013-5190-8
8. Yang, X.-S. (2010). Engineering optimization: An introduction with metaheuristic applications.
9. Rao, R. V., & Kalyankar, V. D. (2014). Optimization of modern machining processes using advanced optimization techniques: A review. *International Journal of Advanced Manufacturing Technology, 73*, 1159–1188. https://doi.org/10.1007/s00170-014-5894-4
10. Rao, R. V. (2016). Jaya: A simple and new optimization algorithm for solving constrained and unconstrained optimization problems. *International Journal of Industrial Engineering Computations, 7*, 19–34. https://doi.org/10.5267/j.ijiec.2015.8.004
11. Agarwal, N., Shrivastava, N., & Pradhan, M. K. (2020). Optimization of relative wear ratio during EDM of titanium alloy using advanced techniques. *SN Applied Science, 2*. https://doi.org/10.1007/s42452-019-1877-2.
12. Gunantara, N. (2018). A review of multi-objective optimization: Methods and its applications. *Cogent Engineering, 5*, 1–16. https://doi.org/10.1080/23311916.2018.1502242
13. Mohanty, A., Talla, G., & Gangopadhyay, S. (2014). Experimental investigation and analysis of EDM characteristics of inconel 825. *Materials and Manufacturing Processes, 29*, 540–549. https://doi.org/10.1080/10426914.2014.901536

14. Rao, K. V., Murthy, P. B. G. S. N., & Vidhu, K. P. (2017). Assignment of weightage to machining characteristics to improve overall performance of machining using GTMA and utility concept. *CIRP Journal of Manufacturing Science Technology, 18*, 152–158. https://doi.org/10.1016/j.cirpj.2016.12.001.
15. Rao, R., Modelling, V. P.-A. M. (2013). *Multi-objective optimization of heat exchangers using a modified teaching-learning-based optimization algorithm*. Elsevier.

Effect of Spiral Pattern of Electrolyte Jet on Material Removal Rate and Surface Roughness During Electrochemical Machining of 100Cr6 in NaCl

S. K. Tiwari and R. K. Upadhyay

1 Introduction

Electrochemical machining (ECM) has got an industrial importance due to its capability of controlled atomic-level metal removal [1, 2]. It is an anodic dissolution process utilizes the principles of electrolysis, where a high-amperage, low-voltage current is used to dissolve the metal and to remove it from workpiece [3, 4]. ECM offers atomic-level material removal which results in higher surface finish with stressed crack-free surface irrespective of hardness of the workpiece materials [5]. Because of these facts, the ECM process is opted as best alternative within the field of machining process to machine alloy steels without generation of thermal stresses with less tool wear, thus suitable for mass production work with low labor requirements [6, 7].

In electrochemical machining, electrolyte is pumped into the interelectrode gap maintained between tool and workpiece, and a low-voltage high-amperage current passes through this electrochemical circuit [8]. Mobility of ions accompanied by electron flow promotes the dissolution of metal according to Faraday's law [9]. Many researchers have optimized the ECM parameters at different angles although the potential of ECM is unexplored. From the literature, it has been found that different jet patterns of tool electrode have significant effect on the electrolyte solution during electrochemical dissolution of workpiece material, which directly influences on the material removal rate and surface roughness of the machined workpiece [10]. The

S. K. Tiwari (✉)
Department of Production Engineering, Birla Institute of Technology, Mesra—Off-Campus Deoghar, Deoghar 814142, India
e-mail: sanjivtiwari@bitmesra.ac.in

R. K. Upadhyay
Department of Mechanical Engineering, Birla Institute of Technology, Mesra—Off-Campus Deoghar, Deoghar 814142, India
e-mail: ritesh.upadhyay@bitmesra.ac.in

R. Agrawal et al. (eds.), *Recent Advances in Smart Manufacturing and Materials*, Lecture Notes in Mechanical Engineering,
https://doi.org/10.1007/978-981-16-3033-0_43

achievement of higher material removal rate and better surface finish in ECM is a strong research base which may be possible by change in composition of electrolyte solution and change in electrolyte flow pattern [11]. Therefore, the present work focused on the flow of electrolyte jet in spiral pattern to improve the machining characteristics.

2 Experimentation

The design resulted in total ten experiments, which were performed at 12 V supply. A brass tool was used as cathode and 100Cr6 as anode. Chemical composition of 100Cr6 is Fe 96.92%, Cr 1.43%, C 0.97%, Si 0.28%, Mn 0.28%, Ni 0.11%, P 0.006%, and S 0.002%. A typical experimental setup of electrochemical machining is shown in Fig. 1.

2.1 *Working Condition*

The machining parameters selected based on performance characteristics are shown in Table 1.

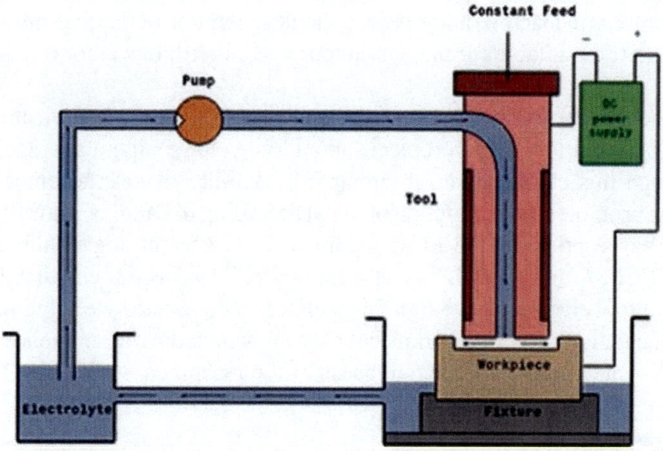

Fig. 1 Experimental setup

Table 1 Working conditions for analysis

S. No.	Parameters	Values
1	Voltage	12 V
2	Current density	10–50 A/cm^2
3	Interelectrode gap	0.005–0.009 (cm)
4	Electrolyte flow rate	10 l//min
5	Electrolyte and its concentration	NaCl aqua solution
6	Tool material	Brass
7	Workpiece	100 Cr6
8	Machining time	2 min

2.2 Estimation of Material Removal Rate

The experiments were performed at different current densities with and without tool rotation, and rest of the parameters are set according to Table 1 for each run. Table 2 portrays the experimental design. The initial weight of the workpiece was taken for by using weighing machine. The electrode was then fed continuously toward the workpiece, and time was recorded. After machining, the cavity was formed on the workpiece. The final weight of the workpiece was taken, and material removal rate was calculated as per the following formula [12].

$$\text{MRR} = \frac{\text{Initial weight} - \text{Final weight}}{\text{Density} \times \text{Total time}} \tag{1}$$

Table 2 Experimental design

S. No.	Current density (A/cm^2)	Tool rotation (rpm)	Feed rate (cm/s)	IEG (cm)
1	10	0	0.0030	0.0090
2	10	300	0.0030	0.0090
3	20	0	0.0058	0.0080
4	20	300	0.0058	0.0080
5	30	0	0.0075	0.0070
6	30	300	0.0075	0.0070
7	40	0	0.0098	0.0060
8	40	300	0.0098	0.0060
9	50	0	0.025	0.0050
10	50	300	0.025	0.0050

2.3 Estimation of Surface Roughness

The quality of the machined surface was measured with Talysurf (Model: Taylor Hobson) with sampling length of 4 mm. The experimental results of surface roughness for machining with two different patterns of jets are shown in Tables 3 and 4.

Table 3 Experimental results of MRR and surface roughness during machining with straight jet in circular pattern

S. No.	Weight before machining (g)	Weight after machining (g)	MRR_{th} (g/s)	MRR_{act} (g/s)	η	SR (μm)
1	360.4	360.2	0.1430	0.0886	62.0	0.922
2	375.6	375.4	0.1430	0.0956	66.9	0.810
3	387.8	387.4	0.2860	0.2010	70.3	0.765
4	398.8	398.3	0.2860	0.2073	72.5	0.650
5	411.5	410.9	0.4290	0.3080	71.8	0.686
6	417.7	417.3	0.4290	0.3372	78.6	0.600
7	431.8	430.9	0.5720	0.4187	73.2	0.630
8	445.5	444.5	0.5720	0.4622	80.8	0.501
9	340.6	339.5	0.7150	0.5412	75.7	0.405
10	310.5	309.3	0.7150	0.6027	84.3	0.290

Table 4 Experimental results of MRR and surface roughness during machining with straight jet in spiral pattern

S. No.	Weight before machining (g)	Weight after machining (g)	MRR_{th} (g/s)	MRR_{act} (g/s)	η	SR (μm)
1	360.2	359.9	0.1430	0.1060	74.1	0.818
2	375.4	375.1	0.1430	0.1151	80.5	0.734
3	387.4	386.9	0.2860	0.2382	83.3	0.605
4	397.5	397	0.2860	0.2482	86.8	0.528
5	410.8	410.1	0.4290	0.3599	83.9	0.567
6	416.3	415.5	0.4290	0.3960	92.3	0.480
7	430.9	429.9	0.5720	0.4873	85.2	0.499
8	444.5	443.4	0.5720	0.5394	94.3	0.387
9	339.4	338.1	0.7150	0.6277	87.8	0.290
10	309.3	307.9	0.7150	0.6921	96.8	0.180

2.4 *Estimation of Current Efficiency*

Current efficiency mainly depends on the current density, workpiece material, electrolyte type, and machining conditions. Here, in this experiment, current efficiency was calculated by using Eq. (2) [13].

$$\text{Current Efficiency}, \eta = \frac{\text{Actual metal removed}}{\text{Theoretical metal removed}} \tag{2}$$

3 Result and Discussion

The variation of current density with varying feed rate and varying interelectrode gap (IEG) is shown in Figs. 2 and 3. It is observed that current density increases with increase in feed rate and decrease with increase in interelectrode gap. The material removal rate during machining is greatly influenced by rotary action of the tool. In this work, the geometrical characteristics of tool electrode is employed to apply the electrolyte solution into the interelectrode gap with circular and spiral patterns of the jet under stationary and rotary conditions of the tool.

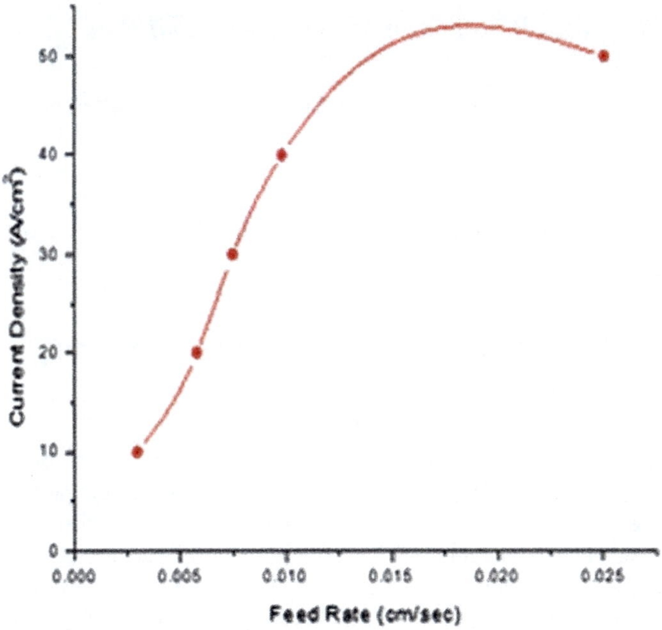

Fig. 2 Plot of current density against feed rate

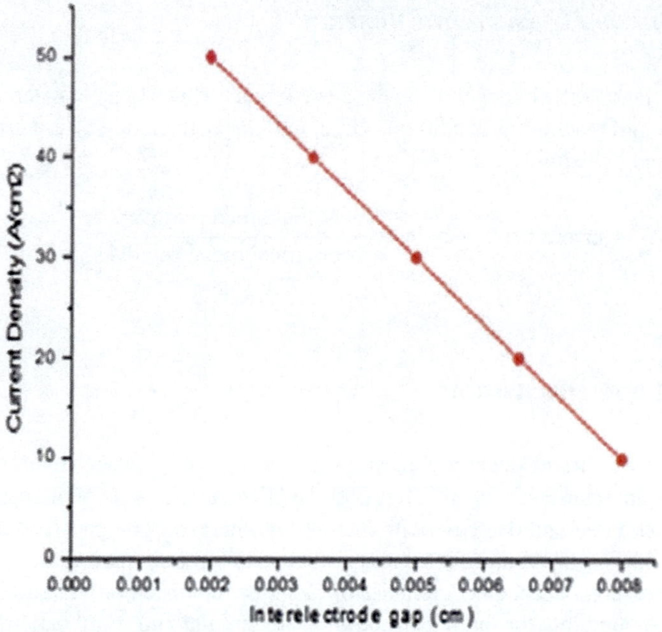

Fig. 3 Plot of current density with respect to IEG

3.1 Effect of Tool Rotation on Material Removal Rate

The results of effects of tool rotation on material removal for two different jet patterns rate are shown in Tables 3 and 4. The experimental result shows an improvement of MRR while machining with rotational tool compared with the stationary tool. It is because rotary action of tool causes more mobility of ions from the metal to the solution which increases the speed of the chemical reactions [14].

3.2 Effect of Straight Jet with Spiral Pattern on Material Removal Rate

The experimental results shown in Table 4 reveal that straight jet in spiral pattern facilitates the better use of electrical current which prevents the generation of short circuits during machining process. Because of better electrolyte distribution pattern of jet in spiral pattern, a maximum MRR of 0.6921 g/s was achieved under rotary condition of tool. This result is 13% higher when compared to the results of MRR using straight jet in circular pattern. The experimental MRR at different current densities is plotted in Figs. 4 and 5 for circular and spiral patterns of jets, respectively. The maximum MRR obtained using rotational tool is 0.6027 g/s for circular pattern

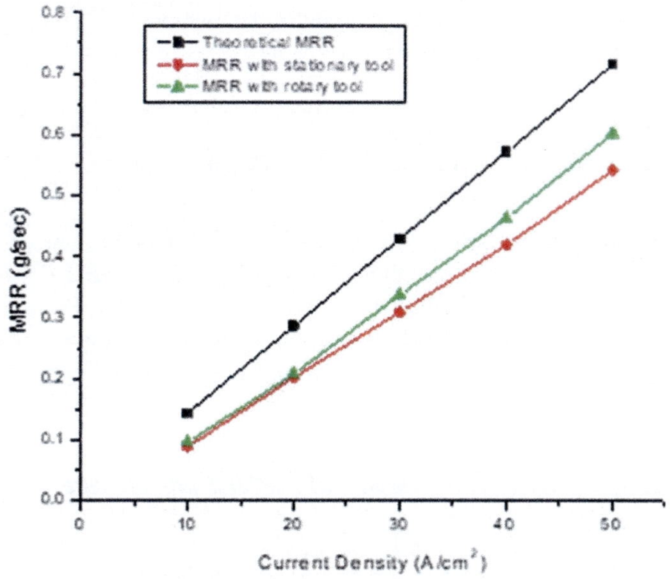

Fig. 4 Plot of current density versus MRR (for straight jet in circular pattern)

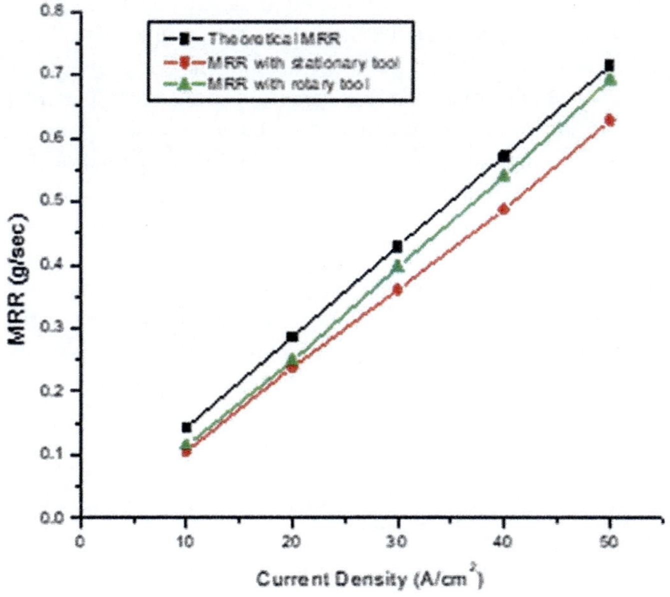

Fig. 5 Plot of current density versus MRR (for straight jet in spiral pattern)

of jets and 0.6921 g/s for spiral pattern of jets which is more than obtained using stationary tool under same set of machining conditions.

3.3 Effect of Tool Rotation on Surface Roughness

From the results of surface roughness under the effect of tool rotation, it was observed that rotating tool gives better surface finish when compared with the stationary tool. The enhancement rates with circular and spiral pattern of jets are 13.2, 15, 12.5, 20.5, 8% and 10.3, 12.8, 15.3, 22.5, 28.4%, respectively, for same values of current density.

3.4 Effect of Straight Jet with Spiral Pattern on Surface Roughness

The results of effect of straight jet in spiral pattern on surface roughness show that spiral pattern of electrolyte jet tends to produce minimum surface roughness. It is due to the fact that the flow of electrolyte admitted into the interelectrode gap provokes a minimum fluid turbulence at workpiece surface when compared with straight jet in circular pattern and consequently promotes the homogeneous condition of metal dissolution during machining of metal [15]. The plot of surface roughness at different current densities is shown in Figs. 6 and 7 for circular and spiral patterns of jets, respectively. The minimum surface roughness achieved using rotational tool was 0.290 μm for circular pattern of jets and 0.180 μm for spiral pattern of jets which is more than obtained using stationary tool under same set of machining conditions.

An improvement in current efficiency was observed by using rotary tool compared with that using a stationary tool for machining with two different patterns of jets. The efficiency is improved by (7.3, 3, 8.7, 9.4, 10.3) % when machining with circular pattern of jets and (8, 4, 9.1, 9.7, 9.3, 9.3) % when machining with spiral pattern of jets for current density values (10, 20, 30, 40, 50) A/cm^2, respectively.

4 Validation of Experimental Observations Through CFD Analysis

Figures 8 and 9 demonstrate the CFD analysis of turbulence kinetic energy for both straight jet in circular and spiral pattern. The analysis reveals the fact that the minimum and maximum turbulent kinetic energy associated with the eddies of the fluid flow in case of straight jet in circular pattern are 9 m^2/s^2 and 33.87

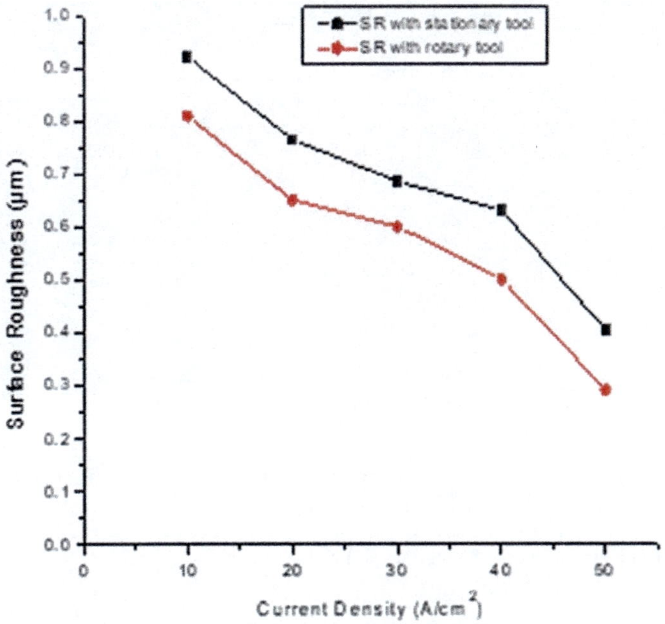

Fig. 6 Plot of current density versus SR (for straight jet in circular pattern)

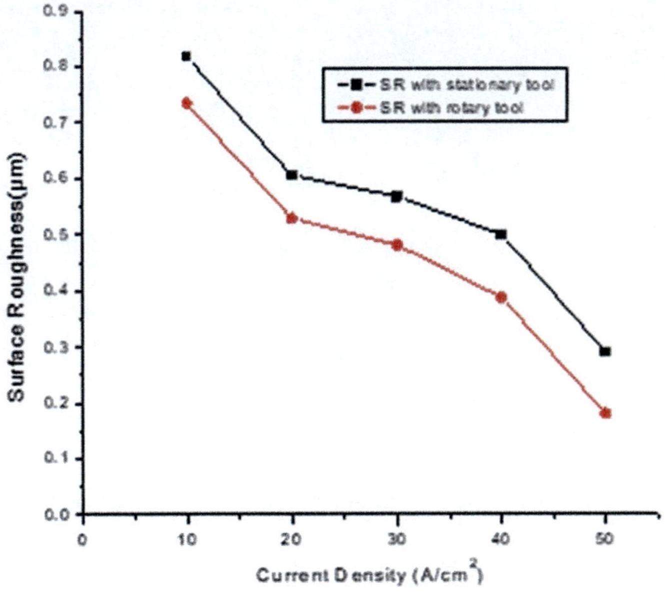

Fig. 7 Plot of current density versus SR (for straight jet in spiral pattern)

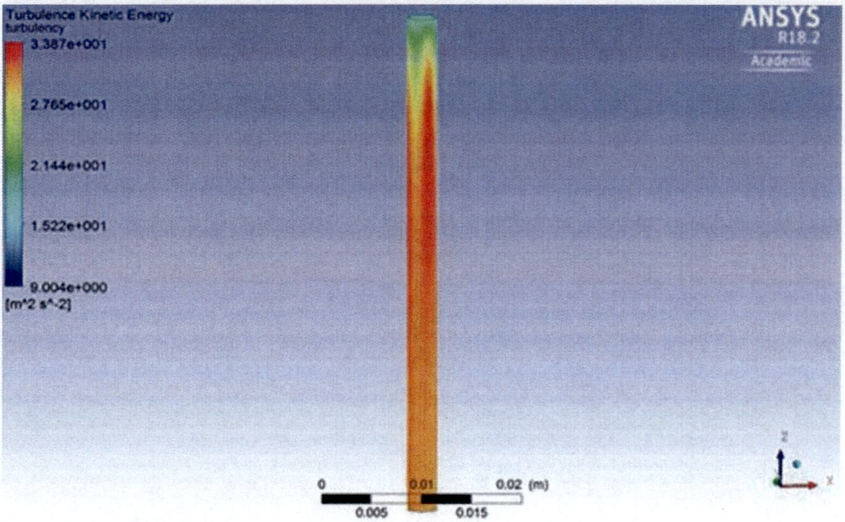

Fig. 8 Turbulence kinetic energy for straight jet in circular pattern

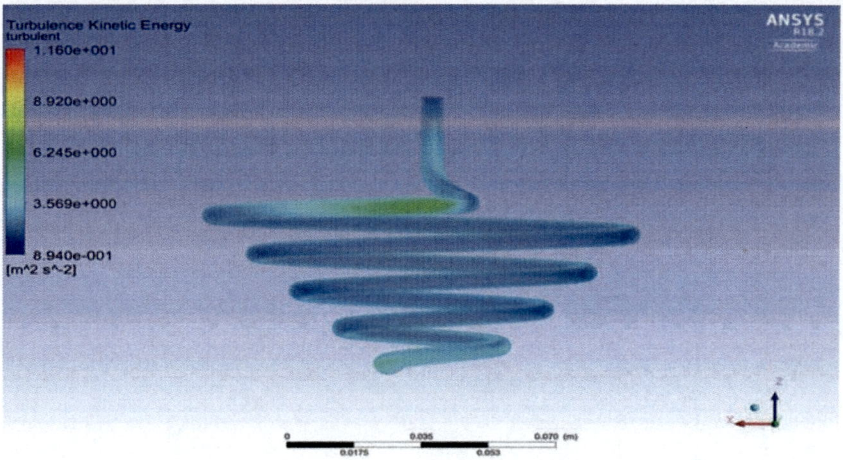

Fig. 9 Turbulence kinetic energy for straight jet in spiral pattern

m^2/s^2, respectively, while in case of spiral pattern, it is 0.89 m^2/s^2 and 11.6 m^2/s^2, respectively.

Higher turbulent kinetic energy leads to a high surface roughness, whereas low turbulent kinetic energy gives better surface finish. CFD analysis of selected jet patterns evidenced for smooth surface finish while electrochemical machining of 100Cr6 in NaCl with spiral pattern of jets. The analysis of percentage improvement

Table 5 Improvement analysis of MRR and SR with reference jet pattern

S. No.	Description	Types of jet patterns	
		Straight jet in circular pattern	Inclined jet in spiral pattern
1	Maximum achieved MRR (g/s)	0.6027	0.6921
2	Minimum achieved SR (μm)	0.290	0.180
3	Working conditions	50 A/cm^2	50 A/cm^2
		0.025 cm/s	0.025 cm/s
		300 rpm	300 rpm
4	Percentage improvement on the selected objective	Reference jet pattern	13% in MRR and 38% in SR

in material removal rate and surface roughness with reference to straight jet in circular pattern is shown in Table 5.

5 Conclusion

The electrochemical characteristics of 100Cr6 in aqueous NaCl solution have been analyzed experimentally to study the influence of jet patterns on MRR and SR under rotary and stationary conditions of tool. The effects of major influencing parameters like feed rate, interelectrode gap, and current densities have been studied for improving the material removal rate using two different jet patterns. The experimental investigations support the conclusion that machining of 100Cr6 in aqueous NaCl solution under rotating condition of tool using straight jet in spiral pattern gives better material removal rate and minimum surface roughness of around 0.6921 g/s and 0.180 μm, respectively, at 50 A/cm^2. The rates of improvement in MRR and SR were found to be 13% and 38%, respectively, for the same sets of machining conditions.

References

1. Sekar, T., & Marappan, R. (2008). Experimental investigations into the influencing parameters of electrochemical machining of AISI 202. *Journal of Advanced Manufacturing Systems, 7*(2), 337–343.
2. Bhattacharyya, B., Malapati, M., Munda, J., & Sarkar, A. (2007). Influence of tool vibration on machining performance in electrochemical micro-machining of copper. *International Journal of Machine Tools and Manufacture, 47*(2), 335–342.
3. Kozak, J., Rajurkar, K. P., & Makkar, Y. (2004). Selected problems of micro-electrochemical machining. *Journal of Materials Processing Technology, 149*(1–3), 426–431.

4. Mount, A. R., & Muir, R. N. (2007). Dissolution characteristics of iron and stainless steels in chloride under electrochemical machining conditions. *Journal of the Electrochemical Society, 154*(3), E57.
5. Upadhyay, R. K., Kumar, A., & Srivastava, P. K. (2017). Experimental investigations of catalytic effect of Cu^{2+} during anodic dissolution of iron in NaCl electrolyte. *Proceedings of the Institution of Mechanical Engineers, Part B: Journal of Engineering Manufacture, 231*(13), 2408–2415.
6. Upadhyay, R. K., Kumar, A., & Srivastava, P. K. (2016). High rate anodic dissolution of stainless steel 316 (SS316) using nano zero valent iron as reducing agent. *Journal of Applied Science and Engineering, 19*(1), 47–52.
7. Upadhyay, R. K. (2016). Thermodynamic feasibility of low valence dissolution of titanium during electrochemical machining of titanium in NaCl. *International Journal of Interdisciplinary Research Centre, 2*(1), 37–51.
8. Bhattacharyya, B., Malapati, M., & Munda, J. (2005). Experimental study on electrochemical micromachining. *Journal of Materials Processing Technology, 169*(3), 485–492.
9. Mukherjee, S. K., Kumar, S., & Srivastava, P. K. (2007). Effect of electrolyte on the current-carrying process in electrochemical machining. *Proceedings of the Institution of Mechanical Engineers, Part C: Journal of Mechanical Engineering Science, 221*(11), 1415–1419.
10. da Silva Neto, J. C., Da Silva, E. M., & Da Silva, M. B. (2006). Intervening variables in electrochemical machining. *Journal of Materials Processing Technology, 179*(1–3), 92–96.
11. Sen, M., & Shan, H. S. (2005). Analysis of hole quality characteristics in the electro jet drilling process. *International Journal of Machine Tools and Manufacture, 45*(15), 1706–1716.
12. Ayyappan, S., & Sivakumar, K. (2014). Experimental investigation on the performance improvement of electrochemical machining process using oxygen-enriched electrolyte. *The International Journal of Advanced Manufacturing Technology, 75*(1–4), 479–487.
13. Ayyappan, S., & Sivakumar, K. (2016). Enhancing the performance of electrochemical machining of 20MnCr5 alloy steel and optimization of process parameters by PSO-DF optimizer. *The International Journal of Advanced Manufacturing Technology, 82*(9–12), 2053–2064.
14. Sekar, T., & Marappan, R. (2007). Experimental studies on effect of tool geometry over metal removal rate in ECM process. *Journal Nonconventional Technologies, 3*, 107–110.
15. Hewidy, M. S. (2005). Controlling of metal removal thickness in ECM process. *Journal of Materials Processing Technology, 160*(3), 348–353.

Soft Modeling of WEDM Process in Prediction of Surface Roughness Using Artificial Neural Networks

H. D. S. Aiyar, G. Chauhan, and N. Gupta

1 Introduction

Modern-day industries find huge application of materials like composites, ceramics and super alloys due to their superior mechanical properties like better fatigue characteristics, high bending stiffness, high strength, good damping capacity and low thermal expansion. These materials are widely used in manufacturing of automotive, aviation and surgical components. Such materials are difficult to machine owing to their superior characteristics. Non-conventional machining processes like WEDM can be used to produce a precise and complex irregular-shaped job [1].

WEDM works by creating a voltage difference between the workpiece and the wire electrode which is immersed in a dielectric medium. The wire electrode serves as the cathode, and the workpiece serves as the anode. Between the workpiece and wire electrode, a certain gap is maintained so that the spark is generated and is sustained due to the breakdown of dielectric medium. The spark produces a very high temperature which erodes the metal surface. Upon cooling, the metal chips are removed by the dielectric fluid due to its flushing effect. There is a huge surge of interest in WEDM process as it produces highly precise and accurate surface quality parts [2, 3].

The input parameters of WEDM include pulse off time, voltage, pulse on time, dielectric pressure, wire tension, wire feed rate and peak current which affects the operating parameters [4–8]. Surface roughness affects the friction, fatigue, corrosion resistance, light reflection and wear resistance. Determining surface roughness by analytical equations is very difficult due to the intricacies involved in the formation mechanism of surface roughness process. Surface profilometer is used in measuring the surface roughness which is a time-consuming and expensive process and also requires trained workers [5–13]. Thus, an attempt has been made in this paper to

H. D. S. Aiyar · G. Chauhan · N. Gupta (✉)
Mechanical Engineering Department, DTU, New Delhi, India

R. Agrawal et al. (eds.), *Recent Advances in Smart Manufacturing and Materials*,
Lecture Notes in Mechanical Engineering,
https://doi.org/10.1007/978-981-16-3033-0_44

develop a mathematical model for measuring the surface roughness based on the process parameters.

Over the years, numerous researches have been performed in predicting the surface roughness using experimental and analytical models. But in recent times, artificial intelligence approaches have become highly popular over conventional approaches. The AI approaches do not require the explicit defining of the machining process. Only a dataset of the process involving the process parameters and operating parameters is sufficient which can be designed by performing the experiments. Artificial neural networks (ANNs) are one of the most popular and widely used artificial intelligence techniques for developing predictive models. ANN is an artificial way to represent a human brain which is used to simulate the learning process. An ANN is a directed graph where nodes perform calculations and are connected with other nodes.

Upon reviewing the literature, a lot of information regarding ANN was acquired. Tarng et al. predicted the two most important operating parameters of the WEDM process, cutting speed and surface roughness, using ANN. Then, the optimal cutting parameters were predicted using a simulated annealing algorithm [14]. Spedding and Wang modeled the WEDM process with very high accuracy using responsive surface methodology and ANN. Autoregressive models were also used to describe the stochastic components of surface roughness [15]. Risboodet et al. predicted dimensional deviation and surface roughness of a mild steel rod during the dry and wet turning process using ANN. Taguchi's experimental technique was used in developing the dataset containing 26 experiments [16]. Caydas et al. worked on building an adaptive neuro-fuzzy inference system which can estimate the surface roughness and white layer thickness [17].

In one of the studies performed by Ramaiah and Rajyalakshmi, the WEDM process parameters—mrr and surface roughness—were optimized by Taguchi's grey relational analysis. Taguchi's L36 orthogonal array was used to prepare the dataset for the experiment [18]. Huan et al. in their study selected the optimal process parameters of WEDM process using grey relational analysis. A Taguchi's L16 dataset was built by performing experiments on the Chromium12 alloy steel [19]. Neeraj et al. in their study used response surface methodology to predict the surface roughness and metal removal rate on a high strength low alloy steel workpiece [20]. Kuppan et al. prepared a dataset with current, power-on, rotational speed and duty factor as the process parameters using Taguchi's experimental design method. Response surface methodology technique was used in the estimation of average metal removal rate and surface roughness and [21]. In another study by UlasCaydas and AhmetHascalik, response surface methodology was used to predict the recast layer thickness and electrode wear. Analysis of variance (ANOVA) pointed out that the output parameters are significantly influenced by the input process parameters [22].

In the present study, the dataset for WEDM was obtained from the article: Process optimization and estimation of machining performances using artificial neural network in wire EDM [23]. CONCORD DK7720C four axes wire electric discharge machine was used for performing the experiment. Bed speed, pulse on, current and pulse off were chosen as the process parameters and surface roughness as the output parameter. Taguchi's L16 orthogonal array was used for designing the experiment. The obtained surface roughness values were predicted using ANN. The ANN model is trained on MATLAB NNTOOLBOX. Feed forward back propagation neural networks are used for building the model. Levenberg–Marquardt algorithm is chosen as the optimizer as it converges the result very quickly with high accuracy and is best suited for small datasets.

2 Experimental Work

The WEDM dataset used was an experimental dataset used in research work [23]. In this work, the WEDM experimentation was conducted on CONCORD DK7720C four axes CNC wire electric discharge machine. According to the workpiece material and desired output, the person who operates the machine has to choose the parameters for input. Pulse off, pulse on, current and bed speed are the four process parameters considered to investigate the surface roughness of the workpiece.

In Fig. 1, schematic setup of the WEDM is shown. High-frequency AC current is ejected from the wire to the workpiece, and a spark is generated. The positioning system which is controlled by the computer maintains a 0.02 mm gap between the wire electrode and workpiece. Figure 2 shows the experimental setup on the CONCORD DK7720C four axes CNC WEDM which uses a molybdenum wire with 0.18 mm diameter as the electrode. Fixed parameters and the control features selected are listed in Table 1. The debris is removed by flushing the dielectric fluid (water is used in this case). Each of the control factors have four levels and are listed in Table 1 [23].

Classical experimental design methods are complex and become highly expensive in terms of time as well as cost. So, Taguchi method is used to conduct experimentation efficiently [23]. Taguchi's L16 orthogonal array is used for designing the experiment. The process parameters are shown in Table 2 with their selected levels.

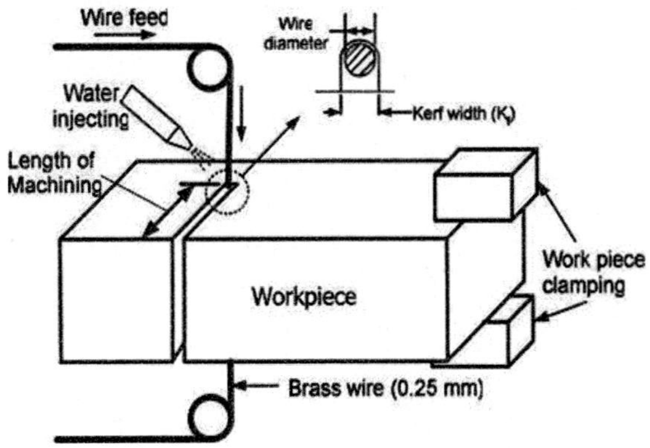

Fig. 1 WEDM schematic diagram

Fig. 2 Experimental setup

Table 1 Input parameters and their levels

S. No.	Control factors	Level 1	Level 2	Level 3	Level 4
1	Pulse on	16	20	24	28
2	Pulse off	4	6	8	10
3	Current	3	4	5	6
4	Bed speed	20	25	30	35

Table 2 Experimental results

Run (μm)	Pulse on (μm)	Pulse off (μm)	Current (A)	Bed speed (μm/s)	Surface roughness (μm)	Predicted surface roughness (μm)
1	16	4	3	20	2.392	2.392000017
2	16	6	4	25	2.384	2.383999979
3	16	8	5	30	2.526	2.511744596
4	16	10	6	35	2.792	2.945800769
5	20	4	4	30	2.631	2.615439361
6	20	6	3	35	2.541	2.541
7	20	8	6	20	2.843	2.842999994
8	20	10	5	25	2.891	2.891000011
9	24	4	5	35	3.317	3.316999992
10	24	6	6	30	3.281	3.280999991
11	24	8	3	25	2.476	2.476000018
12	24	10	4	20	2.253	2.655811313
13	28	4	6	25	3.024	3.024000013
14	28	6	5	20	2.586	2.585999993
15	28	8	4	35	2.482	2.482000025
16	28	10	3	30	2.348	2.347999969

3 ANN Model Development and Training

ANN is an artificial way to represent a human brain which is used to simulate the learning process. An ANN is a directed graph where nodes perform calculations and are connected with other nodes. The connections transfer a signal from one node to another which are linked with some weight. These weights are responsible for the augmentation of the signal. The weights in the graph are randomly initialized, and a learning function tweaks the weights and biases after every epoch such that the predictions come closer to the target values.

In this study, feed forward back propagation neural networks will be used. The ANN was created using the neural network toolbox in MATLAB as shown in Fig. 3. The neurons are arranged in a first-order sequence, and the last layer receives the results. Since four process parameters have been chosen, the input layer consists of four neurons. The output layer has a single neuron with linear activation which predicts the surface roughness. In between the input and the output layers, hidden layers exist. Based upon repeated experiments and extensive literature review, 16 neurons in a single hidden layer with Tanh activation were found to be the most favorable architecture. Tanh activation binds the activation between -1 and $+1$. Continuously fed activation rates from one layer to another in a forward direction allow their names as a feed supply network.

The learning function optimizes the weights and biases so that the predictions come close to the actual values. Levenberg–Marquardt algorithm (trainlm) will be used as the learning function because of its fast convergence and is best suited for small datasets. The metrics as well as the loss function is chosen as mean square error. The dataset was divided randomly into training and testing dataset each with 70% and 30% of the total data, respectively. The networks were trained up to 3000 epochs.

4 Results and Discussions

After the models were trained, tested and evaluated successfully, the model with 4-16-1 architecture was found to be the most optimum in the present work. The neural network toolbox of MATLAB was greatly used in predicting the output parameters, and errors were well within the range of experimental and predicted values of the ANN model. The model was trained for 3000 epochs, and the goal MSE for training was kept at $1e-8$ (Fig. 4).

The performance of the model is calculated using mean squared error method. The mean of squared difference between actual and predicted values is known as MSE. The values close to 0 are desired while using MSE. In Fig. 5, the validation performance shows that the optimum solution is reached at zeroth epoch. The MSE target is achieved in third epoch with training MSE $= 2.01e-16$ and testing MSE $=$

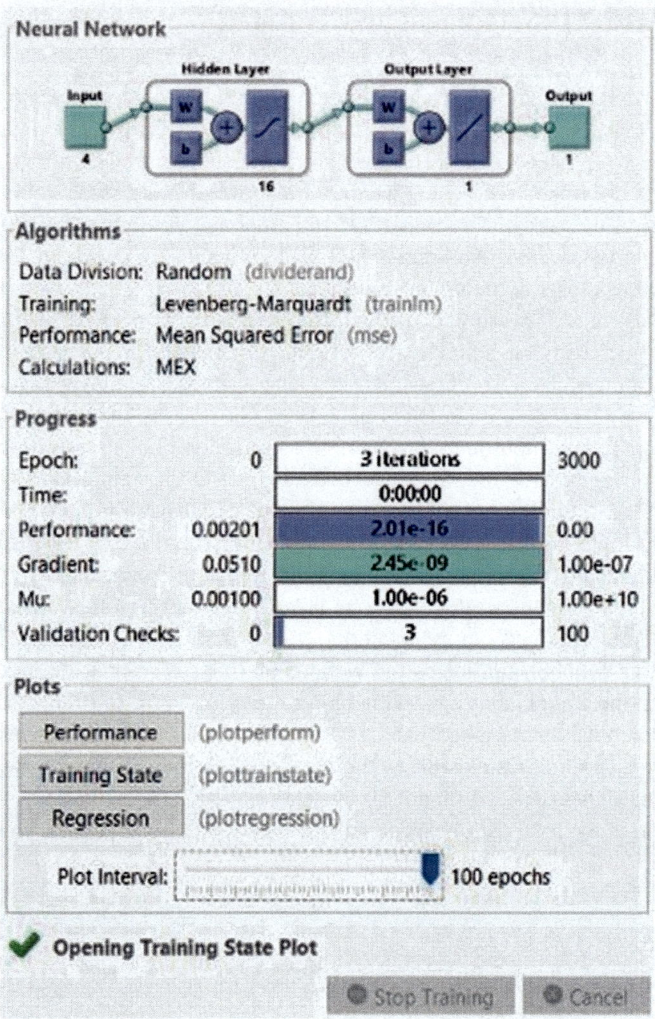

Fig. 3 MATLAB NNTOOLBOX settings

1e − 1. The predicted surface roughness values for the whole dataset are very close to the experimental values.

In Fig. 6, the regression values for training are 0.98903, testing is 1, and validation is also 1. The regression values are closer to unity, indicating that the ANN model is able to very well capture and learn the behavior of the trained dataset.

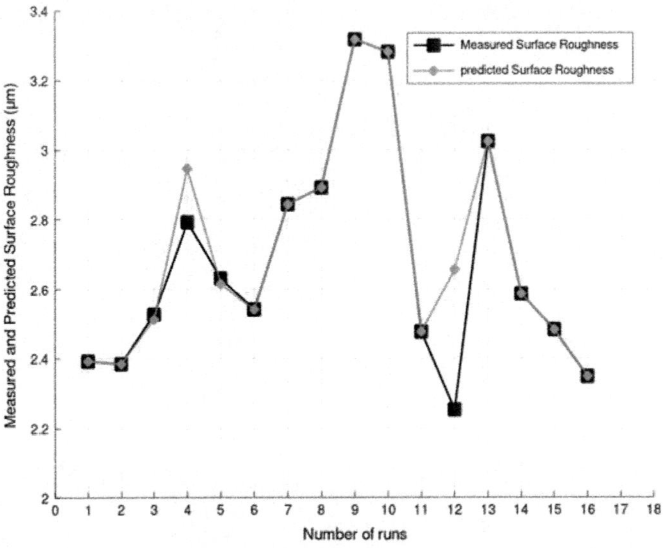

Fig. 4 Comparison of predicted and experimental values

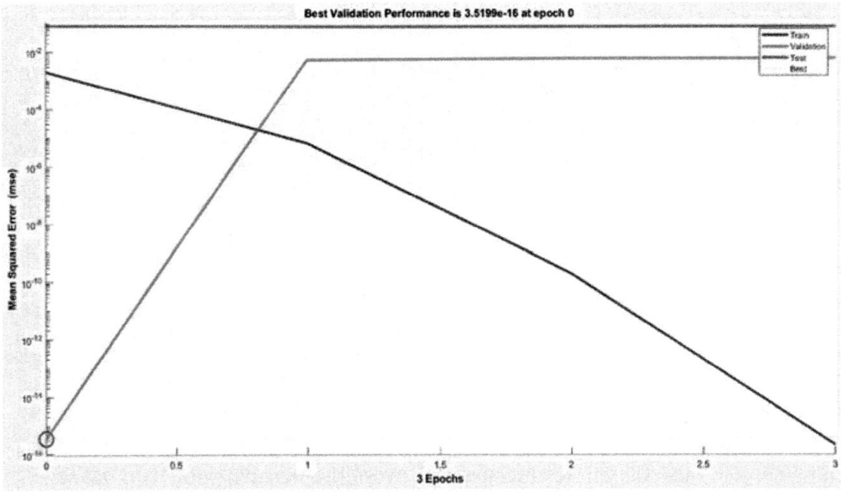

Fig. 5 Error plot of train, test and validation

5 Conclusion

This paper has presented a mathematical model for the prediction of surface roughness using ANN which will lead to massive reduction in production time and cost,

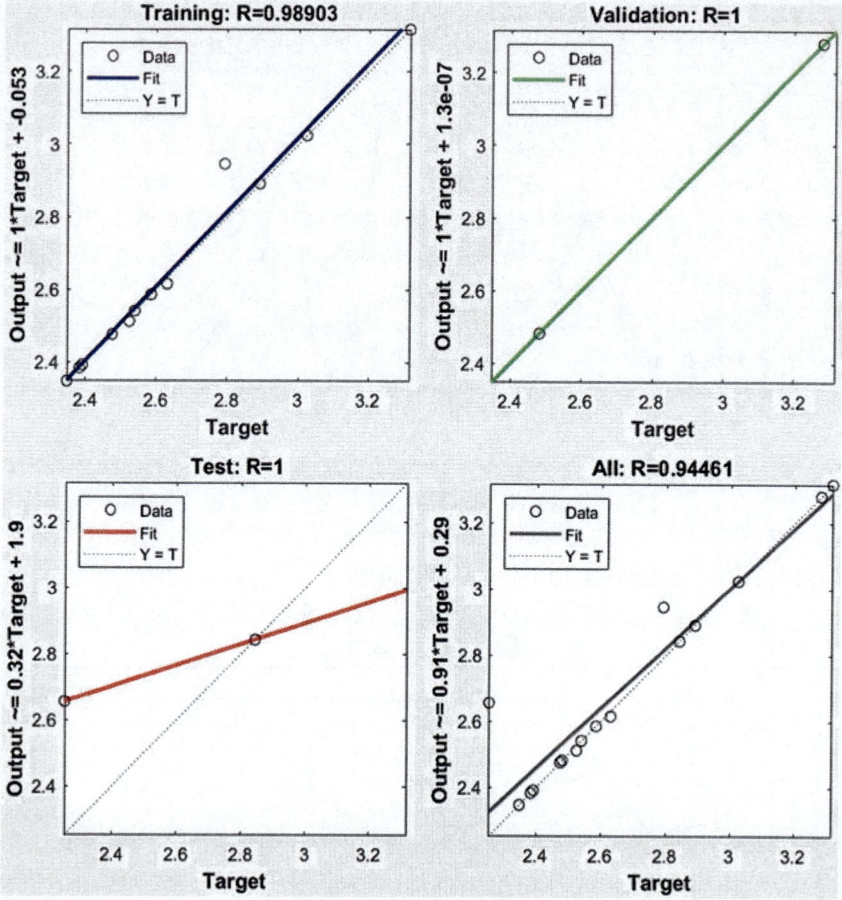

Fig. 6 Regression plots of train, test, validation and overall dataset

eventually making the WEDM process highly efficient and paves the way for automation. The 4-16-1 ANN model is effectively used for prediction of output parameters with training MSE 2.01e − 16 and testing MSE = 1e − 1 on a batch of 70% data in training set and 30% data in testing set. The overall regression value is 0.94 which is very close to 1, indicates that the ANN model has performed exceptionally well and has high efficacy. This technique can be generalized and used to calculate the operating parameters of other non-conventional machining processes, and this leads to the focus of our further research work.

The ANN predicted values are very close to actual experimental values, which makes it accurate and reliable [24]. The soft modeling approach is widely used in the optimization of various manufacturing process problems.

References

1. Mohd Abbas, N., Solomon, D. G., & Fuad Bahari, Md. (2007). A review on current research trends in electrical discharge machining (EDM). *International Journal of Machine Tools and Manufacture, 47*(7–8), 1214–1228. https://doi.org/10.1016/j.ijmachtools.2006.08.026
2. Sharma, V., & Sagar, M. K. (2017). Modeling the wire-EDM process parameters for EN-8 carbon steel using artificial neural networks. *International Journal of Engineering, Science and Technology, 9*(2), 26. https://doi.org/10.4314/ijest.v9i2.3
3. Srinivasa Rao, P., Ramji, K., & Satyanarayana, B. (2016). Effect of wire EDM conditions on generation of residual stresses in machining of aluminum 2014 T6 alloy. *Alexandria Engineering Journal, 55*(2), 1077–1084. https://doi.org/10.1016/j.aej.2016.03.014
4. Surya, V. R., Kumar, K. M. V., Keshavamurthy, R., Ugrasen, G., & Ravindra, H. V. (2017). Prediction of machining characteristics using artificial neural network in wire EDM of Al7075 based in-situ composite. *Materials Today: Proceedings, 4*(2), 203–212. https://doi.org/10.1016/j.matpr.2017.01.014
5. Thankachan, T., Soorya Prakash, K., Malini, R., Ramu, S., Sundararaj, P., Rajandran, S., Rammasamy, D., & Jothi, S. (2019). Prediction of surface roughness and material removal rate in wire electrical discharge machining on aluminum based alloys/composites using Taguchi coupled grey relational analysis and artificial neural networks. *Applied Surface Science, 472*, 22–35. https://doi.org/10.1016/j.apsusc.2018.06.117
6. Tosun, N., & Pihtili, H. (2003). The effect of cutting parameters on wire crater sizes in wire EDM. *The International Journal of Advanced Manufacturing Technology, 21*(10–11), 857–865. https://doi.org/10.1007/s00170-002-1404-1
7. Saha, P., Singha, A., Pal, S. K., & Saha, P. (2007). Soft computing models based prediction of cutting speed and surface roughness in wire electro-discharge machining of tungsten carbide cobalt composite. *The International Journal of Advanced Manufacturing Technology, 39*(1–2), 74–84. https://doi.org/10.1007/s00170-007-1200-z
8. Chaudhari, R., Vora, J., Parikh, D. M., Wankhede, V., & Khanna, S. (2020). Multi-response optimization of WEDM parameters using an integrated approach of RSM–GRA analysis for pure titanium. *Journal of the Institution of Engineers (India): Series D, 101*(1), 117–126. https://doi.org/10.1007/s40033-020-00204-7.
9. Çaydaş, U., & Hasçalık, A. (2008). A study on surface roughness in abrasive waterjet machining process using artificial neural networks and regression analysis method. *Journal of Materials Processing Technology, 202*(1–3), 574–582. https://doi.org/10.1016/j.jmatprotec.2007.10.024
10. Wang, X., Huang, C., Zou, B., Liu, G., Zhu, H., & Wang, J. (2017). Experimental study of surface integrity and fatigue life in the face milling of inconel 718. *Frontiers of Mechanical Engineering, 13*(2), 243–250. https://doi.org/10.1007/s11465-018-0479-9
11. Maher, I., Eltaib, M. E. H., Sarhan, A. A. D., & El-Zahry, R. M. (2014). Cutting force-based adaptive neuro-fuzzy approach for accurate surface roughness prediction in end milling operation for intelligent machining. *The International Journal of Advanced Manufacturing Technology, 76*(5–8), 1459–1467. https://doi.org/10.1007/s00170-014-6379-1
12. Shucai, Y., Chunsheng, H., & Minli, Z. (2018). A prediction model for titanium alloy surface roughness when milling with micro-textured ball-end cutters at different workpiece inclination angles. *The International Journal of Advanced Manufacturing Technology, 100*(5–8), 2115–2122. https://doi.org/10.1007/s00170-018-2852-6
13. Natarajan, C., Muthu, S., & Karuppuswamy, P. (2011). Prediction and analysis of surface roughness characteristics of a non-ferrous material using ANN in CNC turning. *The International Journal of Advanced Manufacturing Technology, 57*(9–12), 1043–1051. https://doi.org/10.1007/s00170-011-3343-1
14. Tarng, Y. S., Ma, S. C., & Chung, L. K. (1995). Determination of optimal cutting parameters in wire electrical discharge machining. *International Journal of Machine Tools and Manufacture, 35*(12), 1693–1701. https://doi.org/10.1016/0890-6955(95)00019-t

15. Spedding, T. A., & Wang, Z. Q. (1997). Parametric optimization and surface characterization of wire electrical discharge machining process. *Precision Engineering, 20*(1), 5–15. https://doi.org/10.1016/s0141-6359(97)00003-2
16. Risbood, K. A., Dixit, U. S., & Sahasrabudhe, A. D. (2003). Prediction of surface roughness and dimensional deviation by measuring cutting forces and vibrations in turning process. *Journal of Materials Processing Technology, 132*(1–3), 203–214. https://doi.org/10.1016/s0924-0136(02)00920-2
17. Çaydaş, U., Hasçalık, A., & Ekici, S. (2009). An adaptive neuro-fuzzy inference system (ANFIS) model for wire-EDM. *Expert Systems with Applications, 36*(3), 6135–6139. https://doi.org/10.1016/j.eswa.2008.07.019
18. Rajyalakshmi, G., & Venkata Ramaiah, P. (2013). Multiple process parameter optimization of wire electrical discharge machining on Inconel 825 using Taguchi grey relational analysis. *The International Journal of Advanced Manufacturing Technology, 69*(5–8), 1249–1262. https://doi.org/10.1007/s00170-013-5081-z
19. Huang, J. T., & Liao, Y. S. (2003). Optimization of machining parameters of Wire-EDM based on grey relational and statistical analyses. *International Journal of Production Research, 41*(8), 1707–1720. https://doi.org/10.1080/1352816031000074973
20. Sharma, N., Khanna, R., & Gupta, R. (2013). Multi quality characteristics of WEDM process parameters with RSM. *Procedia Engineering, 64*, 710–719. https://doi.org/10.1016/j.proeng.2013.09.146
21. Kuppan, P., Rajadurai, A., & Narayanan, S. (2007). Influence of EDM process parameters in deep Hole drilling of Inconel 718. *The International Journal of Advanced Manufacturing Technology, 38*(1–2), 74–84. https://doi.org/10.1007/s00170-007-1084-y
22. Çaydaş, U., & Hasçalik, A. (2007). Modeling and analysis of electrode wear and white layer thickness in die-sinking EDM process through response surface methodology. *The International Journal of Advanced Manufacturing Technology, 38*(11–12), 1148–1156. https://doi.org/10.1007/s00170-007-1162-1
23. Ugrasen, G., Ravindra, H. V., Prakash, G. V. N., & Keshavamurthy, R. (2014). Process optimization and estimation of machining performances using artificial neural network in wire EDM. *Procedia Materials Science, 6*, 1752–1760. https://doi.org/10.1016/j.mspro.2014.07.205
24. Gupta, N., Agrawal, A. K., & Walia, R. S. (2019, May). Soft modeling approach in predicting surface roughness, temperature, cutting forces in hard turning process using artificial neural network: An empirical study. In *International conference on information, communication and computing technology* (pp. 206–215). Springer, Singapore.

Education 4.0 to Industry 4.0 Vision: Current Trends and Overview

Shruti Agrawal, Nidhi Sharma, and Sumedha Bhatnagar

1 Introduction

1.1 Industrial Revolution till 4.0

The term Industry 4.0 was firstly coined in 2011 in Germany that includes the representatives from business class, political and academic field, together they initiate the word known as Industry 4.0 [1]. Industry 4.0 can be considered as recent trend in manufacturing organization which aims at the exchaning automation and data between the physical entities. The motive of the government of Germany was to promote computerization of manufacturing via a research project. The term "Industry 4.0" was then publicly introduced in 2011 at Hannover Fair. It is also known as IIoT that is Industrial Internet of things or smart manufacturing that denotes physical production along with the smart digital technology, machine learning and large data that complies into a more holistic and updated connected ecosystem that focus on manufacturing and supply chain management for the companies. The concept of Industry 4.0 is a transformation that enables faster, flexible and efficient processes that generates high-quality products at minimum cost. The emergence of Industry 4.0 helps the manufactures to overcome current challenges and issues by adopting more flexible approaches that enhance the speed of innovation and give enough focus to customer-centred approaches. Further, different components for Industry 4.0 is represnted in Fig 1.

The revolution in industrial sector has evolved in several steps from almost past 200 years ago [2]. In Table 1, the evolution from phase Industry 1.0 to latest Industry 4.0 is shown.

S. Agrawal (✉) · N. Sharma · S. Bhatnagar
Department of Humanities and Social Sciences, Malaviya National Institute of Technology Jaipur, Jaipur, Rajasthan, India
e-mail: 2019rhs9506@mnit.ac.in

R. Agrawal et al. (eds.), *Recent Advances in Smart Manufacturing and Materials*,
Lecture Notes in Mechanical Engineering,
https://doi.org/10.1007/978-981-16-3033-0_45

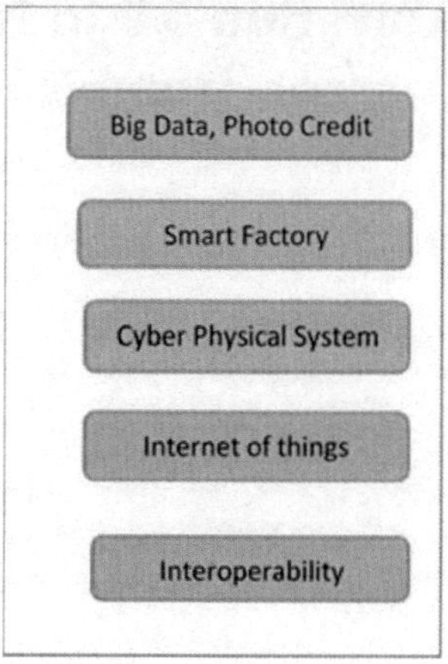

Fig. 1 Components of Industry 4.0 revolution

Table 1 Evolution from Industry 1.0–4.0

Industrial Revolution	Time period	Technology
Industry 1.0	Till 1760	Mechanization, steam power and weaving loom
Industry 2.0	In the beginning of twentieth century	Mass production, assemble lines, electrical power
Industry 3.0	In the beginning of 1970s period	Automation production, computers, IT systems and robotics
Industry 4.0	Present period	The smart factory, autonomous system, IoT, machine learning

Dating back towards 1760, the first phase of Industrial Revolution starts that considers the process of water and steam power machines in the field of manufacturing and mechanical production [3]. The first field that experienced such impact was laid down on the textile industry. The first phase was thoroughly favourable in terms of manufacturing a huge number of goods that creates an improved standard of living. Since after Industry 1.0 has taken place, there has been a massive increase in the production sector. This is then followed by the second revolution of industry which is known as Industry 2.0 and also "The Technological Revolution" as common.

In the beginning of twentieth century, in second revolution, electricity has played a major role in the entire process which results into greater production and suave machines. The first T production process model developed by Henry's Ford proved as a climax factor in the revolution chain along with the establishment of conveyor belts and increased demand of labour in the manufacturing sector.

In the phase of Industry 3.0, the third Industrial Revolution began with the first computer era. Earlier, these computers were often very easy, difficult and incredibly huge relative to the computing power before they were able to supply. In this stage, the use of programmable electronic devices was implemented by the industries. These updated programmable devices are more efficient and flexible in the production of digital automation. In the modern automation system sector, the Third Industrial Revolution is still being used.

Industry 4.0, the Fourth Industrial Revolution, is known as the era of smart machines with more storage systems and feasible production facilities that create autonomous exchange of information, trigger actions and facilitate each other with no human interference. With the support of the Industrial Internet of things (IIoT), the exchange of information is being made possible. In the Industry 4.0, it has been seen the major two key development directions that can be defined in expected future time [2]. Application-pull is the first direction that states the need for changes in way to adoption of dynamic operative framework conditions. Such directions lead to flexibility on the development of products, lessen the development of product time, and customers choose to individualize the products.

On the other hand, the second direction is known as technology-push that leads to the improvement of automation, mechanization, digital technology and networking in the industry sector [4]. The focus of this direction is implied on the technology that is being used in the current time in the industries and manufacturing sectors including the factors on how to improvise it more.

The fundamental paradigm shifts in Industry 4.0 can witness the combination of Internet technologies along with the future-oriented technologies. It could enable the open communication pathway with each other in between the devices, machines, sensors and people by making use of Internet technology known as Internet of Things (IoT) [5]. This entire process is named as cyber-physical system (CPS) in which the whole components in the system, machines-to-machines and machines-to-human, could hold communication with each other and transfer from the production to consumption process [6].

This concurrent revolution in industry may impact various other fields except manufacturing, such as recruitment of labour workforce, information and communication technology (ICT), various ventures, business, education sector and much more [7]. There will be significant change in the recruitment of labour workforce in Industry 4.0 as the talent needed will be completely changed as will the updating technologies. The different talents or skills are required in order to compress some basic technical qualifications in specified industries. The knowledge of information technology (IT), organizational understanding, procession knowledge and new technologies is important for the development of Industry 4.0.

In case of business sector like small–medium enterprises (SME), Industry 4.0 has made it to go through various difficulty as they are not skilled as needed in terms of applications and technical surroundings [8]. The reasons behind this are that the SMEs do not owe a skilled labour class staff who possess the basic skills to cooperate with current technologies, less availability of funds which is required to set up the updated technologies, and hence, it counts as loss to their business.

In the education sector, Industry 4.0 has played a major role and turned the wheel out for a lot of big changes [9]. In the era of the new Industrial Revolution, the industries now demand the more qualified candidates to work with them as it leads to better quality of production and results [10]. In the education sector, it provides the candidates with required training and skills and by the due course makes them prepare as a qualified candidate for the industry sector in upcoming time [5]. Along with this entire concept, a new term has generated in between researchers, education practitioners and policy makers that is Education 4.0.

India, being a developing country, will constantly face the same challenge with Industry 4.0 and Education 4.0. Hence, this paper aims to highlight the overall overview on Indian higher education system towards Industry 4.0. This paper comprises various reported literature amended on Industry 4.0 and Education 4.0 from 2011 to up till the present policies. The use of various online platforms are being used during this study which includes Google Scholar, Science Direct, Web of Science and policies by Indian government. The use of snowballing method is being used so that it helps to gather more sources as maximum as possible also keeping in view the emergence of Education 4.0.

The paper is being categorized in five sections that are as follows:

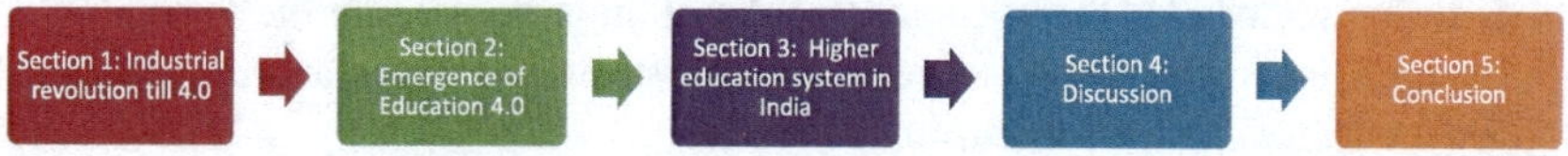

2 Emergence of Education 4.0

Education 4.0 is the term appeared publicly during the uptrend of Industry 4.0 as it inevitably motivates almost every other sector and most specially the education sector. The role of teaching methods now would rather focus on student-centric approach instead of teacher-centric approach along with the updated technology in teaching style. At the time of the pandemic of COVID-19, it resulted into the entire reform of the education system, and thus, it may promote Education 4.0 in almost every country. Adopting such changes at the time of pandemic has given a boost to Education 4.0 in alignment with Industry 4.0.

2.1 *Education Revolution*

The revolution of education is as similar to the Industry 4.0 that could be categorized on the basis of its emergence from the past as shown in Table 2. The evolution of education started decades ago or even centuries ago. In that period, teaching verbally was a mode of transmission of knowledge, teachers act as a medium to share that knowledge, and students just receive the information in the specified manner.

Table 2 Education 1.0–4.0

Education revolution	Methods	Technology used
Education 1.0	Oral dictation instructions One-way communication	No technology is being used at the time of education process
Education 2.0	Progressive A step towards access to Internet	Limited access
Education 3.0	Production of knowledge Co-constructivism	Allow full access for knowledge, constructivism and transmission
Education 4.0	Innovation and invention Virtual classrooms	Blended learning More of Internet and technology driven

Education is just not simply a means of transferring knowledge, but it is about imparting the knowledge in such a manner that may enhance an individuals' perspective and help him to make a standard of living and survive in the world. Hence, with every Industrial Revolution, we can see a subsequent revolution in the education sector too. The vision of Education 4.0 has been modelled a decade ago. The research paper published by Harkins described "leapfrogging" and its practices and paradigm of education [11]. Leapfrogging stands to go forward in the ongoing competition or the current state of art via innovation, saving time and cost.

In the education sector, the teaching learning process has evolved along with the improvement of technology. Education 1.0 is more like the first generation of Web that is one-way process. Teaching–learning activity takes place in the classroom. Teachers are the suppliers of knowledge, and students are the receptors. This method is called as instructivism where the knowledge receivers do not interact much and behave passively [12]. Mainly, it can be said as a teacher-centric approach where the decision of teachers is meant to be of utmost importance.

Education 2.0 initiated with the boom of Web 2.0 that recognized this education revolution [13]. It promotes the communication among individuals and users, and also, it is the initial phase of growth of social media. Education 2.0 is grounded on progressivism where the learning process is not just restricted to teacher to student but also student to student and student to experts. The growth of Education 2.0 gives enough emphasis on the use of the Internet as a medium of learning activity though its access is limited. Also, Education 2.0 focus on the use of skype, wikis, sharing of blogs and use of other social media networking websites in the classrooms for teaching purposes.

Education 3.0 is purely constructed on the belief that content is actively and freely available. This phase is self-directed, interest-based learning in which problem-solving approach along with innovation and creativity is promoted with the education. In this revolution, there are no specified teaching professionals but also those who have required source of materials can conduct the learning activities. It can also be stated as a cross-institutional, cross-cultural way of education opportunity. The entire learning phase is student-centric.

Education 4.0 is the updated recent trend that caters to the needs of society as a whole in the "innovative era". Parallelism, connectivism and visualization are the special characteristics of Education 4.0 [14]. This recent learning behaviour surely motivates the learners to develop critical thinking and cognitive thinking skills along with the application of new technology. Education 4.0 is trusted to empower the learners to come up with innovations and substantiations production of knowledge [15].

2.2 Applied Technology

During the continuous revolution in education, various technologies are used in the entire learning process. Early in the nineteenth and twentieth century, from 1840 to 1950, the correspondence and broadcast courses were the first technology that is being used in the education sector. Sustainable developement and minimizing carbon emissions are the major concerns in industries of developing nations [16–18]. This mentioned technology included mail facility to the community to spread the knowledge content which was further followed by the operations of televisions and radio to transmit the information.

At the period of twentieth century, in this phase of revolution of education, mainly the technologies are based on the computer-based training programs. In 1960, Donal L. Bitzer developed the first software named as Programmed Logic for Automation Operations (PLATO) to the learning-based computers. However, the role of training programs and local awareness program is also important for sustainability practices in industries [19]. After this development, many familiar technologies have been started to develop by the other researchers. In Fig. 2, the timeline of education has been illustrated.

The first educational revolution was initiated in the 1990s in alignment with the procurement of the World Wide Web (WWW). This is known as Education 1.0 revolution which was completely aligned with the first stage of Web 1.0. The first stage of WWW Web 1.0 development is a one-way information and one-way process, but this technology was used too less in the Education 1.0 phase [12].

The Web 2.0 was developed along with the second revolution of known as Education 2.0 [13]. The Web 2.0 is the next stage of WWW with some additional read and write characteristics and also an improvement over the first phase of Web 1.0. The interactions between machine and human are being permitted in this phase. The concept of blended learning in education comes after this technology in the 2000s as a new teaching learning method. Blended learning is a combination of classroom face-to-face learning and e-learning. E-learning is electronic learning where the learning process takes place via use of the Internet. It is also a mode of asynchronous learning in which the education is being transferred even when the instructor and the learner are not present at the same time and same place. The learning can take place according

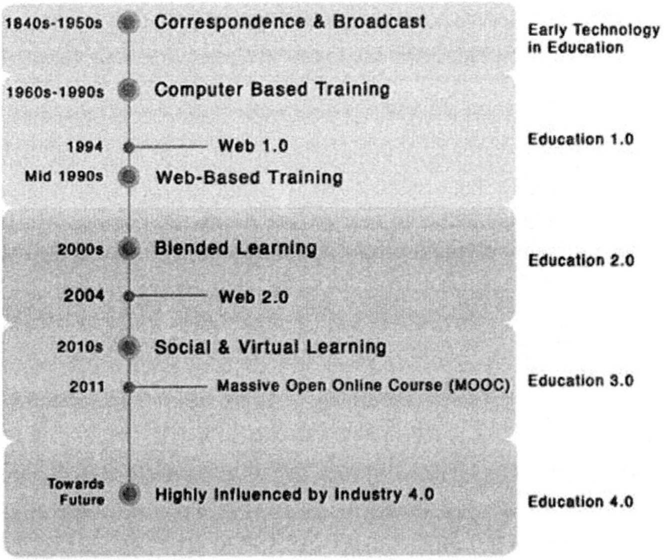

Fig. 2 Timeline of educations

to the convenience of the learner. Such technologies via the Internet can be operated on personal computers (PC), laptop, tablet, kindle, mobile phone, and thus, the production of such is increasing on a daily basis.

"www" has become a great source of providing information via the Internet to the large numbers of users. The sudden growth and the increased demand of the Internet, trend of social media caused the third phase of educational revolution that is Education 3.0. In the year 2010, the concept of virtual learning emerged because of increased demand for education via social media platforms (S. Jeschke, Higher Education 4.0—Trends and Future...—Google Scholar, n.d.). Dave Cormier introduced the concept of Massive Open Online Learning (MOOC) that gives a platform to students to learn anywhere and anytime. It provides huge data of content in an open access to every learner and breaks the traditional form of learning.

Education 4.0 is an emerging concept of education, and importantly, at the time of COVID-19 pandemic, it has given a boost to this phase of revolution. Industry 4.0 include Cyber physical systems and machine learning which requires skilled labour set for industries [20, 21]. It is a non-traditional form of education where face-to-face learning, classroom learning no longer serve. It is a technology-based learning and may call it the future of learning. It moves in a direction that is more progressive, innovative, intellectual, knowledge-driven and future-ready.

2.3 Link Between Education 4.0 and Industry 4.0

The educational sector is already being determined by the cognitive and cloud computing, IoT, cyber-physical system and various other parameters imposed by Industry 4.0. Students have become technological driven in their field. Education sector has left with no choice but instead to adopt changes as specified in industry that are bought due to disruptive technologies. By providing the flexible, self-learning, self-sustainable, problem-solving, solution-oriented, collaborative and mentoring environment to the learners, the future generation will definitely respond to the need of Industry 4.0.

2.4 How Education 4.0 Can Revolutionize the Wave the Industry 4.0

Education 4.0 gives an opportunity to all the learners to grow and learn at their own pace and to engage in the updated technology and software that enhance their skills and knowledge in depth for technology [22]. The art of creative thinking, creative writing will also emerge that allows them to think out of the box and provides innovative and sustainable solutions for the problem [23]. The role of the Internet in all over revolution is foremost and a main source that provides knowledge in the learning process.

2.5 How Education 4.0 Creates Jobs for Industry 4.0

Adopting Education 4.0 may lead to the presence of more skilled, qualified, intelligent and technical-friendly students at the time when smart manufacturing and production reduces the proportion of labour-intensive opportunities. The manufacturing industry are labour-intensive and Industry 4.0 is a skill-intensive which requires an array of skills. The available jobs for students in the upcoming trend include flexible and innovative thinking, emotional and social intelligence, virtual collaboration, designing mind-set, "thinking like a data-scientist" and interdisciplinary [24]. Machines are still unable to replicate judgement, insight, conviction, opinion and communication ability.

3 Higher Education System in India

The education system of India is being divided into two segments that are pre-independence era and post-independence era. Ancient education system has been very wide in India. In ancient times, education was categorized into two systems that are "Vedic and Buddhist". After then, vocational education system was promoted that includes the knowledge of professions and art in education. When ancient education started declining, that was the phase of evolution of higher education in post-independence India.

As per the recent trend in education, Education 4.0 will be the upcoming next happening to the world. India being a developing country will also exhibit the same in future. In the view to accept this new phase of educational revolution, it is needed for India to get ready with the talent management to make a balance with the expected conditions towards Education 4.0.

The Ministry of Education (MoE) which was earlier known as Ministry of Human Resource Development (MHRD) of India has recently revised the National Education Policy 2020 that is basically a blueprint of the education system. This policy is a roadmap of education which is aligned with the global trend [25]. The last educational policy was formed in 1986, and after 34 years, it has been revised. Due to the unprecedented changes in the education system also at the time of lockdown due to the pandemic, this policy is a necessity.

The government of India has given the major priority to access, affordability, equity and accountability. This policy is prepared to re-design the entire education system in India. The higher plays a significant role in promoting well-being of humans as well as of society. It also contributes to the sustainable development of the economy as a whole. The key areas in higher education are to develop character, ethics, values, curiosity, creativity, capability and spirit of service. It comprises accomplishment, enlightenment and productive contribution to the society. The NPE is re-energizing the education system so that it could produce high-quality higher education with equity and inclusion. To make the classroom environment open and independent Government of India has launched several policies [26].

4 Discussion

Indian government has given major emphasis on re-structuring the higher education system, which is believed to make the education system in India to be more authentic and successful. The role of MOOC and non-conventional education is a prime point in the view of Education 4.0. The learning and teaching outcomes in Education 4.0 are purely learner-centric, or it can also be known as reflective level of teaching where the learner occupies primary place and teachers are assumed to be at second place. The aim of this approach is to develop the reflective powers of the learner so that they can solve problems on their own by applying required reasoning, logic and imagination. It is the high level of teaching learning approach that enhances the logical thinking of the student. Collaborative learning is also that necessary in higher education as it promotes the art of soft-skills among the learners (Razali et al., n.d.). It is termed as one of the most important approaches that can be adapted in the Education 4.0 learning process with the collaboration of knowledge and social interactions.

5 Conclusion

The future of education can be seen with Education 4.0 that responds with the need of Industry 4.0 where humans and technologies are aligned to enable new opportunities. The education sector should be well-updated in order to provide training and skills to the students that are talented especially in the higher education sector to meet the need of Industry 4.0. India is in the stage of adapting the Education 4.0 in the higher education system by forming the required programmes and policies. In the concern of future work, the authors can work upon the conceptual framework of Indian higher education system that supports and justifies the importance of Education 4.0 after the pandemic conditions.

References

1. Lasi, H., Fettke, P., Kemper, H. G., et al. (2014). Industry 4.0. *Business & Information Systems Engineering, 6*, 239–242. https://doi.org/10.1007/s12599-014-0334-4
2. Drath, R., & Horch, A. (2014). Industrie 4.0: Hit or hype? [Industry Forum]. *IEEE Industrial Electronics Magazine, 8*, 56–58.
3. Industrial revolution: Definition and inventions. HISTORY.com—HISTORY. http://www.history.com/topics/industrial-revolution. Accessed 18 December 2020.
4. Halili, S. H. (2019). Technological advancements in education 4.0. *The Online Journal of Distance Education and e-Learning, 7*(1), 63–69.
5. Baygin, M., Yetis, H., Karakose, M., & Akin, E. (2016). An effect analysis of industry 4.0 to higher education. In *2016 15th international conference on information technology based higher education and training, ITHET 2016*. Institute of Electrical and Electronics Engineers Inc.
6. Rajkumar, R., Lee, I., Sha, L., & Stankovic, J. (2010). Cyber-physical systems: The next computing revolution. In *Proceedings—Design automation conference*, pp 731–736.
7. Hussin, A. A. (2018). Education 4.0 made simple: Ideas for teaching. *International Journal of Education and Literacy Studies, 6*(3), 92–98.
8. Faller, C., & Feldmúller, D. (2015). Industry 4.0 learning factory for regional SMEs. In *Procedia CIRP*, 88–91.
9. Salmon, G. (2019). May the Fourth Be with You: Creating Education 4.0. *Journal of Learning for Development, 6*(2), 95–115.
10. Mourtzis, D., Vlachou, E., Dimitrakopoulos, G., & Zogopoulos, V. (2018). Cyber-physical systems and education 4.0–the teaching factory 4.0 concept. *Procedia manufacturing, 23*, 129–134.
11. Harkins, A. M. Leapfrog principles and practices: Core components of education 3.0 and 4.0.
12. Gerstein, J. (2014). Moving from education 1.0 through education 2.0 towards education 3.0. *Experiences in Self-Determined Learning.*
13. Yamamoto, G. T., & Karaman, F. (2011). Education 2.0. *Horiz, 19*, 109–117. https://doi.org/10.1108/10748121111138308

14. Goldie, J. G. S. (2016). Connectivism: A knowledge learning theory for the digital age? *Medical Teacher, 38*, 1064–1069. https://doi.org/10.3109/0142159X.2016.1173661
15. Education 4.0: New challenge of learning. Puncreobutr. *St. Theresa Journal of Humanities and Social Sciences*. http://www.stic.ac.th/ojs/index.php/sjhs/article/view/PositionPaper3. Accessed 19 December 2020
16. Jamwal, A., Agrawal, R., Sharma, M., & Kumar, V. (2021). Review on multi-criteria decision analysis in sustainable manufacturing decision making. *International Journal of Sustainable Engineering*, 1–24.
17. Jamwal, A., Agrawal, R., Sharma, M., Kumar, V., & Kumar, S. (2021). Developing A sustainability framework for Industry 4.0. *Procedia CIRP, 98*, 430–435.
18. Jamwal, A., Agrawal, R., & Sharma, M. (2021). Life cycle engineering: past, present, and future. In *Sustainable Manufacturing* (pp. 313–338). Elsevier.
19. Jamwal, A., Agrawal, R., Gupta, S., Dangayach, G. S., Sharma, M., & Sohag, M. A. Z. (2020). Modelling of sustainable manufacturing barriers in pharmaceutical industries of Himachal Pradesh: an ISM-fuzzy approach. In *Proceedings of international conference in mechanical and energy technology* (pp. 157–167). Springer, Singapore.
20. Jamwal, A., Agrawal, R., Sharma, M., & Giallanza, A. (2021). Industry 4.0 Technologies for Manufacturing Sustainability: A Systematic Review and Future Research Directions. Applied Sciences, 11(12), 5725.
21. Jamwal, A., Agrawal, R., Sharma, M., Kumar, A., Kumar, V., & Garza-Reyes, J. A. A. (2021). Machine learning applications for sustainable manufacturing: a bibliometric-based review for future research. *Journal of Enterprise Information Management*.
22. Hariharasudan, A., & Kot, S. (2018). A scoping review on Digital English and Education 4.0 for Industry 4.0. *Social sciences, 7*(11), 227.
23. Agrawal, S., Jamwal, A., & Gupta, S. (2020). Effect of COVID-19 on the India economy and supply chain. Preprints 2020050148. https://doi.org/10.20944/preprints202005.0148.v1.
24. Jedaman, P., Buaraphan, K., & Pimdee, P., et al. (2018). Analysis of sustainable leadership for science learning management in the 21st Century under education THAILAND 4.0 framework. In *AIP conference proceedings* (p. 030062). American Institute of Physics Inc.
25. Ciolacu, M., Tehrani, A. F., Beer, R., & Popp, H. (2017). Education 4.0-Fostering student's performance with machine learning methods. In *2017 IEEE 23rd international symposium for design and technology in electronic packaging, SIITME 2017—Proceedings* (pp. 438–443). Institute of Electrical and Electronics Engineers Inc.
26. National Education Policy. (2020). Ministry of Human Resource Development Government of India.

Mechanical and Tribological Behavior of Al7075/SiC/WS$_2$ Stir Casting Fabricated Composite with Optimization of EDM Parameters

Amit Kumar and M. K. Pradhan

1 Introduction

In recent years, there has been a growing interest in non-conventional materials and including ceramic particle reinforced composites, hybrid composites, metal-ceramic combinations called MMCs for the development of advanced components. Aluminum metal matrix composites (AMMCs) are composites in which aluminum is used as a matrix and several reinforced materials are embedded in a matrix. Hybrid composites consist mostly of one matrix and two or more reinforcements. Hybrid AMMC has superior mechanical properties such as higher specific strength, low thermal coefficient of specific modulus and good wear resistance. [1]. So, the hybrid AMMC is widely used in the various advanced engineering areas such as automotive, aircraft, locomotive industries, marine and defense sector. [2]. There has been an increased and continuous demand for smarter material having high strength to weight ratio, efficient materials with less cost, which literary initiate the people to put the focus on the investigation of composite materials in the last few years. The material like silicon carbide (SiC) and tungsten di-sulfide (WS$_2$) which is used for reinforcement are improved mechanical and tribological properties [4]. With the addition of silicon carbide (cheaper, lightweight and strength provider) with tungsten di-sulfide (harder, wear resistance and self-lubrication) gives optimum properties. The macro and microhardness of the composites were increased with respect to the addition of wt% of SiC [13]. But, due to the presence of these reinforcement materials, these hybrid composite materials become "difficult-to-machine" materials for the conventional machining process. It is difficult to get good surface quality of the conventional machining. Hence, non-conventional machining like EDM, ultrasonic machining, electrochemical machining, etc., are generally preferred to get better machining characteristics [7]. *EDM* is a precise metal removal *process* that is used to take out

A. Kumar (✉) · M. K. Pradhan
Maulana Azad National Institute of Technology, Bhopal 462003, India

R. Agrawal et al. (eds.), *Recent Advances in Smart Manufacturing and Materials*, Lecture Notes in Mechanical Engineering,
https://doi.org/10.1007/978-981-16-3033-0_46

metal through electric spark erosion. The EDM method is one of the most popular and quite extensively used non-traditional machining methods for conductive material like these types of aluminum alloys. Tahat [15] fabricated LM6/SiC/Graphite hybrid metal matrix composites were made of (Al-11.8%Si)/SiC/Gr by dual stir casting method and investigated the mechanical properties and wear behavior of composite. The wear test was conducted and it was observed that with the increase of the SiC/Graphite reinforcement contents in the matrix, and the wear rate of Al/SiC/Gr hybrid metal matrix composites reduces. Ravindranath et al. [12] studied hybrid aluminum metal matrix composites Al2219 is reinforced with 8% boron carbide (B_4C) and 3% graphite particles hybrid aluminum metal matrix composites were fabricated. It is investigated that the wear rate of hybrid aluminum composites decrease with increase graphite reinforcement concentration in the composite. Singh et al. [14] studied hybrid particulate composites of Al/SiC/Al_2O_3/WS_2 (0, 5, 9 wt%) in which fabrication is done by using the powder metallurgy method. The analysis of variance (ANOVA) was used to find the percentage contribution of various parameters on the wear and friction behavior. The result shows that the friction coefficient and the wear rate of WS_2 are minimum for 5% compare to 0 and 9 wt% of WS_2 in hybrid particulate composites. The effect of electrode shape on MRR, electrode wear rate (EWR), wear ratio (WR) and average surface roughness (Ra) has been investigated for mild steel work material and copper electrode [8]. It is investigated the effect of Ton, Toff, Ip, gap control setting on the EDM of cast Al 6063-Sic MMC [5]. To increase productivity with good surface quality, it is very difficult to achieve without setting the optimal parameters on EDM. Henceforth, multi-objective optimization approach like TOPSIS, gray relational, SAW, VIKOR are used to get optimal parameters of EDM responses. In all these SAW, it is the simple and best popular method which is used in this research. Aminesh et al. [3] studied the effect of parameters like speed, electrode gap and abrasive particle size on recast layer thickness and Ra using full factorial analysis method. Multi-response optimization problem has been solved by searching an optimal parametric combination, capable of producing with high productivity and at the same time ensuring the reduced surface quality of the EDMed product [10].

2 Materials and Methods

The aluminum alloy AA7075 was used as the base material and SiC and WS_2 were used as reinforcement. The chemical composition of the hybrid aluminum alloy used in this research is shown in Table 1. SiC (10 wt%) at constant proportion in the hybrid

Table 1 Chemical composition of the AA7075

Alloying element	Cu	Mg	Si	Zn	Cr	Ti	Fe	Mn	Pb
% composition	1.5	2.5	0.5	6	0.5	0.5	0.25	0.2	0.2

Table 2 Specimen dimension

Test	Code	Dimension	No. of specimens
Tensile	IS 1608	6 mm Ø, $L = 15$ mm	4
Compression	ASTM E9	15 mm Ø, $L = 20$ mm	4
Wear rate	IS 1501	10 mm Ø, $L = 30$ mm	4
Hardness	IS 1500	15 mm Ø, $L = 15$ mm	4

composite (Al7075 + SiC + WS_2) and WS_2 at various proportions such as (0.75, 1 and 1.5 wt%) has been used. The SiC powder is lighter as compared to tungsten di-sulfide powder. Moreover, the SiC can withstand at a very high temperature and tungsten di-sulfide (WS_2) have self-lubrication properties.

2.1 Fabrication of Specimen

The hybrid particulate composites (Al7075 + SiC + WS_2) were fabricated by the stir casting method. Al7075 alloy in the required quantity was taken as matrix material in a graphite crucible and heated in an electric furnace. The temperature of the melt was raised from 750 to 900 °C. This molten metal was stirred through a graphite impeller at a speed of 300–400 RPM to generate a vortex. The SiC and WS_2 reinforcement particles were taken separately and mixed with an Al7075 matrix. SiC (10 wt%) at constant proportion in the composite has been used and WS_2 at various proportions such as (0.75, 1 and 1.5 wt%). SiC and WS_2 powders were preheated at 450 °C to drive off the moisture. After the rotating of stir, the vortex formation takes place in the melt. Stirring of melt (hybrid metal matrix alloy with particulate reinforcement) was continued for another 5 min. Even after the end of reinforcement particles feeding to get a homogeneous distribution of the SiC and WS_2 powder particles in the composite material. The molten material was finely poured into the metal mold made up of cast iron material and allowed to solidify at room temperature for 3–4 h. The composite was then cut into pieces of different dimension as shown in Table 2 for various test such as tensile, compression, wear rate and hardness test.

2.2 Experimentation on EDM

The CNC electrical discharge machine is used to experimentation. The hybrid aluminum metal matrix composite workpiece with a 15 mm diameter and an electrolytic pure copper rod is used as a tool material. In current work, the experimental parameters and their levels are shown in Table 3. The experiment was conducted and the effect of various machining parameters; I_p, Ton, T_{au} and V on MRR, TWR, ROC

Table 3 Input variables

Variable	Unit	Levels		
		1	**2**	**3**
Discharge current (I_p)	A	8	10	12
Pulse on time (Ton)	μs	75	100	150
Voltage (V)	V	50	60	70

Table 4 Experimental parameters and responses on EDM machine

Wt. (%)	I_p (A)	Ton (μs)	V (V)	MRR (mm^3/min)	TWR (mm^3/min)	ROC (μm)	Ra (μm)
0.75	8	75	50	38.14446	0.145768	5.6458	7.1
0.75	10	100	60	17.76367	0.090511	5.7056	11.328
0.75	12	150	70	56.06992	0.119039	5.6782	12.51
1.0	8	100	70	18.3321	0.292309	5.6931	8.041
1.0	10	150	50	19.13204	0.378279	5.6258	10.381
1.0	12	75	60	39.21962	0.362711	5.6365	8.844
1.5	8	150	60	18.03273	0.230593	5.6248	10.323
1.5	10	75	70	18.2872	0.046769	5.6782	7.12
1.5	12	100	50	39.71567	0.101573	5.5806	9.723

and Ra on the given materials [11]. The experimental parameters and their responses to the EDM machine were obtained in Table 4.

3 Machining Behavior

3.1 Metal Removal Rate (MRR)

The metal removal rate is the measuring parameter for which how much material is removed from a part in a given period. The formula was given to calculate the metal removal rate:

$$\text{MRR}\ \left(\text{mm}^3/\text{min}\right)\frac{\text{Weight loss (gm)} \times 60}{\text{The density of sample}\ \left(\text{gm/mm}^3\right) \times \text{Machining time (s)}} \tag{1}$$

3.2 Tool Wear Rate (TWR)

In EDM, due to the rotational motion of the tool rises the tool wear rate (TWR) in all stages of current and pulse on time. TWR can be defined as the volumetric loss of tool in unit time. It is expressed as

$$\text{TWR} = \frac{\Delta V}{t} = \frac{\Delta W}{\rho * t} \quad (2)$$

where ΔV is the volumetric loss from the electrode, ΔW is the weight loss from the tool, t is the machining time during the process and $\rho = 8960$ kg/m the density of the electrode.

3.3 Surface Roughness (Ra)

Surface roughness is a measurement of the technical quality of the product, which mostly influences the manufacturing cost of the product. Surface roughness is one of the most important characteristics in the EDM process because it affects the fatigue strength of the component. The measurement of surface roughness was done using a portable stylus type profilometer, Talysurf (Taylor Hobson, Surtronic 3^{+}). The profilometer was set up to a cut-off length of 0.8 mm, filter 2CR, traverse speed 1 mm/s and 4 mm evaluation length. It is calculated as the arithmetic value of the profile from the centerline along the length. This can be express as

$$\text{Ra} = \frac{1}{L} \int |y(x)| \mathrm{d}x|| \quad (3)$$

where L is the sampling length, y is the profile curve and x is the profile direction. The average 'Ra' is determined within $L = 0.8$ mm.

3.4 Radial Overcut (ROC)

The dimensional accuracy of EDM is significantly affected by the ROC resulting from the discharge gap and electrode wear. The input variables such as I_{p}, Ton, the voltage and the workpiece material are greatly influenced by ROC. It increases with the increase of I_{p} but up to a certain limit besides it depends upon the gap between electrode and workpiece, voltage and chip size, which may vary with the amperage used.

$$\text{ROC} = \frac{D1 - D2}{2} \quad (4)$$

where $D1$ = Right side reading on the digital micrometer and $D2$ = Left side reading on the digital micrometer.

4 Analysis Methods

4.1 Taguchi Method

The main aim of this method is to create a good quality product at a low-cost to the manufacturer. Taguchi used the concept of the signal-to-noise (S/N) ratio to find the best levels for each controllable factor. This approach is used for experimental design in this research. The experiment was conducted with three controllable three-level factors of the EDM process and four responses. Three factors were I_p, Ton and V. The four quality characteristics measured were the MRR, TWR, Ra and ROC. In this study, nine experimental runs based on the orthogonal array $L_9(3^4)$ as shown in Table 4.

4.2 Simple Additive Weighting (SAW)

This is the simplest and widest used multi-criteria decision-making process method. In this paper, to solve the personnel selection problem using MCDM, the SAW method is used. The basic concept of the SAW method is to find the weighted sum of performance ratings on each alternative on all attributes. In this method, each element is given a weight and the sum of all weights must be 1. The overall performance score of an alternative is given by Eq. (1).

$$P_i = \sum_{j=1}^{K} WjXi(k) \quad (5)$$

If all the elements consist in the decision table are normalized, then SAW can be used for any type and any number of elements. For that case Eq. (1) becomes;

$$P_i = \sum_{j=1}^{K} WjXi * (k) \quad (6)$$

where $Xi^*(k)$ = The normalized value of $Xi(k)$, P_i = The overall score of the alternative A_i, then the highest value of P_i will be the best alternative in all the alternatives. The advantage of the simple additive weighting method compared with other decision-making models is that to perform decisions more precisely because it is based on pre-defined value and preference weight.

4.3 Data Preprocessing

The data preprocessing is the processing of the data for the analogous arrangement of the original data. In which, the experimental results are normalized in the range of 0 and 1, depending on the responses either higher the better or lower the better. For the higher productivity, the metal removal rate should be high, hence, treated as the "higher is better" (HB) attribute. The experimental data can be normalized as;

$$Xi^*(k) = \frac{Xi(k) - \min Xi(k)}{\max Xi(k) - \min Xi(k)} \tag{7}$$

And for increasing the productivity the TWR, ROC and Ra of the material will be as low as possible. For these three machining parameters, "lower is better" (LB) attribute is used as.

$$Xi^*(k) = \frac{\max Xi(k) - Xi(k)}{\max Xi(k) - \min Xi(k)} \tag{8}$$

where $i = 1, 2, \ldots n$; $k = 1, 2, \ldots., p$; $Xi^*(k)$ is the normalized value of the kth element in the ith sequence; max $Xi^*(k)$ is the largest value of $Xi(k)$; min $Xi^*(k)$ is the smallest value of $Xi(k)$; n is the number of experiments; p is the number of quality characteristics.

4.4 Principal Component Analyses (PCA)

A conversion of a set of observations of possibly correlated variables (various numerical values) into a set of values of uncorrelated variables comes under principal component analysis. It was discovered by Pearson in 1901 and then it was developed by Hotelling in 1933. After invention, it becomes a popular mathematical tool to investigate the data analysis for the formation of predictive models. PCA may be operated by either eigenvalue decomposition of a data covariance matrix or particular value decomposition of a data matrix. It is used for recognizing arrangements of data and conveying the data in such a way as to highlight their similarities and differences. Once the arrangement of data has been recognized, the data can be compacted, i.e.,

Table 5 Eigenvalues and explained variation for principal component

Principal component	Eigenvalue	Proportion	Explained variation (%)
First	1.3471	0.337	33.7
Second	1.2293	0.307	30.7
Third	0.9019	0.225	22.5
Four	0.5217	0.13	13

by decreasing the number of dimensions, without much loss of information. This is one of the principal advantages of PCA. The main aim of PCA is to:

1. Extract the most important information from the data,
2. Squeeze the dimension of the data set by keeping only the significant,
3. Simplify the explanation of the data set,
4. Examine the arrangement of the readings and the variables. The step by step procedure is described as follows [6, 9]. The normalized value is shown in Table 7 by Minitab 18 as a multivariate principal component to determine the eigenvalues and explained variation as in Table 5. Also, it was found the correlation matrix consisting of all four principal components and weights of all the four variables.

4.5 Determination of the Composite Score and Ranking

The weight has been calculated from the principal component method which is further used for calculating the overall performance score (composite score). For each alternative, the normalized values have been used for all the responses as in the table. For example, the values of composite score for the 1st run calculate as: $0.5320 \times 0.546121 + 0.7014 \times 0.023409 + 0.4784 \times 0.114921 + 1.0000 \times 0.315844 = 0.67780$. After determining, the composite score ranking is done from higher to lower as the composite score in descending order as given in Tables 6 and 7.

The detailed analysis method has been done and the composite score with the run order has been presented in Fig. 1 and it can be seen that the run order 1 has the highest composite score.

5 Result and Discussion

5.1 Optimization of EDM Machining Parameters

To get the optimum result of the machining parameters and to reduce the manufacturing cost of the composite product, the SAW and PCA methods were used. For optimization of EDM parameters of machining of hybrid composite (Al7075

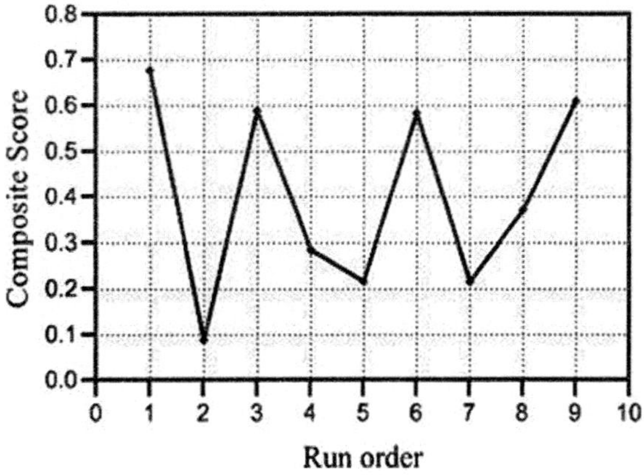

Fig. 1 Response plot for the composite score

Table 6 Eigen analysis of the correlation matrix

Variable	PC1	PC2	PC3	PC4	Contribution
MRR (mm^3/min)	0.739	0.032	0.265	− 0.618	0.546121
TWR (mm^3/min)	0.153	0.706	0.529	0.445	0.023409
ROC (μm)	0.339	− 0.684	0.37	0.529	0.114921
Ra (μm)	− 0.562	− 0.18	0.716	− 0.373	0.315844

Table 7 Normalized decision matrix with composite score and rank

Run	Normalized value				Composite Score	Rank
	MRR (mm^3/min)	TWR (mm^3/min)	ROC (μm)	Ra (μm)		
1	0.5320	0.7014	0.4784	1.0000	0.67780	1
2	0.0000	0.8681	0.0000	0.2185	0.08933	9
3	1.0000	0.7820	0.2192	0.0000	0.58962	3
4	0.0148	0.2593	0.1000	0.8261	0.28657	6
5	0.0357	0.0000	0.6384	0.3935	0.21717	7
6	0.5601	0.0470	0.5528	0.6776	0.58455	4
7	0.0070	0.4455	0.6464	0.4043	0.21623	8
8	0.0137	1.0000	0.2192	0.9963	0.37074	5
9	0.5731	0.8347	1.0000	0.5152	0.61013	2

+ SiC + WS_2), rank has been given to each run order. The rank is given to each run order according to the descending order of the composite score in Table 7. The composite score for each experiment of the L9 orthogonal array was calculated as discussed in the previous section (Table 1). From the all performed experimental run, the experiment No. 1 (Run 1) had the highest composite score. Therefore, optimum machining parameter setting behind experiment No. 1 for the desired output

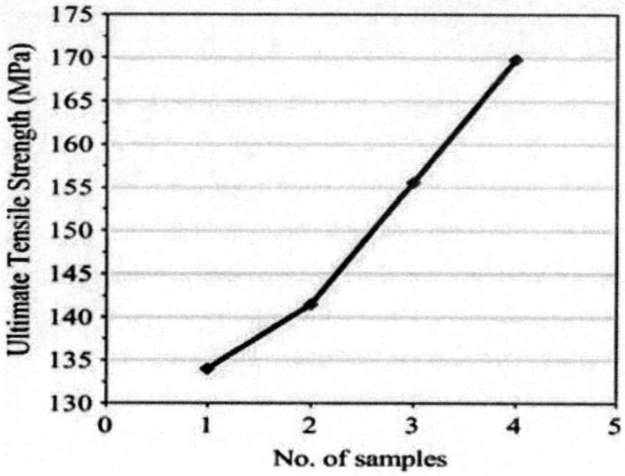

Fig. 2 Ultimate tensile strength of the AMMC

responses which are the best multi-performance characteristics among all nine experiments. The optimum machining performance parameter on EDM from the SAW-PCA optimization method was 8 A discharge current (level 1), Pulse on time, Ton = 75 μs (level 1) and Voltage $V = 50$ V (level 1). The confirmation experiment with the above optimal parameters has been done and as expected the getting of higher MRR with lower TWR, Ra and ROC successfully.

5.2 *Mechanical Behavior*

5.2.1 Ultimate Tensile Strength of Hybrid Composite

The tension test is conducted on the UTM machine for various samples of hybrid composite (Al7075 + SiC + WS_2) at 80 KN load and 3 mm strain speed. The four specimens were tested and the value of ultimate tensile strength evaluated as shown in Figure. Figure 2 shows that the ultimate tensile strength of hybrid composite increases with an increase in wt% of reinforcements in the composite.

5.2.2 Compressive Strength of Hybrid Composite

The compression test is conducted on the UTM machine for various samples of hybrid composite (Al7075 + SiC + WS_2) at 80 KN load and 3 mm strain speed. The specimens were tested and the value of compressive strength plotted in Fig. 3. The figure shows that with the increase in the wt% of reinforcement in the hybrid composite (Al7075 + SiC + WS_2) the compressive strength decreases.

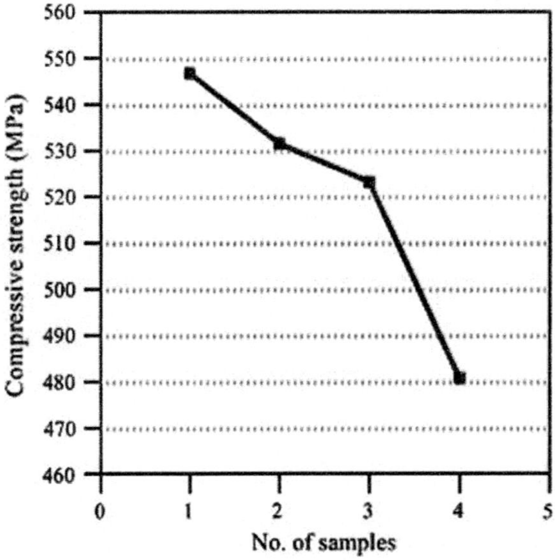

Fig. 3 Compressive strength of the AMMC

5.2.3 The Rockwell Hardness of Hybrid Composite

The hardness test was conducted on a digital Rockwell hardness testing machine for hybrid composite (Al7075 + SiC + WS_2).

The Rockwell hardness of various samples shown of the hybrid composite was evaluated and was plotted. Figure 4 has been plotted between wt% of reinforcement and Rockwell hardness of the hybrid composite. It can be seen that with the increase the wt% of reinforcement particle in the composite than their hardness increases.

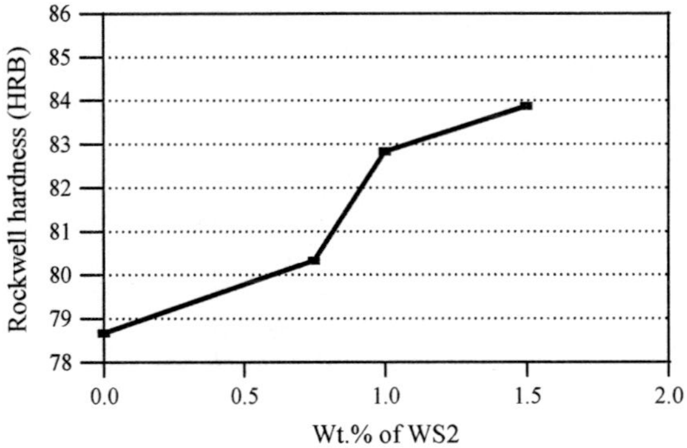

Fig. 4 Rockwell hardness (HRB) of the AMMC

5.3 Tribological Behavior

5.3.1 Wear Test of Hybrid Composite

A pin-on-disk wear testing machine is used to perform the wear test. The test was done on the samples of hybrid composite (Al7075 + SiC + WS_2) at loads of 2 and 4 kg and various sliding distance, with 100 mm track diameter, the different readings have been taken at 10 min interval. Figures 5, 6, 7 and 8 show the relation between wear rate and sliding distance for different samples at different load. It can be seen that with the increase of the load the wear rate increases and also if substantial readings have been taken at one face of the sample at a constant load for the respective sliding distance the wear rate decrease. The effect of reinforcement material can also be concluded on the wear rate of composite that the percentage of reinforcement increase the wear rate decreases.

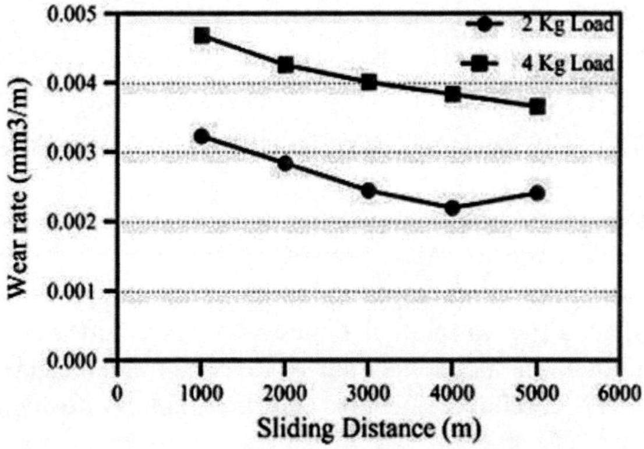

Fig. 5 Wear testing of Al7075 alloy at 2 and 4 kg load

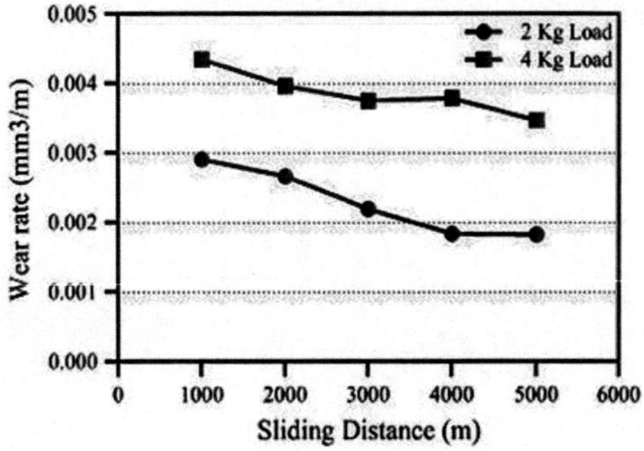

Fig. 6 Wear testing of Al7075 + 10wt%SiC + 0.75wt%WS_2 alloy at 2 and 4 kg load

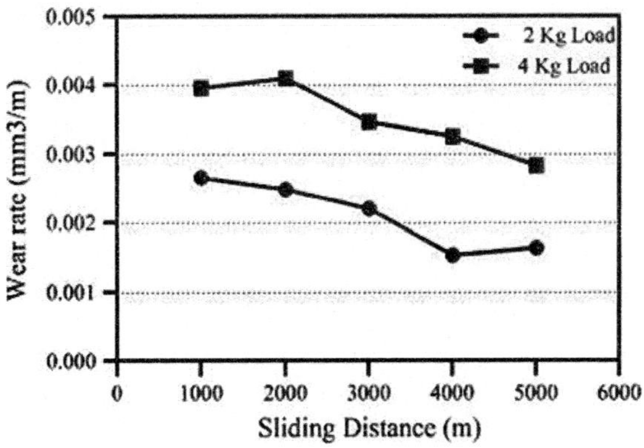

Fig. 7 Wear testing of Al7075 + 10wt%SiC + 1.0wt%WS_2 alloy at 2 and 4 kg load

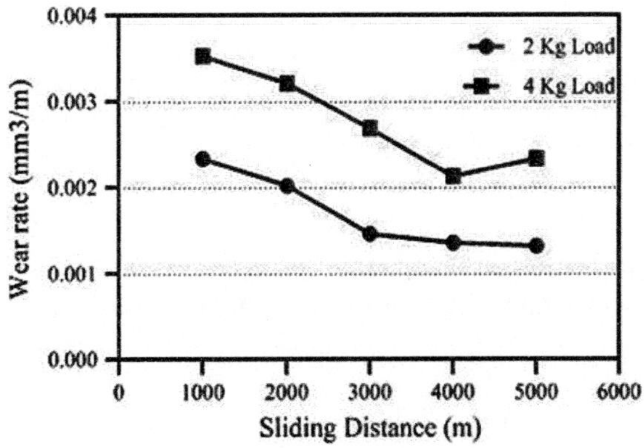

Fig. 8 Wear testing of Al7075 + 10wt%SiC + 1.5wt%WS_2 alloy at 2 and 4 kg load

5.4 The Effect of EDM Parameters on Machining Characteristics of EDM Machine

The EDM process parameters and machining characteristics are observed on CNC EDM as shown in Table 4.

5.4.1 Metal Removing Rate

To analyzing the mean of means S/N rations analysis of variance (ANOVA) technique is used, which is a theoretical approach that involves graphing the special effects and visually making out the important influences of various influencing factors. Figure 9 of S/N was plotted to see the effect of EDM parameters on metal removing rate by ANOVA at Minitab software and it was seen that when the wt% of reinforcements (SiC + WS_2) increases then MRR decreases and when the input current, Ip increases

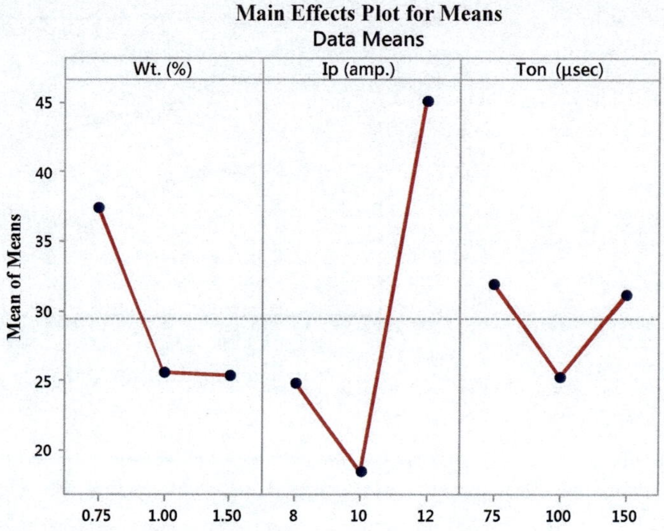

Fig. 9 Effect of process parameters on MRR

then first MRR decreases from 8 to 10 (A) then after increases abruptly. Simultaneously, first increment in Ton and voltage the MRR first decrease from 75 to 100 μs and 50 to 60 V, respectively, then after increases gradually.

5.4.2 Tool Wear Rate

Figure 10 depicts the effect of EDM parameters on tool wear rate and it was seen that when the wt% of reinforcements ($SiC + WS_2$) in hybrid composite sample and voltage (V) increases, and the TWR first increases up to 0.75–1 wt% and 50–60 (V), respectively, then after decreases abruptly. Simultaneously, when the I_p and Ton increases, initially, the TWR decreases up to 10 A and 100 μs, respectively, and then increases gradually.

5.4.3 Radial Overcut

In Fig. 11, the effect of EDM parameters on radial overcut has been depicted and it is seen that when the wt% of reinforcements ($SiC + WS_2$) increases the ROC decrease, when the Ip and Ton increases, the first TWR increases from 8 to 10 (A) and 75 to 100 (V), respectively, then after decreases gradually. Simultaneously increment in voltage, the ROC increases gradually.

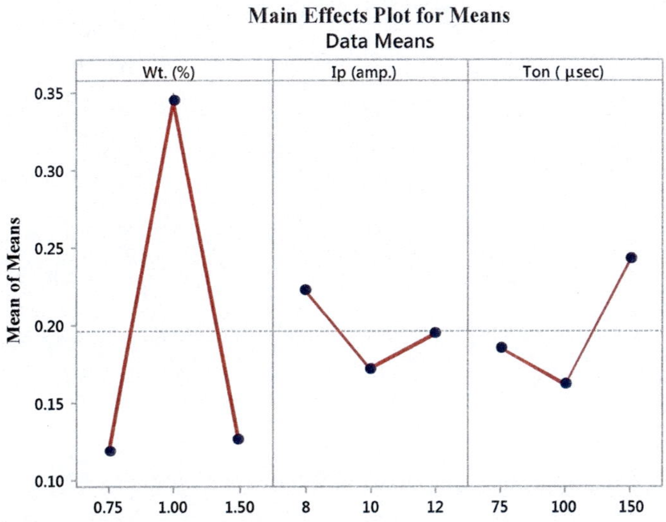

Fig. 10 Effect of process parameters on TWR

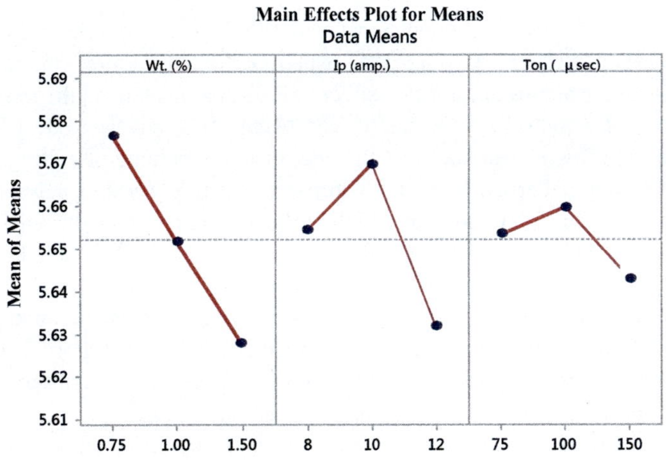

Fig. 11 Effect of process parameters on ROC

5.4.4 Surface Roughness

Figure 12 shows the influence of EDM parameters on surface roughness and it was seen that when the wt% of reinforcements ($SiC + WS_2$) increases, the Ra value decreases. When the Ip and Ton increases, the Ra increases gradually, voltage increases from 50 to 60 V the Ra increases after that it decreases up to 70 V.

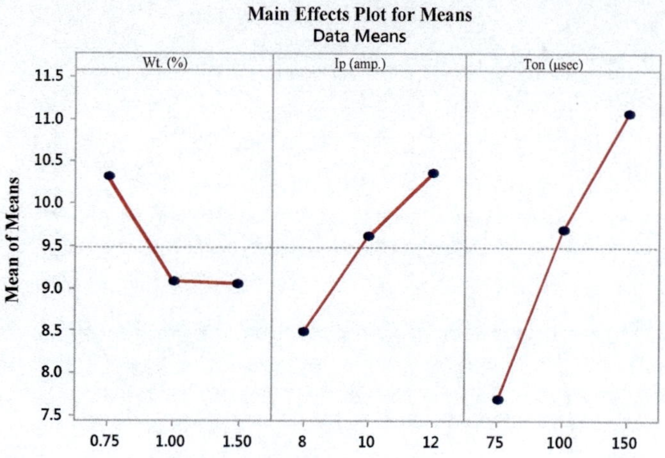

Fig. 12 Effect of process parameters on Ra

6 Conclusion

This research investigates the mechanical and tribological characteristics of the hybrid composite (Al7075 + SiC + WS_2) such as the tensile strength, compressive force, hardness and wear rate. The effect of EDM parameters on the machining characteristics of a hybrid composite surface using a copper electrode has been studied. The experiments were performed under different parametric settings as per the Taguchi orthogonal array. Minitab 18 software was used for the analysis of the ANOVA and Taguchi techniques for modeling the alternatives and attributes of the EDM machining parameters. The best alternative was to use SAW coupled with the PCA method. The best possible machining parameters ($I_p = 8$ A, Ton = 75 μs and $V = 50$ V) for EDM are being found. It is noted that the Ip is the most effective parameter in the EDM machining characteristics compared to Ton and V. For mechanical features, hybrid composite hardness (Al7075 + SiC + WS_2) has been observed to be increasing when the composite matrix wt% of reinforcement particle increases. When the weight of the reinforcement particles increases in the composite matrix, the ultimate tensile strength of the hybrid composite increases. The compressive strength of the hybrid composite is slightly reduced when the wt% of the reinforcement particle increases in the composite matrix. The rate of wear of hybrid composites decreases when the wt% of reinforcement increases in the composite matrix. It is also shown that the rate of wear increases with an increase in load.

References

1. Alaneme, K. K., & Olubambi, P. A. (2013). Corrosion and wear behaviour of rice husk Ash/alumina reinforced al–mg–si alloy matrix hybrid composites. *Journal of Materials Research and Technology, 2*(2), 188–194.
2. Ahlatci, H., Kocer, T., Candan, E., & Çimenoğlu, H. (2006). Wear behavior of Al/(Al_2O_3+ SiC) hybrid composites. *Tribology International, 39*(3), 213–220.
3. Amineh, S. K., & Tehrani, A. F. (2013). Mohammadi Improving the surface quality in wire electrical di charge machined specimens by removing the recast layer using a magnetic abrasive finishing method. *International Journal of Advanced Manufacturing Technology, 66*, 1793–1803.
4. Boopathi, M. M., Arulshri, K., & Iyandurai, N. (2013). Evaluation of mechanical properties of aluminum alloy 2024 reinforced with silicon carbide and fly ash hybrid metal matrix composites. *American Journal of Applied Sciences, 10*(3), 219.
5. Dvivedi, A., Kumar, P., & Singh, I. (2008). Experimental investigation and optimization in EDM of al 6063 SiC metal matrix composite. *International Journal of Machining and Machinability of Materials, 3*(3–4), 293–308.
6. Fung, C. P., & Kang, P. C. (2005). Multi-response optimization in friction properties of PBT composites using Taguchi method and principal component analysis. *Journal of Materials Processing Technology, 170*(3), 602–610.
7. Ho, K. H., & Newman, S. T. (2003). State of the art electrical discharge machining (EDM). *International Journal of Machine Tools and Manufacture, 43*(13), 1287–1300.
8. Khan, A., Ali, M., & Haque, M. (2009). A study of electrode shape configuration on the performance of die-sinking EDM. *International Journal of Mechanical and Materials Engineering, 4*(1), 19–23.
9. Pradhan, M. K. (2012). Determination of optimal parameters with multi-response characteristics of EDM by response surface methodology, grey relational analysis, and principal component analysis. *International Journal of Manufacturing Technology and Management, 26*(1–4), 56–80.
10. Pradhan, M. K. (2011). Modelling and optimization of electrical discharge machining variables using RSM coupled with GRA and PCA. In *AICTE sponsored national conference on emerging trend & its application in engineering*, pp. 26–28.
11. Pradhan, M. K. (2018). Optimization of EDM process for MRR, TWR, and radial overcut of D2 steel: a hybrid RSM-GRA and entropy weight-based TOPSIS approach. *International Journal of Industrial and Systems Engineering, 29*(3), 273–302.
12. Ravindranath, V., Shankar, G. S., Basavarajappa, S., & Kumar, N. S. (2017). Dry sliding wear behavior of hybrid aluminum metal matrix composite reinforced with boron carbide and graphite particles. *Materials Today: Proceedings, 4*(10), 11163–11167.
13. Selvam, J. D. R., Smart, D. R., & Dinaharan, I. (2013). Synthesis and characterization of Al6061-fly ash-sic composites by stir casting and compacting methods. *Energy Procedia, 34*, 637–646.
14. Singh, S., Maheshwari, S., & Pandey, P. (2004). Some investigations into the electric-discharge machining of hardened tool steel using different electrode materials. *Journal of Materials Processing Technology, 149*(1–3), 272–277.
15. Tahat, M. S. (2016). Machinability and tribological properties of stir cast lm6/sic/gr hybrid metal matrix composite. In *MATEC Web of Conferences, 75*, 02001.

Experimental Study on Movement of Crankshaft During Machining

Sunil Pal and Surendra Kumar Saini

1 Introduction

To enhance the productivity of a small-scale industry that manufacture crankshaft for Eicher Tractor, the DMAIC technique is used to identify the problems with the help of flow process and basic observations in terms of material movement and rate of production. In order to achieve required mechanical properties, most of crankshafts for automobile are forged with micro alloyed steel. Usually, the shapes of crankshafts are complex [1, 2]. Forged crankshaft passes through a number of processes and thus prepared a finished crankshaft as per the desired dimensions and design. Subharaj et al. [3] developed a sustainable ecodesign for crankshaft speed sensor assembly line. Li et al. [4] developed a mathematical model to optimize the conventional layout and scheduling (i.e., multi-row workshop layout and flexible job shop scheduling) for crankshaft. Their objective of study was to develop the green manufacturing environment for shipbuilding industry. Rehman et al. [5] employed lean technique to decline the manufacturing waste for productivity improvement.

Jamwal et al. [6] improved productivity of caravan using continuous improvement method. They simulate the entire process and validated to enhance the product life cycle. Ou et al. [7] proposed industry 4.0 applications such as data analytics and cloud technology to upgrade smart manufacturing of automotive crankshaft. Lee [8] studied effectiveness of Internet of things, cloud environment to support the cyber manufacturing. He reveled that these tools have capability to detect real-time changes in ongoing process in advanced industries. Diverse optimization techniques have studied and fuzzy and Taguchi methodology used for elect the optimum parameters for computer numerical control machining of steel [9–12]. Jamwal et al. [13] implemented lean manufacturing technique to obtain zero wastes in small-scale

S. Pal · S. K. Saini (✉)
Mechanical Engineering Department, Poornima College of Engineering Jaipur, Jaipur, Rajasthan 302022, India

R. Agrawal et al. (eds.), *Recent Advances in Smart Manufacturing and Materials*,
Lecture Notes in Mechanical Engineering,
https://doi.org/10.1007/978-981-16-3033-0_47

industry. The purpose of the present study is to reduce the material movement. The time required in transportation and handling is also needed to reduce by suitable modifications in the present process.

2 Material and Methods

Forged crankshaft is machined at "SWATI INDIA PRIVATE LIMITED" Alwar, Rajasthan. Different machining operations performed on the forged crankshafts. It is used in single cylinder of Eicher Engines. The work study examines the work process and eliminates nonproductive processes, which can reduce the number of process, space utilization, and production and operation time. The data that needs to be carried out in this work is flow process of the work, details of each process, distance traveled by the work piece for specific process, number of crankshafts that can be produced in specific time. To improve the productivity, either input is reduced for same output or output is increased with same input resources.

Method study concerned with the way in which work is carried out and used for finding better ways of doing work. It is a procedure for cost reduction. An incredible accept of method study is that there is always a better way for doing a task. The unique thing of method study is that it is a step-by-step process for improvements of methods of work. It proceeds to the collection and recording of the facts, identification of the problem and finally suggested solution. Method study involves Select, Record, Examine, Develop, Define, Install, and Maintain [6]. Plant layout is the most effective physical arrangement, either existing or in plans of industrial facilities, i.e., arrangement of machines, processing equipment, and service department to achieve greatest co-ordination and efficiency of resources.

3 Results and Discussion

Present process is observed in working industry. Distance travel and cycle time are elected for improvement in present process. Cycle time is observed 92.73 min using stop watch with present layout. The total flow of work piece within the unit for machining at various stations is observed as 6574 inches horizontal and 365 inches vertical. To conduct different operations on work, movement of work is done manually. A trolley is used to transport large amount of material in initial stages of machining. By applying method study, the complete movement of work piece is observed moving from one work station to other. There are many point observed in this study where movement of the work piece is very large, and required serious effort to reduce. For this, few alterations are suggested and implemented, so that the flow of work piece is in one direction and back tracking and queuing are also avoided.

In order to reduce the movement of work piece between the work stations, some alterations are applied in the present layout. Initial step of centering-facing is performed on the lathe. Then, it is used to shift in the successive machining shed along with the raw material place. Hence, flow of the material is uni-directional and avoids excess movement of the work piece. The stock of raw material in its new place has one more advantage. Unloading of the work pieces from the truck becomes easier and lesser distance required to travel in order to place it in its desired place. In the initial phase, the distance travel by the work piece is 598 in. horizontally which are reduced to 199 in. by implementing the alteration. Thus, material movement is reduced in the first step.

By shifting the grinding machine, a large effect is observed with the material movement. This results in the accordance of the group technology as the similar processes are performed in the particular area or premises. The distance travel by the work piece before the alteration is 635 in. horizontally and after the alteration is 560 in.

By changing the sequence of buffing and lapping operation, the motion of material is reduced. If lapping and then pin buffing are performed, the material flow is 267 in. horizontal, and if the process is reversed, i.e., if pin buffing and then lapping is applied, the motion of the work piece is observed as 96 in. horizontally keeping the vertical movement same in both the case, and then, no need to mention it separately.

These improvements are implemented because of the observations of present data being recorded and analyzed. Various bottlenecks are observed and efforts are made to eliminate or to reduce them up to a great extent.

The material handling process need to be improved as it required a large amount of material flow manually, and the time required for this process is very high which directly affects the productivity. By doing this, the labor requirement as well as the queued time by the machine tools is reduced in many of the work stations. This helps in the cost reduction of the process.

4 Conclusions

Modified horizontal and vertical distance traveled by work piece is measured. It is found that traveled distance is reduced by 809 in. in horizontal and 128 in. in vertical. These modifications directly affect the cost of the machining system which is reduced to a great extent thus productivity are improved. The cost of machining in present process is Rs. 211.15, and after implementing the modifications, this cost is reduced to Rs. 196.19.

Acknowledgements Thanks to SWATI INDIA PRIVATE LIMITED Alwar, Rajasthan (India) for extending their facilities to conduct this study. Also thanks to Poornima College of Engineering Jaipur for support and guidance.

References

1. Bulut, M., & Cihan, O. (2020). Stress and deformation analysis of a connecting rod by using ANSYS. *International Journal of Automotive Engineering and Technologies, 9*, 154–160.
2. Breyfogle, F. W. *Implementing six sigma: Smarter solutions using statistical methods*. John Wiley & Sons, New York, NY, USA.
3. Subharaj, C., Natarajan, U., & Hyacinth, S. X. (2019). Sustainable eco-design for fixture index mechanism in crankshaft speed sensor assembly line. *Advances in Mechanical Engineering, 11*, 1–12.
4. Li, H., Duan J., Zhang Q. Multi-objective integrated scheduling optimization of semi-combined marine crankshaft structure production workshop for green manufacturing. *Transactions of the Institute of Measurement and Control*. https://doi.org/10.1177/0142331220945917.
5. Rehman, A. U., Usmani, Y. S., Umer, U., & Alkahtani, M. (2020). Lean approach to enhance manufacturing productivity: A case study of Saudi Arabian factory. *Arabian Journal for Science and Engineering, 45*, 2263–2280.
6. Jamwal, A., Agrawal, R., Sharma, M., & Kumar, V. (2021). Review on multi-criteria decision analysis in sustainable manufacturing decision making. *International Journal of Sustainable Engineering*, 1–24.
7. Ou, X., Huang, J., Chang, Q., Hucker, S., & Lovasz, J. G. (2020). First time quality diagnostics and improvement through data analysis: A study of a crankshaft line. *Procedia Manufacturing, 49*, 2–8.
8. Lee, H. (2017). Framework and development of fault detection classification using IoT device and cloud environment. *Journal of Manufacturing Systems, 43*, 257–270.
9. Saini, S. K., & Pradhan, S. K. (2014). Optimization of multi-objective response during CNC turning using taguchi-fuzzy application. *Procedia Engineering, 47*, 141–149.
10. Saini, S. K., & Pradhan, S. K. (2014). Soft computing techniques for the optimization of machining parameter in CNC turning operation. *International Journal of Emerging Technology and Advanced Engineering, 4*, 117–124.
11. Saini, S. K., & Pradhan, S. K. (2014). Optimization of machining parameters for CNC turning of different materials. *Applied Mechanics and Materials, 592*, 605–609.
12. Pradhan, S. K., & Saini, S. K. (2014). Multi-objective optimization of CNC turning machining parameters. *Advanced Materials Research, 1016*, 172–176.
13. Jamwal, A., Aggarwal, A., Gupta, S., & Sharma, P. (2019). A study on the barriers to lean manufacturing implementation for small-scale industries in Himachal region (India). *International Journal of Intelligent Enterprise, 6*, 393–407.

Optimization of EDM Process Parameters for Machining of Steel

Rajesh Kumar Mohanty and Surendra Kumar Saini

1 Introduction

Materials like hard steels, composites, carbides and super alloys find applications in distinct types of industries such as nuclear, tool and die making, aerospace and biomedical, etc. [1]. However, machining of these materials is grim with conventional machining methods. Researchers are exploring the capability of advanced machining processes to machine such materials. Electrical discharge machining process is widely used machining process to machine such materials. In electrical discharge machining process, the metal removal takes place by electric spark or in other words metal erosion takes place because of the electric spark. Here, in this process, the electric spark formed between two electrodes is used to erode material from work piece. The shape of the tool is an actual replica of the shape that is to be produced in the work piece [2]. A high pulsating electric charge of high density is being given to the work piece through the electrode. So there is a continuous erosion of metal which occurs in work piece at a controlled rate [3]. Molineti et al. [4] studied the influence of manganese and silicone powders in two different concentrations of dielectric, while study of EDM on AISI H13 too steel.

They investigated hardness, surface roughness, microstructure and chemical composition. They found surface hardness five times better in comparison of conventional EDM. They concluded that powder suspended in the dielectric was responsible for surface quality improvement. EDM on Inconel X-750 was performed by Pratap et al. [5], and they optimized process parameters. The input parameters are pulse duration, peak current and voltage while the quality responses are material removal rate and surface roughness. The optimum values for parameters such as depth of cut, feed and cutting speed elected for machining of metals using hybrid

R. K. Mohanty · S. K. Saini (✉)
Department of Mechanical Engineering, Poornima College of Engineering Jaipur, Jaipur, Rajasthan, India

R. Agrawal et al. (eds.), *Recent Advances in Smart Manufacturing and Materials*, Lecture Notes in Mechanical Engineering,
https://doi.org/10.1007/978-981-16-3033-0_48

optimization techniques. They also reviewed different techniques of optimization used for machining parameters optimizations [6–9]. Wire electrochemical discharge machining (WECDM) on quartz was investigated by Kumar et al. [10]. They found WECDM has enormous scope to machine hard and brittle materials. They used WECDM process which is basically a combination of ECM and wire EDM and so combines the characteristics of both processes. Taguchi orthogonal array and ANOVA were being used to find the optimum factors and their significant influence on response parameters.

Kumar et al. [11] performed EDM on Inconel 706. They investigated the EDM of Inconel-706 by using a hybrid magnetic field assisted powder mixed process. They observed that by increasing peak current and pulse on time results into a rough surface because of presence of microholes. Machining of aluminum-based SIC reinforced composite was performed using EDM by Nayim et al. [12]. The input parameters are pulse on/off duration and current while response parameters were material erosion rate, microhardness, and surface roughness and recast layer thickness. They observed a higher surface finish with higher pulse off time and lower peak current. In present research work electric discharge machining on steel H13 with copper tool performed. Experimental values are measured and analyzed. Signal-to-noise ratios which are associated with the observed values were determined. Factors that affect the most on response were found out using MINITAB software.

2 Experimentations

Orthogonal array is used to perform EDM experiments on steel (grade H 13). Copper made tool used as electrode and make positive polarity. Photograph of machining setup (EDM with work material) is shown in Fig. 1. Electrode thickness is taken with two levels while pulse on time and current is taken with three levels (Table 1). Orthogonal array (L18) design is used to perform the EDM experiments on H 13 steel (Table 2). Weight of the work piece and tool measured using precision weighing machine. Image of machined work is shown in Fig. 2.

3 Optimization

Equation 1 used to calculate single to noise (S/N) ratio for considered response, i.e., MRR (using larger is better). Table 3 shows the mean values of S/N ratio.

$$S/N = -10 \log 10\,(\text{M.S.D}) \qquad (1)$$

where, M.S.D. = Mean square deviation.

Fig. 1 Photograph of die sinker EDM

Table 1 Variables with level

Variables	Unit	Levels		
		1	2	3
Electrode thickness (*A*)	mm	5.2	6.2	–
Discharge current (*B*)	A	1	3	5
Spark/pulse on time (*C*)	μs	200	300	400

ANOVA for the factors is shown in Table 4. From Table 4, it can be found out that spark on time is the most important factor. The next is electrode thickness and last is discharge current. Main effects of parameters on MRR are shown in Fig. 3. Mean ratio of S/N of MRR is shown in Fig. 4. Optimum level of process parameters for MRR is electrode thickness ($B2 = 6.2$ mm), discharge current ($B3 = 5$ A) and spark time ($C1 = 200$ μs) as shown in Fig. 3.

4 Conclusion

Machining of tool steel (grade H-13) by EDM with copper electrode done successfully. Following inference are found from present study.

- The machining parameters selected are pulse on time, discharge current and thickness of the tool. Orthogonal array L18 used to performed experiments and response

Table 2 Experiment design matrix

S. No.	*A*	*B*	*C*
1	5.2	1	200
2	5.2	1	300
3	5.2	1	400
4	5.2	3	200
5	5.2	3	300
6	5.2	3	400
7	5.2	5	200
8	5.2	5	300
9	5.2	5	400
10	6.2	1	200
11	6.2	1	300
12	6.2	1	400
13	6.2	3	200
14	6.2	3	300
15	6.2	3	400
16	6.2	5	200
17	6.2	5	300
18	6.2	5	400

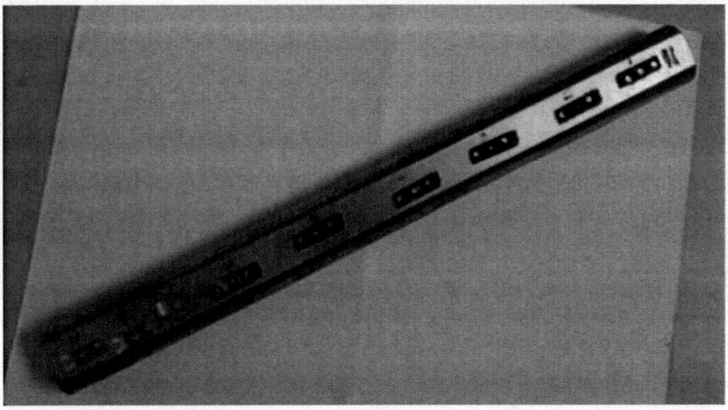

Fig. 2 Image of machined work piece

Table 3 Mean (S/N) ratio for MRR

Parameters	Level 1	Level 2	Level 3
Electrode thickness (*A*)	$A1 = 11.93$	A2 = 19.767	–
Discharge current (*B*)	$B1 = 10.654$	$B2 = 17.169$	B3 = 19.722
Spark on time (C)	C1 = 21.142	$C2 = 15.842$	$C3 = 10.561$

Table 4 ANOVA results for MRR

Parameters	DOF	SS	MS	F_0 = (MS/MS error)	Contribution (%)
Electrode thickness (*A*)	1	276.5	276.5	51.04	29.43
Discharge current (*B*)	2	262.19	130.10	24.017	27.9
Spark on time (*C*)	2	335.878	167.939	31.002	35.75
Error	12	65	5.417		6.92
Total	17	939.568			100

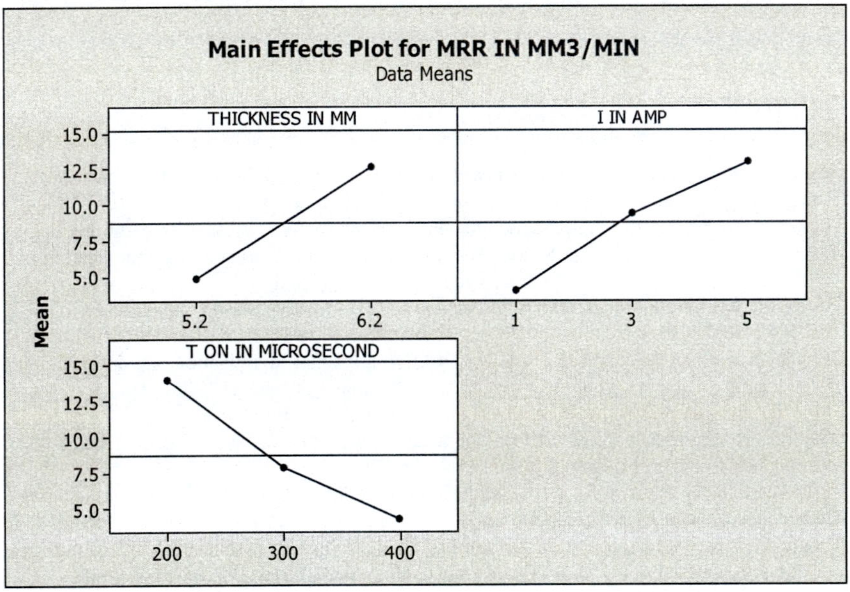

Fig. 3 Main effect plot for MRR

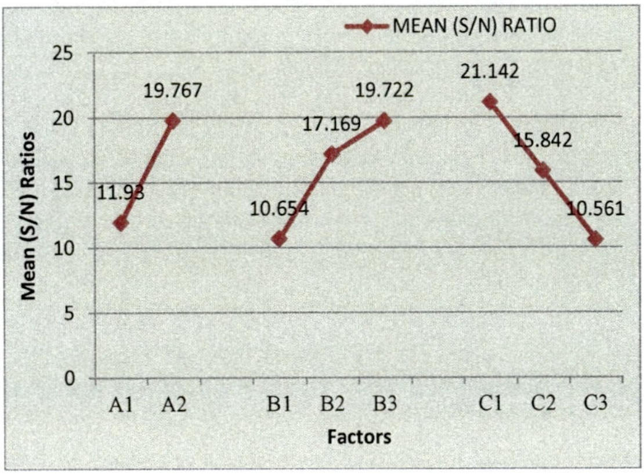

Fig. 4 Graph for mean (S/N) ratio for MRR

material removal rate taken. Optimization of process parameters for EDM of H13 steel done.

- Pulse on time is influencing parameter then electrode thickness and the last is discharge current for elected response.
- Optimum level of process parameters for MRR are electrothickness (6.2 mm), discharge current (5 A) and spark time (200 μs).

Acknowledgements Thanks to PMC, Ghaziabad Utter Pradesh (India), for extending their facilities to conduct this study. Authors also thank to Poornima College of Engineering Jaipur for support and guidance.

References

1. Callister, W. D. *Materials science and engineering—An introduction*. John Wiley and Sons.
2. Jain V. K. (2009). *Advanced machining processes*. Allied Publishers.
3. Ghosh, A., & Mallik, A. K. (2005). Manufacturing science by West Press Private Limited 2005.
4. Molinetti, A., Amorim, F. L., Soares, P. C., Czelusniak, T.(2016). Surface modification of AISI H13 tool steel with silicon or manganese powders mixed to the dielectric in electrical discharge machining process. *International Journal of Advanced Manufacturing Technology, 83*, 1057–1068 (2016).
5. Pratap, P., Kumar, J., & Verma, R. K. (2020). Experimental investigation and optimization of process parameters during electric discharge machining of Inconel X-750. *Multiscale and Multidisciplinary Modeling, Experiments and Design, 3*, 161–171.
6. Saini, S. K., & Pradhan, S. K. (2014). Optimization of multi-objective response during CNC turning using taguchi-fuzzy application. *Procedia Engineering, 97*, 141–149.
7. Pradhan, S. K., & Saini, S. K. (2014). Multi-objective optimization of CNC turning machining parameters. *Journal of Advanced Materials Research, 1016*, 172–176.
8. Saini, S. K., & Pradhan, S. K. (2014). Optimization of machining parameters for CNC turning of different materials. *Journal of Applied Mechanics and Materials, 592–594*, 605–609.
9. Saini, S. K., & Pradhan, S. K. (2014). Soft computing techniques for the optimization of machining parameter in CNC turning operation. *International Journal of Emerging Technology and Advanced Engineering, 4*, 117–124.
10. Kumar, S., Goud, M., & Mohan, N. (2020) Experimental investigation of magnetic-field-assisted electric discharge machining by silicon-based dielectric of Inconel 706 superalloy. *Indian Academy of Science (Sadhana), 45.*
11. Singh, B. P., Sidhu, S. S., Payal, H. S., Kaur, S. (2019). Magnetic field influence on surface modifications in powder mixed EDM. *Silicon, 11*, 415–423.
12. Nayim, S. T. I., Hasan, M. Z., Jamwal, A., Thakur, S., & Gupta, S. (2019, September). Recent trends and developments in optimization and modelling of electro-discharge machining using modern techniques: A review. *AIP Conference Proceedings, 2148*(1), 030051.

Can Industry 4.0 Revolutionize the Wave of Green Finance Adoption: A Bibliometric Analysis

Sumedha Bhatnagar, Dipti Sharma, and Shruti Agrawal

1 Introduction

Industry 4.0 or the fourth industrial revolution is considered as a significant step toward the integration and upgradation of industries through new technologies [1, 2]. The Industry 4.0 practices in recent times supports the objectives of circular economy and cleaner production along with optimum production efficiency [3]. The idea of Industry 4.0 aims at output optimization and reduce waste generation thus supporting the vision of circular economy and sustainability dimensions [4]. The aim of circular economy further converges with the concept of green economy in terms of economic and environmental dimensions [5]. The transition to green economy involves utilization of natural resources, energy and new technologies with cleaner and efficient production methods that can further result into economic growth and generation of job opportunities. Adoption of green economic model can be a success when it involves global commitment and strategically designed framework for global investment involving both developing and developed countries. Industry 4.0, through digitization of processing and innovation, can play integral role in sustainable growth [6, 7]. It can also be considered relevant for effective growth of green financing. Its components can ensure efficient processing in the industries thus can play a key role in mitigating the financing risks to environmentally friendly industries. It can further aid in development of the technology for the financial system for improved transparency in the system.

Present study attempts to identify the link between Industry 4.0 and green financing for effective and efficient economic growth and transition to green economy. The study presents the overview of both the concepts and the common parameters between them for addressing the challenges in their adoption.

S. Bhatnagar (✉) · D. Sharma · S. Agrawal
Department of Humanities and Social Sciences, Malaviya National Institute of Technology Jaipur, Jaipur, Rajasthan, India

R. Agrawal et al. (eds.), *Recent Advances in Smart Manufacturing and Materials*, Lecture Notes in Mechanical Engineering,
https://doi.org/10.1007/978-981-16-3033-0_49

1.1 Green Economy and Green Finance

The concept of green economy was first discussed by Pearce et al. [8] in their report "Blueprint of Green Economy." The study reviewed the research and policy recommendation on various approaches for the valuation of environmental services. Since then not only environment protection has gained attention, the focus has transformed toward more efficient utilization and sophisticated consumption of resource [9]. The drivers for the green economy include specific policies directed toward elimination or reducing environmentally harmful subsidies, creating a market for ecosystem goods and services, providing market-based incentives, opportunities and enabling institutions through an appropriate regulatory framework. Babonea and Joia [10] stated that the concept of green economy is a multidimensional concept that pays particular attention to economy and environment. It is based on the four basic components, namely equity, ecological scarcities, environmental risks and human well-being [11].

Green finance is based on the motive for transition toward green economy. G20 study group referred green finance as "financing of investments that provide environmental benefits in the broader context of environmentally sustainable development. These environmental benefits include: reduction in air, water and land pollution, reduction in greenhouse gas (GHG) emissions, improved energy efficiency while utilizing existing natural resources, as well as mitigation of and adaptation of climate change and their co-benefits." It is a ladder to address multiple environmental challenges and converge the economies on a sustainable pathway [12].

UNEP Inquiry [13] explained the terminologies related to green finance in the manner shown in Fig. 1. The figure explains that investment under sustainable

Fig. 1 Concept of sustainable finance, green finance and climate finance

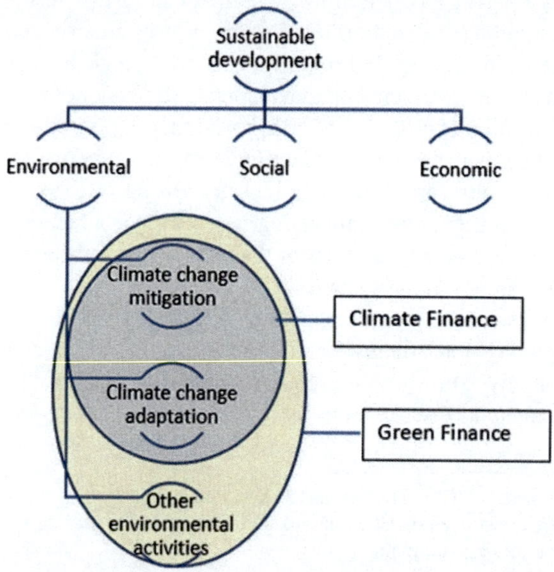

finance addresses development that is inclusive of economic, social and environmental aspects. Whereas, green finance includes investments that has focus on environmental objectives with particular emphasis on biodiversity and resource conservation [14]. Lastly, climate finance focuses on financing that supports mitigation and adaptation activities that addresses climate change. Implementation of green finance requires innovation in the financial institutions of the country. It requires integration of financial industry with the environmental factors for long-term economic growth. The innovation includes technical innovation as well as development of human capital for the adaptability to the economic-friendly environmental practices. Babonea and Joia [10] stated that the transition to green economy requires emphasis on knowledge creation, research and innovation.

1.2 Industry 4.0

Industry 4.0 or fourth industrial revolution refers to the strategic initiative that was first undertaken by Germany for the rapid advancement and generation of manufacturing technologies [15, 16]. These technologies are related to cyber-physical systems, Internet of things (IoT), cloud computing, industrial integration, enterprise architecture, service-oriented architecture, business process management, industrial information integration, etc. [2, 15, 16]. The concept of Industry 4.0 is to converge new and emerging technologies to efficient productivity and digital solutions for higher industrial performance [1, 17]. The concept of Industry 4.0 is an improvement in manufacturing and service system that enable faster, flexible and efficient processes that generates high-quality products at minimum cost [2, 18].

Since the inception of industrial revolution, the strategy for the adoption of technology has evolved. Figure 2 shows the key technology adoption of previous industrial revolutions.

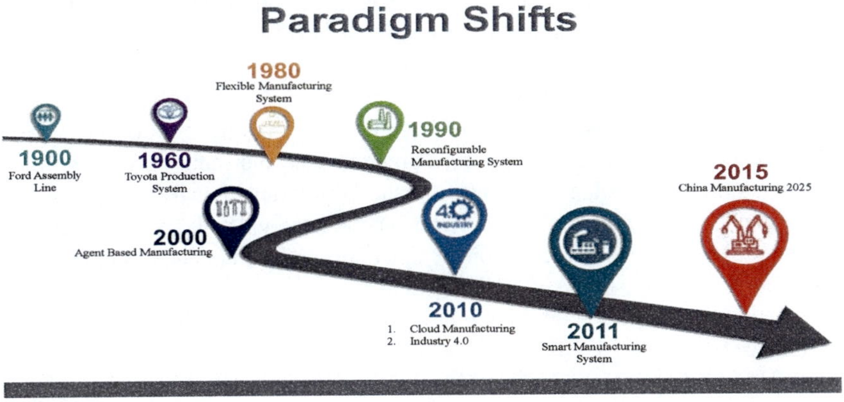

Fig. 2 Evolution of industrial revolution

The first phase of industrial revolution dates back to 1760s, encouraged the process of water and steam power machines in the field of manufacturing and mechanical production. It focused on manufacturing of a huge number of goods that creates an improved standard of living. Following to that, the second industrial revolution which is known as Industry 2.0 and also "The Technological Revolution" marks its beginning in twentieth century. During this period, electricity played an important role in the entire process which results into massive production, assemble lines and electric power processing. The third industrial revolution began with the use of programmable electronic device was implemented by the industries. These updated devices were much more efficient and flexible in the production of digital automation. Industry 4.0 is the fourth industrial revolution that also refers to the era of smart machines that aims for improved industrial performance [17, 19]. It has potential to significantly impact various field other than manufacturing, such as financial sector, information and communication technology (ICT), various ventures, business, employment generation, etc. [20, 21].

Industry 4.0 explicitly does not mention the ecological dimension of sustainability but improved production technology can significantly impact the environment through reduced pollution, greenhouse gases emissions, decrease in energy consumption and increase in profits simultaneously. Erol, (n.d.), stated that the smart technologies can further improve the product life cycle. Further improving the production efficiency through digital transformation can support sustainable or green transition by significantly reducing the consumption of natural resources. Increased energy efficiency can also be a key element that is the common link between Industry 4.0 and transition to green economy [22]. Financing the implementation of Industry 4.0 practices and transitioning to green economy is one of the challenges to their implementation. Financing green economy depends on multiple factors such as technological upgradation, availability of skilled manpower, robust and adaptable financial system aligned with the economic regulations and policies and availability of investment scenario [23]. Present study focuses on the nexus between the green financing and Industry 4.0 for the efficient and environmentally resilient development of the economy. It further attempts to provide an overview of both the concepts and how they can be integrated for the long-term economic growth. The study also addresses the question "how can adoption of Industry 4.0 and green financing together support the transition toward green economy?".

The structure of the paper is as follows: section one gives the brief overview of the three concepts namely; green economy, green finance and Industry 4.0. Section two presents the methodology employed for the bibliographic analysis. Section 3 gives the main results of the analysis and Sect. 4 concludes and makes suggestion for future researches. This study focuses on the link between green finance and Industry 4.0 and address the following research question:

RQ1: How adoption of Industry 4.0 can play an integral role in the growth of green finance?

RQ 2: How implementation of Industry 4.0 and green finance can ensure transition to green economy?

2 Literature Review: Data and Methods

A systematic literature search has been conducted with the high level of inclusiveness [24]. Further bibliometric analysis has been applied to obtain full, thematic overview of the existing knowledge on Industry 4.0, green finance and green economy. The bibliographic records including keywords and citation are key elements of the scientific research. They are also the reliable methods to study the scope and dynamics of scientific fields in financing and industrial revolution. Bibliometric analysis also facilitates objective and reliable analysis by providing broad picture of research and science mapping [25]. It also reveals the dynamic and structural aspects of scientific research.

2.1 Article Search

Search query has been constructed using the Web of Science (WoS) database. Since the focus of the study is green finance, similar terms have been used for the exhaustive list of articles. The term that is often used in interchanging with green finance is, green finance, climate finance and carbon finance [26]. For further increasing the coverage of literature, the keywords used with variation are "green financing," "climate financing," "carbon financing," "low-carbon finance," "environment finance," "green investment" and "green bond."

The keyword search included "green finance" or "carbon finance" or "climate finance" or "low-carbon finance" or "environment finance" or "green financing" or "climate financing" or "carbon financing" or "low-carbon financing" or "environment financing" or "green investment" or "green bonds" and "Industry 4.0" or "industrial revolution." Since we are interested in the literature that focuses on the role of Industry 4.0 in successful implementation of green finance practices, we searched in the subject fields related to engineering and technology, management, business, economic literature and the related domains, which identified 89 documents. Papers included in the subject areas: meteorology atmospheric sciences, ecology, geography, public environmental occupational health, chemistry physical, electrochemistry, geosciences multidisciplinary, food science technology, urban studies and water resources are excluded in the present study.

These documents were further filtered and peer reviewed academic journal articles, review papers, proceeding papers and early access are selected for analysis. Lastly to ensure the relevance of the articles, the abstracts, keywords and introductions were checked manually. This allowed elimination of false-positive articles or the articles not related to the topic of interest. The overall search process resulted in a sample of 89 bibliographic records, covering more than 150 cited references and 200 unique keywords.

2.2 Methods

To undertake literature review, present study has applied co-citation of author keywords analysis. Co-citation analysis is used to identify the conceptual roots of the article that is, the papers and/or authors that are more frequently are co-cited by the articles in the sample of articles selected [27]. To undertake both the bibliographic analysis, the work of [28] is followed.

3 Findings

The results show that based on the selected list of documents, the first article was published in 2003. The rise in academic interest toward environment protection and the need for efficient development of industrial sector can be seen with the rise in number of publications this area. The maximum publication of 23 articles is recorded in 2020. Nearly, 75% (63 articles) of articles were published during the period of 2016–2020. Table 1 shows top authors working in the area of green finance and industrialization. It is found that "Journal of Cleaner Production" has the maximum number of publications which is followed by sustainability.

Table 1 Top authors and sources

S. No.	Top authors		Top sources	
	Author name	TP	Journal/source name	TP
1	Springmann, M.	2	Journal of Cleaner Production	9
2	Sueyoshi, T.	2	Sustainability	8
3	Tsai, S. B.	2	Climate Policy	4
4	Wang, K.	2	Agricultural Economics ZEMEDELSKA EKONOMIKA	3
5	Zhang, X.	2	Climate and Development	3
6	Abro, M. M. Q.	1	Energy for Sustainable Development	2
7	Aglietta, M.	1	Environmental Resource Economics	2
8	Aldakhil, A. M.	1	Environmental Science and Pollution Research	2
9	Ali, M.	1	International Environmental Agreements Politics Law and Economics	2
10	Ali, S.	1	Climate and Development	2

3.1 Analysis of the Research Trends

The analysis suggests that most of the articles focus on adoption of efficient processing practices that accounts for 35% of the articles. The analysis also shows that the focus of the studies is more of engineering related journals rather than management and economic outlets. Moreover, the limited studies have focused on the role of innovation in financial institutions and industrial sector and the role of Industry 4.0 in supporting the greener growth of the economy. The concept of green finance is a relatively new concept that is much established in developed economies and developing economies still have immense scope for its growth. Similarly, implementation of Industry 4.0 approach is much robust in developed economies than in developing economies. The studies also show that both the concepts ensure productivity and efficient utilization of the natural capital.

Table 2 shows that most of the research in the given research area has been undertaken by the developed countries. People Republic of China has the maximum contribution in this area. In recent years China has taken multiple initiatives for promoting green finance and increase industrial performance of the country. On the basis of keyword analysis 7 major clusters can be formed that merges the clusters that include similar terms. Table 3 shows the major clusters that can be formed through the bibliometric analysis. The clusters have been named based on the commonality between the keywords. The interlink between industrial processing and green finance is included in cluster 5.

Figure 3 shows the network formed on the basis of the keywords. Application of Industry 4.0 can improve the productivity of the sector thus reducing the cost of production and increase in contribution of the sector. Green supply chain is another important component that is the interlink between green finance and Industry 4.0. The integration of Industry 4.0 with environment dimension requires utilization of

Table 2 Top countries, subject areas and keywords

Country	TP	Top subject areas	TP	Top keywords used	Occurrence
People Republic of China	26	Environmental Science and Technology	39	Climate Finance	6
USA	13	Business Economics	25	Green Finance	6
Germany	11	Science Technology other topics	20	Industry 4.0	5
France	7	Engineering	15	Climate change	4
Canada	6	Energy Fuels	7	Green bond	4
England	6	Computer Science	4	Green Economy	4
Italy	4	Public Administration	5	Economy	4
Australia	3	Government Law	4	Industry	4
Japan	3	Development Studies	3	Green FDI	3
Singapore	3	Agriculture	2	Green Patent	2

Table 3 Cluster on basis of keywords

Cluster	Elements
Cluster 1: green economy	Environmentally adjusted multifactor productivity, ESG information disclosure, green FDI, green growth, green patent, renewable energy, green investment, green finance
Cluster 2: public policies	Environmental regulations, business policies, green policies, green governance, political, corporate governance, green operation
Cluster 3: decision making	ANP, AHP, MCDM models, criteria, DEMATEL, drivers, MCDM approach
Cluster 4: energy	Carbon emissions, component analysis, financial analysis, labor, panel data analysis, energy consumption
Cluster 5: Industry 4.0	4th Industrial revolution, bitcoin, index, technology, Industry 4.0, artificial intelligence, green bonds, green supply chain
Cluster 6: financial market	Bank ownership, economics, enterprises, markets, increasing returns, capital budgets, prices, public investments, long run uncertainties, interest rate
Cluster 7: sustainable development	Barriers, health, economic growth, mental health, education, generation, credit policy, CO_2 emission, generation, Kuznets curve

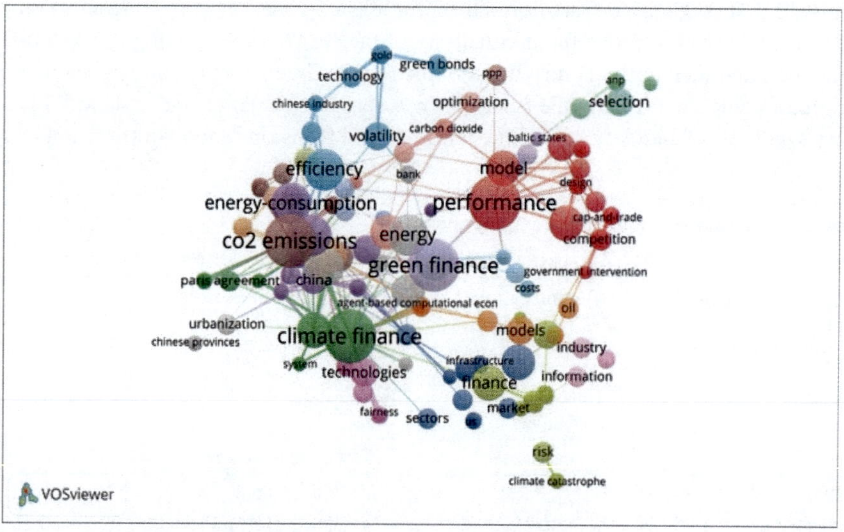

Fig. 3 Network map of keywords

resources in closed loop supply chain that also includes reuse and recovery, technology transfer to developing and least developed countries and achieve sustainable development through eco-innovation [29].

4 Concluding Remarks

In conclusion, the paper highlights that implementation of Industry 4.0 technologies enables the growth of green financing to achieve the transition toward green economy. Through bibliometric analysis, the main focus of current literature linking green finance, Industry 4.0 practices and green economy has been identified. The study reveals that limited literature available interlinks the concept of green finance and implementation of Industry 4.0. Both the concepts are relatively new for the developing economies and both requires innovation in technology and development of human capital for their successful implementation. Industry 4.0 ensures efficiency in industrial sector of the country. Whereas, green finance ensures adoption of energy efficient and environment friendly practices. The green financing instruments that include green bonds, green loans encourage the growth in investment for the green transition. The adaptation of green finance requires adoption of innovation and upgraded technology. This is where industrial revolution can play a pivotal role. Through the given cluster, it can also be inferred that through the adoption of the fourth industrial revolution, investment risk involved in green financing can be mitigated. Adoption of Industry 4.0 practices can improve industrial performance and ensure efficient utilization of resources. It can support in reducing the risk involved in financing green products. It can also support in generation of green job and skilled workforce in the economy that again contributes toward transition to green economy. Innovation in technology also reduces the production cost and production time in the industry. Adoption of smart manufacturing, artificial intelligence and other emerging technologies of Industry 4.0 has the potential to fill the technological gap and attract investors participation in green investment thus supporting transition to green economy.

5 Future Scope of Study

Several studies have been conducted on the concept of Industry 4.0, green finance and green economy. The studies have identified technology innovation as one of the key challenges in increasing green financing. This study addresses the role of Industry 4.0 in increasing green finance and transition to green economy. Further research can be conducted to empirically study the role of Industry 4.0 for improving green finance. Further, a conceptual framework can be developed that highlights the approaches through which Industry 4.0 can fill the technological gap in green investment and green economy transition.

References

1. Jamwal, A., Agrawal, R., Sharma, M., & Kumar, V. (2021). Review on multi-criteria decision analysis in sustainable manufacturing decision making. *International Journal of Sustainable Engineering*, 1–24.
2. Jamwal, A., Agrawal, R., Manupati, V. K., Sharma, M., Varela, L., & Machado, J. (2020, December). Development of cyber physical system based manufacturing system design for process optimization. *IOP Conference Series: Materials Science and Engineering, 997*(1), 012048.
3. Rajput, S., & Singh, S. P. (2019). Connecting circular economy and industry 4.0. *International Journal of Information Management, 49*, 98–113. https://doi.org/10.1016/j.ijinfomgt.2019.03.002
4. Jamwal, A., Agrawal, R., Sharma, M., & Giallanza, A. (2021). Industry 4.0 Technologies for Manufacturing Sustainability: A Systematic Review and Future Research Directions. *Applied Sciences, 11*(12), 5725.
5. D'Amato, D., Droste, N., Allen, B., et al. (2017). Green, circular, bio economy: A comparative analysis of sustainability avenues. *Journal of Cleaner Production, 168*, 716–734. https://doi.org/10.1016/j.jclepro.2017.09.053
6. Popkova, E. G., & Sergi, B. S. (2018) Will Industry 4.0 and other innovations impact russia's development? In *Exploring the future of Russia's economy and markets* (pp. 51–68). Emerald Publishing Limited.
7. Ibarra, D., Ganzarain, J., Igartua, J. I. (2018). Business model innovation through Industry 4.0: A review. *Procedia Manufacturing*, pp 4–10.
8. Pearce, D., Markandya, A., & Barbier, E. (1989). *Sustainable development: The implications of sustainable development for resource accounting, project appraisal and integrative environmental policy*. London Environment Economics Centre.
9. Turner, R. K. (1990). Report on reports: Blueprint for a green economy. *Environment, 32*, 25–26. https://doi.org/10.1080/00139157.1990.9929030
10. Babonea, A.-M., & Joia, R.-M. (2012). Transition to a green economy-a challenge and a solution for the world economy in multiple crisis context.
11. UNEP (2015). The financial system we need: Aligning the financial system with sustainable development policy summary. *UNEP Inquiry Reports, 112.*
12. Falcone, P. M., Morone, P., & Sica, E. (2018). Greening of the financial system and fuelling a sustainability transition: A discursive approach to assess landscape pressures on the Italian financial system. *Technol Forecast Soc Change, 127*, 23–37. https://doi.org/10.1016/j.techfore.2017.05.020
13. United Nations Environment Programme (2019). Sustainable finance progress report, 26.
14. Dörry, S., & Schulz, C. (2018). Green financing, interrupted. Potential directions for sustainable finance in Luxembourg. *Local Environment, 23*(7), 717–733.
15. Da, Xu. L., Xu, E. L., & Li, L. (2018). Industry 4.0: State of the art and future trends. *International Journal of Production Research, 56*, 2941–2962. https://doi.org/10.1080/00207543.2018.1444806
16. Lu, Y. (2017). Industry 4.0: A survey on technologies, applications and open research issues. *Journal of Industrial Information Integration, 6*, 1–10.
17. Dalenogare, L. S., Benitez, G. B., Ayala, N. F., & Frank, A. G. (2018). The expected contribution of Industry 4.0 technologies for industrial performance. *International Journal of Production Economics, 204*, 383–394. https://doi.org/10.1016/j.ijpe.2018.08.019
18. Bartodziej, C. J., & Bartodziej, C. J. (2017). The concept Industry 4.0. In *The concept Industry 4.0* (pp. 27–50). Springer Fachmedien Wiesbaden.
19. Rojko, A. (2017). Industry 4.0 concept: Background and overview. *International Journal of Interactive Mobile Technologies, 11*, 77–90. https://doi.org/10.3991/ijim.v11i5.7072
20. Gorecky, D., Schmitt, M., Loskyll, M., & Zühlke, D. (2014). Human-machine-interaction in the industry 4.0 era. In *Proceedings—2014 12th IEEE international conference on industrial informatics, INDIN 2014* (pp. 289–294). Institute of Electrical and Electronics Engineers Inc.

21. Jamwal, A., Agrawal, R., Sharma, M., Kumar, V., & Kumar, S. (2021). Developing A sustainability framework for Industry 4.0. *Procedia CIRP, 98*, 430–435.
22. Ejsmont, K., Gladysz, B., & Kluczek, A. (2020). Impact of Industry 4.0 on sustainability—Bibliometric literature review. *Sustainability, 12*, 5650. https://doi.org/10.3390/su12145650.
23. Burritt, R., & Christ, K. (2016). Industry 4.0 and environmental accounting: A new revolution? *Asian Journal of Sustainability and Social Responsibility, 1*, 23–38. https://doi.org/10.1186/s41180-016-0007-y
24. Petticrew, M., & Roberts, H. (2008). Systematic reviews in the social sciences: A practical guide. John Wiley & Sons.
25. Jamwal, A., Agrawal, R., Sharma, M., Kumar, A., Kumar, V., & Garza-Reyes, J. A. A. (2021). Machine learning applications for sustainable manufacturing: A bibliometric-based review for future research. *Journal of Enterprise Information Management.*
26. Zhang, D., Zhang, Z., & Managi, S. (2019). A bibliometric analysis on green finance: Current status, development, and future directions. *Finance Research Letters, 29*, 425–430. https://doi.org/10.1016/j.frl.2019.02.003
27. Boyack, K. W., & Klavans, R. (2010). Co-citation analysis, bibliographic coupling, and direct citation: Which citation approach represents the research front most accurately?. *Journal of the American Society for information Science and Technology, 61*(12), 2389–2404.
28. Jamwal, A., Agrawal, R., Sharma, M., Dangayach, G. S., & Gupta, S. (2020). Application of optimization techniques in metal cutting operations: A bibliometric analysis. *Materials Today: Proceedings.*
29. Ben, Y. A. (2020). How can Industry 4.0 contribute to combatting climate change? *Rev d'Economie Ind, 169*, 161–193. https://doi.org/10.4000/rei.8911
30. Erol, S. Where is the Green in Industry 4.0? or How information systems can play a role in creating intelligent and sustainable production systems of the future 1 Industry 4.0-the last chance for truly sustainable production.